Proceedings of the 18th International Meshing Roundtable

T0180851

Brett W. Clark (Ed.)

Proceedings of the 18th International Meshing Roundtable

 Springer

Brett W. Clark
CUBIT Project Lead
Computational Simulation Infrastructure (1543)
Sandia National Laboratories
P.O. Box 5800
Albuquerque, NM, 87185-1321
USA
E-mail: bwclark@sandia.gov

ISBN 978-3-642-42570-7   ISBN 978-3-642-04319-2 (eBook)

DOI 10.1007/978-3-642-04319-2

Typesetting: Data supplied by the authors

Production: Scientific Publishing Services Pvt. Ltd., Chennai, India

Cover Design: WMX Design GmbH, Heidelberg.

Printed in acid-free paper

9 8 7 6 5 4 3 2 1

springer.com

Preface

The papers in this volume were selected for presentation at the 18th International Meshing Roundtable (IMR), held October 25-28, 2009 in Salt Lake City, UT, USA. The conference was started by Sandia National Laboratories in 1992 as a small meeting of organizations striving to establish a common focus for research and development in the field of mesh generation. Now after 18 consecutive years, the International Meshing Roundtable has become recognized as an international focal point annually attended by researchers and developers from dozens of countries around the world.

The 18th International Meshing Roundtable consists of technical presentations from contributed papers, research notes, keynote and invited talks, short course presentations, and a poster session and competition. The Program Committee would like to express its appreciation to all who participate to make the IMR a successful and enriching experience.

The papers in these proceedings were selected from more than 40 paper submissions. Based on input from peer reviews, the committee selected these papers for their perceived quality, originality, and appropriateness to the theme of the International Meshing Roundtable. We would like to thank all who submitted papers. We would also like to thank the colleagues who provided reviews of the submitted papers. The names of the reviewers are acknowledged in the following pages.

We extend special thanks to Jacqueline Hunter for her time and effort to make the 18th IMR another outstanding conference.

August 2009   18th IMR Program Committee

18th IMR Conference Organization

Committee

Brett Clark (Committee Chairman)
Sandia National Laboratories, Albuquerque, NM
bwclark@sandia.gov

Steve Karman
University of Tennessee, Chattanooga, TN steve-karman@utc.edu

Joseph Walsh
Simmetrix, Inc., Clifton Park, NY
jwalsh@simmetrix.com

Suzanne Shontz
Pennsylvania State University, University Park, PA
shontz@cse.psu.edu

Yasushi Ito
University of Alabama, Birmingham, AL
yito@uab.edu

John Verdicchio
Rolls-Royce, London, UK
john.verdicchio@rolls-royce.com

Laurent Anne
Distene, Bruyeres le Chatel, FR
laurent.anne@distene.com

Paresh Patel
MSC Software, Santa Ana, CA
paresh.patel@mscsoftware.com

Bill Jones
NASA Langley Research Center, Hampton, VA
w.t.jones@nasa.gov

Coordinators	**Jacqueline A. Hunter**
	Sandia National Laboratories, Albuquerque, NM
	jafinle@sandia.gov
Web Designer	**Jacqueline A Hunter**
	Sandia National Laboratories, Albuquerque, NM
	jafinle@sandia.gov
Web Site	**http://www.imr.sandia.gov**

Reviewers

Alauzet, Frederic	Inria, Sophia-Antipolis, France
Alleaume, Aurelien	Distene S.A.S., France
Benzley, Steven	Brigham Young Univ., USA
Blacker, Ted	Sandia Nat. Labs, USA
Blades, Eric	Mississippi State U., USA
Canas, Guillermo	Harvard University, USA
Cavallo, Peter	Craft Tech, USA
Chernikov, Andrey	C. of William and Mary, VA, USA
Chrisochoides, Nikos	C. of William and Mary, VA, USA
Clark, Brett	Sandia Nat. Labs, USA
Cortes Aguirre, Javie	Universidad Nacional Autonoma de Mexico, Mexico
Cuilliere, Jean-Christophe	Universit du Quebec a Trois-Rivieres, Canada
Dannenhoffer, John	Syracuse University, NY, USA
Diachin, Lori	Lawrence Livermore National Laboratory, USA
Ebeida, Mohamed	U. of California, Davis, USA
Erten, Hale	U. of Florida, USA
Francois, Vincent	Université du Québec à Trois-Rivières, Canada
Garimella, Rao	Los Alamos Nat. Labs, USA
Guoy, Damron	U. of Illionois, Urbana-Champaign, USA
Iannetti, Anthony	NASA Glenn Research Center, USA
Ito, Yasushi	U. of Alabama, Birmingham, USA
Jiao, Xiangmin	State U. of New York, Stony Brook, USA
Jones, Bill	NASA Langley Research Center, USA
Karman, Steve	U. of Tennessee, Chattanooga, USA
Knupp, Patrick	Sandia Nat. Labs, USA
Leon, Jean-Claude	Institut National Polytechnique de Grenoble, France
Lipnikov, Konstantin	Los Alamos Nat. Labs, USA
Loseille, Adrien	George Mason University, USA
Masters, James	Arnold Engineering Development Center, USA
Ollivier-Gooch, Carl	U. of British Columbia, Canada
Owen, Steve	Sandia Nat. Labs, USA

Patel, Paresh	MSC Software, USA
Pav, Steven	U. of California, San Diego, USA
Phillips, Todd	Carnegie Mellon University, USA
Plaza, Angel	Universidad de Las Palmas de Gran Canaria, Spain
Quadros, William	Sandia National Laboratories, USA
Remacle, Jean-Francois	Universite Catholique de Louvain, UK
Samareh, Jamshid	NASA Langley Research Center, USA
Sarrate, Jose	Universitat Politecnica de Catalunya, Spain
Schneiders, Robert	MAGMA Giessereitechnologie GmbH, Germany
Sekar, Balu	USAF Wright Patterson AFB, USA
Shewchuk, Jonathan	U. of California at Berkeley, USA
Shih, Alan	U. of Alabama, Birmingham, USA
Shontz, Suzanne	Pennsylvania State Univ., USA
Si, Hang	Weierstrass Institute of Applied Analysis and Stochastics, Germany
Simpson, Bruce	U. of Waterloo, CA
Staten, Matthew	Sandia Nat. Labs, USA
Steinbrenner, John	Pointwise, USA
Thompson, David	Mississippi State University, USA
Thornburg, Hugo	USAF Wright Patterson AFB, USA
Tournois, Jane Maud	Inria, Sophia-Antipolis, France
Ungor, Alper	U. of Florida, USA
Vaughn, Ed	US Army Aviation & Missile RDEC,USA
Vavasis, Stephen	U. of Waterloo, Canada
Verdicchio, John	Rolls-Royce, UK
Walsh, Joe	Simmetrix, Inc., NY, USA
Weill, Jean-Christophe	CEA, France
Wyman, Nick	Pointwise, USA
Xia, Hao	U. of Cambridge, UK
Yamakawa, Soji	Carnegie Mellon U., PA, USA
Yvinec, Mariette	Inria-Prisme, France
Yu, Zeyun	U. of Wisconsin, USA
Zhang, Yongjie (Jessica)	U. of Texas, Austin, USA

Contents

Session 3A: Parallel & Hybrid

Session 3B: Applications

Session 4: Tetrahedral Meshing

Session 5: Adaptivity

Size Function Smoothing Using an Element Area Gradient

John Howlett and Alan Zundel

Department of Civil and Environmental Engineering
Brigham Young University
john.d.howlett@byu.net, zundel@byu.edu

Abstract. This paper presents a method to improve element size transitions when using a size function to govern the mesh generation. The method modifies the size function to meet a user specified adjacent element area change limit. The method can be used to either refine or coarsen the resulting mesh. Two sample meshes generated using the method are presented.

1 Introduction

Size functions consist of spatially distributed points with an associated value used to specify the desired spacing of finite element nodes. Size functions should contain only positive, nonzero values. In addition, size functions should contain values for the entire domain.

Sources of size functions vary based on what numerical model the finite element mesh being generated will be used with. They may come from manual specification by the user, analysis performed on a previous calculation [6], or from geometric data such as depth. For example, the size function may be specified based on the wavelength when using CGWAVE a wave prediction model for simulating the propagation and transformation of ocean waves in coastal regions and harbors [5]. In any case, the size function specifies how far apart nodes should be in the finite element mesh.

Figure 2 shows a wave shape represented by a different number of elements. The quality of the representation varies as the number of nodes changes. To include more nodes, the size of each element must be a smaller fraction of the wavelength. Since wavelength is related to the depth of the water, which varies spatially over the domain, a size function can be created based on the wavelength in the area being modeled. Shallow areas, with short wavelengths, require closer node spacing than deeper areas, with long wavelengths [2].

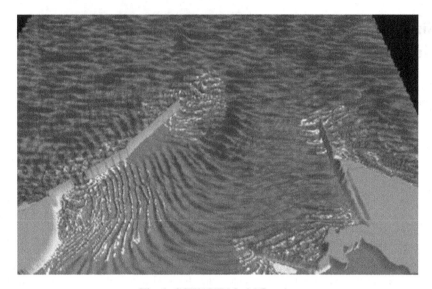

Fig. 1. CGWAVE Model Results

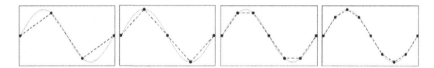

Fig. 2. Wave Shape as Represented by Different Number of Elements

2   Size Function Smoothing Using an Element Area Gradient

If the size function changes too quickly, it will be impossible to space the mesh nodes at a distance close to the distance specified by the size function. Size functions that change too quickly are said to be poorly conditioned. Size functions that change at an acceptable rate are said to be well conditioned [3][7][8].

Size function smoothing using an element area gradient significantly improves the quality of meshes produced by size function based mesh generators. By creating high quality initial meshes, the amount of manual mesh editing required is greatly reduced or eliminated.

To understand why size function smoothing is needed, consider the case of a one dimensional grid between $x = 0$ and $x = 20$ with a desired node spacing varying from one unit at $x = 0$ to five units at $x = 20$. The creation of intermediate nodes to match the desired spacing could proceed, starting at $x = 0$, as shown in Figure 3.

The first node is located at $x = 0.0$ and has a size value of 1.0. The next node is offset one unit to $x = 1.0$ and has a size function value of 1.2. The third node is offset 1.2 units to $x = 2.2$ and has a size function value of 1.44. This process continues until the entire one dimensional grid is filled in.

The creation of intermediate nodes to match the desired spacing could also proceed, starting at $x = 20$, as shown in Figure 4.

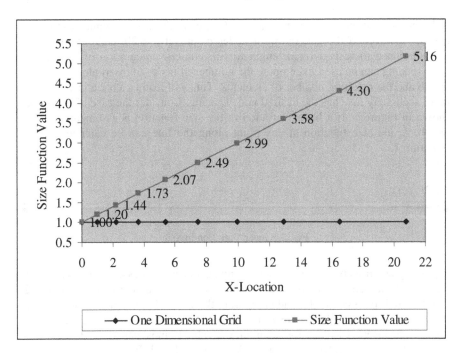

Fig. 3. One Dimensional Grid Node Distribution – Case 1A

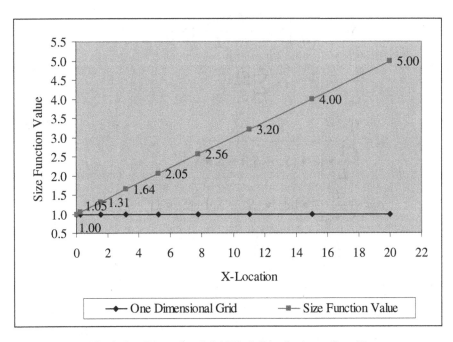

Fig. 4. One Dimensional Grid Node Distribution – Case 1B

If the generation of nodes begins $x = 0$, the resulting grid contains nine elements. However, if the generation of nodes begins at $x = 20$, eight elements are created. Since there is a fixed length to fill, numeric integration of the desired node spacing can be used to compute the number of nodes to insert along the fixed length and the resulting number of elements. This will lead to a node insertion that more accurately matches the desired node spacing. Consider the case of two nodes shown in Figure 5. If a linear variation in the size function is assumed along the length, L, the size function at any point along the line can be calculated using Equation 1.

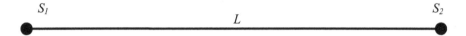

S_1 L S_2

Fig. 5. Linear Variation in Size

Taking small steps along L, each segment, dx, has an associated size S. The number of elements, n, to fill the length L can be calculated using Equation 2. Combining Equation 1 and 2 and computing the integral yields the number of elements required to fill the length L as shown in Equation 3. Since the number of nodes to be inserted will be not be an integer value in most cases, the desired node spacing values are scaled using the scale calculated with Equation 4. The scaled node spacing values produce an integer value for the number of nodes to insert.

$$S(x) = \left(\frac{x}{L}\right) \bullet S_2 + \left(\frac{L - x}{L}\right) \bullet S_1 \tag{1}$$

$$n = \sum_{i=0}^{n} \frac{dx}{S_i} \tag{2}$$

$$n = \int_0^L \frac{dx}{\dfrac{1}{L} \bullet (x \bullet (S_2 - S_1) + L \bullet S_1)} \tag{3}$$

$$n = \frac{L}{S_2 - S_1} \bullet \ln\left|x \bullet (S_2 - S_1) + S_1 \bullet L\right|_0^L$$

$$n = \frac{L}{S_2 - S_1} \bullet \ln\left(\frac{S_2}{S_1}\right)$$

$$Scale = \frac{\updownarrow n}{n} \tag{4}$$

This method is able to place points in an acceptable manner unless the fixed length is not long enough to match the desired node spacing values. Consider the same one dimensional grid presented earlier, but this time with a desired node spacing varying from one unit at $x = 0$ to one hundred units at $x = 20$ as shown in Figure 6.

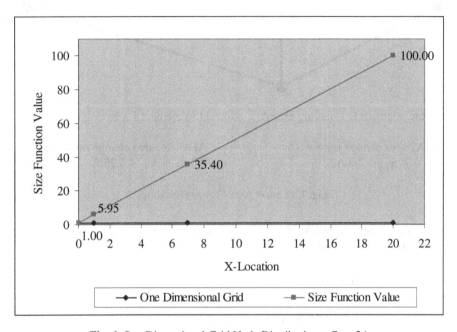

Fig. 6. One Dimensional Grid Node Distribution – Case 2A

Using this desired node spacing, if the generation of nodes begins $x = 0$, and the same offset method as described earlier is used, the resulting grid contains three elements. However, if the generation of nodes begins at $x = 20$, one element is created. So the question becomes, what is a reasonable variation of the desired node spacing? The desired node spacing should be a function of the fixed length to fill. We will present an algorithm capable of determining the minimum, $S_{min,}$ and maximum, S_{max}, size values for a node located a distance, L, from the starting node, whose desired node spacing is known.

To simplify the algorithm, we eliminate the need to compute S_{min} by always computing S_{max} for the smaller of the two desired node spacings and adjusting the larger desired node spacing value to be less than or equal to S_{max}.

To compute S_{max}, we must define how fast the desired node spacing is allowed to change. One guideline for the RMA2 model from the USACE-ERDC states that the area of an element should fall between 50 and 200 percent of its neighbor's area [4]. For example, if an element has an area of ten square meters, the adjacent elements should have an area between five and twenty square meters. ADCIRC

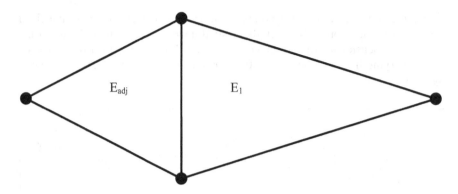

For any element E_{adj}, having area A_{adj}, adjacent to element E_1, having area A_1:

RMA2 area change ratio = 0.5   ADCIRC area change ratio = 0.8
$0.5\ A_1 < A_{adj} < 2.0\ A_1$ $0.8\ A_1 < A_{adj} < 1.25\ A_1$

Fig. 7. Element Area Change Guidelines

model developers recommend even more restrictive guidelines that limit area change to between 80 and 125 percent [9].

$$P_g = \sqrt{\frac{1}{r}} \qquad (5)$$

P_g = Maximum percent growth

r = area change ratio

$$L_e' = L_e \bullet P_g$$

The area change ratio, r, is specified as a value between zero and one. The minimum area of an adjacent element is the current element's area times the area change ratio. The maximum area of an adjacent element is the current element's area times one over the area change ratio. Since the area of a triangle is proportional to the nodal spacing squared, the allowable change in the desired node spacing is the square root of the user specified area change ratio.

Fig. 8. Maximum Nodal Spacing Growth Rate

Given a segment with a target size of S_1 at one end and a length of L, we want to compute the maximum size of a segment, S_{max}. We begin by determining how far down the segment one element extends. Since S_1 is the minimum size, it will extend farther than S_1. This distance can be computed by finding the length to get one element using Equation 6.

$$1 = \frac{L_e}{S_2 - S_1} \bullet \ln\left(\frac{S_2}{S_1}\right) \text{ where } S_2 = P_g \bullet S_1 \tag{6}$$

$$1 = \frac{L_e}{(P_g \bullet S_1) - S_1} \bullet \ln\left(\frac{P_g \bullet S_1}{S_1}\right)$$

$$1 = \frac{L_e}{(P_g - 1) \bullet S_1} \bullet \ln(P_g)$$

Solving for L_e gives :

$$L_e = \frac{(P_g - 1) \bullet S_1}{\ln(P_g)}$$

$$S_f \equiv \frac{P_g - 1}{\ln(P_g)}$$

$$L_e = S_1 \bullet S_f$$

The next segment length can be computed in the same fashion with Equation 7.

$$S_1 = P_g \bullet S_1 \tag{7}$$

$$S_2 = (P_g)^2 \bullet S_1$$

$$1 = \frac{L_{e_2}}{(P_g)^2 - P_g \bullet S_1} \bullet \ln\left(\frac{(P_g)^2 \bullet S_1}{P_g \bullet S_1}\right)$$

$$1 = \frac{L_{e_2}}{(P_g - 1) \bullet P_g \bullet S_1} \bullet \ln(P_g)$$

Solving for L_{e_2} gives :

$$L_{e_2} = \frac{(P_g - 1) \bullet P_g \bullet S_1}{\ln(P_g)}$$

$$L_{e_2} = P_g \bullet S_1 \bullet S_f$$

$$L_{e_3} = (P_g)^2 \bullet S_1 \bullet S_f \tag{8}$$

Similarly, the third segment length can be computed with Equation 8. The resulting nodal size values and segment lengths are shown in Figure 9.

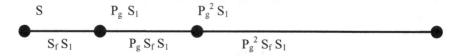

Fig. 9. Segment Lengths and Associated Nodal Size Values

Therefore, the total length of n segments can be computed using Equation 9 and the nodal size value using Equation 10.

$$L_{e_n} = S_1 \bullet S_f \bullet \sum_{i=0}^{n-1} P_g^{\ i} \tag{9}$$

$$S_n = P_g^{\ n} \bullet S_1 \tag{10}$$

The number of terms to complete length L_e can be found using Equation 11.

$$L_e = S_1 \bullet S_f + P_g \bullet S_1 \bullet S_f + \left(P_g\right)^2 \bullet S_1 \bullet S_f + ... + \left(P_g\right)^{n-1} \bullet S_1 \bullet S_f \tag{11}$$

$$L_e = S_1 \bullet S_f \bullet \left(1 + P_g + P_g^{\ 2} + ... + P_g^{\ n-1}\right)$$

$$\frac{L_e}{S_1 \bullet S_f} = \frac{1 - P_g^{\ n+1}}{1 - P_g}$$

$$\text{Let } S_n = \frac{L}{S_1 \bullet S_f} = \frac{1 - P_g^{\ n+1}}{1 - P_g}$$

$$P \equiv P_g - 1$$

$$P_g = P + 1$$

$$S_n = \frac{1 - P_g^{\ n+1}}{-P}$$

$$-PS_n = 1 - P_g^{\ n+1}$$

$$PS_n = -1 + P_g^{\ n+1}$$

$$PS_n + 1 = P_g^{\ n+1}$$

$$Log_{P_g}\left(PS_n + 1\right) = n + 1$$

$$n = Log_{P_g}\left(PS_n + 1\right) - 1$$

The number of terms, n, from Equation 11 can be used in Equation 10 to determine the maximum size value for a node located a distance, L, from the node with size value S_1, given the maximum percent growth, P_g.

3   Size Function Smoothing Tools in SMS

The Surface-water Modeling System (SMS) [1] contains a size function smoothing tool based on the method presented in this paper. The smoothing tool modifies size function values to honor the user specified maximum area change limit (see Figure 7). The tool allows the user to control whether size function values are increased (maximum value anchored) or decreased (minimum value anchored). Increasing size function values results in a less refined mesh, while decreasing them results in a more refined mesh. A minimum size function value can also be specified.

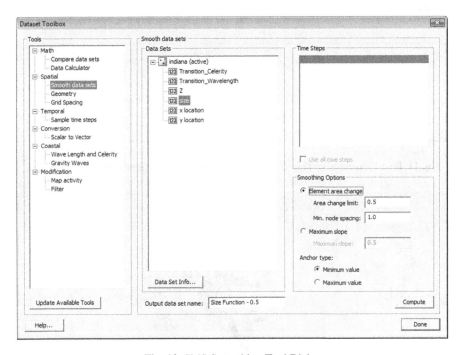

Fig. 10. SMS Smoothing Tool Dialog

The smoothing tool operates on a size function stored as a triangulated irregular network (TIN) by performing the following steps for the minimum value anchored example:

1. Sort TIN nodes from smallest to largest size function value
2. Set the size function value to the larger of the current size function value or the specified minimum size function value

3. For each sorted TIN node, T_i:
4. For each TIN node adjacent to T_i:
5. Calculate the distance, L, from the adjacent TIN node to T_i
6. Compute n using Equation 11
7. Compute S_{max} using Equation 10
8. Set the size function value at T_i to the smaller of the current size function value or S_{max}
9. If the size function value at T_i was changed, resort T_i in the list of TIN nodes

4   Examples

Consider a two dimensional grid, similar to the one dimensional grid presented earlier, with a desired node spacing varying from one unit at x = 0 to one hundred units at x = 20 with no variation in the y direction. The resulting mesh is shown in Figure 11.

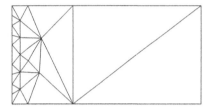

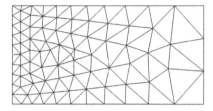

Fig. 11. Poor Element Size Transitions   **Fig. 12.** Good Element Size Transitions

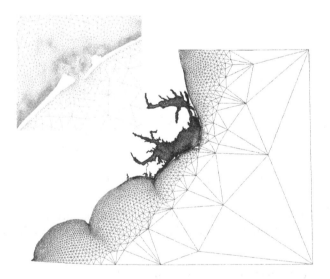

Fig. 13. Beaufort, North Carolina Mesh Generated Using Original Size Function

By applying the smoothing algorithm with an area change limit of 0.8, the revised size function has a desired node spacing varying from one unit at x = 0 to 3.23 units at x = 20 with no variation in the y direction. The resulting mesh is shown in Figure 12.

Figure 13 shows a mesh generated using a size function of 10 elements per 12 hour period (semi-diurnal tidal constituents) wave. The size function varies from 4 to 827,685. Applying the smoothing algorithm with an area change limit of 0.8 results in a new size function varying from 4 to 38,283. The resulting mesh from this new size function is shown in Figure 14.

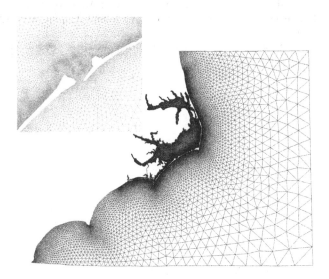

Fig. 14. Beaufort, North Carolina Mesh Generated Using Smoothed Size Function

5 Conclusion

Size function smoothing using an element area gradient has been proposed as a new method for improving element size transitions when using a size function to govern the mesh generation. The method can be applied as a pre-process to a sizing function generated from any criteria regardless of the mesh generator used. Future work will involve application to three dimensions.

References

1. Aquaveo LLC. Reference manual: The Surface water Modeling System (2009)
2. Aquaveo LLC. Tutorial document: CGWAVE Analysis (2009)
3. Borouchaki, H., Hecht, F., Frey, P.J.: Mesh Gradation Control. In: Proceedings, 6th International Meshing Roundtable, pp. 131–141. Sandia National Laboratories (1997)

4. Donnel, B.P., Letter, J.V., McAnally, W.H.: Users Guide for RMA2 Version 4.5. U.S. Army Engineer Research and development Center, Vicksburg MS (2001)
5. Demirbilek, Z., Panchang, V.: CGWAVE: A Coastal Surface Water Wave Model of the Mild Slope Equation, USACE Technical Report CHL-98-xx, U.S. Army Engineer Research and Development Center, Vicksburg MS (1998)
6. Hagen, S.C., Michael Parrish, D.: Unstructured Mesh Generation for the Western North Atlantic Tidal Model Domain. In: Proceedings, 11th International Meshing Roundtable, pp. 87–98. Sandia National Laboratories (2002)
7. Howlett, J.D.: Size Function Based Mesh Relaxation. Depatment of Civil and Environmental Engineering, Brigham Young University (2005)
8. Li, X., Remacle, J.-F., Chevaugeon, N., Shephard, M.S.: Anisotropic Mesh Gradation Control. In: Proceedings, 13th International Meshing Roundtable, pp. 401–412. Sandia National Laboratories (2004)
9. Luettich Jr., R.A., Westerink, J.J.: Reference manual: ADCIRC, University of North Carolina at Chapel Hill (2004)

Removing Self Intersections of a Triangular Mesh by Edge Swapping, Edge Hammering, and Face Lifting

Soji Yamakawa and Kenji Shimada

The Department of Mechanical Engineering
Carnegie Mellon University

Abstract. This paper describes a computational method for removing self intersections of a triangular mesh. A self intersection is a situation where a part of a surface mesh collides with another part of itself, i.e., two mesh elements intersect each other. It destroys the integrity of the mesh and makes the mesh unusable for certain applications. A mesh generator often creates a self intersection when a relatively large element size is specified over a region with a narrow clearance. There has been no automated method that automatically removes self intersections, and such self intersections needed to be corrected by manually editing the mesh. The proposed method automatically resolves a self intersection by re-connecting edges and adjusting node locations. This technique removes a typical self intersection and recovers the integrity of the triangular mesh. Experimental results show the effectiveness of the proposed method.

1 Introduction

This paper describes a computational method for removing self intersections of a triangular mesh. A self intersection is a situation where a part of a surface mesh collides with another part of itself, i.e., at least two mesh elements intersect each other. A self intersection destroys the integrity of the surface mesh and makes the surface mesh unusable for certain applications. The proposed method removes such self intersections and recovers the integrity of the mesh.

A triangular mesh is used for many applications such as finite element analysis, visualization, and so on. Many of those applications require the mesh to be free of self intersections. A self intersection in a triangular mesh could cause failure of the finite element analysis or make unwanted artifacts in the visualization.

A triangular mesh is also used as a boundary of a tetrahedral mesh, and such a triangular mesh must not include a self intersection. A tetrahedral mesh is usually created by first creating a triangular mesh of the boundary of the target volume and then filling the inside of the triangular mesh with tetrahedral elements. If the boundary triangular mesh includes a self intersection, the mesh no longer defines a legitimate volume, and the tetrahedral mesh generator may create severely distorted elements or even fail to create a mesh at all. Therefore, a self intersection in a triangular mesh is very problematic and needs to be removed.

Theoretically, the best approach for removing self intersections is to mesh the original CAD model with more appropriate mesh sizing. However, it is often impossible to choose adequately short edge length. For example, some types of analyses require minimum edge length longer than certain threshold. For those analyses, minimum edge length condition is more important than the boundary fidelity. The original CAD model is sometimes not available to the analyst due to confidentiality issues. When an analysis needs to be performed on a very old geometry, or a legacy model, the original CAD model may no longer exist, and the model may be available only in the form of a mesh. Therefore, self-intersections need to be removed by modifying the mesh in many real-world problems.

There has been, however, no automated method for removing self intersections of a triangular mesh. Hence, the appropriate care must have been taken before a triangular mesh was created to avoid self intersections. The input geometric model needed to be de-featured, geometric constraints needed to be appropriately added or removed, and a non-uniform sizing function was often necessary. If those preconditioning measures were not adequate, a surface mesh generator would yield self intersections, which needed to be removed by manually editing the mesh.

The proposed method effectively removes typical self intersections created due to a defective CAD model or a very narrow clearance and recovers integrity of the triangular mesh. The proposed method takes as input a triangular mesh, open or closed, and applies to the mesh a sequence of three types of operations: (1) edge swapping, (2) edge hammering, and (3) face lifting. The edge-swapping operation has been used mainly for mesh-quality improvement and is used for reducing self intersections in this context. The edge-hammering and face-lifting operations adjust locations of the nodes used by an intersecting triangular element. These operations calculate new node locations based on the neighboring nodes and the locations of intersections. The node is moved to the new location if the move reduces the number of intersecting elements without making a new intersecting element.

The organization of the paper is as follows. Section 2 presents previous work. Section 3 describes typical sources of self intersections. Section 4 explains details of the proposed method, and Section 5 discusses some potential expansions and discussions. Some experimental results are presented in Section 6. Section 7 concludes the paper.

2 Previous Work

The proposed method pertains to facet-repair techniques, which repair defects included in a surface mesh. A surface mesh may include two types of defects, geometric defects and topological defects. A self intersection is one of the common geometric defects, and a common topological defect is a gap or a hole located where the mesh needs to be closed.

Since facet-repair techniques have been of great interest in the industry, substantial research has been done. Nonetheless, most of the attention has been paid to the topological aspect of the problem.

The majority of the published facet-repair techniques are gap-closure and hole-filling techniques. If a triangular mesh has a gap or hole located where the mesh needs to be closed, such a gap or hole needs to be closed. Barequet and Sharir have presented a method for closing gaps and holes of a polyhedral surface [1]. Barequet and Kumar have presented a method for repairing a geometric model defined by a triangular mesh [2]. Gueziec et al. have presented a method for converting a polygonal surface to a manifold surface [3]. Branch et al. have presented a method for filling holes of a triangular mesh [4]. Li et al. have presented a method for repairing holes of a triangular mesh [5]. Those gap-closure and hole-filling techniques identify gaps and holes that need to be closed and then insert triangular elements and/or stitch the edges together so that gaps and holes are filled by triangular elements.

Gap closure and hole filling are important to recover the topological integrity of a triangular mesh. However, it is also necessary to remove self intersections and recover the geometric integrity. Self-intersection avoidance has been relying mainly on *a priori* methods. In other words, adequate pre-conditioning needed to be performed before a triangular mesh was created so that the mesh generator would not yield a self intersection.

One of the common sources of a self intersection is a very small feature, which is usually irrelevant to the purpose of the mesh. Ribelles et al. have presented a method that automatically removes features of a geometric model [6]. Lee et al. also have presented another approach for small-feature suppression [7].

A self-intersection can also be avoided by identifying features and appropriately adding geometric constraints near the features. Numerous researches have been done for feature-line identification [8-12]. Jiao has presented a method for identifying features of a geometric model and preserves them during the mesh-generation process [13].

Alternatively, a self-intersection can be avoided by creating smaller elements near small features. Quadros et al. have presented a method for automatically creating an element-sizing function based on the input geometry [14].

Self intersections can be avoided by embedding intersection curves on a mesh and then dividing the mesh into a set of non-intersecting sub-meshes [15]. This method, however, divides a mesh into multiple meshes and can be utilized only in limited types of applications.

Despite the extensive research on automatic feature identification and removal techniques, those techniques are not always adequate for avoiding all self intersections. If the mesh generator could not avoid creation of self intersections in a triangular mesh, the user needed to manually edit the mesh to remove those self intersections.

The proposed method modifies a triangular mesh and reduces self intersections. It is useful for reducing the manual mesh editing when the mesh generator can not avoid self intersections.

3 Typical Sources of a Self Intersection

Even when the CAD model itself does not include a self intersection, there is a possibility that the mesh generator creates a self intersection when an excessively large element size is specified. A large element size yields a large error (in distance) between the mesh surface and the original surface. In two dimensions, when a curve is discretized into line segments, an expected error between a curve and its discretization is calculated as:

$$e=\frac{1}{\kappa}-\sqrt{(1/\kappa)^2-(h/2)^2} \; ,$$

where h is the length of a segment of the discretization and κ is the curvature of the curve as shown in Figure 1.

In three dimensions, an expected error between a surface and its triangulation is calculated as:

$$e=\frac{1}{\kappa}-\sqrt{(1/\kappa)^2-(h^2/3)} \; ,$$

where h is the edge length, and κ is the principal curvature of the surface. The assumption is the curvature does not change rapidly and the triangular element is equilateral.

Assume that an equilateral triangle with edge length of h is lying on X-Y plane as shown in Figure 2. The coordinates of the three corners are $(0,0,0)$, $(h,0,0)$, and $(h/2,0,(\sqrt{3}/2)h)$, and the center of the triangle is $(h/2,0,(\sqrt{3}/6)h)$. The center of one of its circumscribed spheres with radius of $r=(1/\kappa)$ is located at $(h/2,0,(\sqrt{3}/6)h)$. Since the distance from the center to the origin, where one of the triangle corners is located, is equal to the radius, the following equation is derived:

$$r^2=\left(\frac{h}{2}\right)^2+y^2+\left(\frac{\sqrt{3}}{6}h\right)^2=\frac{h^2}{3}+y^2.$$

It can be solved for y as:

$$y=\sqrt{r^2-\frac{h^2}{3}} \; .$$

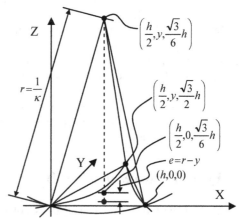

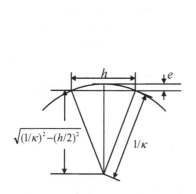

Fig. 1. Expected error between a curve and its discretization

Fig. 2. Expected error between a curved surface and its discretization

Since the expected error can be calculated as the difference between the radius and the distance between the element to the center of the sphere, the expected error becomes:

$$e = \frac{1}{\kappa} - \sqrt{(1/\kappa)^2 - (h^2/3)} \; . \tag{1}$$

Therefore, when two surfaces are separated with a small clearance d, the edge length h should be small enough so that the expected error is smaller than d.

However, in practice, element size is often decided based on the application in which the mesh is used, and it can be way too large for the surface clearance of the original geometric model. Such a large element size yields self intersections in the mesh. For example, Figure 3 (a) shows a CAD model that includes two co-axial cylindrical surfaces. The radii of the outer and inner cylindrical surfaces are 5.0 and 4.8, respectively. Therefore, the clearance between the two surfaces is 0.2. The curvature of the outer surface is 1.0/5.0=0.2. From equation (1), the edge length of the mesh needs to be less than 2.42. Figure 3 (b) shows a triangular mesh created from this CAD model with an average edge length of 2.42, and it does not include a self intersection. However, if the same geometry is meshed with average edge length of 3, some self intersections are clearly visible as shown in Figure 3 (c). In this particular case, self intersections can be avoided by identifying cylindrical surfaces and adding some mesh constraints on the surface. Figure 3 (d) shows some new constrained edges added by the polygon-crawling method [12]. The mesh generator preserves those constrained edges, and the output triangular mesh does not include a self intersection as shown in Figure 3 (e).

Ideally, a sufficiently small element size needs to be chosen or appropriate mesh constraints need to be added for the regions of small clearance. However, some of such ill conditions are often overlooked. Particularly, such ill conditions tend to be left untreated if they are:

1. contained within a very small feature, or
2. away from the point of interest for the application in which the mesh is used.

Such less-visible ill conditions are most likely left untreated before the model is given to the mesh generator. Even if all of those ill conditions are detected, a complex model could have hundreds of them and could take substantial manual operations to correct or add constraints to all of them. As discussed in Section 2, if such an ill condition was overlooked and the mesh generator created self intersections, the user needed to manually correct those self intersections.

The next section explains an automated method for removing self intersections included in a triangular mesh. The method substantially reduces the manual operations required to eliminate self intersections to recover the integrity of the mesh.

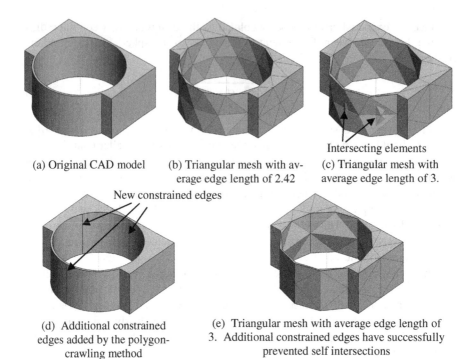

(a) Original CAD model

(b) Triangular mesh with average edge length of 2.42

Intersecting elements

(c) Triangular mesh with average edge length of 3.

New constrained edges

(d) Additional constrained edges added by the polygon-crawling method

(e) Triangular mesh with average edge length of 3. Additional constrained edges have successfully prevented self intersections

Fig. 3. A narrow clearance causing self intersections

4   Detail of the Proposed Method

4.1   Improvement Criteria

The proposed method uses three types of operations, edge swapping, edge hammering and face lifting, which are explained later in this section. Although the three operations are effective in reducing self intersections, those operations cannot always guarantee the successful reduction of self intersections. In the worst case, those operations could make the situation worse. Therefore, the proposed method needs criteria to test the improvement, and the proposed method will not apply the operations if the criteria cannot be met.

The improvement criteria used by the proposed method are as follows:

1.   The number of intersecting triangles decreases.
2.   No new intersecting triangle is created.

The first criterion guarantees that the process terminates within finite computational time. The mesh has a finite number of intersecting triangles, and the edge-swapping, edge-hammering and face-lifting operations do not change the total number of triangles. Therefore, the process terminates when no more reduction of the intersecting triangles is possible.

The second condition prevents the effect of the edge-swapping, edge-hammering, and face-lifting operations from propagating through the mesh. If the number of intersecting triangles is reduced by creating a new intersecting triangle, the new intersection needs to be removed by a subsequent edge-swapping, edge-hammering, or face-lifting operation. As a result, the effect of those operations could propagate through the mesh. Such propagation can be prevented by the second condition.

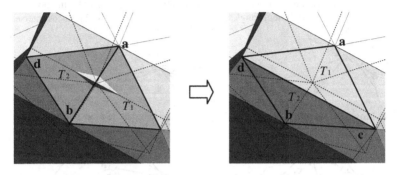

Fig. 4. Removing a self intersection by edge swapping

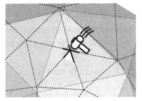 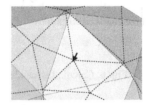

(a) Five triangles are intersecting with another triangle. Dashed lines are behind the outside layer of the triangular mesh.

(b) Intersections are removed by Edge Hammering

Fig. 5. An Example of edge hammering

4.2 Edge Swapping

Edge swapping reconnects an edge shared by two triangles. When an edge connecting nodes **a** and **b** is shared by two triangles T_1 and T_2, and if triangle T_1 is using nodes **a**, **b**, and **c**, and T_2 is using **b**, **a**, and **d**, triangles T_1 and T_2 are replaced with triangles **cad** and **dbc** as shown in Figure 4. Edge swapping removes edge **ab** and creates edge **cd**. The proposed method applies edge swapping to an edge used by an intersecting triangle if it satisfies the improvement criteria.

4.3 Edge Hammering

Edge Hammering moves a node and corrects an edge sticking out of another triangle. When a triangular mesh is created from two curved surfaces with a small clearance, a self intersection can be created as shown in Figure 5 (a). In this

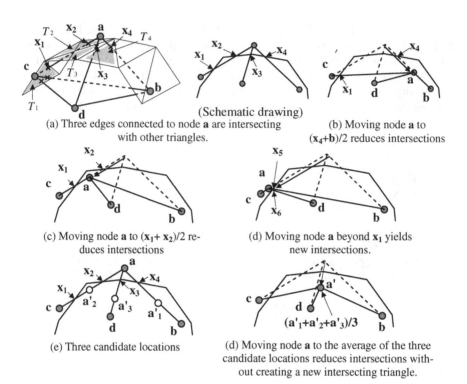

(Schematic drawing)
(a) Three edges connected to node **a** are intersecting with other triangles.

(b) Moving node **a** to $(x_4+b)/2$ reduces intersections

(c) Moving node **a** to $(x_1+x_2)/2$ reduces intersections

(d) Moving node **a** beyond x_1 yields new intersections.

(e) Three candidate locations

(d) Moving node **a** to the average of the three candidate locations reduces intersections without creating a new intersecting triangle.

Fig. 6. Calculating candidate points to which a node is moved

example, five edges from the inside-layer triangles are sticking out of an outside-layer triangle. Such a self intersection can be removed by moving the node of the intersecting edges as shown in Figure 5 (b).

If an edge is intersecting with another triangle at one point, the intersection on the edge will disappear by moving one of the edge nodes toward the other node beyond the intersecting point. For example, in Figure 6 (a), edge **ab** is intersecting with triangle T_4 at x_4. Edge **ab** will no longer have an intersection if node **a** is moved toward node **b** beyond x_4 as shown in Figure 6 (b). In this example, this move also makes triangles T_2, T_3, T_4, and **abd** free of intersection. The new location of node **a** can be anywhere between x_4 and **b** to make edge **ab** intersection-free. The mid point, $(x_4+b)/2$, is advantageous because it does not make edge **ab** too short and makes a reasonable clearance from x_4.

If an edge is intersecting with other triangles at more than one point, the intersection may be reduced by moving one of the edge nodes toward the other node beyond the first intersecting point before reaching the second. For example, edge **ac** in Figure 6 (a) is intersecting with triangles T_1 and T_2 at x_1 and x_2, respectively. If node **a** is moved toward node **c** beyond x_2 before x_1, intersection x_2 disappears, and triangles T_2, T_3, T_4, and **abd** become free of intersection as shown in Figure 6 (c). However, if node **a** is moved beyond x_1, a new intersection x_5 will be created as shown in Figure 6 (d). From this observation, if an edge intersects with other triangles at more than one point, one of the edge nodes should be moved toward the other node beyond the first intersection to reduce the number of intersections. However, moving the node beyond the second intersection will create a new intersection. Node **a** can be moved to anywhere between the first and second closest intersections. The simplest choice of the point is the mid point of the first two intersections.

Since a node can be connected to more than one intersecting edge, the node can have multiple candidate points. If edge $\mathbf{ab_n}$ is the nth intersecting edge connected to node **a**, the nth candidate point $\mathbf{a'_n}$ is calculated as:

(a) If edge $\mathbf{ab_n}$ is intersecting with another triangle at a single point **x**, candidate point $\mathbf{a'_n}$ is calculated as $\mathbf{a'_n} = (\mathbf{x} + \mathbf{b_n})/2$, and

(b) If edge $\mathbf{ab_n}$ is intersecting with other triangles at multiple points $\mathbf{x_1}$, $\mathbf{x_2}$, ..., and $\mathbf{x_k}$, and $\mathbf{x_1}$ and $\mathbf{x_2}$ are the first and second closest intersection to node **a**, candidate point $\mathbf{a'_n}$ is calculated as $\mathbf{a'_n} = (\mathbf{x_1} + \mathbf{x_2})/2$.

The proposed method moves node **a** to the average of the candidate points:

$$\mathbf{a'} = \sum \mathbf{a'_n}/n \,,$$

if the move satisfies the improvement criteria described in Section 4.1.

In the example shown in Figure 6, three candidate locations, $\mathbf{a'_1}$ and $\mathbf{a'_2}$ are calculated as:

$$\mathbf{a'_1} = (\mathbf{x_4} + \mathbf{b})/2$$

$$\mathbf{a'_2} = (\mathbf{x_1} + \mathbf{x_2})/2$$

$$\mathbf{a'_3} = (\mathbf{x_3} + \mathbf{d})/2,$$

as shown in Figure 6 (e), and node **a** is moved to $\mathbf{a'} = (\mathbf{a'_1} + \mathbf{a'_2} + \mathbf{a'_3})/3$ because triangles T_2, T_3, T_4, and **abd** become intersection free without making a new intersecting triangle. Although triangles T_1, **acb**, and **acd** are still intersecting, the move yields no new intersecting triangles, and the three remaining intersecting triangles can be made intersection free by subsequent edge hammering and face lifting operations.

4.4 Face Lifting

The face lifting operation moves a triangle at a time to reduce the number of intersections. For example, a highlighted triangle in Figure 7 (a) is intersecting with multiple edges. The face-lifting operation moves the triangle to its normal direction and removes the intersections as shown in Figure 7 (b).

Assume that triangle T_1, consisting of three nodes $\mathbf{t_1}$, $\mathbf{t_2}$, and $\mathbf{t_3}$, is intersecting with one or more edges, and the nodes $\mathbf{p_1}$, $\mathbf{p_2}$, ..., $\mathbf{p_n}$ are the nodes used by the intersecting edges. The unit normal vector of T_1 is $\mathbf{n_1}$, and ε is the given clearance requirement between a triangle and a node.

The minimum and maximum of the signed distances of nodes $\mathbf{p_i}$ $\{i=1 \text{ to } n\}$ relative to T_1, denoted as d_{min} and d_{max} respectively, are then calculated as:

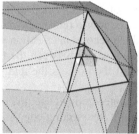 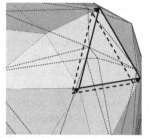

(a) A triangle intersecting with multiple edges

(b) Face-lifting operation removes the intersections

Fig. 7. Face-lifting operation

$$d_{min} = \min((\mathbf{p}_i - \mathbf{t}_i) \cdot \mathbf{n}_i),$$

$$d_{max} = \max((\mathbf{p}_i - \mathbf{t}_i) \cdot \mathbf{n}_i).$$

The proposed method moves triangle T_1 to two possible new locations, by offsetting $\mathbf{n}_i(d_{max} + \varepsilon)$ and $\mathbf{n}_i(d_{min} - \varepsilon)$ from the original location. And for each new location, the proposed method tests if the move satisfies the improvement criteria described in Section 4.1.

If only one of the two new locations satisfies the improvement criteria, triangle T_1 is moved to the new location that satisfies the improvement criteria. If both

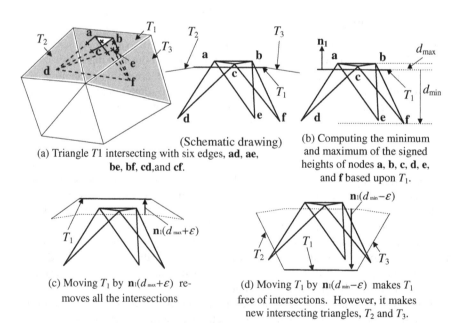

(Schematic drawing)

(a) Triangle $T1$ intersecting with six edges, **ad, ae, be, bf, cd**, and **cf**.

(b) Computing the minimum and maximum of the signed heights of nodes **a, b, c, d, e**, and **f** based upon T_1.

(c) Moving T_1 by $\mathbf{n}_i(d_{max} + \varepsilon)$ removes all the intersections

(d) Moving T_1 by $\mathbf{n}_i(d_{min} - \varepsilon)$ makes T_1 free of intersections. However, it makes new intersecting triangles, T_2 and T_3.

Fig. 8. Calculating possible new locations of an intersecting triangle

moves satisfy the improvement criteria (implies that the intersecting edges are disconnected from the other part of the mesh that T_1 belongs to,) T_1 is moved to the new location that makes the move smaller.

For example, in Figure 8 (a), triangle T_1 intersects with six edges, **ad**, **ae**, **be**, **bf**, **cd**, and **cf**. The minimum and maximum of the signed heights of the nodes of the intersecting edges are calculated as shown in Figure 8 (b). If the triangle T_1 is moved by $\mathbf{n}_1(d_{max}+\varepsilon)$, triangle T_1 becomes free of intersection without making a new intersecting triangle as shown in Figure 8 (c). Therefore, it satisfies the improvement criteria described in Section 4.1. However, if T_1 is moved by $\mathbf{n}_1(d_{min}-\varepsilon)$ as shown in Figure 8 (d), triangles T_2 and T_3, which were intersection free before the move, intersect with some of the edges. Therefore, it violates the improvement criterion. In this case, triangle T_1 is thus moved by $\mathbf{n}_1(d_{max}+\varepsilon)$ from the original location.

4.5 Order-Dependency Issues

The result of edge-swapping, edge-hammering, and face-lifting operations depends on the order in which the operations are applied. If the operations are applied to the mesh in an inappropriate order, they may either make an unnecessarily large deformation to the mesh or simply fail to reduce the self intersections. However, it is virtually impossible to find the ideal order of the operations.

In this research, we have tested several different measures in an attempt to reduce the adverse effect caused by the order dependency. In the first approach, we have applied edge swapping first, edge hammering second, and face lifting last. Since edge swapping does not move nodes, it seemed to have the least impact on the geometry. Edge hammering moves one node at a time while face lifting moves three nodes at a time. Edge hammering thus seems to have less impact on the geometry compared to face lifting. Although this approach gave somewhat good results, it often gave less than satisfactory results. That led to the two-phase approach described below.

To reduce the adverse effect caused by the order-dependency, the proposed method applies a sequence of edge-swapping, edge-hammering, and face-lifting operations in two phases: (1) cost-calculation phase and (2) implementation phase.

In the cost-calculation phase, a sequence of edge-swapping, edge-hammering, and face-lifting operations are applied to all applicable edges, nodes, and elements. After each operation, the cost of the operation is calculated, and then the modifications made to the mesh by the operation are retracted back to the original state of the mesh. Therefore, the mesh does not change before and after the cost-calculation phase.

The cost of an operation is a measurement that evaluates the magnitude of deformation of the mesh made by the operation, which can be measured as a change of volume made by the operation. The cost of the edge-swapping operation is the volume of a tetrahedron enclosed by the two triangles deleted by the operation and the two new triangles created by the operation as shown in Figure 9 (a).

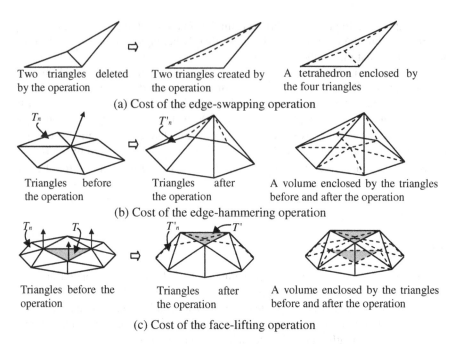

Two triangles deleted by the operation | Two triangles created by the operation | A tetrahedron enclosed by the four triangles

(a) Cost of the edge-swapping operation

Triangles before the operation | Triangles after the operation | A volume enclosed by the triangles before and after the operation

(b) Cost of the edge-hammering operation

Triangles before the operation | Triangles after the operation | A volume enclosed by the triangles before and after the operation

(c) Cost of the face-lifting operation

Fig. 9. Cost of the edge-swapping, edge-hammering, and face-lifting operations

The cost of the edge-hammering operation is calculated as follows. Let T_1, T_2, ..., T_n be the triangles sharing the node to be moved by the operation. The operation moves the node, and T_n becomes T'_n. The cost of the edge-hammering operation is the volume enclosed by T_1, T_2, ..., T_n, and T'_1, T'_2,..., T'_n as shown in Figure 9 (b).

The cost of the face-lifting operation is calculated as follows. Let T be the triangle to be moved by the operation and T_1, T_2, ..., T_n be the triangles sharing at least one node with T. The operation moves T to T', and T_n becomes T'_n. The cost of the edge-hammering operation is the volume enclosed by T, T', T_1, T_2, ..., T_n, and T'_1, T'_2,..., T'_n as shown in Figure 9 (c).

Then, in the implementation phase, the edge-swapping, edge-hammering, and face-lifting operations are applied to the mesh in the order of ascending cost.

5 Potential Expansions and Discussions

5.1 Expanding the Proposed Method for a Quadrilateral Mesh

The proposed method can easily be expanded so that it can deal with a quadrilateral mesh by the following steps:

1. Tessellate every quadrilateral by adding a diagonal edge while keeping track of which triangle comes from which quadrilateral.

2. Apply the proposed method.

3. Merge triangles to re-construct the quadrilaterals.

However, there are two remaining issues that need to be addressed for this expansion as follows.

(a)   Edge preservation vs. effectiveness

Some quadrilaterals may not be re-constructed after applying the proposed method due to edge swapping. To guarantee the re-construction, edge swapping needs to be carefully applied so that edges of the original quadrilaterals are preserved. However, this limitation could reduce the effectiveness of the proposed method.

(b)   Non-planar quadrilateral

Four nodes of a quadrilateral element may not be co-planar, and the geometry of a quadrilateral element is a bi-linear surface. Therefore, tessellating a quadrilateral into two triangles changes the geometry, and some self intersections may not be detected after the tessellation. Or, even if the triangular mesh is made self-intersection free by the proposed method, it does not guarantee that the quadrilateral mesh is also self-intersection free after quadrilateral elements are re-constructed.

Further research is needed for addressing these issues.

5.2 Locally Adjusting the Clearance Requirement

The proposed method takes the parameter ε, a clearance requirement for the face-lifting operation, as input. The face-lifting operation moves an intersecting triangular element in its normal direction so that it will have a clearance of ε.

However, in some cases, it is ideal to vary ε over the domain. For example, if the thickness of the geometry substantially changes over the domain, ε should be proportional to the local thickness of the domain.

The proposed method can be adapted to such requirement with a small modification. When triangle T_1 is being moved by the face-lifting operation, instead of taking ε as a user input, it can be calculated as follows.

Assume triangle T_1 consists of nodes p_1, p_2, and p_3, and its normal is n. t_n is the distance that a ray travels from p_n in the direction $-n$ until it hits another triangular element as shown in Figure 10. If the ray does not hit another triangle, t_n is zero. Each of t_n gives a rough estimation of the thickness at p_n. If at least two of the three t_n s are non-zero, ε is taken from the median of t_n s. If only one t_n is non-zero, ε is taken from the non-zero t_n. If all three t_n s are zero, the face-lifting operation is not attempted, and this intersection is deferred to the edge-swapping and edge-hammering operations.

In some applications, the clearance can also be specified relative to the edge length of the triangle to be moved. For example, if the triangular mesh is used as the boundary of a tetrahedral mesh, and if the maximum aspect ratio of the tetrahedral mesh needs to be smaller than γ, the required clearance ε must be $1/\gamma$ times the longest edge length of the triangle. In this application, a similar adaptation is also possible for the edge-hammering operation.

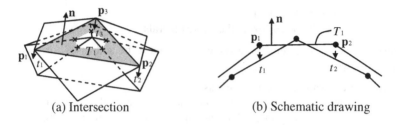

(a) Intersection   (b) Schematic drawing

Fig. 10. Estimating the clearance requirement from the local thickness

5.3 Boundary Fidelity Issue

The edge-hammering and face-lifting operations move nodes of the input triangular mesh. The mesh generator usually places nodes exactly on the original surfaces. However, the nodes may be moved away from the original surfaces by the proposed method. In other words, edge-hammering and face-lifting operations trade the boundary fidelity for removal of self intersections. It raises an obvious question: if it is allowed to sacrifice boundary fidelity to remove self intersections.

A surface mesh needs to satisfy multiple requirements including edge length, clear of self-intersection, element quality, in addition to boundary fidelity. As presented in Section 6, self intersections in a boundary triangular mesh yield a tetrahedral mesh that is utterly unusable in the finite element simulation. It implies that self intersections can be more problematic than loss of the boundary fidelity. If the boundary fidelity is very important, another option is to make smaller elements so that no self intersection is created as discussed in Section 3.

In summary, those mesh requirements, edge length, clear of self intersection, element quality, and boundary fidelity, are in the relation of trade offs and may not be satisfied all together. Priorities of those requirements must be considered based on the best outcome of the application in which the mesh is used. It is possible that boundary fidelity is not the top priority, and then diverting the nodes away from the original surface to remove self intersections can be justified.

Loss of the boundary fidelity could impact the outcome of the application in which the mesh is used, and therefore the diversion should be kept as small as possible. Nonetheless, a certain loss of boundary fidelity should be tolerated in order to obtain the best outcome in the application.

6 Examples

This section presents some examples to demonstrate the effectiveness of the proposed method. Figure 11 (a) shows a thin-walled solid with ribs and a screw hole. Such a shape frequently appears in a geometric model of a plastic-casing. Figure 11 shows a triangular mesh with an average edge length of 1.5. This mesh does not include self intersections. However, the number of elements can be too many for the application in which the mesh is used. The number of elements can be reduced by specifying larger element size. Figure 11 (c) shows a triangular mesh with an average edge length of 3mm. The middle section of the outer surface of the screw hole

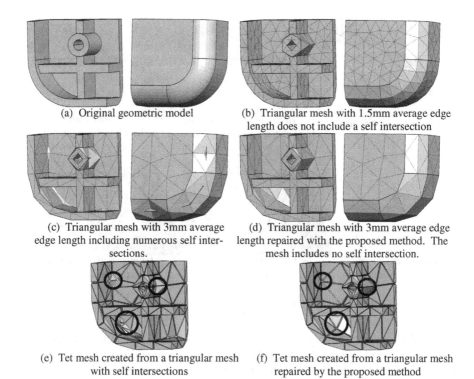

(a) Original geometric model

(b) Triangular mesh with 1.5mm average edge length does not include a self intersection

(c) Triangular mesh with 3mm average edge length including numerous self intersections.

(d) Triangular mesh with 3mm average edge length repaired with the proposed method. The mesh includes no self intersection.

(e) Tet mesh created from a triangular mesh with self intersections

(f) Tet mesh created from a triangular mesh repaired by the proposed method

Fig. 11. Sample model with ribs and a screw hole

is collapsed to a single edge, and it intersects with the inside wall. Some inside-wall elements also collide with the outer-wall of the fillet. The proposed method successfully removes all self intersections as shown in Figure 11 (d). Figure 11 (e) shows a tetrahedral mesh created from the triangular mesh shown in Figure 11 (c). Due to the self intersections in the input triangular mesh, the output tetrahedral mesh has some exterior edges that are shared by more than two exterior triangular elements. Figure 11 (f) shows a tetrahedral mesh created from the triangular mesh shown in Figure 11 (d). Since no self intersection is included in the triangular mesh, the output tetrahedral mesh is a valid and usable in the finite element analysis.

Figure 12 (a) shows a CAD model of a plastic casing of a telephone. Figure 12 (b) is a triangular mesh of the model meshed with an average edge length of 4mm. With this edge length, the mesh generator does not create a self intersection. Figure 12 (c) is a triangular mesh with 10mm average edge length. This edge length yields numerous self intersections as shown in Figure 12 (d). Note that the majority of the elements are intersection free. Only a small fraction of the elements are intersecting. Despite the small percentage of intersecting elements, these self intersections could make the mesh utterly unusable for the application. The proposed method successfully removes all the intersections. The changes made by the

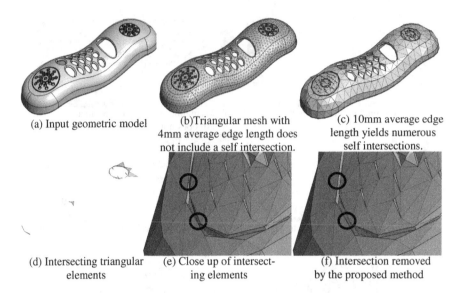

(a) Input geometric model

(b)Triangular mesh with 4mm average edge length does not include a self intersection.

(c) 10mm average edge length yields numerous self intersections.

(d) Intersecting triangular elements

(e) Close up of intersecting elements

(f) Intersection removed by the proposed method

Fig. 12. Sample model of a plastic casing of a telephone

proposed method are hardly visible without zooming into the specific locations. Figure 12 (e) shows a close up look of two of the self intersections included in the 10mm mesh, and the proposed method removes them as shown in Figure 12 (f).

7 Conclusions

This paper has presented a computational method for removing self intersections included in a triangular mesh. The method systematically applies edge-swapping, edge-hammering, and face-lifting operations to the input triangular mesh. The proposed method has been applied to many test cases, and the results showed the effectiveness of the proposed method.

References

[1] Barequet, G., Sharir, M.: Filling Gaps in the Boundary of a Polyhedron. Computer Aided Geometric Design 12, 207–229 (1995)
[2] Barequet, G., Kumar, S.: Repairing CAD Models. In: Proceedings of IEEE Visualization 1997, pp. 363–370 (1997)
[3] Gueziec, A., Taubin, G., Lazarus, F., Horn, B.: Cutting and Stitching: Converting Sets of Polygons to Manifold Surface. IEEE Transactions on Visualization and Computer Graphics 7, 136–151 (2001)
[4] Branch, J., Prieto, F., Boulanger, P.: A Hole-Filling Algorithm for Triangular Meshes Using Local Radial Basis Function. In: Proceedings of 15th International Meshing Roundtable, pp. 411–431 (2006)

[5] Li, F., Chen, B., Leng, W.-h.: A Hole Repairing Method for Triangle Mesh Surfaces. In: Proceedings of International Conference on Computational Intelligence and Security, pp. 972–975 (2006)

[6] Ribelles, J., Heckbert, P., Garland, M., Stahovich, T., Srivastava, V.: Finding and Removing Features from Polyhedra. In: Proceedings of ASME Design Automation Conference (2001)

[7] Lee, K.Y., Armstrong, C.G., Price, M.A., Lamont, J.H.: A Small Feature Suppression/Unsuppression System for Preparing B-Rep Models for Analysis. In: Proceedings of ACM Symposium on Solid and Physical Modeling, pp. 113–124 (2005)

[8] Nomura, M., Hamada, N.: Feature Edge Extraction from 3D Triangular Meshes Using a Thinning Algorithm. In: Proceedings of SPIE - Vision Geometry X, pp. 34–41 (2001)

[9] Jiao, X., Heath, M.T.: Feature Detection for Surface Meshes. In: Proceedings of 8th International Conference on Numerical Grid Generation in Computational Field Simulations, pp. 705–714 (2002)

[10] Baker, T.J.: Identification and Preservation of Surface Features. In: Proceedings of 13th International Meshing Roundtable, pp. 299–309 (2004)

[11] Yoshizawa, S., Belyaev, A., Seidel, H.-P.: Fast and Robust Detection of Crest Lines on Meshes. In: Proceedings of ACM Symposium on Solid and Physical Modeling, pp. 227–232 (2005)

[12] Yamakawa, S., Shimada, K.: Polygon Crawling: Feature-Edge Extraction from a General Polygonal Surface for Mesh Generation. In: Proceedings of 14th International Meshing Roundtable, pp. 257–275 (2005)

[13] Jiao, X.: Volume and Feature Preservation in Surface Mesh Optimization. In: Proceedings of 15th International Meshing Roundtable, pp. 359–373 (2006)

[14] Quadros, W.R., Vyas, V., Brewer, M., Owen, S.J., Shimada, K.: A Computational Framework for Generating Sizing Function in Assembly Meshing. In: Proceedings of 14th International Meshing Roundtable, pp. 55–72 (2005)

[15] Glimm, J., Simanca, S.R., Tan, D., Tangerman, F.M., Vanderwoude, G.: Front Tracking Simulations of Ion Deposition and Resputtering. SIAM Journal on Scientific Computing 20, 1905–1920 (1999)

Conformal Refinement of Unstructured Quadrilateral Meshes

Rao Garimella

MS B284, Los Alamos National Laboratory, P.O. Box 1663,
Los Alamos, NM 87544
rao@lanl.gov

Abstract. A multilevel adaptive refinement technique is presented for unstructured quadrilateral meshes in which the mesh is kept conformal at all times. This means that the refined mesh, like the original, is formed of only quadrilateral elements that intersect strictly along edges or at vertices, i.e., vertices of one quadrilateral element do not lie in an edge of another quadrilateral. Elements are refined using templates based on 1:3 refinement of edges. It is demonstrated that by careful design of the refinement and coarsening strategy, high quality elements can be maintained in the refined mesh. The method is demonstrated on a number of examples with dynamically changing refinement regions.

1 Introduction

Adaptive mesh refinement is a well-known and widely employed technique for accurately capturing special features of the solution in steady and unsteady simulations. In such simulations, adaptive refinement enables the capturing of complex solution features by focusing refinement in critical areas without having to refine the mesh everywhere. Adaptive mesh refinement is now standard practice in simplicial meshes (triangular and tetrahedral) in a wide variety of applications. The unique topological properties of simplices allow the refinement in such meshes to be confined to fairly local regions while maintaining a high element quality [10] and keeping the mesh conforming. Conformity of the mesh implies that the intersection of a pair of elements, if not null, is strictly a lower dimensional mesh entity such as a face, an edge or a vertex. Non-conformity of mesh is commonly interpreted to mean that a lower order boundary entity (e.g. a vertex) of one element lies on a higher order boundary entity (e.g. an edge) of another element.

For quadrilateral meshes, the most common approach to adaptation is to refine elements in a non-conformal way. This allows the refinement to remain local but introduces non-conformal nodes which lie on the edges of neighboring elements. However, mesh non-conformity necessitates augmentation of the PDE solution algorithm to deal with the special nodes. Non-conformity is typically dealt with by constraining the solution at the non-conformal nodes to be dependent on the solution at the nodes of the edge it lies on using constraint equations [15] or Lagrange multipliers [4] or by the use of mortar elements to link the non-matching elements [1].

In this research, a technique is described to refine an unstructured quadrilateral mesh such that the result is also a hierarchically refined, conforming mesh of only quadrilaterals with high quality albeit a little worse than the parent mesh quality.

2 Previous Work

There has been considerable research on conformal triangular refinement for adaptive simulations since termination of refinement for simplices is very easy (see, for example, [12]). However, for quadrilateral meshes most researchers choose to use non-conformal quadtree type refinement with specialized code to handle non-conformal nodes (see, for example, [3]). There have been only a few articles describing conformal quadrilateral mesh refinement and coarsening, and even fewer that deal with the issue in a dynamic setting, i.e., conformally refining and coarsening a quadrilateral mesh that has been previously refined.

One of the best known papers on the issue of conformal quadrilateral refinement is by Schneiders [14]. In the paper, Schneiders discusses 2-refinement (bisection of edges) and 3-refinement (trisection of edges). He chooses the trisection of edges because it simplifies the algorithm. The refinement information is propagated from elements to nodes and refinement templates are defined based on the number of marked nodes (See Figure 1). The refinement templates are chosen such that the scheme is stable, i.e., the quality of elements does not deteriorate with increasing refinement levels. However, even though in this research, uniformly refined quadrilaterals have trisected edges and are split into 9 child quadrilaterals, templates used in adjacent elements to terminate the refinement have bisected edges as seen in the figure. In general, Schneiders scheme is more complicated to implement than the scheme presented here. Still, it is a valid scheme for conformal quadrilateral refinement and has been used by other researchers such as Zhang and Bajaj [18]. Schneiders has extended the work to hexahedral refinement as well but correctly points out that certain refinement patterns for the faces of hexahedra may not admit a valid decomposition of the parent hexahedron. Ito et al. have also used Schneiders' approach for octree based hexahedral refinement templates [8].

Tchon et al. have proposed a quadrilateral refinement strategy in which they find layers of elements, shrink the layers of elements and reconnect the shrunk layer with the surrounding mesh [17]. Clearly this strategy assumes certain structure to the mesh and specific refinement patterns while ignoring the issues of multiple levels of refinement, mesh quality and dynamic adaptation. Hence, the approach is of limited utility.

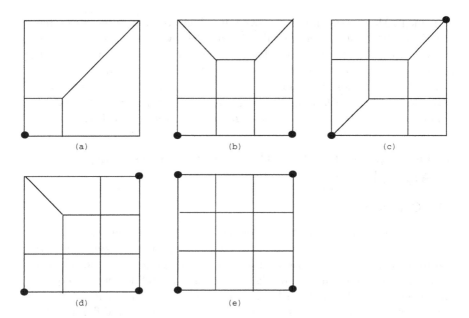

Fig. 1. Schneiders' subdivision templates for quadrilateral refinement (refinement vertices are marked with circles) (a) unrefined quadrilateral (b) one vertex marked (c) two adjacent vertices marked (d) two diagonally opposite vertices marked (e) three vertices marked (f) all vertices marked (uniform refinement of quadrilateral).

Several researchers have proposed a quadrilateral refinement strategy where the end result is a mixture of quadrilaterals and triangles, for example [5]. Similarly, others have proposed hexahedral refinement strategies which result in a combination of hexahedra and prisms. However, this conflicts with the stated goal of achieving a conforming all-quadrilateral or all-hexahedral mesh.

Benzley et al. have proposed quadrilateral mesh coarsening strategies that are quite general and do have an advantage over nested refinement strategies in that they can coarsen beyond the original resolution of the mesh [16].

The research that is closest to the presented is the work by Sandhu et al. [13] although this work was developed without knowledge of this earlier research. In this work, Sandhu et al. use node marking and trisection of edges to define templates for refining elements and terminating the refinement. They define one less than the number of templates used in this work. Similar to this work, they also recommend undoing non-uniform refinement of quadrilaterals before further refinement to maintain quality. However, all their examples show only static refinement and aspects of dynamic refinement such as coarsening, remapping etc. are not explored.

This paper discusses a dynamic mesh adaptation strategy for quadrilateral meshes that results in a conformal all-quadrilateral mesh with nested refinement. Moreover, while not proved, it is believed that in piecewise linear complexes the resulting mesh quality is bounded by the quality of the parent mesh regardless of the number of levels of refinement at each time step or the number of time steps in the mesh. The adapted mesh is suitable for use in a wide range simulations without any special procedures since it is composed of only conformal quadrilaterals. Finally, the nested refinement allows for easy remapping of cell based quantities from one time step to another.

3   Description of Mesh Refinement/Coarsening Algorithm

3.1   Overview

The adaptive mesh modification algorithm starts with tagging elements that must be refined because they do not adequately represent some geometric feature or because the solution error in these elements is deemed to be too high. These elements and their edges are tagged for refinement (or coarsening), if necessary, to multiple levels below (or above) their current level of refinement[1]. When an edge is adjacent to two elements with different refinement levels, it is refined to the higher of the two levels. Once the appropriate elements have been tagged by the application, the mesh is coarsened wherever the application requests the elements to be larger than they currently are. After coarsening, the mesh is refined wherever the application requests elements to be smaller than they currently are. During both coarsening and refinement, the target refinement levels of elements are adjusted so that they are consistent with their siblings (children of their parents) and such that the target refinement levels of two adjacent elements do not differ by more than one. The one-level difference rule ensures that the number of templates required to make the mesh conforming is limited to a manageable number and that the mesh is smoothly graded.

3.2   Subdivision Templates

When some elements in the mesh get uniformly refined, one or more edges of adjacent quadrilaterals are also refined. To make the mesh strictly conforming, these adjacent elements must also be subdivided into quadrilaterals such that the refinement terminates. To facilitate conformal subdivision of elements that are not uniformly refined, edges are trisected instead of being bisected as in triangular meshes. The reason for choosing trisection over bisection is that if an odd number of edges of a quadrilateral were bisected, the resulting polygon would have an odd number of edges and could not be subdivided into quadrilaterals in a self contained manner. The templates used

---

[1] Regardless of whether an element is being coarsened or refinement, it will always be referred to its target level in the heirarchy of meshes below the coarsest mesh as its target level of refinement.

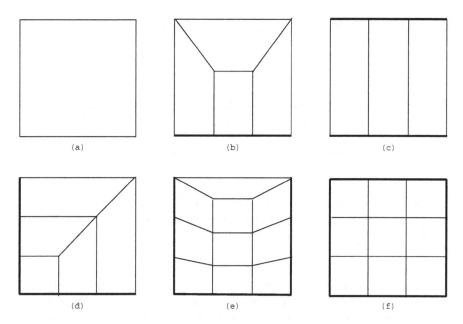

Fig. 2. Subdivision templates for quadrilateral refinement (thick edges are refined edges) (a) unrefined quadrilateral (b) one edge refined (c) two opposite edges refined (d) two adjacent edges refined (e) three edges refined (f) all edges refined (uniform refinement of quadrilateral).

for subdividing quadrilaterals with different edges refined are shown below in Figure 2. Some of these templates have been described in previous works [13] and some are new.

The quadrilaterals that result from uniform refinement of a parent quadrilateral are called *regular* elements while quadrilaterals resulting from refinement of one, two or three edges of the parent quadrilateral are called *irregular* quadrilaterals.

It must be pointed out that the templates described above are different from the templates in Schneiders' work. In that work, refinement tags are transmitted to vertices of elements and templates derived from the combinations of vertices tagged for refinement. Those templates are shown in Figure 1. As can be seen from the picture, the only template the two approaches have in common is the uniform refinement template. In the remaining cases, edges of elements that adjacent to uniformly refined elements are refined using an irregular 1:2 pattern. Also, even if only one edge of an element adjacent to a uniformly refined element is refined, the template proposed by Schneiders refines two other edges of the element. This in turn forces refinement of other elements. Figure 3 shows a simple example of this over-refinement as a consequence of uniform refinement of the central element in a 3x3 mesh of quadrilaterals. As can be seen in the picture, Schneiders' scheme modifies

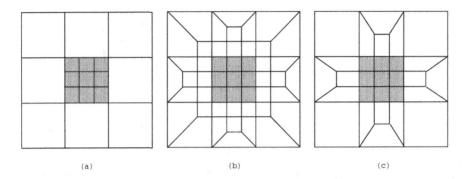

Fig. 3. Comparison of Schneiders' scheme and the proposed scheme of refinement on a 3x3 grid of quadrilaterals (a) Central element refined uniformly (b) Surrounding mesh made conforming by Schneiders' scheme (c) Surrounding mesh made conforming by proposed scheme.

every element in the 3x3 mesh while the proposed scheme affects only the edge connected neighbors. This leads us to the conclusion that the Schneiders' scheme is a little more complex to implement that the current scheme in which termination requires refinement of only edge neighboring faces. In terms of numbers of elements, the number of extra elements created by both schemes seems comparable for large problems.

3.3   Coarsening

In the mesh adaptation method presented here, coarsening of elements is done first before refinement. In this approach, coarsening is performed strictly using the knowledge of the hierarchical structure of the adapted mesh, i.e., if an element is to be deleted then its siblings are also deleted simultaneously and the parent element is restored. For this reason, the coarsening strategy of this paper cannot coarsen beyond the original mesh. Coarsening is performed on elements whose current refinement level is higher than the target refinement level. Before actual deletion of elements, however, the target levels of elements are adjusted to ensure that there is not more than one level of difference between two adjacent elements and that the target levels of siblings are consistent.

Consider an element whose current refinement level is L_c and target refinement level is L_t. Assume the maximum refinement level of all of its edge connected neighbors (and therefore, all of its edges) is L_a. Then, if the target refinement level of this element is greater than one level less than the maximum target level of its edges, then set the target level to be exactly one less than the maximum target level of its edges. Algorithmically, this can be expressed more succinctly as: if $L_t < L_a - 1$, then $L_t = L_a - 1$ (See Figure 4a). For example, if for a particular element $L_c = 5$, $L_t = 1$ and the neighbors have targets of 1, 3, 1, 2. Then $L_a = 3$ and L_t is set to 2.

For making the refinement levels consistent between siblings, a conservative approach is taken and the element and its siblings are marked for coarsening only up to the maximum level (smallest size), L_s, allowed by the element and all its siblings. So, if $L_t <= L_s < L_c$ then $L_t = L_s$. For example, if $L_c = 5$, $L_t = 1$ for an element, but $L_s = 3$, i.e., one of the siblings of the element has a target level of 3 (Figure 4c). Then the current element cannot be coarsened to a level lower than $L_t = 3$. On the other hand if $L_t < L_c < L_s$, then $L_t = L_c$. For example, $L_c = 5$, $L_t = 1$ as before, but $L_s = 7$, i.e., a sibling wants to be refined from the current level while the element wants to be coarsened. Then the element cannot be coarsened above the current level, $L_t = L_c$ (Figure 4d).

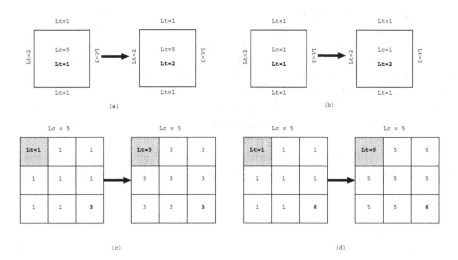

Fig. 4. Level adjustment during coarsening and refinement (a)(b) Refinement level adjusted due to maximum level of edges (c) Coarsening level adjusted due to target level of sibling (d) No coarsening allowed above current level because sibling wants to be further refined.

Next the elements are coarsened one level at a time starting from the highest level. Every time an element and its siblings are deleted we transmit the target refinement level to its parent. After coarsening the mesh at a particular level, the level adjustment is redone before coarsening at the next lower level.

3.4   Refinement

The most important rule imposed during refinement is that *irregular elements are never refined* as their repeated subdivision can lead to unbounded deterioration of quality. Instead, whenever an irregular element is tagged for refinement, the element and its siblings are deleted and its parent element is tagged for uniform refinement upto the maximum level requested by the element and its siblings. This rule ensures that the quality of the refined mesh is always bounded by the quality of the parent mesh. Schneiders defines refinement schemes with this property as being stable [14].

Refinement of the mesh and adjustment of levels before refinement is bit more complex than coarsening because regular and irregular elements have to be dealt with separately. On the other hand, level adjustment has to be done only once before multilevel refinement as opposed to the doing it at each level for coarsening.

To do level adjustment for refinement, the algorithm looks at each element whose target refinement level L_t is higher than its current refinement level, L_c. Then it gets the maximum refinement level, L_a of all its edge connected neighbors. As before, if its target refinement level is greater than one level lower than the maximum target level of its edges, i.e. $L_t < L_a - 1$, then the target level of this element is adjusted as $L_t = L_a - 1$. Also, if the element is irregular and one of its edges is to be refined, then the element and its siblings are marked for deletion and its parent is marked for refinement to L_t. Finally, if two adjacent elements are to be subdivided irregularly, it is ensured that the common edge of the two is also to be subdivided. This ensures better element quality as shown in Figure 5.

Then the algorithm deletes the irregular faces and subdivides the remaining faces according to the templates based on the number of edges that are

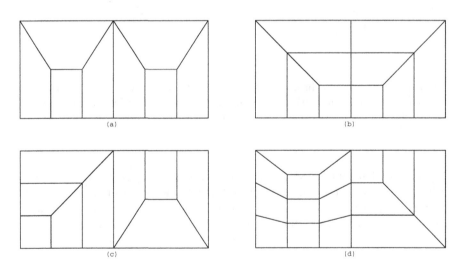

Fig. 5. Refining the common edge between two irregularly refined elements (a) Two adjacent elements with one edge refined (b) same elements with their common edge refined (c) One element with two edges refined next to an element with one edge refined (d) same elements with their common edge refined

refined. Every time an element is refined, all its children are marked with the target refinement level. The procedure continues to iterate over the mesh elements until all elements have reached their target level of refinement.

4   Remapping or Solution Transfer

Unlike mesh adaptation for capturing geometry, adaptation to reduce solution error for solving a PDE is tightly coupled with the issue of remapping or transfer of quantities from the base mesh to the adapted mesh. There are two types of solution quantities that must be remapped between meshes - integral quantities such as mass or energy and pointwise quantities such as diffusivity. The remapping of both quantities must be done accurately and remapping of integral quantities must be done in a conservative manner (for example, the densities of the child elements must be assigned such that the total mass of the parent element is conserved). When a group of elements is coarsened, one can just sum up integral quantities such as mass (or energy) over the child elements and assign it to the parent. For pointwise quantities, one can take an average of the values for the children weighted by their areas. On the other hand, when an element is refined uniformly, then one can equidistribute the mass over the children (less accurate) or do a linear reconstruction of the density function over the parent element and integrate over each child to get its mass (more accurate) [2]. Likewise pointwise properties such as diffusivity can also be linearly reconstructed over the parent's neighborhood and an accurate value derived for the child. Field variables (such as velocity) can be obtained by either evaluating an interpolant over the parent at nodes of the child or by solving a local problem over the refined elements using the solution over the base mesh to impose boundary conditions for the local problem. Also, in the proposed algorithm, special care must be taken for remapping when irregular elements are targeted for refinement since the mesh is coarsened back to the parent element and refined down uniformly. Using a summation of masses of the irregular children to get a mass for the parent element and then redistributing it to the regular children can be a poor choice and will lead to lower order accuracy remapping. Rather, it is better to use an intersection based remapping routine locally to get second order accuracy [6].

5   Results

First a static example of refinement is presented of a structured mesh mesh adapted to a superimposed line in the mesh. Any element that is intersected by the line is refined up to level 3 (level 0 is the original mesh). The super-imposed line goes from $(-0.308207, 1.106007)$ to $(1.106007 - 0.308207)$. The quality of the mesh before and after the refinement is also compared in terms of the average condition number of the element, $\bar{\kappa}$, defined as the mean of the condition numbers [9, 7] at all corners of the element. One can see from the

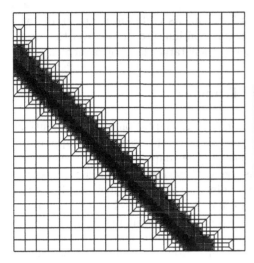

$\bar{\kappa}$	Original	Refined
1.0 – 1.5	400	12412
1.5 – 2.0	0	1396
2.0 – 3.0	0	0
4.0 – 5.0	0	0
5.0 –	0	0

Fig. 6. A 20x20 structured mesh refined using distance from center as the refinement criterion

histogram of the refined mesh that it is not shifted dramatically from the ideal case and that the worst quality element has an average condition number of only 1.69. In fact, in simulations where a line is moved diagonally across the domain and the mesh refined around it, the worst element condition number stays at 1.69. Also, the worst element quality stays at 1.69 regardless of what refinement level is applied to elements intersected by the line.

Next refinement induced by the same line in an unstructured quadrilateral mesh is demonstrated. Figure 7a shows the original mesh with the elements marked for refinement to level 3 due to intersection with the line (also shown). Figure 7b shows the refined mesh after the levels have been adjusted to enforce a one-level difference between adjacent elements. Also included is a table showing the distribution of condition numbers before and after refinement. The worst condition number goes from 3.12 to 3.79 after refinement.

In the following example, several snapshots from an dynamic adaptation procedure are shown where a circle of radius 0.1 is moved along a circular path in the domain. The center of the circle traces a circle of radius 0.2 centered at $(0.5, 0.5)$. The starting point of the circle center is $(0.7, 0.5)$. The target size for the elements to be refined is $0.05d$ where d is the distance between the centroid of the element and the center of the circle. As the circle moves, previously refined parts of the mesh are coarsened and new parts are refined with considerable overlap between the coarsened and refined regions. As expected, the worst element quality stays at 1.69 throughout the dynamic adaptation procedure.

Finally, several snapshots of a dynamic adaptation procedure are shown in which elements intersecting one of two expanding circles are refined to

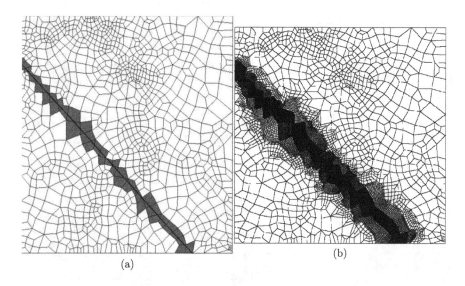

$\bar{\kappa}$	Original	Refined
1.0 – 1.5	1007	42491
1.5 – 2.0	15	1776
2.0 – 3.0	1	504
3.0 – 4.0	1	15
4.0 – 5.0	0	0
5.0 – 7.5	0	0

(a) (b) (c)

Fig. 7. Refinement of an unstructured mesh along a line (a) Target refinement levels (b) Refined mesh (c) Histograms of condition numbers

level 3 and elements intersecting both circles are refined to level 4. One circle is centered at $(0.0, 0.0)$ and the other circle is centered at $(1.0, 0.25)$. Both circles start with a radius of 0.11 with their radii increasing in increments of 0.05. As the circles grow, they intersect each other and eventually grow out of the domain. Elements that intersect one or the other circle are refined to a level of 3 while elements that intersect both circles are refined to a level of 4. Again the worst quality is stays fixed at 1.69 throughout the adaptation process.

6 Discussion

This paper presented a comprehensive mesh adaptation procedure for quadrilaterals that results in conformal meshes with nested refinement. The refinement is based on templates devised from a consistent 1:3 refinement of element edges. It also presented algorithms for adjustment of refinement levels of elements, both for coarsening and for refinement, such that there is never

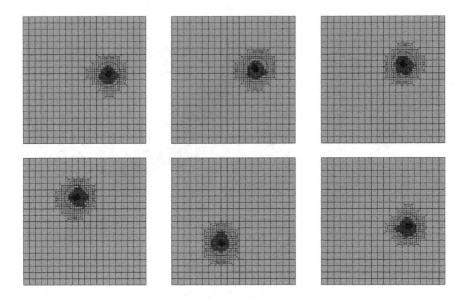

Fig. 8. Snapshots of dynamic mesh adaptation of a 20x20 structured mesh with respect to a circle rotating about the center of the domain

more than a one level difference between the refinement levels of adjacent elements. The quality of the refined mesh is kept high by never refining irregular elements used to bridge refined and coarse regions of the mesh. Instead if irregular elements must be refined, they are deleted and their parent is uniformly refined instead. Using several dynamic mesh adaptation examples, it was shown that the procedure effectively refines the mesh where necessary and coarsens it where it is not.

Although one other paper discussing similar templates and strategy was found after this algorithm was devised, that paper does not discuss dynamic mesh adaptation and mesh coarsening explicitly although it too suggests that irregular elements not be refined.

Compared to the algorithm proposed by Schneiders and the templates in his papers, this algorithm produces fewer elements and is simpler due to the consistent use of 1:3 edge refinement. Also, Schneiders does not discuss the issue of mesh quality when forced to refine irregular elements. Finally, the issue of solution transfer or variable remapping is addressed in the current paper which is often ignored in most conformal quadrilateral refinement papers.

In 3D, the combinatorial complexity of the current algorithm could be more complex than that of Schneiders' algorithm. That is because this algorithm tags edges instead of vertices for refinement, thereby resulting in $\sum_{i=0}^{12} {}^{12}C_i = 4096$ possible combinations. Of course, many of these can be eliminated due to symmetry of rotation and inversion. Even so, the number is expected to be higher than in Schneiders' algorithm. Also, it is possible, just

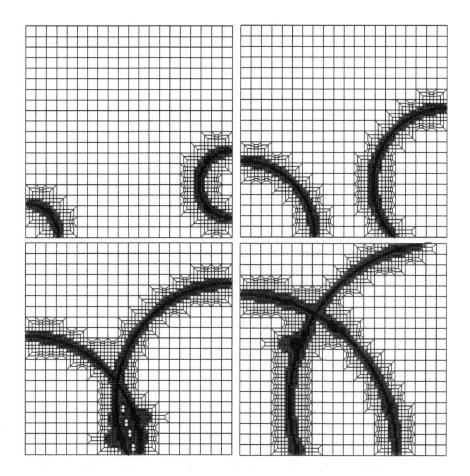

Fig. 9. Snapshots of dynamic mesh adaptation of a 20x20 structured mesh with respect to two expanding circles

like in Schneiders' algorithm, that some subdivisions of the hexahedron faces may not admit a subdivision into hexahedra. In such a case, one can refine additional edges of such hexahedra to be able to mesh them and propagate the refinement further. In such a case, one can only hope that the refinement does not consume the entire mesh. Alternatively, one can use the modified templates proposed by Parrish et al. to terminate the refinement [11]. This topic will be addressed in future work.

References

1. Arbogast, T., Cowsar, L.C., Wheeler, M.F., Yotov, I.: Mixed finite element methods on nonmatching multiblock grids. SIAM Journal on Numerical Analysis 37(4), 1295–1315 (2000)

2. Barth, T.J., Jespersen, D.C.: The design and application of upwind schemes on unstructured meshes. In: Proceedings of the 27th Aeroscience Meeting, Reno, NV, AIAA (January 1989)
3. Bass, J.M., Oden, J.T.: Adaptive finite element methods for a class of evolution problems in viscoplasticity. International Journal of Engineering Science 25(6), 623–653 (1987)
4. Bathe, K.J., Bouzinov, P.A.: On the constraint function method for contact problems. Computers and Structures 64, 1069–1085 (1997)
5. Botkin, M.E., Wentorf, R., Karamete, B.K., Raghupathy, R.: Adaptive refinement of quadrilateral finite element shell meshes
6. Garimella, R.V., Kucharik, M., Shashkov, M.J.: An efficient linearity and bound preserving conservative interpolation (remapping) on polyhedral meshes. Computers and Fluids 36(2), 224–237 (2007)
7. Garimella, R.V., Shashkov, M.J., Knupp, P.M.: Triangular and quadrilateral surface mesh quality optimization using local parametrization. Computer Methods in Applied Mechanics and Engineering 193(9-11), 913–928 (2004)
8. Ito, Y., Shih, A.M., Soni, B.K.: Octree-based reasonable quality hexahedral mesh generation using a new set of refinement templates. International Journal for Numerical Methods in Engineering 77(13), 1809–1833 (2009)
9. Knupp, P.M.: Achieving finite element mesh quality via optimization of the jacobian matrix norm and associated quantitites. Part I - a framework for surface mesh optimization. International Journal for Numerical Methods in Engineering 48(3), 401–420 (2000)
10. Liu, A., Joe, B.: On the shape of tetrahedra from bisection. Mathematics of Computation 63(207), 141–154 (1994)
11. Parrish, M., Borden, M.J., Staten, M.L., Benzley, S.E.: A selective approach to conformal refinement of unstructured hexahedral finite element meshes. In: Brewer, M., Marcum, D.L. (eds.) Proceedings of the 16th International Meshing Roundtable, Seattle, WA, Octopber 2007, pp. 251–268. Sandia National Laboratories (2007)
12. Rivara, M.-C.: Algorithms for refining triangular grids suitable for adaptive and multigrid techniques. International Journal for Numerical Methods in Engineering 20, 745–756 (1984)
13. Sandhu, J.S., Menandro, F.C.M., Liebowitz, H., Moyer Jr., E.T.: Hierarchical mesh adaptation of 2d quadrilateral elements. Engineering Fracture Mechanics 50(5/6), 727–735 (1995)
14. Schneiders, R.: Refining quadrilateral and hexahedral element meshes. In: Proceedings of the 5th International Conference on Numerical Grid Generation in Computational Fluid Simulations, pp. 679–688. Mississippi State University (1996)
15. Shephard, M.S.: Linear multipoint constraint applied via transformation as part of a direct assembly process 20, 2107–2112 (1984)
16. Staten, M., Benzley, S., Scott, M.: A methodology for quadrilateral finite element mesh coarsening. Engineering with Computers 24(3), 241–251 (2008)
17. Tchon, K.-F., Dompierre, J., Camarero, R.: Automated refinement of conformal quadrilateral and hexahedral meshes. International Journal for Numerical Methods in Engineering 59, 1539–1562 (2004)
18. Zhang, Y., Bajaj, C.: Adaptive and quality quadrilateral/hexahedral meshing from volumetric data. Computer Methods in Applied Mechanics and Engineering 195, 942–960 (2006)

Guaranteed-Quality All-Quadrilateral Mesh Generation with Feature Preservation

Xinghua Liang, Mohamed S. Ebeida, and Yongjie Zhang*

Department of Mechanical Engineering, Carnegie Mellon University, USA
`jessicaz@andrew.cmu.edu`

Abstract. In this paper, a quadtree-based mesh generation method is described to create guaranteed-quality, geometry-adapted all-quadrilateral meshes with feature preservation for arbitrary planar domains. Given point cloud, our method generates all-quad meshes with these points as vertices and all the angles are within $[45°, 135°]$. For given planar curves, quadtree-based spatial decomposition is governed by the curvature of the boundaries and narrow regions. 2-refinement templates are chosen for local mesh refinement without creating any hanging nodes. A buffer zone is created by removing elements around the boundary. To guarantee the mesh quality, the angles facing the boundary are improved via template implementation, and two buffer layers are inserted in the buffer zone. It is proved that all the elements of the final mesh are quads with angles between $45° \pm \varepsilon$ and $135° \pm \varepsilon$ ($\varepsilon \leq 5°$) with the exception of badly shaped elements that may be required by the specified geometry. Sharp features and narrow regions are detected and preserved. Furthermore, boundary layer meshes are generated by splitting elements of the second buffer layer. We have applied our algorithm to a set of complicated geometries, including the Lake Superior map and the air foil with multiple components.

Keywords: Guaranteed quality, all-quadrilateral mesh, quadtree, sharp feature, narrow region, boundary layer.

1 Introduction

Provably good-quality triangular mesh generation methods were well developed for planar and curved surfaces. However, quadrilateral elements are preferred in finite element analysis due to their superior performance. Although it was proved that any planar n-gon can be meshed by $O(n)$ quads with all the angles bounded between $45° - \varepsilon$ and $135° + \varepsilon$ [5], fewer algorithms exist in the literature for all-quad mesh generation, and most of these algorithms are heuristic. Circle-packing techniques have been developed to construct quads with no angles larger than $120°$ for polygon interiors, but with no bound on the smallest angle [4]. Later, a balanced quadtree was utilized to generate a quad mesh with bounded minimum angle $18.43°$, but the

---
* Corresponding author.

maximum angle bound is 180° [1]. In this paper, we present an approach to generate guaranteed-quality all-quad meshes for given point cloud or planar curves, in which all the angles of any element are within $[45° \pm \varepsilon, 135° \pm \varepsilon]$, where $\varepsilon \leq 5°$, except badly shaped elements that may be required by the specified geometry.

Our algorithm generates unstructured adaptive all-quad meshes. For given point cloud, we firstly define a size function based on the relative location of these points, and then generate adaptive quadtree using 2-refinement templates. Three cases are considered to produce all-quad meshes conforming to the given point cloud, with all the element angles within $[45°, 135°]$.

For planar curves, six steps are adopted to construct guaranteed-quality all-quad meshes. Firstly we decompose each curve into a set of line segments based on its curvature; Secondly, a strongly balanced quadtree is constructed and hanging nodes are removed using 2-refinement templates. The element size is governed by the curvature of the boundary and narrow regions; Thirdly, elements outside and around the boundary are removed to create the quadtree core mesh and a buffer zone; Next, we design four categories of templates to adjust the boundary edges and therefore improve the angles facing the boundary in the quadtree core mesh; Then the angular bisectors are used to construct the first buffer layer; Finally the points generated during the first layer construction are projected to the boundary and form the second buffer layer. It is proved that all the angles in the constructed mesh are within $[45° \pm \varepsilon, 135° \pm \varepsilon]$ ($\varepsilon \leq 5°$) for any planar smooth curves. A few bad angles may be required to preserve sharp features such as small angles on the boundary. Boundary layers are generated by splitting the second buffer layer.

We have applied our algorithm to a set of complicated geometries, including the Lake Superior map and the air foil with multiple components. Our robust algorithm efficiently deals with curves in large-scale size, and generates meshes with guaranteed quality while minimizing the number of elements.

The reminder of this paper is organized as follows: Section 2 reviews the related work on quad mesh generation; Section 3 describes the guaranteed-quality meshing of point cloud; Section 4 explains the detailed algorithm for guaranteed-quality meshing of smooth curves; Section 5 talks about sharp feature and boundary layer generation; Section 6 shows two application results; Section 7 presents our conclusion and future work.

2   Previous Work

There are three main categories for direct unstructured all-quad mesh generation [13]: domain decomposition, advancing front and grid-based methods.

Domain Decomposition Methods: The domain is divided into simpler convex or mappable regions, and then template-based, mapping or geometric algorithms are utilized to generate the mesh for each of these regions. Domain decomposition can be achieved by various techniques. Tam and Armstrong

[18] introduced medial axis decomposition. Joe [10] decomposed the domain based on geometric algorithms. Quadros *et al.* [14] introduced an algorithm that couples medial axis decomposition with an advancing front method. In general, these methods produce high quality meshes but they are not robust and may require a great deal of user interaction especially if the domain has non-manifold boundaries.

Advancing Front Methods: This approach starts with the initial placement of nodes on the boundaries of the domain. Quad elements are then formed by projecting each edge on the front towards the interior and a new front is formed using edges on the new boundary. This process is repeated recursively until the domain is completely covered with quads. Zhu [25] is among the first to propose a quadrilateral advancing front algorithm. In his approach, two triangles are created using the traditional advancing front methods then combined to form a single quad. Blacker and Stephenson [6] introduced the paving algorithm in which they place a complete row of quads next to the front toward the interior. White and Kinney [19] enhanced the robustness of the paving algorithm through creating individual quads rather than a complete row. The advancing front methods generate near-boundary elements with high quality. However, the closure algorithms for the elements at the interior are still unstable, especially if the two overlapping elements have large difference in size. In such instance, heuristic decisions are made and these usually generate elements with poor quality in this region. Moreover, the detection and resolution of the closure regions can be very time consuming and sensitive to floating point errors.

Grid-Based Methods: A grid-based method starts with a uniform Cartesian background grid or a quadtree structure generated using the local feature sizes. Quads, conforming to the domain boundaries, are then fitted into that grid. Baehmann *et al.* [2] modified a balanced quadtree to generate a quad mesh for an arbitrary domain. Zhang *et al.* developed an octree-based iso-contouring method to generate adaptive quadrilateral and hexahedral meshes [22, 21, 24]. Schneiders *et al.* [16] used an isomorphism technique to conform an adaptive tree structure to the object boundaries. Grid-based algorithms are robust but often generate poor quality elements at the boundary.

Quality Improvement: In finite element analysis, small angles within the mesh usually lead to ill-conditioned linear systems. Further problems are caused due to elements with angles close to $180°$. Therefore, a post-processing step is crucial to improve the overall quality of the elements. Smoothing and clean-up methods are the two main categories of mesh improvement. Smoothing methods relocate vertices without changing the connectivity [7]. These methods are simple and easy to implement. However, they are heuristic and sometimes invert or degrade the local elements. To solve this problem, optimization-based smoothing methods were proposed [8]. In this approach, a node is relocated at the optimum location based on the local gradient of the surrounding element quality. Optimization-based methods provide much better mesh quality but they are not practical due to excessive amount of computations. For this reason such methods are usually combined with a Laplacian smoothing technique [9]. Surface feature preservation represents another challenging problem. Methods based on local curvature and volume preserving geometric flows are presented in [3, 23] to identify and preserve the main surface features. Clean-up methods for quad meshes [17, 11] are utilized to improve the node valence. Pillowing [12] is used to ensure that any two adjacent quads share at most one edge.

All of these quality improvement techniques do not guarantee any bounds for the element angles in the final mesh. In this paper, we will present an approach to generate all-quad meshes with guaranteed quality.

3   Guaranteed-Quality Meshing of Point Cloud

Given N points in a planar domain Ω, we aim to find a quad mesh M that includes the given points as vertices and all the elements have the minimum and maximum angle bounds. Let $\mathcal{X}$ denote the set of input points, we start by the spatial decomposition using 2-refinement templates [15] and the strongly balanced quadtree algorithm [20], which means the quadtree level difference between any two neighboring cells is ≤ 1. Compared to 3-refinement, 2-refinement provides a gradual transition and preserves element angles. It is well-known that any quad mesh has an even number of boundary nodes. This fact is utilized in eliminating each pair of the hanging nodes in the quadtree structure using 2-refinement templates. This algorithm considers the vertex sequence $\{V_1, V_2, \ldots, V_{2k}\}$ on the boundaries. Half of the vertices in this list are set to be active and the other half are set to be inactive such that each active vertex is followed by an inactive vertex. Any element that contains only one active node is refined using the transition template in Fig. 1(b) while any element containing two or more active nodes is refined using the refinement template shown in Fig. 1(c).

During the spatial decomposition, an additional requirement is applied so that each cell, c, containing a point $P \in \mathcal{X}$ is surrounded by eight empty cells of the same size. We then subdivide c into 16 identical regions, B_i $(i = 1, 2, \ldots, 16)$. One region of these will contain the point P. Here we consider three regions only as shown in Fig. 2(a). The other possibilities can be obtained by symmetry. If $P \in B_1$, we set the grid node $V_1 = P$ as shown in Fig. 2(b). When $P \in B_2 \cup B_3$, the local refinement shown in Fig. 2(c) is carried out first. If $P \in B_2$, we set the grid node $V_2 = P$ and adjust V_3 vertically as shown in Fig. 2(d). Similarly, if $P \in B_3$ as shown in Fig. 2(e), we set $V_3 = P$ and adjust V_2 correspondingly. As proved in Lemma 1, this algorithm guarantees all the angles in the final mesh are within $[45°, 135°]$, and the maximum aspect ratio is 1.5 (*The aspect ratio* of a quad is defined as the longest edge length over the shortest edge length). Fig. 1(d-e) show two testing cases.

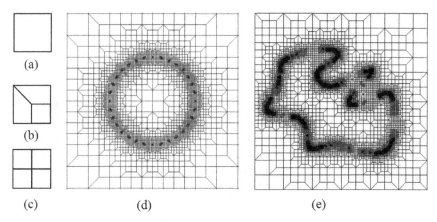

Fig. 1. 2-refinement templates and mesh generation conforming to a set of points. (a) A quadtree cell at Level i; (b) The transition template; (c) The refinement template; (d) A quad mesh of 32 points distributed uniformly along a circle; and (e) A quad mesh of 215 points distributed non-uniformly along a curve.

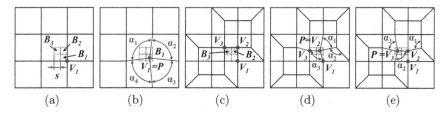

Fig. 2. All-quad mesh generation conforming to a set of points. (a) The cell containing a point is subdivided into 16 regions, supposing $P \in B_1 \cup B_2 \cup B_3$; (b) When $P \in B_1$, $V_1 = P$; (c) Local refinement for $P \in B_2 \cup B_3$; (d) When $P \in B_2$, $V_2 = P$ and V_3 is adjusted; and (e) When $P \in B_3$, $V_3 = P$ and V_2 is adjusted.

Lemma 1. *The method above results in a quad mesh with angles between* $45°$ *and* $135°$ *with the maximum aspect ratio of 1.5.*

Proof. In Fig. 2(b), the maximum displacement that V_1 can move is the diagonal of the square region B_1, $\sqrt{2}s$. The resulting angles α_1, α_2, α_3 and α_4 due to this displacement have a minimum value of $67°$ and a maximum value of $127°$. The maximum aspect ratio of the quad around P is 1.3. When $P \in B_2 \cup B_3$, the local refinement in Fig. 2(c) guarantees that both V_2 and V_3 are surrounded initially by four angles of $90°$. In Fig. 2(d-e), V_2 and V_3 are adjusted. One node is moved to P, and the other node is adjusted to maintain the slope between V_2 and V_3. The worst case here is similar to moving a corner of a square cell of size s a distance of $\sqrt{2}s$. The resulting angles α_1, α_2, and α_3 due to this displacement have a minimum value of $53°$ and a maximum value of $108°$. The maximum aspect ratio of the quads around P is 1.5. Therefore,

our algorithm produces all-quad meshes with angles between 45° and 135°
with the maximum aspect ratio of 1.5.   ◇

4 Guaranteed-Quality Meshing of Smooth Curves

Given a planar domain Ω and closed smooth curves $\mathcal{C}$ represented by cubic
splines, a set of points or polygons, we aim to generate a guaranteed-quality
all-quad mesh for the regions enclosed by $\mathcal{C}$. Six steps are applied as shown in
Fig. 3, including (1) curve decomposition, (2) adaptive quadtree construction,
(3) buffer zone clearance, (4) template implementation, (5) first buffer layer
construction, and (6) second buffer layer construction.

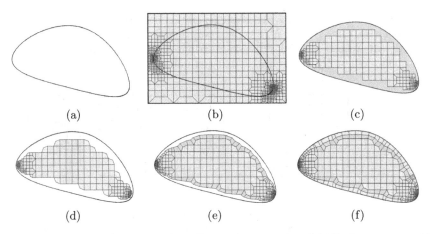

(a)   (b)   (c)

(d)   (e)   (f)

Fig. 3. Flow chart of guaranteed-quality mesh generation. (a) The input curve; (b)
Adaptive quadtree construction; (c) Buffer zone clearance; (d) Template implementation; (e) First buffer layer construction; and (f) Second buffer layer construction.

4.1 Curve Decomposition

Given closed smooth curves $\mathcal{C}$ represented by cubic splines, we first decompose
$\mathcal{C}$ into a set of piecewise-linear line segments using three criteria: (1) The angle
between two neighboring line segments is $\leq 5°$; (2) The approximation error
of each line segment is less than a given threshold (i.e., 0.01); and (3) Each
curve in $\mathcal{C}$ is represented by line segments in the clockwise direction. As a
result, non-uniform points $\mathcal{X}$ are created based on the curve local curvature.

4.2 Adaptive Quadtree Construction

For a set of points $\mathcal{X}$, a size function is first defined. For example, we define the
size function as $s_i = min(d_{ij})$, where d_{ij} is the distance between two points
i and j ($i, j \in \mathcal{X}$ and $i \neq j$). The basic concept of a spatial representation

consists of enclosing the points $\mathcal{X}$ in a bounding box, denoted as $\mathcal{B}(\Omega)$, which is corresponding to the root of the spatial decomposition quadtree. This box is either a square with one cell or a rectangle with one row of square cells. Each cell is recursively subdivided based on the size function and the strongly balanced quadtree algorithm. A cell, c, is *crowded* if its size is greater than the size function at any point within c or if the quadtree level difference around c is more than one. The spatial decomposition is achieved by splitting any crowded cell recursively until there is no crowded cell, and 2-refinement templates are used to remove all hanging nodes, see Fig. 3(b).

Narrow region is an important feature of the input curve. A region is defined as *narrow* if it contains only none/one/two quadtree cells in one direction after the domain decomposition. The detected narrow region will be refined until all directions contain more than two quadtree cells, which guarantees the topology will be preserved during the mesh generation.

4.3 Buffer Zone Clearance

After generating the adaptive quadtree, we remove elements near the boundary curves so that there is enough space to construct guaranteed-quality quad elements. Such a process is called *buffer zone clearance*. Here are definitions used in the algorithm description:

Buffer zone: *Any zonal area that serves the purpose of keeping the quad mesh distant from boundaries, e.g., the blue region in Fig. 3(c).*
Boundary edge: *A boundary edge is contained in only one element, e.g., the edges $\overline{AB}$, $\overline{BC}$ and $\overline{CD}$ in Fig. 4(a).*
Boundary point: *The two vertices of each boundary edge are named boundary points, e.g., the points A, B, C and D in Fig. 4(a).*
Boundary angle: *The angle formed by two neighboring boundary edges facing the nearest boundary, e.g., the angles α, β, and γ in Fig. 4(a).*
Boundary edge angle: *The angle formed from the boundary edge to the boundary. The angle must be inside the buffer zone, and can not intersect with the quadtree core mesh. The boundary edge angle has a range of $[-180°, 180°]$, and the counterclockwise direction is positive. For example, the angle ψ in Fig. 4(a) is negative.*

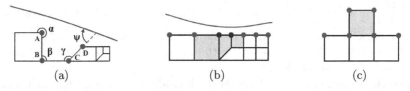

(a)   (b)   (c)

Fig. 4. Definition and criteria in Buffer Zone Clearance. (a) The red curve is the boundary. A, B, C, and D are boundary points. $\overline{AB}$, $\overline{BC}$, and $\overline{CD}$ are boundary edges. α, β, and γ are boundary angles. ψ is a boundary edge angle; (b) One exception of Criterion 3; and (c) Single element removal (the blue quad).

Three criteria are used in the buffer zone clearance: Criterion 1: all elements outside the regions to be meshed are deleted; Criterion 2: all elements intersecting with curves $\mathcal{C}$ are deleted; and Criterion 3: if the distance of any vertex to the nearest boundary is less than or equal to a pre-defined threshold ε_s, all elements sharing this vertex are deleted. Here we choose $\varepsilon_s = 0.5 * max(s_i)$, where s_i is the size of the ith element sharing this vertex. However, there are two exceptions. For example in Fig. 4(b), the two blue boundary points are very close to the boundary, so they need to be deleted according to Criterion 3. However, we choose to keep them because the boundary edge formed by them are almost parallel to the boundary. In addition, we also need to remove all the single elements. A *single element* is an element whose four vertices are all boundary points, and it has only one edge shared by another element as shown in Fig. 4(c). After the buffer zone clearance, the quadtree core mesh and a buffer zone are created, see Fig. 3(c). The buffer zone will be filled with guaranteed-quality quad elements.

4.4   Template Implementation

If all the boundary edges are parallel to the boundary, it is easy to construct good quality meshes. However, after the buffer zone clearance, the range of the boundary angles is $[45°, 315°]$, and the boundary edge angles are within $[-180°, 180°]$. In order to generate good quality elements around the boundary, we design templates and use them to improve the boundary angles and the boundary edge angles, keeping all the element angles in the quadtree core mesh $\in [45°, 135°]$.

Uniform Grid: If the quadtree core mesh is uniform, there are only three possible boundary angles after the buffer zone clearance: 90°, 180°, and 270°. Only 90° and 270° will possibly introduce bad elements when we fill the buffer zone. To guarantee the element quality, we use Templates 1(a-b) to modify these quad elements. In Tab. 1, the left column shows the original configurations, and the middle column shows the modified templates. For each template, the boundary angles are denoted in the figure and the boundary angle sequence is provided. For example, in Template 1(a), the boundary angle sequence is $(\frac{180°}{270°})$-90°-$(\frac{180°}{270°})$, which indicates $\varphi = (\frac{180°}{270°})$, $\alpha = 90°$, $\psi = (\frac{180°}{270°})$, where "$(\frac{180°}{270°})$" means "$[180°, 270°]$". By using Templates 1(a-b), the possible boundary angles in the uniform mesh become 112.5°, 180°, 202.5°, and 225°. It is obvious that the boundary angle range is improved from $[90°, 270°]$ to $[112.5°, 225°]$.

The right column of Tab. 1 shows the boundary edge angle for each template. For example in Templates 1(a-b), the grey points and dash lines indicate the edges deleted during the buffer zone clearance. We draw two solid-line circles at the starting point S and the ending point T, and draw a dash-line circle at one of the deleted points, denoted as B''. The solid-line circle means

Table 1. Template implementation of Categories 1-4.

Template	Original template	Modification	Boundary edge angles
1(a-b) Uniform: Adaptive:	or $(^{180°}_{270°})$-90°-$(^{180°}_{270°})$ or $(^{90°}_{180°})$-270°-$(^{90°}_{180°})$ $(^{180°}_{315°})$-90°-$(^{180°}_{315°})$ or $(^{90°}_{225°})$-270°-$(^{90°}_{225°})$	$(^{112.5°}_{202.5°})$ - 225° - $(^{112.5°}_{202.5°})$ $(^{112.5°}_{247.5°})$ - 225° - $(^{112.5°}_{247.5°})$	
2(a)	$(^{225°}_{315°})$ - 45° - 225° - 180°	$(^{161.5°}_{251.5°})$ - 153.5°	
2(b)	$(^{225°}_{315°})$ - 45° - 225° - 135° - 225°	$(^{161.5°}_{251.5°})$ - 153.5° - 180°	
2(c)	$(^{225°}_{315°})$ - 45° - 225° - 180° - 90°	$(^{161.5°}_{251.5°})$ - 153.5° - 180° - 180°	
3(a)	$(^{225°}_{315°})$ - 90° - $(^{225°}_{315°})$	$(^{135°}_{225°})$ - 225° - $(^{135°}_{225°})$	
3(b)	$(^{225°}_{315°})$ - 90° - 135° - 225°	$(^{153.5°}_{243.5°})$ - 161.5°	
3(c)	$(^{225°}_{315°})$ - 90° - 180° - 90° - 270°	$(^{153.5°}_{243.5°})$ - 180° - 161.5°	
3(d)	$(^{225°}_{315°})$-90°-180°-180°-90°-270°	$(^{146°}_{236°})$ - 180° - 180° - 169°	
4(a)	$(^{225°}_{270°})$ - 135° - 90° - 225°	$(^{161.5°}_{206.5°})$ - 153.5°	
4(b)	225° - 90° - 225°	180°	
4(c)	$(^{90°}_{180°})$ - 270° - $(^{90°}_{180°})$	$(^{135°}_{225°})$ - 225° - $(^{90°}_{180°})$	

Note: The boundary angle sequence below each template starts from left to right. In 1(a), $(^{180°}_{270°})$ -90°-$(^{180°}_{270°})$ indicates $\varphi = (^{180°}_{270°})$, α=90°, $\psi = (^{180°}_{270°})$. "$\varphi = (^{180°}_{270°})$" means "$\varphi \in [180°, 270°]$".

the boundary line must be on or outside the circle since the center is an existing point, while the dash-line circle means the boundary line must be inside the circle since the center is a deleted point. The radius of each circle is half size of the maximum element surrounding its center. By balancing the three circles, we can determine two boundary lines L_1 and L_2, which give the range of possible boundary edge angles for each template. In Templates 1(a-b), the slope of L_1 is $-90°$, and the slope of L_2 is $0°$. Since the boundary edge $\overline{SB'}$ has a slope of $22.5°$ and the boundary edge $\overline{B'T}$ has a slope of $67.5°$, the boundary edge angle becomes $[-67.5°, 67.5°]$. As a final result, we obtain that the range of the boundary edge angles in Templates 1(a-b) is $[-67.5°, 67.5°]$.

Adaptive Grid: For the adaptive mesh, the range of boundary angles after the buffer zone clearance is $[45°, 315°]$. Here Templates 1(a-b) are also used, but the boundary angles φ and ψ have larger ranges. In addition, three other categories of templates are designed to improve the boundary angles as listed in Tab. 1: the boundary angle $\alpha = 45°$ in Category 2, $\alpha = 90°$ in Category 3, and $\alpha = \{135°, 225°, 270°, 315°\}$ in Category 4. All the other possible configurations can be derived from these basic templates. Meanwhile, templates beginning with $\alpha = 180°$ are not listed in Tab. 1, because they are either categorized in other templates or good enough and therefore need not to be improved.

Here we use Template 2(a) as an example to explain how to derive these templates based on the 2-refinement algorithm (Fig. 1(a-c)). Fig. 5(a) shows a row of three uniform cells, where the black points indicate the boundary nodes. We then use 2-refinement templates to refine these cells, and Fig. 5(b) is one possible result. Here we always keep the left element and the corresponding boundary points in Fig. 5(b). The red points are boundary points which are very close to the boundary. These points and their neighboring cells (grey cells) are removed during the buffer zone clearance, which results in Fig. 5(c). The cells now have new boundary points and boundary edges. However, the boundary edges $\overline{AB}$, $\overline{BC}$ and $\overline{CD}$ form a sharply concave geometry with the minimum boundary angle of $45°$. To guarantee the mesh quality, we must eliminate the concaveness by modifying the boundary edges as shown in Fig. 5(d), and finally we obtain Template 2(a) in Tab. 1. Changing the number of cells in Fig. 5(a) or/and applying 2-refinement templates to these cells differently, we can obtain the rest templates in Tab. 1 in similar ways. It is noticed that the same template can be derived from different combinations, and some combinations are impossible and never appear in the mesh. Moreover, the number of cells in Fig. 5(a) can be restricted to be between 1 and 4. If the number of cells is greater than 4, they can always be split so that each part contains no more than 4 cells.

From the right column of Tab. 1, we can obtain the range of the boundary edge angles for each template. We also take Template 2(a) as an example. L_1 is horizontal and its slope is $0°$. Since the dash-line circle at the center B'' is

Fig. 5. An example of deriving a template. (a) Given uniform cells; (b) Cells after applying 2-refinement templates; (c) Cells after the buffer zone clearance; and (d) Cells after the template implementation.

double size of the solid-line circle at the center T, and L_2 is tangent to both circles, we can obtain the slope of L_2 is $39.5°$. The boundary edge angles of the two horizontal boundary edges are within $[-39.5°, 0°]$. However, the boundary edge $\overline{SC}$ has a slope of $26.5°$, so the boundary edge angle becomes $[-13°, 26.5°]$. As a final result, the range of the boundary edge angles in Template 2(a) is $[-39.5°, 26.5°]$.

Lemma 2. *After the template implementation, all the element angles in the quadtree core mesh are within $[45°, 135°]$, all the boundary angles are within $[112.5°, 251.5°]$, and all the boundary edge angles are within $[-90°, 90°]$.*

Proof. It is obvious that all the element angles in the quadtree core mesh are within $[45°, 135°]$. From the modified templates in Tab. 1, we can see that the boundary angles are within $[112.5°, 251.5°]$, noticing that the possible boundary angle $90°$ in Template 4(c) will be further improved by Template 1(a)/3(a). We check the boundary edge angles of each template in Tab. 1 and list them in Tab. 2. From the table, it is obvious that all the boundary edge angles are in the range of $[-67.5°, 90°]$. Due to the symmetry of these templates, we release the range of the boundary edge angle to $[-90°, 90°]$. ◇

Discussion: During the template implementation, the boundary angles and the boundary edge angles are improved by filling with new quads, changing the shape of quads or changing the connectivity of quads. All these improvements will help to guarantee the element quality during the following first and second buffer layer construction. Moreover, we classify all the templates into three priorities as shown in Tab. 2: high, medium and low. Templates in lower priority will not be implemented until all the higher priority templates are implemented.

4.5   First Buffer Layer Construction

For each boundary point, we use angular bisectors to calculate a corresponding point inside the buffer zone called *the first buffer point*. Then, each pair of neighboring boundary points and their first buffer points construct a quad. All these new quads form the first buffer layer as presented in Fig. 3(d).

Here we develop an algorithm to calculate the first buffer points using angular bisectors. As shown in Fig. 6(a), the boundary point B is shared by two boundary edges $\overline{AB}$ and $\overline{BC}$ with the boundary edge angles of φ and ψ.

Table 2. Boundary edge angles of each template in Tab. 1.

Template	Slope of L_1	Slope of L_2	α	Boundary edge angle	Priority
1(a)	$-90°$	$0°$	$22.5°$	$[-67.5°, 67.5°]$	Low
1(b)	$-90°$	$0°$	$22.5°$	$[-67.5°, 67.5°]$	Medium
2(a)	$0°$	$39.5°$	$26.5°$	$[-39.5°, 26.5°]$	High
2(b)	$0°$	$28.1°$	$26.5°$	$[-28.1°, 26.5°]$	High
2(c)	$0°$	$28.1°$	$26.5°$	$[-28.1°, 26.5°]$	High
3(a)	$-63.4°$	$45°$	$45°$	$[-63.4°, 90°]$	Low
3(b)	$0°$	$26.5°$	$18.5°$	$[-26.5°, 18.5°]$	High
3(c)	$0°$	$26.5°$	$18.5°$	$[-26.5°, 18.5°]$	High
3(d)	$0°$	$14°$	$11.3°$	$[-14°, 11.3°]$	High
4(a)	$0°$	$26.5°$	$26.5°$	$[-26.5°, 26.5°]$	High
4(b)	$-26.5°$	$26.5°$	-	$[-26.5°, 26.5°]$	High
4(c)	$-90°$	$0°$	$45°$	$[-45°, 90°]$	Medium

Suppose $\psi \geq \varphi$, we draw the angular bisectors of the larger angle ψ and the boundary angle β. Finally, the two angular bisectors intersect at B' which is the corresponding first buffer point of B. By using all the boundary points and their first buffer points, a set of quad elements are generated to form the first buffer layer.

There are two special cases for the first buffer point calculation. If both boundary edges of a boundary point are approximately parallel to the boundary, or the boundary edge angles are less than $10°$, we create the first buffer point as the middle of each boundary point and its projection on the boundary. For example in Fig. 6(b), since the two boundary edges $\overline{AB}$ and $\overline{BC}$ are almost parallel to the boundary, we project B to the boundary, and take the middle point of B and its projection as the corresponding first buffer point B'. Similarly, A' and C' are calculated. These boundary points A, B, C and the first buffer points A', B', C' construct two quads for the first buffer layer.

As shown in Fig. 6(c), the second special case is introduced by Templates 1(a-b) in Tab. 1. The boundary angles at boundary points B and C are $\alpha = 112.5°$ and $\beta = 180°$, and the angular bisectors are l_1 and l_2. B' is the first buffer point of B. It can be observed that B' is close to l_2, which may

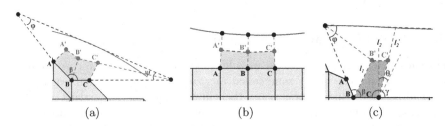

Fig. 6. First buffer layer construction. (a) The general case; (b) The parallel case; and (c) The special case.

introduce a quad with bad angles or even self-intersection. Suppose l_1 and l_2 form an angle $\theta = (\beta - \alpha)/2 = 33.75°$, l_2 is rotated by $\theta/2$ so that it is away from B'. This new angular bisector l_2' will be used to calculate C'.

With these three cases in Fig. 6, we can get all the corresponding first buffer points and generate the first buffer layer. After the construction of the first buffer layer, the original boundary points have now become inner points, and new boundary points are on the first buffer layer. The following Lemma 3 proves the angle bounds of the first buffer layer.

Lemma 3. *All the four angles of any element in the first buffer layer are in the range of $[45° \pm \varepsilon, 135° \pm \varepsilon]$, where $\varepsilon \leq 5°$.*

Proof. To simplify the proof, we first assume that the boundary curve around each boundary edge is a straight line. In Fig. 7(a), B' and C' are the first buffer points corresponding to the boundary points B and C, and the boundary edge angle of $\overline{BC}$ is ψ. Suppose B' and C' are on the angular bisector of ψ. There are two possible cases we need to consider in order to prove this lemma.

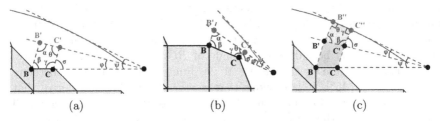

(a)   (b)   (c)

Fig. 7. Guaranteed-quality proof. (a) The first buffer layer (Case 1); (b) The first buffer layer (Case 2); and (c) The second buffer layer.

The first case is shown in Fig. 7(a). The boundary angle of B is $\leq 180°$ and the boundary angle of C is $\geq 180°$. From Lemma 2, we can obtain $\psi \in [-90°, 90°]$. Here we choose $\psi \in [0°, 90°]$ due to the symmetry, therefore $\varphi = \psi/2 \in [0°, 45°]$. After the template implementation, all the boundary angles are in the range of $[112.5°, 251.5°]$. Therefore, the boundary angle of B is in the range of $[112.5°, 180°]$ and the boundary angle of C is in the range of $[180°, 251.5°]$. Since $\overline{BB'}$ and $\overline{CC'}$ are angular bisectors, $\beta \in [56.25°, 90°]$ and $\gamma \in [90°, 125.75°]$. Likewise, we can get $\sigma = 180° - \gamma \in [54.25°, 90°]$, $\theta = \varphi + \sigma \in [54.25°, 135°]$, $\alpha = 180° - \beta - \varphi \in [45°, 125.75°]$. Therefore, all the angles of the quad $BCC'B'$ are within $[45°, 135°]$.

As shown in Fig. 7(b), the second case is that both of the boundary angles are greater than $180°$. This case is introduced by Templates 1(a-b). From Tabs. 1-2, we can obtain the two boundary angles are $202.5°$ at B and $225°$ at C, and the boundary edge angle $\psi \in [-67.5°, 67.5°]$. Here we choose $\psi \in [0°, 67.5°]$ due to the symmetry. Therefore, we have $\beta = 202.5°/2 = 101.2°$, $\gamma = 225°/2 = 112.5°$, $\sigma = 180° - \gamma = 67.5°$, $\varphi = \psi/2 \in [0°, 33.8°]$,

$\theta = \sigma + \varphi \in [67.5°, 101.3°]$, and $\alpha = 180° - \beta - \varphi \in [45°, 78.8°]$. It is obvious that all the four angles of the quad $BCC'B'$ are within $[45°, 112.5°]$.

Following the similar ways, we can easily prove that all the angles in the two special cases (Fig. 6(b-c)) are also within $[45°, 135°]$. In summary, all the four angles of any element in the first buffer layer are within $[45°, 135°]$. Considering that the boundary curve has a small perturbation $\varepsilon \leq 5°$, we relax the angle range to $[45° \pm \varepsilon, 135° \pm \varepsilon]$. $\diamond$

4.6   Second Buffer Layer Construction

The second buffer layer construction is simple. As shown in Fig. 7(c), we project the first buffer points to the boundary and obtain the corresponding second buffer points. Then we construct a quad using these first and second buffer points, and all these new quads form the second buffer layer, see Fig. 3(f). We use Lemma 4 to prove its angle bounds.

Lemma 4. *All the four angles of any element in the second buffer layer are in the range of* $[45° \pm \varepsilon, 135° \pm \varepsilon]$, *where* $\varepsilon \leq 5°$.

Proof. Similar to Lemma 3, we assume that the boundary curve is a straight line around each boundary edge. In Fig. 7(c), B' and C' are the first buffer points of the boundary points B and C. B' and C' are on the angular bisector of ψ, which is the boundary edge angle of $\overline{BC}$. From Lemma 2, we have $\psi \in [-90°, 90°]$. Here we choose $\psi \in [0°, 90°]$ due to the symmetry, therefore $\varphi = \psi/2 \in [0°, 45°]$. It is obvious that $\gamma = \theta = 90°$. We can derive that $\alpha = \sigma = 90° - \varphi \in [45°, 90°]$, and $\beta = 180° - \sigma \in [90°, 135°]$. Therefore, all the four angles of the quad $B'C'C''B''$ are within $[45°, 135°]$. Considering that the boundary curve may have a small perturbation $\epsilon \leq 5°$, the angular range is relaxed to $[45° \pm \varepsilon, 135° \pm \varepsilon]$. $\diamond$

Remark: After applying the designed six steps in Fig. 3, the element angles in the quadtree core mesh, the first buffer layer and the second buffer layer are all in the range of $[45° \pm \varepsilon, 135° \pm \varepsilon]$ $(\varepsilon \leq 5°)$.

5   Sharp Feature and Boundary Layer

To preserve sharp features, we keep the quadtree core mesh nearby each sharp feature uniform during the adaptive quadtree construction. An additional template in Fig. 8(a) is implemented, and the boundary angle range is improved from $[90°, 270°]$ to $[116.6°, 206.6°]$. Suppose P is the corner and α is the sharp angle as shown in Fig. 8(b-d), we develop different meshing algorithms for three various cases: $\alpha \in (0°, 135°]$, $(135°, 270°]$, and $(270°, 360°)$.

When $\alpha \in (0°, 135°]$ as shown in Fig. 8(b), we first draw two lines l'_1 and l'_2, where $l'_1 \perp l_1$ and $l'_2 \perp l_2$. In the green area enclosed by l'_1 and l'_2, we find the closest point M after constructing the first buffer layer, then project it to the boundary and obtain two corresponding points L and N. The four

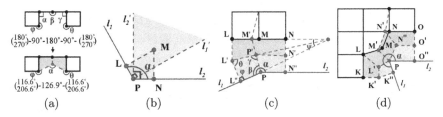

Fig. 8. A new template and three algorithms for sharp features. (a) An additional template; (b) $\alpha \in (0°, 135°]$; (c) $\alpha \in (135°, 270°)$; and (d) $\alpha \in [270°, 360°)$.

points then form a quad $PNML$. Since L and N are the projections of M, if $\alpha \geq 45°$, obviously all the four angles will be within $[45°, 135°]$; otherwise if $\alpha < 45°$, we just keep α, and the range of all the four angles is $[\alpha, 180° - \alpha]$.

When $\alpha \in (135°, 270°]$ as shown in Fig. 8(c), after the template implementation, we use the angular bisector of α to modify the quadtree core mesh close to the sharp feature. First we find that the angular bisector intersects the boundary edge $\overline{LM}$ at M', so we move M to M'. Then we create the first buffer points L' and N'. Suppose the boundary edge angle of $\overline{LM'}$ is ψ, the first buffer point P' of M' is calculated as the intersection of the angular bisector of ψ with $\overline{PM'}$. The constructed yellow quads form the first buffer layer, its quality is guaranteed by Lemma 3. Finally, we obtain the second buffer layer by projecting L' and N' to the boundary. To prove the second buffer layer is also in good quality, we take the quad $PP'L'L''$ as an example and check all its four angles. It is obvious that $\psi \in [0°, 90°]$ and $\varphi = 90°$. If $\alpha \in [180°, 270°]$, it is straightforward to obtain $\beta = \alpha/2 \in [90°, 135°]$, $\theta = 90° - \psi/2 \in [45°, 90°]$, and $\gamma = 180° - \beta + \psi/2 \in [45°, 135°]$. If $\alpha \in (135°, 180°)$, ψ can be restricted to be within $[0°, 67.5° - (180° - \alpha)]$ from Templates 1(a-b), then we have $\beta = \alpha/2 \in (67.5°, 90°)$, $\theta = 90° - \psi/2 \in [45°, 90°]$, and $\gamma = 180° - \beta + \psi/2 \in [90°, 123.25°]$. Now all the four angles of the quad $PP'L'L''$ are within $[45°, 135°]$. Similarly, we can prove that the quad $P'PN''N'$ is of good quality.

When $\alpha \in (270°, 360°)$ as shown in Fig. 8(d), we also use the angular bisector of α to modify the local quadtree core mesh. M' is the intersection point of the angular bisector of α with $\overline{LM}$. Then we move M to M', and move N to N' so that $M'N'//MN$. The angular bisector of $\angle M'PO''$ intersects the angular bisector of $\angle ON'M'$ at point N''; likewise, we create L' in the similar way. Later, we use points K', L', P, N'' and O' to construct the first buffer layer, and use K'', O'' to obtain the second buffer layer. It is noticed that no second buffer layer is created at the sharp feature. This method may not always guarantee good angles, therefore in some cases we have to rotate the bisectors $\overline{PM'}$, $\overline{PN''}$ and $\overline{PL'}$ so that all the angles near the sharp feature are of good quality.

In order to apply a mesh to Computer Fluid Dynamics (CFD) simulations, we must generate one or more boundary layers. In our algorithm, the boundary layer construction is conveniently obtained by splitting the elements of

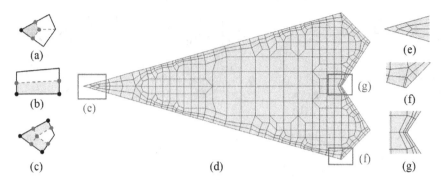

Fig. 9. Templates for boundary layer generation and the delta wing with a small sharp angle of 31°. (a-c) Templates with one, two and three boundary points; (d) Final mesh with all angles $\in [31°, 135°]$; (e-g) Zoom-in pictures of (d).

the second buffer layer. Only three templates are needed as shown in Fig. 9(a-c). The black points are boundary points, the red points are new points, and the blue quads are elements constructed for the boundary layer. It is obvious that Fig. 9(b-c) preserve all the angles. Fig. 9(a) splits an angle using the angular bisector, resulting in a change of the angle range to $[22.5°, 157.5°]$. This only happens at the sharp feature with an angle $> 270°$. Fig. 9(d-g) show the delta wing with 6 sharp angles and a boundary layer.

Remark: Our algorithm guarantees all the angles in the final mesh are within $[45° \pm \varepsilon, 135° \pm \varepsilon]$ $(\varepsilon \leq 5°)$, with the exception of bad elements that may be required by sharp features with an angle $< 45°$ or $> 270°$.

6   Results

We have applied our algorithm to two complicated models: the Lake Superior map and the air foil with multiple components. Our results were computed on a PC equipped with an Intel Q6600 CPU and 4GB DDR-II Memories. As shown in Fig. 10, the Lake Superior map consists of seven closed smooth curves, which has narrow regions. The constructed mesh conforms to the input curves accurately and all the elements are quads with angles within $[43°, 135°]$. The mesh adaptivity is controlled by a size function based on the boundary curvature and narrow region. It took 82 seconds to generate the mesh. The final mesh has 32789 nodes and 30321 quads, the quadtree has 10 levels, and the maximum aspect ratio is 10.

In Fig. 11, the air foil consists of three components, all of them contain sharp angles. It took 15 seconds to generate the mesh. Before generating a boundary layer, the mesh has 24514 nodes and 22929 quads, the quadtree has 14 levels, the angle range in the final mesh is $[45°, 135°]$, and the maximum aspect ratio is 4.7. After generating one boundary layer, the angle range becomes $[27°, 153°]$ due to the template in Fig. 9(a). This result shows that

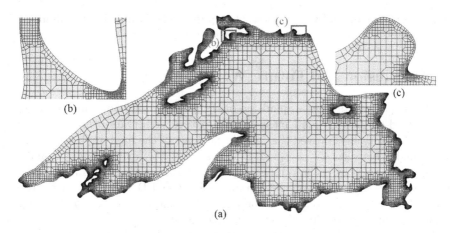

Fig. 10. The Lake Superior map. (a) Final guarantee-quality all-quad mesh with all angles $\in [43°, 135°]$; and (b-c) Zoom-in pictures of (a).

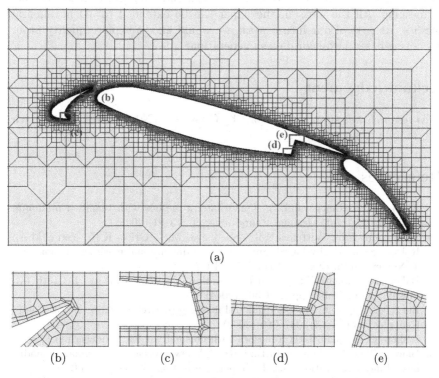

Fig. 11. The air foil with multiple components. (a) All-quad mesh with all angles $\in [45°, 135°]$ before boundary layer generation, and $[27°, 153°]$ after generating one boundary layer; (b-e) Zoom-in pictures of (a).

our algorithm can deal with large-scale inputs while minimizing the number of elements, which are important for aerodynamics simulations.

7 Conclusion and Future Work

In this paper, we present a quadtree-based meshing method, which creates guaranteed-quality all-quad meshes with feature preservation for arbitrary planar domains. It is proved that all the elements of the final mesh are quads with angles between $45° \pm \varepsilon$ and $135° \pm \varepsilon$ ($\varepsilon \le 5°$), except badly shaped elements required by the specified geometry. Our algorithm can conveniently generate boundary layers on the final mesh for the CFD simulation. We have applied our algorithm to a set of complicated geometries, including the Lake Superior map and the air foil with multiple components.

In the current algorithm, the meshes generated inside and outside the curves do not conform to each other. In the future, we will improve our algorithm so that it can be used to generated interior and exterior meshes at the same time. We are also planning to extend our algorithm to dynamic re-meshing. All the ideas in the paper can be extended to 3D hexahedral meshes. As part of our future work, we would like to explore quality proof for octree-based hexahedral mesh generation.

Acknowledgement

This research was supported in part by ONR grant N00014-08-1-0653, which is gratefully acknowledged.

References

1. Atalay, F.B., Ramaswami, S.: Quadrilateral meshes with bounded minimum angle. In: 17th Int. Meshing Roundtable, pp. 73–91 (2008)
2. Baehmann, P.L., Wittchen, S.L., Shephard, M.S., Grice, K.R., Yerry, M.A.: Robust geometrically based, automatic two-dimensional mesh generation. Int. J. Numer. Meth. Eng. 24, 1043–1078 (1987)
3. Baker, T.J.: Identification and preservation of surface features. In: 13th Int. Meshing Roundtable, pp. 299–309 (2004)
4. Bern, M., Eppstein, D.: Quadrilateral meshing by circle packing. Int. J. Comp. Geom. & Appl. 10(4), 347–360 (2000)
5. Bishop, C.J.: Quadrilateral meshes with no small angles (Manuscript) (1991)
6. Blacker, T.D., Stephenson, M.B.: Paving: A new approach to automated quadrilateral mesh generation. Int. J. Numer. Meth. Eng. 32, 811–847 (1991)
7. Canann, S.A., Tristano, J.R., Staten, M.L.: An approach to combined laplacian and optimization-based smoothing for triangular, quadrilateral, and quad-dominant meshes . In: 7th Int. Meshing Roundtable, pp. 211–224 (1998)
8. Freitag, L., Jones, M., Plassmann, P.: An efficient parallel algorithm for mesh smoothing. In: 4th Int. Meshing Roundtable, pp. 47–58 (1995)

9. Freitag, L.A.: On combining Laplacian and optimization-based mesh smoothing techniques. Trends in Unstructured Mesh Generation, ASME 220, 37–43 (1997)

10. Joe, B.: Quadrilateral mesh generation in polygonal regions. Comput. Aid. Des. 27(3), 209–222 (1991)

11. Kinney, P.: CleanUp: Improving quadrilateral finite element meshes. In: 6th Int. Meshing Roundtable, pp. 437–447 (1997)

12. Mitchell, S.A., Tautges, T.J.: Pillowing doublets: Refining a mesh to ensure that faces share at most one edge. In: 4th Int. Meshing Roundtable, pp. 231–240 (1995)

13. Owen, S.: A survey of unstructured mesh generation technology. In: 7th Int. Meshing Roundtable, pp. 26–28 (1998)

14. Quadros, W.R., Ramaswami, K., Prinz, F.B., Gurumoorthy, B.: LayTracks: A new approach to automated geometry adaptive quadrilateral mesh generaton using medial axis transform. Int. J. Numer. Meth. Eng. 61, 209–237 (2004)

15. Schneiders, R.: Refining quadrilateral and hexahedral element Meshes. In: 5th Int. Meshing Roundtable, pp. 383–398 (1996)

16. Schneiders, R., Schindler, R., Weiler, F.: Octree-based generation of hexahedral element meshes. In: 5th Int. Meshing Roundtable, pp. 205–216 (1996)

17. Staten, M.L., Canann, S.A.: Post refinement element shape improvement for quadrilateral meshes. Trends in Unstructured Mesh Generation, ASME 220, 9–16 (1997)

18. Tam, T., Armstrong, C.G.: 2D finite element mesh generation by medial axis subdivision. Adv. Eng. Software 13, 313–324 (1991)

19. White, D.R., Kinney, P.: Redesign of the paving algorithm: Robustness enhancements through element by element meshing. In: 6th Int. Meshing Roundtable, pp. 323–335 (1997)

20. Yerry, M.A., Shephard, M.S.: A modified quadtree approach to finite element mesh generation. IEEE Computer Graphics Appl. 3(1), 39–46 (1983)

21. Zhang, Y., Bajaj, C.: Adaptive and quality quadrilateral/hexahedral meshing from volumetric Data. Comput. Meth. Appl. Mech. Eng. 195, 942–960 (2006)

22. Zhang, Y., Bajaj, C., Sohn, B.-S.: 3D finite element meshing from imaging data. Comput. Meth. Appl. Mech. Eng. 194, 5083–5106 (2005)

23. Zhang, Y., Bajaj, C., Xu, G.: Surface smoothing and quality improvement of quadrilateral/hexahedral meshes with geometric flow. Commun. Numer. Meth. Eng. 25, 1–18 (2009)

24. Zhang, Y., Hughes, T., Bajaj, C.: An automatic 3D mesh generation method for domains with multiple materials. Comput. Meth. Appl. Mech. Eng. (in press, 2009)

25. Zhu, J.Z., Zienkiewicz, O.C., Hinton, E., Wu, J.: A new approach to the development of automatic quadrilateral mesh generation . Int. J. Numer. Meth. Eng. 32, 849–866 (1991)

Advances in Octree-Based All-Hexahedral Mesh Generation: Handling Sharp Features*

Loïc Maréchal

Gamma project, I.N.R.I.A., Rocquencourt, 78153 Le Chesnay, France
`loic.marechal@inria.fr`

Abstract. Even though many methods have been suggested to meet the challenge of all-hexahedral meshing, octree-based methods remain the most efficient from an engineering point of view. As of today, its robustness, speed and robustness are still unmatched. This paper presents advances made in the *Hexotic* project, especially in terms of sharp angles meshing, non-manifold geometries and adaptation.

Keywords: octree, meshing, hexahedra, adaptation.

1 Introduction

1.1 Motivation

Engineers facing the need for hexahedral meshes can access today only two kinds of software:

- Semi-automated block decomposition based methods like super block mapping, extrusion, sweeping algorithms, etc... These produce good quality meshes and enable fine grain control over the mesh, such as element position, orientation and size, but require a big human time cost. To begin with, a good deal of time is needed to learn how to use the software and acquire some knowhow. Afterwards, meshing each object requires a sizable amount of time, from hours to months depending on its complexity and the user's skill.
- Fully automated software, mostly based on the octree method. These products offer high speed, robustness and ease of use at the cost of mesh quality.

Consequently, researchers eager to improve an engineer's condition have but two ways to do so:

- the first is to reduce the human intervention required by semi-automated methods,

---

* This work was funded by the Pôle system@tic under the project E.H.P.O.C.

- the second is to improve the quality of meshes produced by octree-based methods.

This paper addresses the second alternative: what can be done to improve the capabilities of octree methods?

1.2   Limitations of Available Octree-Based Software

Most available products suffer from one ore more of these limitations:

- Non-conformity: hanging vertices are left in the final mesh, adding a burden to the solvers.
- Hybrid meshes: tets, pyramids and prisms are generated along with hexes. These software are often boasted as being hex-dominant, because they produce meshes made of 51% hexes... which is far from satisfactory for many solvers.
- Invalid elements: some products favor geometry accuracy over elements quality and as such tolerate some degenerated hexes (concave, negative volume).
- Smooth geometry: sharp features are smoothed out. No right or acute angles can be represented in the hex mesh. Such rough geometrical accuracy may be enough when dealing with M.R.I based data, but proves to be unsatisfactory for mechanical simulations.
- Constant sized-elements: all hexes are of the same, user-defined, size. Such a constraint generally improves quality but requires too many elements in order to capture small features.

1.3   Objective

We aimed at high objectives when starting this project, even though we knew that, in the end, we would have to water them down. Ideally, we would like to design a new approach to octree which would enforce the following features:

- all-hexahedral meshes,
- full automation, no intervention or knowledge should be required from the user,
- variable element size to capture each geometrical feature with the lowest amount of elements,
- valid meshes, 100% of hexes must have a positive jacobian,
- multiple subdomains and non-manifold subdomains, required in many complex mechanical devices,
- sharp angle meshing,
- adaptive meshing.

To achieve that goal, we chose to combine a modified octree, dual-mesh generation, extensive use of buffer-layers and a new vertex smoothing scheme. These efforts were undertaken under a project named *Hexotic*.

The overall scheme unfolds as:

1. Octree:
   a) analyze and check the input surface mesh
   b) compute the object's bounding box as a starting octant for the octree
   c) refine the octree according to a set of geometrical, physical or user-set criteria,
   d) balance and pair-up the octants,
   e) intersect octants with the triangulated surface,
2. Conforming:
   a) connect octree hanging vertices with the help of arbitrary polyhedra,
   b) build the dual-mesh from the polyhedral mesh,
3. Subdomains
   a) color topological sub-domains,
   b) control topological sub-domain thickness and topological pinches,
   c) insert a buffer layer of hexes around each volume subdomain,
   d) smooth vertices to improve elements quality,
4. Ridge meshing
   a) capture sharp edges on surface elements,
   b) group sharp-edged elements in surface pseudo-subdomains,
   c) insert a buffer layer of hexes around each surface subdomain,
   d) perform a final vertex smoothing while projecting surface vertices on the real geometry.

This paper will guide you through the whole set of meshing steps and present you with some results.

2 Octree Building

2.1 An Almost Regular Octree

Hexotic is based on a regular octree with little, but crucial, modifications:

- compute the object bounding box and use it as first octant,
- recursively subdivide octants according to a set of geometric and computational criteria,
- eventually, subdivide some octants to follow additional topologic rules (balancing, pairing).

Since the scope of this paper is not about fundamentals of octree meshing, reading [13] and [14] should be considered for more information.

2.2 Subdivision Criteria

Two subdivision criteria have been used:

Geometric diameter
Octants are split until they match the local geometry thickness.

Indeed, half this size is required to ensure that each feature is two-element thick. Meshing small features (shell or beam-shaped) with two hexes across allows for much less element distortion in the final boundary recovery process.

Furthermore, it gives the smoothing algorithm much more flexibility, since elements have one to four free vertices inside the volume. One-element thick meshes produce all-vertices locked hexes, which make the mesh extremely stiff in the final projection of surface vertices onto the geometry.

On the other hand, splitting octants once more than required increases the number of elements by a factor of eight in thin areas.

Size-map

In a mesh adaptation scheme, element size is driven by *a priori* or *a posteriori* error estimates. Defining the right size in each area of the mesh is achieved through a so called size-map.

It is made of a set of vertices mapping the whole mesh area, each vertex being associated to a target size. It is up to the solver to derive this size-map from the computation results. Once this size-map is generated, it should be provided, along with the surface mesh, to the hex-mesher. A very simple example is shown in section 10, fig. 16.

Other criteria could be used such as curvature angle or surface mesh triangles size. They are called "geometrical criteria" since they basically associate a size to any area within the bounding-box ($size = min_{i=1..n}(f_i(x, y, z))$, where $f_i(x, y, z)$ is a size criterion) and split octants until their sizes match the target.

2.3   Balancing and Pairing Rules

These rules are also called topological subdivision criteria, as opposed to geometrical criteria presented in the previous section. They set octants sizes depending on their neighbors sizes, enforcing topological rules, hence their name.

Hexotic applies two such criteria to the octree:

Balancing rule

This makes sure that no octant is more than twice or less than twice smaller than neighboring elements (sharing one or more vertices). In other words, the difference between two neighboring octants subdivision level cannot be greater than one.

Pairing rule

This one is quite unusual and its purpose will be fully explained in the following section. Basically, if an octant's son is to be split, its brothers (octants belonging to the same father element) should be split along with it.

Figure 1 shows the four steps an octree mesh goes through: single element bounding-box, octree subdivided according to geometrical rules, then after balancing rule, and finally, after pairing rule has been applied.

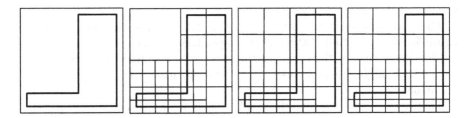

Fig. 1. (From left to right) Object inside its bounding-box; octree matching local features size; octree after balancing; octree after pairing.

After building the octree, the mesh is made only of hexes and each element size matches the size of geometry features in the corresponding area. Still, only the whole bounding-box is meshed and neither subdomains nor geometry boundaries are represented in this hex mesh. Furthermore, size variations between octants leave hanging vertices throughout the mesh, making it non-conformal. The next section will deal with this last issue.

3 Polyhedral Cutting

3.1 Connecting Hanging Vertices

Octree generation leaves us with non-conformal elements, that is, some octants have vertices hanging in the middle of neighboring faces or edges.

Connecting these hanging vertices using all-hex patterns has been a problem for many years.

Several solutions have been suggested (see [15] and [16]) some of them quite satisfactory, but we could do even better using a polyhedral and dual-mesh scheme.

The main idea is to connect hanging vertices using polyhedral elements. Any type of polyhedron and polygonal face can be used, which offers a much greater range of patterns to fill in non-conformal octants. The only restriction stands in the connectivity: the degree of vertices must be eight (there must be only eight elements sharing the same vertex), likewise, each edge can be shared by only four elements. Such a constraint is easily enforced if a cutting process is used to connect hanging vertices (see [2] and [6]).

Furthermore, the cutting process is split into as many steps as there are geometric dimensions. It is known as directional refinement (see [16]).

3.2   2-Dimensional Case

As a starting point, hanging vertices must be tagged according to the dimension they belong to. As depicted in figure 3, vertices hanging at the end of a vertical line are tagged with a small black dot, conversely, horizontal hanging vertices are tagged with a bigger white dot.

Then, we proceed with cutting elements with vertical tags as shown in figure 3 (left). This first cutting step can produce triangles and pentagons.

Finally, elements with horizontal tags are cut, like in figure 3 (right). This second step might cut previously cut polygons, thus cutting hexagons from pentagons.

Thanks to the pairing rule, the number of configurations is reduced to two in 2-dimension. Complex case like the one on the right in fig. 2 are avoided. An octant can only have one subdivided (non-conformal) edge (octants 1 and 3 in fig. 3) or two adjacent subdivided edges (octant 2 in fig. 3).

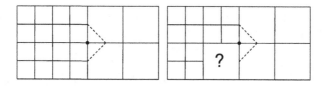

Fig. 2. (left) Correctly paired situation; **(right)** unauthorized unpaired situation, element "?" cannot be properly cut.

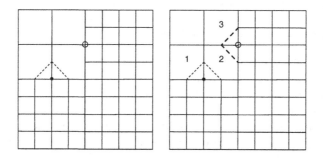

Fig. 3. (left) Polygonal mesh after vertical cutting; **(right)** and after horizontal cutting.

After the two cuttings steps, the mesh is conformal and it is made of polygons, ranging from triangles to hexagons. Vertices are shared by four elements (2-dimensional case).

3.3   3-Dimensional Case

The 3-dimensional case unfolds in the same way, but adds a third cutting step.

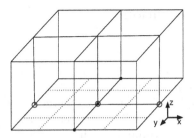

Fig. 4. Octree mesh before cutting: yz plane hanging vertices are tagged with small back dots and xz ones are tagged with bigger white dots. Dotted lines belong to neighboring smaller octants. Note that the vertex in the middle of the non-conformal face belongs to both planes.

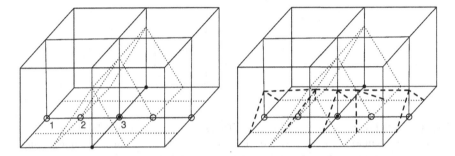

Fig. 5. (**left**) Polyhedral mesh after cutting along the yz plane. Vertices cut in the middle of an edge whose two vertices belong to the same cut plane, inherits the same plane tag. Since vertices 1 and 3 belong to the xz plane (white dots), the vertex 2, cut in the middle of edge 1-3, inherits a white dot tag. (**right**) Polyhedral mesh after two directional cuttings. Fine dotted lines come from the first cut (yz plane), and thick ones come from the second cut (xz plane).

Vertices are first tagged according to the plane they hang from: xy, xz or yz as shown in figure 4.

In 3-dimensions, a new rule must be set: when a vertex falls into the middle of an edge whose vertices have the same tag, then the newly cut vertex inherits this tag (see fig. 5).

The three cuttings steps produce a conformal mesh which is made of polyhedra.

These polyhedra range from four to sixteen polygonal faces, which range from triangles to hexagons.

Thanks to the cutting process, vertices are shared by eight polyhedra, twelve faces and six edges. Likewise, edges are shared by four polyhedra and four faces.

After this step, the mesh is conformal, but made of polyhedra, the next step will make it all hexahedral.

4   Dual-Mesh Generation

4.1   2-Dimensional Case

Generating a dual-mesh consists in creating vertices from primal-mesh elements, and elements from primal-mesh vertices.

A dual vertex is simply built in the barycenter of each primal element.

Creating a dual-element is a little more complicated. For each primal vertex, the ball of elements is computed (that is, the list of primal elements sharing this vertex). From this list of primal elements, the list of their dual-vertices is built. Finally, these dual-vertices are connected to form a dual-element. The process is shown in fig. 6.

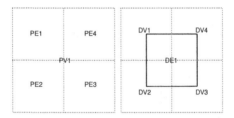

Fig. 6. (left) Four primal-mesh elements (PE) sharing a primal-mesh vertex (PV1); (**right**) the primal-mesh is in dotted lines and the dual one in thick lines. Dual vertices (DV) comes from PE, and the dual element DE comes from PV.

Consequently, the nature of elements in the dual-mesh depends on the vertex connectivity of the primal-mesh. Since primal vertices have a degree of four, then dual-mesh elements will be quadrilaterals. This way, an all-quad mesh is derived from a (not so) arbitrary polygonal mesh.

Figure 7 shows the only two possible configurations in the 2-dimensional case.

Remark

Having a look at the left figure, it appears that we just introduced a new kind of size transition. Well-known quadtree size transition rules are $2 \leftrightarrow 1$ and $3 \leftrightarrow 1$. The dual-mesh generation introduces a $4 \leftrightarrow 2$ transition as shown in fig. 8.

4.2   3-Dimensional Case

Dual vertices

They are set at the barycenter of primal-mesh polyhedra. A vertex is the dual of an element.

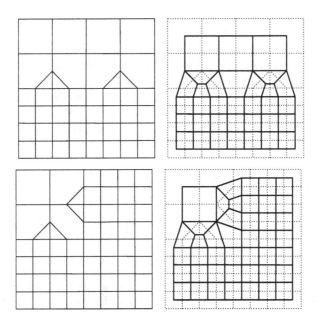

Fig. 7. (left) Primal-meshes; (right) primal-meshes in dotted lines and dual-meshes in thick lines.

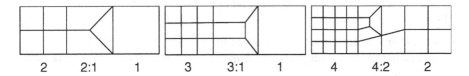

Fig. 8. (left) 2:1 size transition using mixed elements; (center) Schneider's 3:1 transition; (right) 4:2 dual-mesh transition.

Dual edges

They are derived from primal-mesh polygonal faces. A dual edge connects two dual vertices derived from two primal polyhedra sharing the same face. Thus, an edge is the dual of a face.

Dual faces

They are generated from primal edges. The shell (the set of elements which share a common edge) of each primal edge is built, then a dual face is made from the dual vertices associated to these elements. A face is the dual of an edge.

Dual elements

Conversely, they are made from the primal vertices. The ball of polyhedra sharing the same primal vertex is built and the dual element is made from the polyhedra-associated dual vertices. An element is the dual of a vertex.

Since the cutting steps produced only degree-four edges, corresponding dual faces have four vertices. Likewise, every primal vertices have a degree of six, producing dual-elements with six quadrilateral faces: hexahedra!

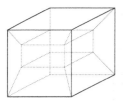 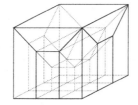 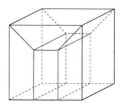

Fig. 9. (**left**) Pattern to mesh a non-conformal octant with one subdivided edge; (**center**) With one subdivided face; (**right**) And with two subdivided edges.

Not surprisingly, the resulting dual-mesh patterns look like Schneiders' original size transitions (see [15]). Only three different patterns are needed to mesh all transition cases (see fig. 9). The advantages over Schneiders' original scheme are:

- smoother size transition, edges are split in two instead of three,
- ability to mesh all configurations, even concave ones, which was impossible with the original method,
- fewer patterns needed.

After this step, the mesh is conformal and hex-only. Still, the mesh fills the whole bounding box and no boundaries nor subdomains are represented. The following steps are about subdomains recovery and boundaries meshing.

5   Subdomains Coloring

The subdomain recovery process begins with coloring concentric main domains. Then tries to color potential non-manifold subdomains within main domains. Finally, some filtering is made to remove one-element thin subdomains and various domain-pinches problems.

Each dual vertex derived form a primal hex intersected by a surface triangle is a boundary one. Conversely, each dual quad which has four boundary vertices is a boundary face.

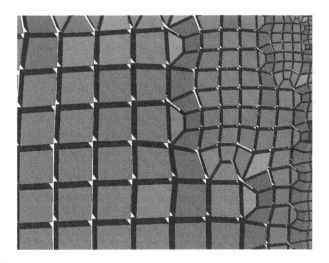

Fig. 10. Cut through a volume mesh: a regular Schneiders' size transition pattern set can mesh any configuration.

6 Geometry and Features Detection

In this step, the hex-mesh boundary quads are to be mapped on the real geometry.

Alongside, some edges of the hex-mesh should be mapped in order to capture sharp angles. These sharp angle edges, called ridges, are set via a user-defined sharp-angle threshold.

Setting objective triangles

Each newly set boundary quad should now be assigned a triangle from the input mesh onto which it will be projected. This process defines a mapping between the generated boundary-quad mesh and the triangular input mesh.

Setting ridges

If two neighboring quads are to be projected on triangles which make and angle greater than a user defined threshold, then their common edge is tagged as ridge. The input triangulated mesh is searched for the closest ridge edge to be assigned as objective feature.

Setting corners

Conversely, if two ridges sharing a common vertex make and angle greater than the threshold, this vertex is tagged as a corner. The closest corner is retrieved from the real geometry to be assigned as objective feature.

7   Buffer-Layer Insertion

Now, we are provided with an all-hexahedral mesh whose hexes are assigned subdomain numbers. Boundary quads, ridges and corners are assigned target geometry position to be projected on. Unfortunately, these boundary vertices cannot be projected yet on the real surface. This is caused by the fact that many boundary hexes have two or three faces to be projected on the same plane, which would make them invalid (see fig. 11). A layer of elements needs to be inserted around boundary hexes to tackle this problem.

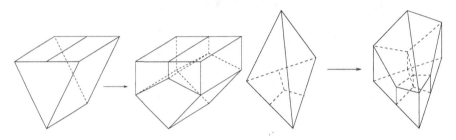

Fig. 11. Two cases of degenerated hexahedra near the boundary: before and after buffer layer insertion.

Inserting volume layers

Each subdomain is surrounded by a boundary layer. A new element is inserted between each boundary quad and the boundary hex it belongs to. This way, some hexes may be surrounded by up to three such boundary elements. Boundary hexes are topologically identical: they have one boundary face, and five internal ones. Thus, they can be freely projected on the real surface while preserving the element quality.

Inserting surface layers

Unfortunately, another problem was left unsolved by the boundary layer insertion. Indeed, some boundary quads have two-ridge edges to be projected onto the same line, making the element invalid (see fig. 12).

This problem is solved by inserting another boundary layer inside the first one. In this case, the subdomains do not come from the geometry, but are built on purpose. First, two-ridge faces are tagged. Then untagged faces standing between tagged ones are also tagged. Finally, adjacent tagged faces are packed together to form the so-called surface subdomains.

Eventually, each surface subdomain is wrapped in a layer a hexes, these new hexes having no more than one boundary face and one-ridge edge.

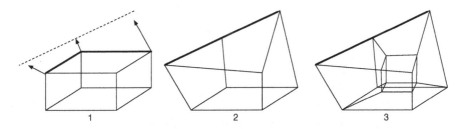

Fig. 12. (**1**) Initial cube with two edges to be projected on a ridge; (**2**) Degenerated element after projection; (**3**) Each hex of the buffer-layer has no more than one ridge-edge.

8   Smoothing

After the insertion of all these layers of elements, the mesh is topologically suitable for projection.

Projection is done on the fly while smoothing the vertices to increase the elements quality. Indeed, surface vertices are not explicitly projected on the geometry. The surface model appears as a constraint when computing the shape of optimal elements in the smoothing scheme.

During the 30 smoothing steps, internal vertices will move so that the quality of inner hexes gradually improves. Surface vertices will slowly be projected on the surface while preserving the quality of boundary hexes. Doing so, it is possible to lower the quality of an element below an acceptance threshold. In this case, we decided to favor element quality over geometry accuracy. Some surface vertices, especially those to be projected on sharp edges, may get stuck during the projection process.

Quality criterion

The quality criterion used is very simple:

$$q = 24\sqrt{3}\frac{V_{min}}{(\sum_{1 \le i \le 12} l_i)^{3/2}}$$

Where V_{min} is the minimum volume among the two sets of five tetrahedra corresponding to the hex (see fig. 13), and l_i are the lengths of the twelve edges.

Since there are as many hex quality criteria as there are solvers, it was pointless to try to find an ideal criterion. Instead, we came up with this basic one, which ranges from 0 (flat element) to 1 (perfect cube) and is negative in case of invalid element (negative volume).

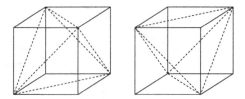

Fig. 13. Two sets of five tetrahedra cut from a hex.

The minimum quality accepted by hexotic is 0.01, which may be unsuitable for some solvers. Setting a higher target may result in poorer geometry accuracy.

A new element based smoothing scheme

This new optimizing scheme loops over elements instead of vertices. That is, for each hexahedron, an optimal element is computed, and its coordinates are added to a set of optimal vertex coordinates. Finally vertices are relaxed toward these optimal positions.

Such a scheme is faster since each optimal hex needs to be computed once. When looping over vertices, a hex optimal shape is computed for each of its vertices, that is, eight times. Furthermore, in this scheme, vertices move all together instead of one by one. When it comes to grid smoothing, a global displacement is preferable.

Computing the optimal element

For each hex, a target optimal element is computed. The main idea is to find the perfect cube, of which vertices are the closest to the original hex one. Thus, minimizing global vertices displacement when moving from initial positions to the optimal one. Here is the 2-dimensional scheme:

- compute the initial quad barycenter (C) and initial base vectors $\vec{V_1}$ and $\vec{V_2}$
- find the pair of orthogonal vectors $\vec{W_1}$ and $\vec{W_2}$, minimizing the sum:

$$\sum_{1 \leq d \leq 2} (\vec{V_d} \cdot \vec{W_d})^2$$

- build a perfect square made up from the center point C and base vectors $\vec{W_1}$ and $\vec{W_2}$ (its size being the average length of the original element, see fig. 14).

If a hex has a boundary face, then the perfect cube should be built so that its boundary quad lies on the constrained surface. In this case, one of the base vectors is set with the surface normal. Likewise, the center point C is moved to a position M (see fig. 15) so that the distance between M and its orthogonal projection on the constrained surface is half the size of the target element.

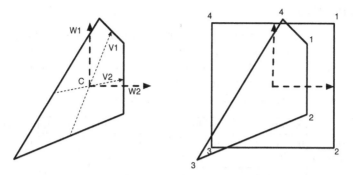

Fig. 14. (**left**) Finding the optimal orthogonal base vectors (in thick dotted lines); (**right**) Building a perfect square in these base.

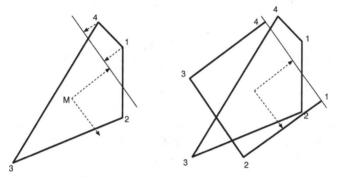

Fig. 15. The edge $1 - 4$ is to be projected on the geometry (thin line). Thus, one of the base vectors is set with the surface normal vector. Likewise, the origin of the new base is positioned so that the boundary edge of the optimal quad lies perfectly on the real surface (right).

9 Conclusion

Let's be honest, *Hexotic* is still far from the hex-meshing Holy Grail.

It is robust, fast (2 million elements per minute on a 2.4 ghz core2 duo) and generates all-hexahedral, conformal and valid meshes while capturing (not so) sharp angles greater than 30°.

It still behaves poorly with thin geometries (the octree generates too many elements) and cannot mesh accurately angles sharper than 30°. Thus it may not be considered as a true general purpose hex-mesher.

However, such a software is able to mesh many real life mechanical geometries as is shown in the last section examples.

Two directions are being investigated to overcome these limitations: using anisotropic smoothing to help handling thin geometries and the insertion of a limited number of tetrahedra and pyramids in very sharp angles.

10 Some Results

Here are some sample results, ranging from simple, but somewhat tricky, geometric pieces, to real life mechanical parts and adapted meshes. All tests have been run on a 2.8 ghz octo-core mac xserve. The smoothing step is multithreaded and takes advantage of all eight cores.

These stats are summed up in the following table:

Figure	Name	# of hexes	Meshing time	Hex quality	Min. q.
Fig. 17, top	nasty cheese	3,556,915	67.11 sec.	0.718	0.010
Fig. 17, bottom	TetTet7	32,995	0.78 sec.	0.735	0.013
Fig. 18, top	anc101_a1	166,280	3.67 sec.	0.620	0.004
Fig. 18, bottom	asm007	121,532	3.12 sec.	0.633	0.011
Fig. 19, top	bm1_90base	818,780	22.74 sec.	0.671	0.011
Fig. 19, bottom	gps25	1,427,535	24.37 sec.	0.721	0.010

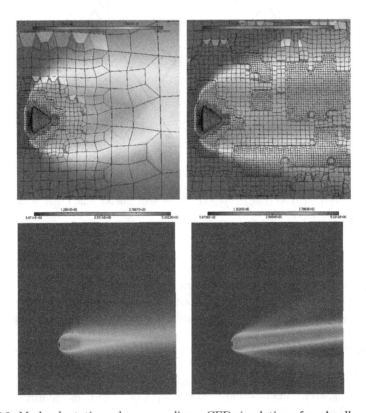

Fig. 16. Mesh adaptation scheme coupling a CFD simulation of an Apollo capsule entering atmosphere (Euler inviscid solver *Wolf* [1]) and *Hexotic*: (**top left**) cut through the initial mesh (66,000 elements); (**top right**) final adapted mesh (538,000 elements); (**bottom left**) initial pressure field; (**bottom right**) The final iteration pressure field is much better captured.

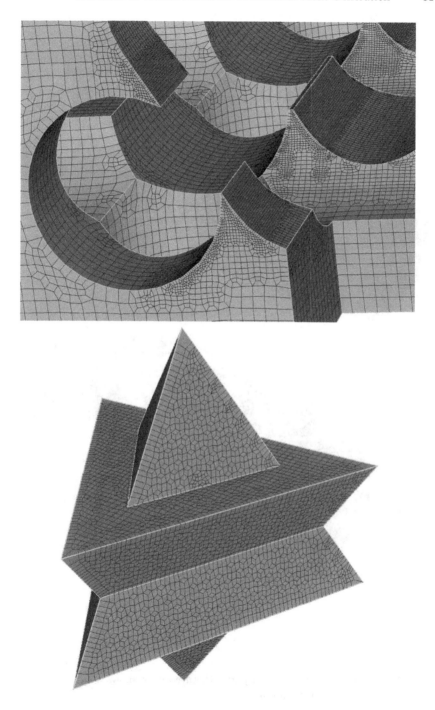

Fig. 17. Test case geometries. The upper picture is a close-up view of nasty cheese, a well known test-case featuring 30° dihedral angles.

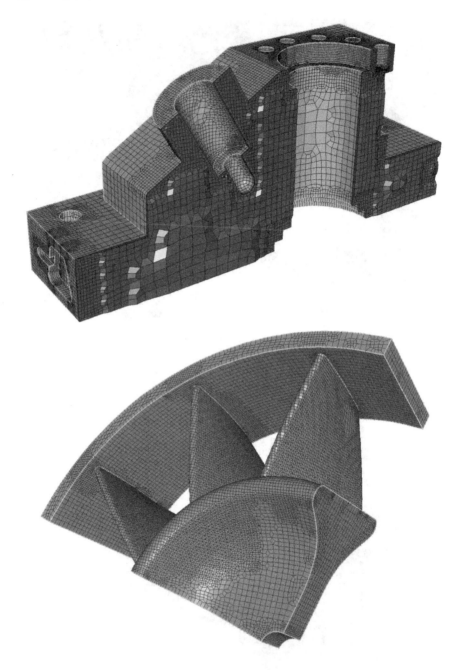

Fig. 18. More complex mechanical parts. Picture above shows a cut through the volume mesh (dark elements are hex faces).

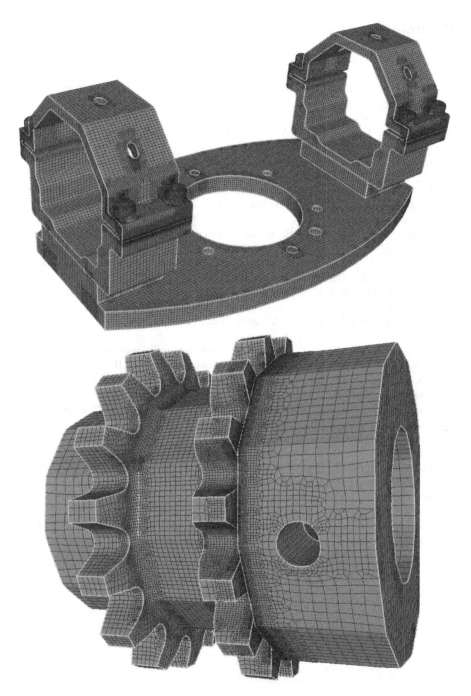

Fig. 19. (**top**) Small features capturing required 818,780 hexes; (**bottom**) A very fine gap between upper part and the top gear forced the octree up to level 9 subdivision. Thus generating 1,427,535 hexes.

References

1. Alauzet, F., Loseille, A.: High Order Sonic Boom Modeling by Adaptive Methods. INRIA Research Report RR-6845 (2009)
2. Armstrong, C.G., Li, T.S., Mckeag, R.M.: Hexahedral meshing using midpoint subdivision and integer programming. Comp. Meth. Appl. Mech. Engng. 124, 171–193 (1995)
3. Armstrong, C.G., Price, M.A.: Mat and associated technologies for structured meshing. ECCOMAS (2000)
4. Blacker, T.: Meeting the challenge for automated conformal hexahedral meshing. IMR 9, 11–19 (2000)
5. Calvo, N.A., Idelsohn, S.R.: All-hexahedral element meshing: Generation of the dual-mesh by recurrent subdivision. Comp. Meth. Appl. Mech. Engng. 182, 371–378 (2000)
6. Dhont, G.: A new automatic hexahedral mesher based on cutting. Int. Jour. Numer. Meth. Eng. 50, 2109–2126 (2001)
7. Folwell, N.T., Mitchell, S.A.: Reliable whisker weaving via curve contraction. IMR 7, 365–378 (1998)
8. Frey, P.J., Maréchal, L.: Fast adaptive quadtree mesh generation. IMR 7, 211–224 (1998)
9. Frey, P.J., George, P.L.: Mesh Generation, ch. 5. Hermes Science publishing (1999)
10. Harris, N.J., Benzley, S.E., Owen, S.J.: Conformal refinement of all-hexahedral element meshes based on multiple twist plane insertion. IMR 13 (2004)
11. Müller-Hannemann, M.: Hexahedral mesh generation by successive dual cycle elimination. IMR 7, 379–393 (1998)
12. Shepherd, J., Benzley, S., Mitchell, S.: Interval assignment for volumes with holes. Int. J. Numer. Meth. Engng. 49, 277–288 (2000)
13. Yerry, M.A., Shephard, M.S.: A modified-quadtree approach to finite element mesh generation. IEEE Computer Graphics Appl. 3, 39–46 (1983)
14. Shephard, M.S., Georges, M.K.: Automatic three-dimensional mesh generation by the finite octree technique. SCOREC Report (1) (1991)
15. Schneiders, R., Bünten, R.: Automatic generation of hexahedral finite element meshes. Computer Aided Geometric Design 12, 693–707 (1995)
16. Schneiders, R.: Octree-based hexahedral mesh generation. Int. J. of Comp. Geom. & Applications 10(4), 383–398 (2000)

Conforming Hexahedral Mesh Generation via Geometric Capture Methods*

Jason F. Shepherd

Sandia National Laboratories
jfsheph@sandia.gov

Abstract. An algorithm is introduced for converting a non-conforming hexahedral mesh that is topologically equivalent and geometrically similar to a given geometry into a conforming mesh for the geometry. The procedure involves embedding geometric topology information into the given non-conforming base mesh and then converting the mesh to a fundamental hexahedral mesh. The procedure is extensible to multi-volume meshes with minor modification, and can also be utilized in a geometry-tolerant form (i.e., unwanted features within a solid geometry can be ignored with minor penalty). Utilizing an octree-type algorithm for producing the base mesh, it may be possible to show asymptotic convergence to a guaranteed closure state for meshes within the geometry, and because of the prevalence of these types of algorithms in parallel systems, the algorithm should be extensible to a parallel version with minor modification.

Keywords: Hexahedral, Mesh, Generation, Dual, Topology Modification.

1 Introduction

In this paper we explore more fully how to capture geometric boundary features using elements in the dual. We build on work completed in [1, 2, 3] and demonstrate conversion of non-geometry conforming hexahedral meshes to conforming hexahedral meshes followed by a mesh conversion to the fundamental mesh. The final quality of these meshes is largely dependent on the quality of the original non-conforming mesh; however, once a reasonable mesh is developed within a geometry it should be feasible to perform mesh conversions to optimize the structures within the mesh to improve geometric quality and mesh topology. The method described in this paper is extensible

---

* Sandia is a multiprogram laboratory operated by Sandia Corporation, a Lockheed Martin Company for the United States Department of Energy's National Nuclear Security Administration under contract DE-AC04-94AL85000.

to multi-volume meshes and geometry tolerant meshing paradigms. Additionally, because the concepts relating to fundamental meshes effectively localize mesh modification to geometric features, parallelizing the algorithm should be relatively straight-forward.

2 Background

The basic concepts that will be utilized throughout this paper utilize a 'dual' representation of a hexahedral mesh. The concept of the hexahedral mesh dual is foundational to many hexahedral mesh modification techniques that have been developed in recent years [4, 1, 5]. Defined by sheets and columns, the dual provides an alternate representation of a conforming hexahedral mesh. This alternate representation has supplied greater understanding about hexahedral mesh topology and has led to the creation of some basic mesh operations. Although the capture techniques introduced in this paper are novel, there is some similarity in the results produced by this method with grid-based methods. The reader is encouraged to review work in [6, 7, 8].

2.1 Dual Sheets and Columns

A hexahedral element contains three sets of four topologically parallel edges, as shown in Figure 1. Topologically parallel edges provide the basis for hexahedral sheets. The formation of a sheet begins with a single edge. Once an edge has been chosen, all elements which share that edge are identified. For each of these elements, the three edges which are topologically parallel to the original edge are also identified. These new edges are searched iteratively to find all connected hexahedra and the topologically parallel edges for each of these elements. This iterative procedure continues until no new adjacent elements are found. The set of all elements which are traversed during this process results in a layer of hexahedra, also known as a hexahedral sheet. Figure 2 shows a hexahedral mesh with a single hexahedral sheet highlighted.

Fig. 1. A hexahedral element has three sets of four topologically parallel edges.

A hexahedral element also contains three pairs of topologically opposite quadrilateral faces, as shown in Figure 3. Topologically opposite faces provide the basis for hexahedral columns. The formation of a column begins with a single face. Once a face has been chosen, the two elements which share that face are identified. For each of these elements, the face which is topologically

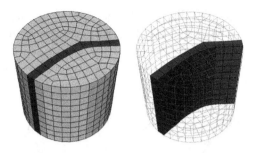

Fig. 2. A hexahedral mesh with one sheet highlighted.

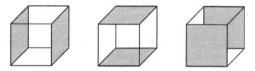

Fig. 3. A hexahedral element has three pairs of topologically opposite faces.

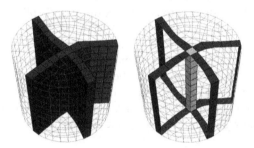

Fig. 4. The intersection of the two sheets shown on the left is defined by the column shown on the right.

opposite of the original face is also identified. These new faces are then used to find the incident hexahedra and topologically opposite faces on these adjacent elements. This process is repeated iteratively until no new adjacent hexahedra can be found. The set of all hexahedra which are traversed during this process makes up a hexahedral column. An important relationship between sheets and columns is that a column defines the intersection of two sheets. This relationship is illustrated in Figure 4.

2.2   Sheet Operations

The dual description of a hexahedral mesh is essentially an arrangement of surfaces satisfying specific criterion. It is, therefore, possible to modify an existing mesh simply by modifying the underlying arrangement of surfaces describing the original mesh. The simplest form of modifying the mesh would involve adding or removing a surface from the arrangement of surfaces. This concept is often referred to as sheet insertion or sheet extraction [9]. Sheet extraction removes a sheet from a mesh by simply collapsing the edges that define the sheet, as shown in Figure 5.

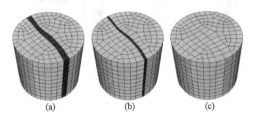

(a)   (b)   (c)

Fig. 5. Sheet extraction: (a) A sheet is selected. (b) The edges that define the sheet are collapsed. (c) The sheet is entirely removed from the mesh.

Using an inverse approach to sheet extraction, it is also possible to insert new sheets into an existing hexahedral mesh. The most common method for inserting a generalized sheet into a hexahedral mesh is pillowing [10]. Unlike sheet extraction, which removes an existing sheet from a mesh, pillowing inserts a new sheet into a mesh. As demonstrated in Figure 6, pillowing is performed on a set of hexahedral elements which make up a 'shrink' set. These elements are pulled away from the rest of the mesh and a new sheet is inserted by reconnecting each of the separated nodes with a new edge and creating new hexahedra utilizing all of the new created edges to fill in the gap. The new sheet surrounds the shrink set and maintains a conforming mesh.

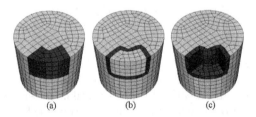

(a)   (b)   (c)

Fig. 6. Pillowing: (a) A shrink set is defined. (b) The shrink set is separated from the rest of the mesh and a sheet is inserted to fill in the gap. (c) The newly inserted pillow sheet.

2.3   Fundamental Hexahedral Meshes

Another concept of importance to the methods outlined in this paper is the notion of a fundamental hexahedral mesh. The definition of a fundamental sheet is relative to the geometric object associated with the mesh [1]. The principle of a fundamental mesh is to have a single sheet for every surface, and a single chord for every curve on a surface (see examples in Figure 7). Such a mesh captures the geometry in such a way that every n-dimensional geometric cell is captured by one or many n-dimensional dual cells with particular restrictions. There is always at least one fundamental mesh for a given geometry, but the fundamental mesh definition allows for many different arrangements which satisfy the definition. If you translate this description in the primal mesh, you get the following definition given in [2].

Definition 2.1 *Let G and M be respectively a 3-dimensional geometric object and a hexahedral mesh. A hexahedral mesh M is a* **fundamental mesh** *with respect to G if and only if:*

1. *M is a strictly geometry-valid hexahedral mesh with respect to G;*
2. *For every geometric surface G_k^2, the number of hexahedral elements incident to G_k^2 is equal to the number of quadrilaterals classified on G_k^2;*
3. *For every curve G_k^1 and every surface $G_{k'}^2$, the number of quadrilaterals on $G_{k'}^2$ incident to edges on G_k^1 is equal to the number of edges classified on G_k^1.*

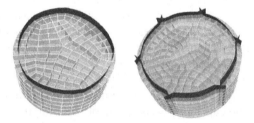

Fig. 7. Fundamental and non-fundamental meshes of a cylindrical geometric object. In the mesh on the right, multiple sheets are utilized to capture a single surface producing a non-fundamental mesh. The image on the left has a single sheet associated with the cylindrical surface and is fundamental.

Complementary explanations about this definition are given [2]. Note, for a geometric object, there is not one unique fundamental mesh, and, in fact, many geometry-valid hexahedral meshes will exist which are fundamental. These meshes are different for two reasons: there are many permutations of boundary sheets which satisfy the fundamental mesh requirements, and the number and configurations of non-boundary sheets within the hexahedral mesh is not restricted.

3 Theory and Assertions

In [1], the assertion is made that any non-fundamental hexahedral mesh can be converted to a fundamental hexahedral mesh. A proof of this assertion was later given in [3]. In this paper, the goal is to show that a non-conforming hexahedral mesh whose boundary is topologically equivalent and geometrically similar to the composite boundary of a given solid geometry can be converted to a fundamental hexahedral mesh of the geometry and maintain reasonable quality metrics for the resulting hexahedra. We will discuss these two conjectures (i.e., conversion and quality) in more detail in this section.

3.1   Non-conforming Meshes to Fundamental Meshes

In [3], it was shown that a non-fundamental hexahedral mesh of a geometric object can always be converted to a fundamental hexahedral mesh of the geometric object. Therefore, it remains to be shown how to convert a non-conforming hexahedral mesh of a given object to a non-fundamental mesh of the same object.

The process of converting a non-conforming mesh to a conforming hexahedral mesh begins with two assumptions regarding the non-conforming mesh and the geometric object to be captured. The first assumption is that the boundary (composite) of the geometry and the boundary of the initial hexahedral mesh must be equivalent. That is, if the geometry is topologically spherical, then the boundary of the initial hexahedral mesh should also be spherical (i.e., it is not possible to convert a hexahedral mesh whose boundary is an n-toroid into a mesh of a spherical geometry). This assumption almost goes without saying, but depending on the method utilized to form the initial base mesh, may be commonly encountered.

The second assumption is more subjective. That is, the boundary of the initial base mesh should be geometrically similar to the boundary (composite) of the geometry to be captured. Satisfaction of this assumption is not binary (i.e., yes or no) like the topology equivalence assumption, and can involve a continuous range of satisfactory values. However, high geometric similarity between the base mesh and the geometry results in decreased modification to the base mesh and an associated higher probability that the quality of the resulting modified mesh will match the quality of the initial base mesh.

Given a base mesh that is non-conforming to the geometry, but topologically equivalent and geometrically similar, the process for converting the mesh to a conforming mesh involves the following steps:

1. Find an embedding of the geometric topology into base mesh.
2. Map the embedded geometric topology in the base mesh to the geometry

For the first item - finding an embedding of the geometric topology into the base mesh - we treat the boundary of the base mesh (e.g., the quadrilaterals, edges and nodes on the boundary of the mesh) as a graph (the mesh graph) and the geometric topology of boundary of the geometric object as a second graph (the geometric graph), and work to find an embedding of the second graph into the first graph. In some cases, this may require an enrichment of the the mesh graph when embedding the curves at high valent vertices from the geometric graph. This enrichment can be accomplished utilizing a pillowing operation in the hexahedral mesh that will be described in the next section.

The second item - Map the embedded geometric topology in the base mesh to the geometry - is accomplished by finding an appropriate location for each of the nodes on the boundary of the base mesh to the appropriate geometric location determined from the embedding in the first step. At this point the mesh is now conforming to the geometry, although it likely has very poor quality (especially, near the new mesh boundary where most of the nodal movement took place in the last step). This quality will be improved during the conversion from the conforming, non-fundamental mesh to a fundamental mesh, and will be described in the next section.

3.2   Assertion on Hexahedral Quality

The second conjecture is in regards to the hexahedral element quality resulting from the conversion process outlined above. While it is very difficult to provide guarantees on potential hexahedral element quality, we refer back to an observation made in [11]. The ideal isotropic mesh contains perfectly planar sheets. Sheet curvature induces 'keystoning' of the hexahedral element (where the edges on one side of the hex are shorter than the opposite edges which are lengthened by the curvature). Sheets that do not intersect each other orthogonally induce element 'skewing' (see Figure 8). However, if only one of the three sheets defining the hex is subject to increased curvature or if there is only a single sheet with non-orthogonal intersections to the other sheets, then the feasible region for non-positive Jacobians remains large and it is likely that a smoothing or mesh optimization algorithm will be able to find a satisfactory nodal placement resulting in suitable hexahedral quality.

Fig. 8. Non-orthogonal intersections between sheets results in element 'skewing'.

We attempt to capitalize on this conjecture. That is, if the base mesh utilized consists of sheets with very low curvature and the sheets intersect each other with near orthogonality, then the sheets which are inserted during the conversion from non-fundamental to fundamental mesh will be the only sheets with any curvature or potential for non-orthogonality. If these inserted sheets are not interacting with each other, then the probability for creating a mesh which cannot be optimized to have reasonable quality is very low.

4 Algorithm

In this section, we describe an algorithm for performing the conversion of the non-conforming mesh to a conforming, non-fundamental mesh, and then finally to a fundamental mesh. We will also provide discussion on optimizing the mesh following the fundamental conversion. The algorithm described is presented for a generalized approach and various alterations can be made which limit generality but improve quality or desired mesh topology. These differences will be discussed in a later section.

The algorithm has four basic steps:

1. Establish the 2-manifolds for volumetric capture.
2. Convert the non-conforming mesh to a conforming mesh through geometric topology capture.
3. Convert the non-fundamental mesh to a fundamental mesh.
4. Mesh optimization to improve mesh quality.

4.1 Establishing 2-Manifolds

The first step in the algorithm is to capture a geometric solid from a pre-existing base mesh. We start by considering the boundary of the geometric solid as a closed, 2-manifold surface. We desire a base mesh whose boundary is topologically equivalent to the boundary of the geometric solid, as well as minimizing the geometric dissimilarities between these two boundaries. We only restrict this base mesh to topological equivalence, except to say, that the resulting quality of the final mesh will be heavily dependent on the geometric similarity between the base mesh and the geometric solid. Therefore, tailoring a base mesh (topologically and geometrically) to improve similarities between the base mesh and the geometric solid can provide dramatic quality differences in the final mesh.

There are several options for establishing the initial base mesh. Perhaps the easiest method is to create a structured grid in the bounding box representation of the solid geometry and eliminating hexahedra outside the boundary of the solid geometry (see Figure 9. This type of approach also allows base meshes to be created using standard octree-meshing techniques.

A similar approach can be utilized by creating simplified solid models and using standard hexahedral meshing algorithms on the simplified geometries and then eliminating hexahedra in the simplified geometry that do not match the original geometry (see example in Figure 10).

This approach can also be used to establish bounding 2-manifolds for multi-volume meshing. The only difference in multi-volume is to ensure that the topology of the all the volumetric base-meshes conform equivalent to the volumes in the solid model. Some initial work by Zhang, et al., demonstractes a method for establishing the base meshes for multi-volume biological models. A similar methodology can be utilized for solid models, although in some cases where the edges in the mesh do not have sufficient topology to match

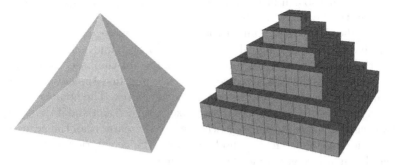

Fig. 9. An original geometry (on left) can be embedded in a regular mesh and the elements contained in the geometry is one alternative for defining a initial base mesh.

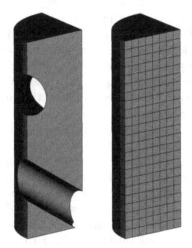

Fig. 10. A base mesh for an original geometry (on left) can also be created using standard hex primitive algorithms, including sweeping. The elements not contained in the mesh on the right can be removed and the resulting mesh would be suitable for a base mesh.

the curves in the solid model, a mesh enrichment step may be needed to provide the additional topology such that the base meshes are equivalent to the solid model.

4.2 Capturing Geometric Topology

Once the base mesh for the solid model has been established, the process of converting the non-conforming base mesh into a conforming base mesh can begin. Essentially, what we want to do in this step is to embed the curves and vertices from the boundary of the solid model into the boundary of the base mesh. This also can be done in one of several ways. We will discuss one approach in this section.

Because the desire is to embed the geometric topology of the boundary into the base mesh, the logical first step is to first find an embedding of each of the geometric vertices into the boundary of the base mesh. In our case, we have done this by a geometric search, and find the closest node in the boundary of the base mesh to the geometric vertex. The number of edges emanating from each of the nodes is also taken into account and if the number of curves emanating from the geometric vertex is greater than the number of edges at the node and a nearby node captures the geometric topology more adequately, then a re-assignment may be made.

If the number of edges emanating from the closest node to a geometric vertex is fewer than the number of curves emanating from the vertex, a mesh enrichment procedure may be utilized to increase the nodal valence. Increasing the nodal valence consists of pillowing a collection of hexes in the neighborhood of the node resulting in additional edges emanating from the node (see Figure 11). This process can be repeated multiple times as needed to increase the nodal valence; however, care should be taken as possible, since the insertion of a small pillow will have reasonably high local curvature (the pillow in this case is essentially hemispherical). The curvature of the pillowed sheet may result in element quality reductions in the final mesh based on the conjecture discussed in Section 3.2.

Once an embedding of the geometric vertices is found and the nodal valence is equal to or greater than the vertex valence, we can begin to work on embedding each of the geometric curves into the mesh. If the quadrilateral boundary of the base mesh is treated as a graph, and the vertices have been embedded in this graph, then this problem can be viewed as similar to a collision-free network search. That is, we want to find a path between the embedded nodes which minimizes the geometric distance from the geometric curve and contains no collisions with the paths for each of the other curves. We demonstrate an example of this process in Figure 12.

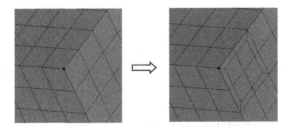

Fig. 11. Nodal valence can be increased using a simple pillowing operation. The original mesh is shown on the left and the resulting nodal valence increase after pillowing is displayed on the right.

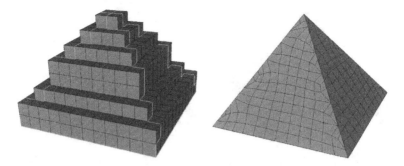

Fig. 12. The geometric graph is embedded into the base mesh using a conflict-free search of the mesh edges on the boundary of the base mesh to ensure topologic equivalence of the new embedding. Once the embedding has been established, the bounding mesh elements are 'snapped' to the appropriate geometry (image on right).

Once an embedding of the vertices and curves has been accomplished, the surface embedding is calculated by finding all of the quadrilaterals contained within the boundary of the embedded curves. This should complete the embedding of the geometric boundary into the boundary of the base mesh. At this point, all of the nodes on the boundary of the base mesh are moved to the correct geometric entity based on the previous embedding. This results in a conforming, but non-fundamental mesh of the geometry, from the previously non-conforming base mesh (see example in Figure 12).

4.3   Fundamental Conversions

Given a the embedding of the vertices and curves into the base mesh, we can begin the process of converting the base mesh to a fundamental mesh of the given geometry. A proof that this conversion can always occur is given in [2], and a construction of this proof can be utilized as follows in two steps:

1. Add a single sheet for each closed shell of boundary surfaces in the geometry. Assuming that the shell is manifold, this operation can be accomplished with a simple pillowing operation.
2. Add a single sheet for the collection of hexahedra contained by quadrilaterals associated with each surface of the geometry. Again, assuming that the collection of quadrilaterals is topologically equivalent to the associated geometric surface and the boundary of this collection of hexahedra if manifold, a simple pillowing operation will suffice for this sheet insertion. It should be noted that the first sheet insertion guarantees that there is a single hexahedron associated with each boundary quadrilateral, which is critical for this construction. Additionally, the embedding step described earlier can be used to guarantee topological equivalence of the quadrilateral collection with the geometric surface.

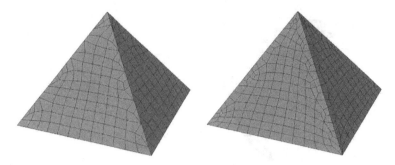

Fig. 13. The pyramid mesh following before (left) and after (right) conversion to a fundamental mesh.

The mesh following the fundamental conversion should be free of flattened (triangle-shaped) quadrilateral elements, and similar improvements will be noted in the interior hexahedra. The mesh for the pyramid example is shown in Figure 13 with the improved quality elements around the boundary.

This particular method for converting the mesh to fundamental is generalized, or in other words, it will work in all cases. However, it should be noted that fundamental meshes are not always required for satisfactory quality, and alternative manipulations of the mesh may be possible and desirable for improved quality. We will discuss this further in Section 6. Additionally, if the mesh in the second step is already fundamental, then no sheet insertion is required and the pillowing operation can be skipped.

4.4   Mesh Optimization

The conversion of the mesh to fundamental has the advantage of giving improved flexibility for mesh optimization near the boundary of the mesh. This is seen quite markedly in earlier papers [12, 1, 8], where dramatic improvements in the scaled Jacobian values are realized following the sheet insertion process and mesh smoothing/optimization. The pillowing process involves non-uniformly scaling elements in order to allow space for the newly inserted elements to occupy. This scaling often causes element inversions; however, the new topology introduced by the sheet insertion allows for greater flexibiliy by increasing the positive quality feasible regions for each of the elements near the boundary. In order to take advantage of this flexibility, mesh optimization algorithms with L2 and L-inf guarantees are recommended. In particular, we have heavily utilized the results of Knupp in mesh untangling [13], condition number optimization [14], and mean-ratio optimization [15, 16]. Additionally, it has been shown [17] that we can dramatically improve the speed of these algorithms by first utilizing a relaxation-based smoother (e.g. Laplacian smoothing), or using a focused-smoothing operation to reduce the number of elements being optimized. A recipe typically utilized following the conversion to fundamental is as follows:

1. Centroidal smoothing [18] each of the surfaces on the mesh boundary.
2. Untangling/optimization for each surface on the boundary (as needed).
3. Laplacian smoothing of the volumetric mesh.
4. Focused untangling of any pockets of mesh with inverted elements.
5. Focused L-inf optimization for any pockets of mesh elements with scaled Jacobian less than 0.2.
6. Additional mesh optimization as desired.

5   Examples

In this section, we demonstrate the process for one more example with slightly increased complexity (Figures 14 and 15). Following this example, we

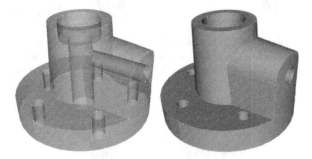

Fig. 14. The original 'sbase' geometric model shown in shaded and transparent modes. (Model provided courtesy of Ansys [19].)

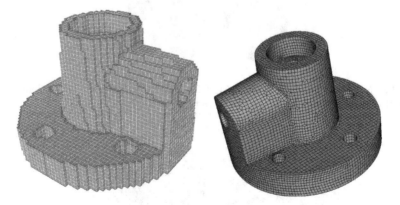

Fig. 15. A base mesh created for the 'sbase' geometry. The base mesh was created by creating a regular mesh in the bounding box of the geometry and eliminating all hexahedra located exterior of the geometry. This image also shows an embedding of the geometric graph in the boundary edges of the base mesh (the blue edges). The final mesh after conversion of the base mesh to a fundamental mesh is shown on the right.

Fig. 16. A mesh of a 6 generation airway of a human lung. The mesh on the right is a close-up view showing the sharp curves generated at the end caps of the model. (Model provided courtesy of Kwai Lam Wong, Oakridge National Laboratories.)

Fig. 17. Mesh of the valve model (Model courtesy of Kyle Merkley, Elemental Technologies, Inc.), and hook model (on right).

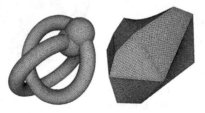

Fig. 18. Mesh of the 'a027' and 'ucp' models ('A027 Model courtesy of Ansys [19]).

demonstrate several other meshes that have been generated using this methodology (Figures 16, 17, and 18).

6   Alternate Methods

The method outlined in the previous sections of this paper are meant to be generalized. That is, these methods should work for all geometries with manifold boundaries. However, alternate conversions may produce better mesh quality, but may not be suitable for general cases. We will discuss some of these cases here, although it should be noted that the list provided below is not exhaustive, and additional methods may be developed and utilized.

6.1   Tri-valent Vertices

As indicated before, the method described in this paper is a generalized method. However, the price for generalization is decreased quality in some cases that could be improved with different sheet configurations that may not work generally. A tri-valent geometric vertex is a good example of this trade-off. In Figure 19(left) a mesh is shown that might be developed using the method outlined in this paper. At the trivalent vertex in the forefront of the image, the fundamental sheets are drawn (red, green, and yellow) for capturing the curves associated with this vertex. This configuration of sheets produces 2*v hexahedra, where v is the vertex valence. So, in this case, there are six hexahedra produced at this vertex location (i.e. the node at the vertex is contained in six hexahedra). An alternate sheet configuration (shown in Figure 19 (right) where the sheets are allowed to intersect one another is also shown. This configuration of sheets produces a single hexahedron at the vertex, and still fundamentally captures all of the curves associated with the

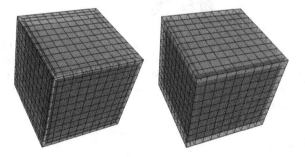

Fig. 19. Allowing changes in the sheet structure can improve the mesh quality. In the image on the left, the three sheets do not intersect resulting in six hexahedra at the vertex in the forefront of the image. On the right, the three sheets intersect and a single hexahedron is produced offering better opportunity for improved mesh quality at the vertex.

vertex. The resulting mesh has a higher potential for quality because of the change in the sheet topology.

Allowing the sheets to intersect one another is, in many cases, an improved strategy for inserting the sheets; however, the disadvantage to this alternate strategy is that it will only work if every vertex is trivalent. Therefore, applying the generalized algorithm is a better option for the original mesh generation algorithm, but it may be possible to 'clean-up' the mesh by re-arranging the sheet topology to improve the mesh quality locally at some of these tri-valent vertices. Additionally, it may be possible to develop similar recipes to reduce the element count and improve the quality potential while still maintaining a valid hexahedral mesh at each of the vertices by allowing some local sheet reconfigurations.

6.2   Regularizing Mesh Near Boundaries

Multiplying the number of sheets can also be utilized to increase the regularity of the mesh near the boundaries (see Figure 20). Balancing the arrangement of the sheets, can have a dramatic effect on the quality of the mesh. It will be advantageous to re-arrange the sheet topology following the initial mesh generation in order to improve regularity of the mesh and quality of the elements while still maintaining conformity with the geometry.

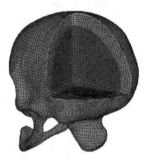

Fig. 20. Allowing for changes to sheet arrangements can improve the quality of the final mesh. In this image, several additional sheets were added near the interior cranial boundary to improve the regular structure of the mesh in this region. Other such additions, rearrangements, and removals of sheets may provide improved quality and topology of the final mesh.

7   Conclusion

We have outlined a method for building all-hexahedral meshes in arbitrary geometries. The procedure involves embedding geometric topology information into the given non-conforming base mesh and then converting the mesh to a fundamental hexahedral mesh. The procedure is extensible to multi-volume meshes with minor modification, and can also be utilized in a geometry-tolerant form (i.e., unwanted features within a solid geometry can be ignored with minor penalty by the meshing procedure). Utilizing an octree-type algorithm for producing the base mesh, it may be possible to show asymptotic convergence to a guaranteed closure state for meshes within the geometry. Due to the prevalence of octree algorithms in parallel systems, the algorithm should also be extensible to a parallel version with minor modification.

References

1. Shepherd, J.F.: Topologic and Geometric Constraint-Based Hexahedral Mesh Generation. Published Doctoral Dissertation, University of Utah (May 2007)
2. Ledoux, F., Shepherd, J.: Topological and geometrical properties of hexahedral meshes. Submitted to Engineering with Computers (2008)
3. Ledoux, F., Shepherd, J.: Topological modifications of hexahedral meshes via sheet operations: A theoretical study. Submitted to Engineering with Computers (2008)
4. Murdoch, P.J., Benzley, S.E.: The spatial twist continuum. In: Proceedings, 4th International Meshing Roundtable, pp. 243–251. Sandia National Laboratories (October 1995)
5. Merkley, K., Ernst, C.D., Shepherd, J.F., Borden, M.J.: Methods and applications of generalized sheet insertion for hexahedral meshing. In: Proceedings, 16th International Meshing Roundtable, pp. 233–250. Sandia National Laboratories (September 2007)
6. Schneiders, R.: An algorithm for the generation of hexahedral element meshes based on an octree technique. In: Proceedings, 6th International Meshing Roundtable, pp. 183–194. Sandia National Laboratories (October 1997)
7. Shephard, M.S., Georges, M.K.: Three-dimensional mesh generation by finite octree technique. International Journal for Numerical Methods in Engineering 32, 709–749 (1991)
8. Zhang, Y., Hughes, T.J.R., Bajaj, C.: Automatic 3D meshing for a domain with multiple materials. In: Proceedings, 16th International Meshing Roundtable. Sandia National Laboratories (October 2007)
9. Borden, M.J., Benzley, S.E., Shepherd, J.F.: Coarsening and sheet extraction for all-hexahedral meshes. In: Proceedings, 11th International Meshing Roundtable, pp. 147–152. Sandia National Laboratories (September 2002)
10. Mitchell, S.A., Tautges, T.J.: Pillowing doublets: Refining a mesh to ensure that faces share at most one edge. In: Proceedings, 4th International Meshing Roundtable, pp. 231–240. Sandia National Laboratories (October 1995)
11. Shepherd, J., Johnson, C.: Hexahedral mesh generation constraints. Journal of Engineering with Computers (2007)
12. Shepherd, J.F., Zhang, Y., Tuttle, C., Silva, C.T.: Quality improvement and boolean-like cutting operations in hexahedral meshes. In: Proceedings, 10th Conference of the International Society of Grid Generation (September 2007)
13. Knupp, P.M.: Hexahedral mesh untangling and algebraic mesh quality metrics. In: Proceedings, 9th International Meshing Roundtable, pp. 173–183. Sandia National Laboratories (October 2000)

14. Knupp, P.M.: Achieving finite element mesh quality via optimization of the Jacobian matrix norm and associated quantities: Part II - a framework for volume mesh optimization and the condition number of the Jacobian matrix. International Journal for Numerical Methods in Engineering 48, 1165–1185 (2000)
15. Brewer, M., Freitag-Diachin, L., Knupp, P., Leurent, T., Melander, D.J.: The MESQUITE mesh quality improvement toolkit. In: Proceedings, 12th International Meshing Roundtable, pp. 239–250. Sandia National Laboratories (September 2003)
16. Diachin, L.F., Knupp, P., Munson, T., Shontz, S.: A comparison of inexact Newton and coordinate descent mesh optimization techniques. In: Proceedings, 13th International Meshing Roundtable, pp. 243–254. Sandia National Laboratories (September 2004)
17. Freitag, L.: On combining Laplacian and optimization-based mesh smoothing techniques. AMD Trends in Unstructured Mesh Generation, ASME 220, 37–43 (1997)
18. Jones, T.R., Durand, F., Desbrun, M.: Non-iterative, feature-preserving mesh smoothing. ACM Transactions on Graphics 22(3), 943–949 (2003)
19. ANSYS. ANSYS (January 2007), http://www.ansys.com

Efficient Hexahedral Mesh Generation for Complex Geometries Using an Improved Set of Refinement Templates

Yasushi Ito, Alan M. Shih, and Bharat K. Soni

Enabling Technology Laboratory
Department of Mechanical Engineering
University of Alabama at Birmingham, Birmingham, AL, U.S.A.
{yito,ashih,bsoni}@uab.edu

Abstract. We proposed a new, easy-to-implement, easy-to-understand set of refinement templates in our previous paper to create geometry-adapted all-hexahedral meshes easily [Ito *et al.*, "Octree-Based Reasonable-Quality Hexahedral Mesh Generation Using a New Set of Refinement Templates," *International Journal for Numerical Methods in Engineering*, Vol. 77, Issue 13, March 2009, pp. 1809-1833, DOI: 10.1002/nme.2470]. This paper offers an extension of the template set to refine concave domains more efficiently. In addition, two new options, temporal rotation and temporal local inflation, are introduced to create hexahedral meshes with fewer elements for long objects or objects with thin regions. Several examples are shown to discuss the improvements of the mesh generation method.

1 Introduction

All-hexahedral mesh generation using an octree data structure is a promising approach to automatically create meshes for complex geometries. However, there are still several issues that need to be addressed for practical applications.

The refinement of a hexahedral mesh with a concave refinement domain is a challenging problem because a resulting mesh must have only hexahedral elements without hanging nodes and the number of elements to be added must be minimized outside of the refinement domain. Several solutions have been presented before, using a combination of templates, and/or pillowing (also known as sheet insertion) techniques [1-7]. Schneiders *et al.* proposed the first template-based method in 1996, which has been used widely because of its simplicity to implement [1]. However, it can be applied to convex refinement domains only. To refine concave refinement domains, many extra elements are needed outside of the domains. Although other template-based methods can be applied to concave refinement domains more efficiently, their implementation is neither easy nor clear, especially for multiple levels of refinement [2, 3]. The pillowing methods are not easy to implement because generally sheets cannot be added locally.

To overcome this problem, we have recently proposed a new, easy-to-implement, easy-to-understand set of refinement templates as part of octree-based, geometry-adapted hexahedral mesh generation [8]. Two new templates were developed for three-node (three nodes of one of the six quadrilaterals) and two-face (two quadrilaterals sharing an edge) refinement. The new template set allows refining hexahedra individually and concave refinement domains can be refined efficiently. In this paper, we propose adding one more new template to the refinement template set to improve the efficiency.

Another challenging problem is how to keep the number of hexahedra minimum for complex geometries while their geometrical fidelity is kept. The alignment of hexahedra in an octree-based mesh mostly depends on the position of the initial octree. The easiest way to select it is by the use of an axial aligned bounding box around the input geometry. However, it tends to create more elements for long objects that do not align with the xyz axes. Objects with thin regions, such as skulls and hearts, need many elements to be represented by octree-based hexahedra without holes. We propose two mesh generation options to relieve this problem.

The reminder of the paper is organized as follows. In Section 2, the new template set is introduced. In Section 3, new options added to our hexahedral mesh generator are introduced to create meshes with fewer elements for complex geometries. In Section 4, examples are shown to demonstrate our mesh generation method. Finally, conclusions are given in Section 5.

2 Refinement Templates

Figure 1 shows the new set of templates for edge (Figure 1a), face (Figure 1b), volume (Figure 1c), three-node (Figure 1d), two-face (Figure 1e) and opposite-edge (Figure 1f) refinement. The templates for edge, face, volume, three-node, two-face and opposite-edge refinement are represented as $T_2(\underline{ABCD}EFGH)$, $T_4(\underline{ABCD}EFGH)$, $T_8(\underline{ABCDEFGH})$, $T_3(\underline{ABC}\underline{D}EFGH)$, $T_6(\underline{ABCDEFGH})$, and $T_{4o}(\underline{ABCD}EF\underline{GH})$, respectively. The underlines indicate nodes to be refined, which correspond to the black points with white letters in Figure 1 (hereafter referred as marked nodes). Only the main structures of T_3, T_6 and T_{4o} are shown in Figure 1 to simplify the drawing. T_2, T_4 and T_8 are from pillowing [4, 5]. T_3 and T_6 are proposed in [8]. T_{4o} is a new template proposed in this paper.

To obtain T_3, the hexahedron is first divided into six base hexahedra as shown in Figure 1d. T_2 is then applied to two hexahedra: $T_2(\underline{ABJIEFNM})$ and $T_2(\underline{DAILHEMP})$. The other four hexahedra, $CDLKGHPO$, $BCKJFGON$, $IJKLMNOP$ and $MNOPEFGH$, are used as is.

To obtain T_6, the hexahedron is first divided into five base hexahedra as shown in Figure 1e. T_2 is applied to two hexahedra: $T_2(\underline{BFGC}JNOK)$ and $T_2(\underline{HDCG}PLKO)$. T_4 is applied to two other hexahedra: $T_4(\underline{AEFB}IMNJ)$ and $T_4(\underline{DHEA}LPMI)$. The hexahedron in the center, $IJKLMNOP$, is used as is.

To obtain T_{4o}, the hexahedron is first divided into seven base hexahedra as shown in Figure 1f. T_2 is applied to four hexahedra: $T_2(\underline{ABJI}EFNM)$, $T_2(\underline{ABCD}IJKL)$, $T_2(\underline{GHP}OFEMN)$ and $T_2(\underline{GHD}COPLK)$. The hexahedron in the center, $IJKLMNOP$, is used as is.

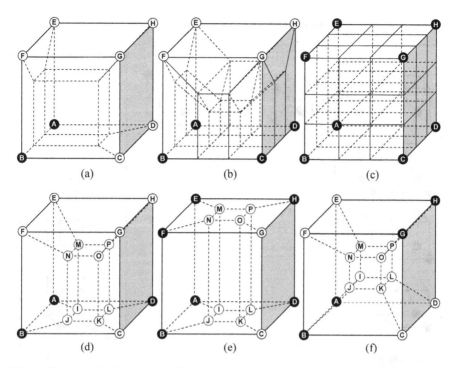

Fig. 1. New set of refinement templates for (a) edge (T_2), (b) face (T_4), (c) volume (T_8), (d) three-node (T_3), (e) two-face (T_6) and (f) opposite-edge refinement (T_{4o}). Note that d, e and f only show the main structures of our new templates.

Table 1. Number of hexahedra in each of the six refinement templates

Templates	T_2	T_4	T_8	T_3	T_6	T_{4o}
# of hexahedra	5	13	27	14	37	23

We can consider that those three templates, T_3, T_6 and T_{4o}, are created by multiple application of sheet insertion. Table 1 shows the number of hexahedra in each of the six templates. Table 2 shows a conversion table for possible refinement patterns and corresponding refinement templates. This conversion table is based on Candidate 2 in [8] to use the new opposite-edge refinement template. Let us think about default templates for quadrilaterals to understand the conversion table better. If a quadrilateral has only one marked node (Figure 2a) or two marked nodes that are in opposite corners (Figure 2b), it is not refined. If a quadrilateral has two neighboring marked nodes (Figure 2c), three marked nodes (Figure 2d) or four marked nodes (Figure 2e), it is refined using $T_{Q2}(\underline{ABCD})$, $T_{Q3}(\underline{ABCD})$ or $T_{Q4}(\underline{ABCD})$, respectively. Patterns 1, 2B, 2C, 3C and 4F shown in Table 2 are not refined, and one of the marked nodes in Pattern 3B or 4D is ignored, accordingly. Although the marked nodes are not always refined, the conversion table works

Table 2. Refinement patterns and corresponding refinement templates

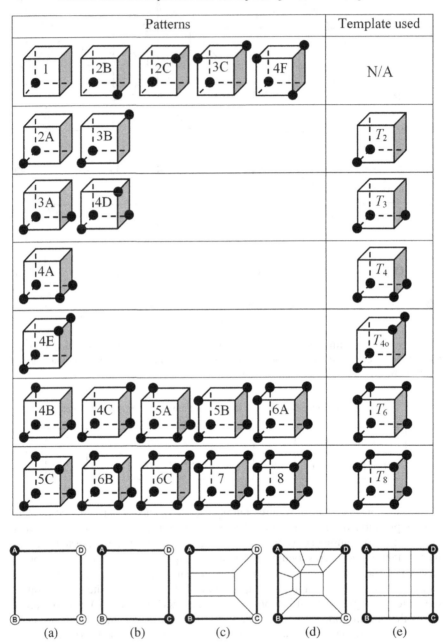

Fig. 2. Quadrilateral refinement templates for (a) node, (b) opposite-node, (c) edge (T_{Q2}), (d) three-node (T_{Q3}) and (e) full refinement (T_{Q4})

well most of the time because those refinement patterns appear only on the verge of the domain to be refined.

All the templates need new nodes. Their positions must be determined consistently and easily such that subdivided hexahedra are in good quality. Note that initial hexahedra are not necessarily cubes but can have low minimum scaled Jacobian values, e.g., the base hexahedra in T_3, T_6 and T_{4o} to which T_2 or T_4 is applied. Let us think about how to calculate the positions of the new nodes in T_8 because those in the other templates are calculated in the same way. Let $\mathbf{x}_i$ be the coordinates of node i. Two nodes added between any edge are calculated as its trisection points. For example, the two nodes between edge AB are obtained as $\frac{2\mathbf{x}_A+\mathbf{x}_B}{3}$ and $\frac{\mathbf{x}_A+2\mathbf{x}_B}{3}$. Each of four nodes added on any quadrilateral is calculated as the center of its any three nodes. For example, the four nodes on quadrilateral $ABCD$ are obtained as $\frac{\mathbf{x}_A+\mathbf{x}_B+\mathbf{x}_C}{3}$, $\frac{\mathbf{x}_B+\mathbf{x}_C+\mathbf{x}_D}{3}$, $\frac{\mathbf{x}_C+\mathbf{x}_D+\mathbf{x}_A}{3}$ and $\frac{\mathbf{x}_D+\mathbf{x}_A+\mathbf{x}_B}{3}$. Eight nodes added inside T_8 are calculated as the trisection points of the four diagonal lines: AG, BH, CE and DF.

3 Options for Hexahedral Mesh Generation

To create hexahedral meshes efficiently for complex geometries, the following two options have been added to our mesh generation method: temporal rotation and temporal local inflation.

3.1 Temporal Rotation

Surface models are not always axially aligned, i.e., their oriented bounding boxes (OBB) do not always align with the xyz axes. This means that the coordinate system of an octree needs to be selected carefully. The temporal rotation option is useful to create a mesh with fewer elements with the same resolution for a long object.

Figure 3 shows an example of a pediatric left humerus model. There are two bounding boxes in Figure 3a: the axial aligned bounding box (AABB; black) and the oriented bounding box (OBB; red) calculated based on the covariance matrix from vertex coordinates [9]. Our mesh generation method creates octants aligned with the AABB by default (Figure 3b). When the temporal rotation option is turned on, octants are automatically aligned with the OBB (Figure 3c). Fewer elements are needed with the same resolution. The quality of the meshes is almost the same. The lowest minimum scaled Jacobian of the hexahedra in the two meshes is 0.37. No adaptation is applied in this case.

3.2 Temporal Local Inflation

The temporal local inflation is a helpful option to represent thin regions with a fewer number of hexahedra. Suppose the local thickness of a surface model is τ_l

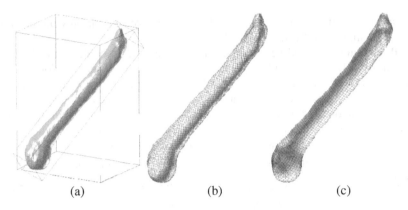

Fig. 3. Hexahedral meshes for a pediatric left humerus: (a) Surface model with AABB (black) and OBB (red); (b) Hexahedral mesh based on AABB (10.5K hexahedra); (c) Hexahedral mesh based on OBB (6.5K hexahedra)

and the local size of the octants (cubes) is ε_l. To represent the thin region without holes, $\varepsilon_l < k\tau_l$ ($k > 2$)) because the octants intersecting with the surface model are removed in our approach during the octree-based core mesh generation process. k is a constant, and the default value is 3.5. If $\varepsilon_l > k\tau_l$, the surface is locally inflated in the local outwardly normal direction so that the thickness temporarily becomes greater than $k\tau_l$. The inflation method is similar to the one used in near-field mesh generation for viscous flow simulations because the local outwardly normal direction and the moving distance may need to be slightly changed during the inflation to avoid creating intersections of the surface [10, 11]. Note that the initial ε_l should not be too big to represent certain details of the input geometry.

Figure 4 illustrates an example of a pediatric skull model. Figure 4a shows the original triangulated surface model. The size of octants can be automatically adjusted based on local thickness of the skull [12] and other user-specified parameters [8] during the octree-based hexahedral mesh generation process. Without the temporal local inflation option, the user needs to specify the parameters so that the size of leaf octants can be small enough to resolve all the thin regions. This induces two problems. If the size of the smallest leaf octants is too small, the resulting mesh may have too many hexahedra to perform computational simulations in limited computational resources (Figure 4b). If the size of the smallest leaf octants is not small enough, the resulting hexahedral mesh can have many holes (Figure 4c).

The temporal local inflation option resolves these problems at some level. When this option is turned on, octants are created inside the surface model after it is temporarily inflated like a balloon only at thin regions as shown in Figure 4d (*cf.* the nose and teeth in Figures 4a and d). The boundary surface of the hexahedral mesh is then recovered based on the original triangulated surface as usual. The resulting mesh shown in Figure 4d has 649K hexahedra. Although the dense mesh created with the temporal local inflation option off (Figure 4d) has 3.08M hexahedra, more elements are needed because there are still holes, especially around the eye sockets, where bones are very thin.

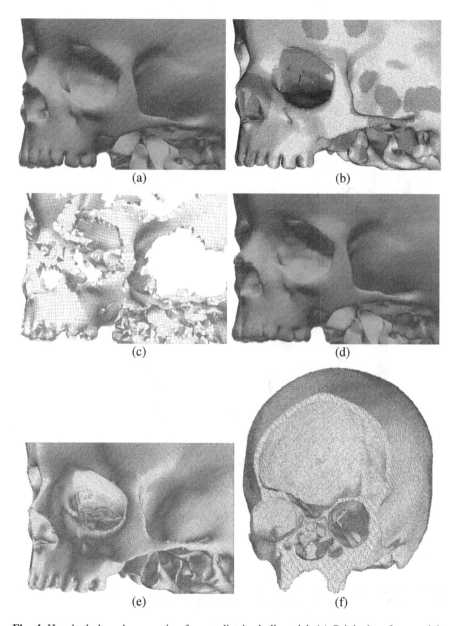

Fig. 4. Hexahedral mesh generation for a pediatric skull model: (a) Original surface model; (b-c) Hexahedral mesh with the temporal local inflation option off – two and three times of local refinement in b and c, respectively; (d) Locally inflated surface model; (e) Hexahedral mesh with the temporal local inflation option on (two times of local refinement); (f) Another view of e with a cross-section (yellow)

(a)

(b)

(c)

Fig. 5. Hexahedral mesh for a Ramses model (a-b) using the new set of the refinement templates (2.14M elements) and (c) the previous set (2.20M elements)

4 Applications

In this section, four geometries are shown to demonstrate our hexahedral mesh generation capability. The quality of hexahedral meshes is also discussed based on the scaled Jacobian metric.

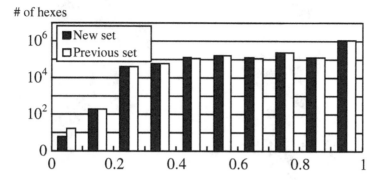

Fig. 6. Element quality distribution by the scaled Jacobian metric for the Ramses meshes

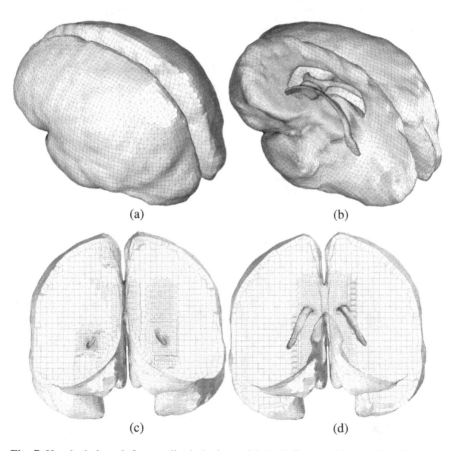

Fig. 7. Hexahedral mesh for a pediatric brain model: (a-b) Front and back sides of surface mesh; (c-d) Cross-sections of the mesh (yellow)

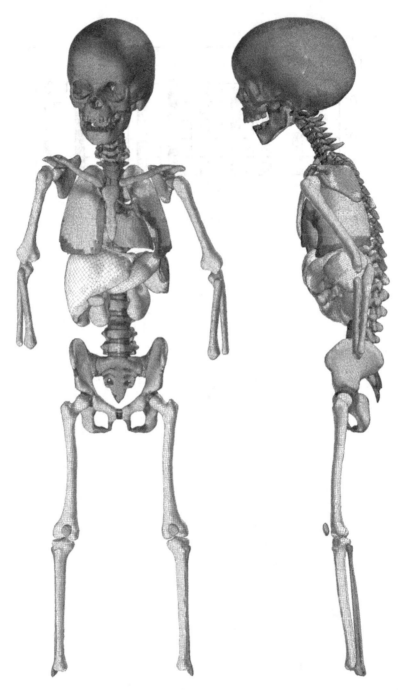

Fig. 8. Hexahedral mesh for a three-year-old child model

4.1 Ramses Model

Figure 5 shows hexahedral meshes for a Ramses model from the AIM@SHAPE Shape Repository. The initial octree was refined equally so that the size of octants becomes approximately 23 and then was locally refined three times based on α proposed in [8] ($\alpha = 50°$). Figures 5a and b show a mesh using the new template set, which has 2.14M elements. Figure 5c shows a mesh using the previous template set (without T_{4o}), which has 2.20M elements (2.8% more). Compared to the previous template set, the new set usually reduces the number of elements by a few percent while keeping almost the same or slightly better mesh quality. The actual reduction depends on each case, especially on the complexity of the refinement domain. The meshes shown in Figures 5b and c are similar except around the right eye.

Figure 6 shows the distribution of element quality of the two hexahedral meshes by the scaled Jacobian metric. The lowest minimum scaled Jacobian values are 0.015 for the mesh using the new set and 0.00079 for the mesh using the previous set. The two meshes have a small number of low quality hexahedra, but their overall quality is good.

4.2 Pediatric Brain Model

Figure 7 shows a hexahedral mesh for a pediatric brain model using the new set of the refinement templates. The initial octree was refined equally so that the size of octants becomes approximately 2.7, and then was locally refined twice based on α = 60°. The mesh has 340,917 hexahedra and is nicely adapted to the geometry. The lowest minimum scaled Jacobian of the hexahedra is 0.10. If the previous template set is used, a mesh with 355,904 hexahedra (4% more) is created.

4.3 Three-Year-Old Child Model

Figure 8 shows a hexahedral mesh for a three-year-old child model. Part of the components, left humerus, skull and brain, is already shown in Sections 3.1, 3.2 and 4.2, respectively. A hexahedral mesh is created for each of the body components separately using the new set of the refinement templates to apply the temporal rotation option to long bones and/or to apply the temporal local inflation option to components with thin regions. The final mesh has 2.79M hexahedra.

5 Conclusions

An octree-based mesh generation method has been developed to automatically create all-hexahedral meshes for complex geometries. An improved set of refinement templates allows the generation of geometry-adapted meshes with fewer elements while keeping almost the same or slightly better mesh quality. Two new options, temporal rotation and temporal local inflation, are useful to create meshes for long objects or thin regions. The capability of our mesh generation method is demonstrated using a few triangulated surface models. The resulting meshes do not have any negative Jacobian elements even for complex geometries, and the quality of the meshes is good.

Acknowledgments

The initial triangulated surface model of the three-year-old child is provided courtesy of Dr. King Yang and Ms. Christina Huber of Wayne State University, Detroit, MI, USA. The Ramses model is provided courtesy of IMATI and INRIA by the AIM@SHAPE Shape Repository. This research is supported in part by the Southern Consortium for Injury Biomechanics (SCIB) No. DTNH22-01-H-07551.

References

1. Schneiders, R., Schindler, R., Weiler, F.: Octree-Based Generation of Hexahedral Element Meshes. In: Proceedings of the 5th International Meshing Roundtable, pp. 205–216. Sandia National Laboratories (1996)
2. Zhang, H., Zhao, G.: Adaptive Hexahedral Mesh Generation Based on Local Domain Curvature and Thickness Using a Modified Grid-Based Method. Finite Elements in Analysis and Design 43, 691–704 (2007)
3. Parrish, M., Borden, M., Staten, M., Benzley, S.: A Selective Approach to Conformal Refinement of Unstructured Hexahedral Finite Element Meshes. In: Proceedings of the 16th International Meshing Roundtable, Seattle, WA, pp. 251–268 (2007) doi: 10.1007/978-3-540-75103-8_15
4. Tchon, K., Dompierre, J., Camarero, R.: Automated Refinement of Conformal Quadrilateral and Hexahedral Meshes. International Journal for Numerical Methods in Engineering 59, 1539–1562 (2004)
5. Harris, N., Benzley, S., Owen, S.: Conformal Refinement of All-Hexahedral Element Meshes Based on Multiple Twist Plane Insertion. In: Proceedings of the 13th International Meshing Roundtable, Williamsburg, VA, Sandia National Laboratories, SAND #2004-3765C, pp. 157–168 (2004)
6. Shepherd, J.F., Zhang, Y., Tuttle, C.J., Silva, C.T.: Quality Improvement and Boolean-like Cutting Operations in Hexahedral Meshes. In: Proceedings of the 10th International Conference on Numerical Grid Generation in Computational Field Simulations, Crete, Greece (2007) (CD-ROM)
7. Merkley, K., Ernst, C., Shepherd, J., Borden, M.J.: Methods and Applications of Generalized Sheet Insertion for Hexahedral Meshing. In: Proceedings of the 16th International Meshing Roundtable, Seattle, WA, pp. 233–250 (2007), doi:10.1007/978-3-540-75103-8_14.
8. Ito, Y., Shih, A.M., Soni, B.K.: Octree-Based Reasonable-Quality Hexahedral Mesh Generation Using a New Set of Refinement Templates. International Journal for Numerical Methods in Engineering 77(13), 1809–1833 (2009)
9. Gottschalk, S., Lin, M., Manocha, D.: OBBTree: A Hierarchical Structure for Rapid Interference Detection. Computer Graphics (SIGGRAPH 1996) 30, 171–180 (1996)

10. Ito, Y., Nakahashi, K.: Improvements in the Reliability and Quality of Unstructured Hybrid Mesh Generation. International Journal for Numerical Methods in Fluids 45(1), 79–108 (2004)
11. Ito, Y., Shih, A.M., Soni, B.K., Nakahashi, K.: Multiple Marching Direction Approach to Generate High Quality Hybrid Meshes. AIAA Journal 45(1), 162–167 (2007)
12. Ito, Y., Shum, P.C., Shih, A.M., Soni, B.K., Nakahashi, K.: Robust Generation of High-Quality Unstructured Meshes on Realistic Biomedical Geometry. International Journal for Numerical Methods in Engineering 65(6), 943–973 (2006)

Embedding Features in a Cartesian Grid

Steven J. Owen and Jason F. Shepherd

*Sandia National Laboratories, Albuquerque, New Mexico, U.S.A.
sjowen@sandia.gov, jfsheph@sandia.gov

Abstract. Grid-based mesh generation methods have been available for many years and can provide a reliable method for meshing arbitrary geometries with hexahedral elements. The principal use for these methods has mostly been limited to biological-type models where topology that may incorporate sharp edges and curve definitions are not critical. While these applications have been effective, robust generation of hexahedral meshes on mechanical models, where the topology is typically of prime importance, impose difficulties that existing grid-based methods have not yet effectively addressed. This work introduces a set of procedures that can be used in resolving the features of a geometric model for grid-based hexahedral mesh generation for mechanical or topology-rich models.

Keywords: grid-based, overlay grid, hexahedral mesh generation, topological equivalence, topology embedding.

1 Background

The general problem of mesh generation involves discretizing a domain into simple shapes such as tetrahedra and hexahedra. In most cases the problem begins with a three-dimensional domain that is represented with topology and geometry. The topology representation can be as simple as a single surface or as complex as a mechanical assembly with hundreds of interconnected volumes, surfaces, curves and vertices related by means of a Boundary-representation (B-Rep) graph structure. In contrast the geometry representation defines the core mathematical foundation of the curves and surfaces and may be defined as a set of non-uniform rational b-splines or as a simple connected set of triangles. The B-Rep graph usually provides the frame on which the geometry is defined. Both geometry and topology should be considered when developing a mesh.

---

* Sandia is a multiprogram laboratory operated by Sandia Corporation, a Lockheed Martin Company for the United States Department of Energy's National Nuclear Security Administration under contract DE-AC04-94AL85000.

Most modern software tools that provide mesh generation capabilities enforce a requirement of *geometry-mesh* ownership. This provides a convenient method by which the user can apply physical properties and attributes to the topological features of the domain rather than dealing with the mesh itself. While attributes must ultimately be represented on the nodes and elements of the mesh for analysis, from a user's perspective, it is more convenient to assign attributes to the B-Rep entities rather than on individual mesh entities that may change over the course of a study.

To accomplish this, individual mesh entities, including, nodes, faces, edges and elements, must have a *child-parent* relationship with the B-Rep entities in the model: vertices, curves, surfaces and volumes. Likewise, B-Rep entities must have a *parent-child* relationship with the mesh entities that they own. This association, in most cases is a *one-to-many* link; that is, a single B-Rep entity may own multiple mesh entities, but a mesh entity can have only one unique parent. This one-to-many relationship has driven most of the modern meshing algorithms, where independent mesh entity groups can be generated for each individual B-Rep entity in the model and subsequently joined to form a contiguous mesh. Underlying the assumption of the B-Rep→mesh one-to-many relationship is that each B-Rep entity does indeed provide important information to the analysis and that a discrete representation of every B-Rep entity in the model will be represented in the mesh. Unfortunately, geometry creation procedures often developed in solid modeling tools frequently generate anomalous curves and surfaces that are not significant to the analysis. Using these models in a mesh generation system that enforces one-to-many ownership can lead to poor quality elements where the B-Rep will routinely over-constrain the resulting mesh. This work assumes that all entities in the geometric model are important to the analysis and that defeaturing or model simplification procedures such as those illustrated in the overview by Thakur et. al. [14] have already been accomplished.

The development of general-purpose unstructured hexahedral mesh generation procedures that effectively capture both geometry and topology for an arbitrary domain have been a major challenge for the research community. A wide variety of techniques and strategies have been proposed for this problem. It is convenient to classify these methods into two categories: *geometry-first* and *mesh-first*. In the former case, a topology and geometry foundation is used upon which a set of nodes and elements is developed. Historically significant methods such as plastering [1], whisker weaving [12] and the more the recent unconstrained plastering [11] can be considered *geometry-first* methods. These methods begin with a well defined boundary representation and progressively build a mesh that ensures that properties of mesh ownership are adhered to. Because these methods use the B-Rep entities themselves as the foundation for the algorithm, the parent-child ownership of mesh entities generated is easily built in to the procedures. Most of these methods define some form of advancing front procedure that requires resolution of an interior void and have the advantage of conforming to a prescribed boundary mesh.

Although work in the area is on-going, the ability to generalize these techniques for a comprehensive set of B-Rep configurations has proven a major challenge and has yet to prove successful for a broad range of models.

In contrast, the mesh-first methods start with a base mesh configuration. Procedures are then employed to extract a topology and geometry from the base mesh. These methods include grid-overlay or octree methods. In most cases these methods employ a Cartesian or octree refined grid as the base mesh. Because a complete mesh is used as a starting point, the interior mesh quality is high, however the boundary mesh produced cannot be controlled as easily as in geometry-first approaches. As a result the mesh may suffer from reduced quality at the boundary and can be highly sensitive to model orientation. In addition, grid-overlay methods may not accurately represent the topology and geometric features as defined in the geometric model. In spite of these inherent deficiencies, mesh-first methods have proven a valuable contribution to mesh generation tools for modeling and simulation. In contrast to geometry-first techniques, fully automatic mesh-first methods have been developed for some applications where boundary topology is simple or is not critical to the simulation. In particular, bio-medical models [17] [18] [3], metal forming applications [8] [4], and viscous flow [13] methods have utilized these techniques with some success. Automating and extending mesh-first methods for use with general B-Rep topologies would provide an important advance in hexahedral meshing technology.

As one of the first to propose an automatic overlay-grid method, Schneiders [8] developed techniques for refining the grid to better capture geometry. He utilized template-based refinement operations, later extended by Ito [3] and H. Zhang [16] to adapt the grid so that geometric features such as curvature, proximity and local mesh size could be incorporated. Y. Zhang [17] [18] and Yin [15] independently propose an alternate approach known as the Dual-contouring method that discovers and builds sharp features into the model as the procedure progresses. This is especially effective for meshing volumetric data where a predefined topology is unknown and must be extracted as part of the meshing procedure.

The dual contouring method for establishing a base mesh described by Y. Zhang [17] begins by computing intersections of the geometry with edges in the grid. Intersection locations are used to approximate normal and tangent information for the geometry. One point per intersected grid cell is then computed using Hermite interpolation from tangents computed at the grid edges. The base mesh in this case is defined as the dual of the Cartesian grid, using the cell centroids and interpolated node locations at the boundary. While attractive as a method for extracting features from volumetric data, it does not guarantee capture of a pre-existing topology such as that contained in a CAD solid model.

Recent work on mesh-first approaches have focused more on the capturing of features of the geometry. A common thread among many of these methods [17] [3] [9] is the introduction of a buffer layer of hex elements to improve element quality near the boundary. This approach, while effective, still relies on a base mesh that is topologically equivalent to the features of a B-Rep. A drawback of many of these methods is that they tend to neglect the parent-child geometry-to-mesh ownership principals which are important for meshing algorithms to effectively engage with CAD-based modeling tools. With the assumption that all features of a B-Rep are indeed important in modeling the domain, the focus of this study is to propose a procedure whereby topological features can be accurately represented in a finite element mesh using a mesh-first method.

Shepherd [10] describes an approach to mesh-first hexahedral mesh generation utilizing geometric capture procedures. This work utilizes theory and assertions developed in [5] [6] . He asserts that a mesh must be topologically equivalent and geometrically similar to its geometry and B-Rep definition in order to develop a valid conformal mesh of the domain. To be topologically equivalent there must be a consistent correlation between the graph of the mesh and the graph of the B-Rep. This can be accomplished by establishing a one-to-many parent-child relationship between B-Rep and mesh entities. B-Rep entities of dimension r must contain a set of one or more contiguous mesh entities of dimension r, where $r = 0, 1, 2, 3$. Although not strictly required, geometric similarity between the mesh and the geometry of the domain is desired in order to maintain reasonable mesh quality. For example, aligning the base mesh with the principal orientations of curves and surfaces of the model will minimize the characteristic stair-step effect and increase mesh quality once mapped to the original geometry.

2 Feature Embedding

For convenience we have limited the base mesh for this study to a Cartesian grid. While it is often desirable to begin with an enriched octree grid or an aligned swept mesh, the Cartesian grid offers simplicity and automation that is easy to generalize for any model. For implementation purposes, a Cartesian grid is very light-weight and fast, avoiding full unstructured mesh data structures required for more general methods. While it is inevitable that mesh enrichment will be needed to more accurately capture small features and high valence vertices and curves, there is value in understanding the principles needed to embed topology through the use of a Cartesian grid. Indeed, it is expected that through careful application of topology embedding, that the need for mesh enrichment will be reduced.

Shepherd [10] outlines an algorithm which is convenient to use as the context for this work. The overall procedure is illustrated in figure 1. Beginning from a CAD model, a Cartesian grid is defined enclosing the model. A 2-manifold is then established from an inside-out procedure on which a topology capture algorithm is performed and subsequently projected to the geometry. A series of one or more buffer layers known as fundamental sheets are then inserted at the boundary, followed by a series of mesh optimization

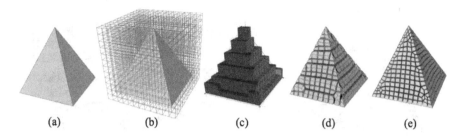

(a)   (b)   (c)   (d)   (e)

Fig. 1. Mesh generation procedure. (a) Initial CAD B-Rep model. (b) Enclosing Cartesian grid is established. (c) Base grid established and topology embedded. (d) Base grid projected to geometry. (e) Fundamental sheets inserted on boundaries.

steps to improve mesh quality. This work will focus specifically on the topology capture procedure illustrated in figure 1(c) and leave the sheet insertion and mesh optimization procedures for a future study.

Establishing a 2-manifold on which topology is captured implies developing a base mesh and using the bounding set of quadrilaterals from a contiguous set of hexahedra on which vertex and curve topology is embedded. Shepherd suggests that tailoring a base mesh to the features and characteristics of the domain is advantageous. For example, where the bulk outline of a model is generally cylindrical, then a base mesh constructed from the sweep of a bounding cylinder would yield better results than using a Cartesian grid. However, this procedure is not easily generalized nor automated and would also be a valuable topic for future study. Instead, most current literature indicates that a base mesh is established from a Cartesian grid. In these cases individual hexes are tested for inclusion or exclusion based on common *in-out* procedures. In-out procedures classify each cell in the grid based upon its centroid's relative position with respect to the boundary of the domain. The continuous set of hexes that are completely contained within the geometry and optionally combined with hexes that are intersecting the domain are used. In many cases, mesh enrichment procedures using octree decomposition are used at this stage to ensure the geometry is effectively represented. In many cases special procedures are used to ensure non-manifold connections and non-contiguous regions within the hex mesh are eliminated.

Whether a base grid is defined using traditional in-out procedures or through a dual contouring procedure, topologic equivalence may not be adequately taken into account. Figures 2 to 4 show examples where standard methods for defining a base grid may be inadequate. The single hexahedra at the apex of the pyramid in figure 2 provides a maximum of a 3-valent node on which to embed a 4-valent vertex. Without subsequent mesh enrichment, topology capture would be impossible at this location. Although Shepherd [10] proposes local pillowing operations to enrich the valence, it also requires an unstructured mesh data representation consequently reducing the

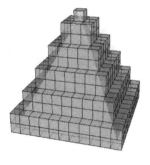

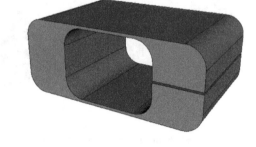

Fig. 2. Base grid defined from a pyramid geometry. Resulting 2-manifold does not provide rich enough mesh to capture 4-valent vertex at pyramid apex.

Fig. 3. B-Rep used to generate the base grids shown in figure 4.

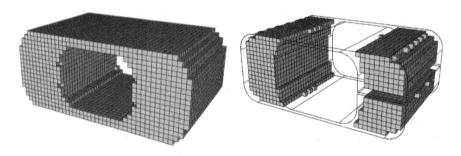

Fig. 4. Two different base grids defined from a C-shaped geometry (figure 3) using common in-out procedures. The grid on the left is defined from the inside and intersected hexes while the grid on the right is defined from only the hexes on the inside of the geometry. Neither satisfies the requirement of topologic equivalence.

efficiency of the proposed procedures and can also result in marginal quality hexahedra to achieve the required valence.

Figures 3 and 4 illustrate another potential issue with traditional base mesh definitions for mesh-first procedures. Neither the cells intersecting the geometry, nor the cells on the interior of the geometry produce a set of hexes that can build a topologically equivalent mesh without special procedures to combine the results from the two cases. In examples such as these, mesh enrichment strategies can be employed to locally refine the grid using template-based refinement techniques [3]. In order to provide complete generality, these methods typically require a $1 \rightarrow 27$ refinement strategy. That is, each hex in the refinement region is decomposed into 27 hexahedra. In addition to introducing artificially high gradients on the local mesh size, transition elements, typically of marginal quality, are needed to ensure a conformal mesh between coarse and fine regions.

In an attempt to better control some of these issues, this work proposes using the full three dimensional Cartesian grid on which the B-Rep topology is progressively extracted. Initially limiting the topology extraction problem to a 2-manifold, as originally proposed by Shepherd [10] and others, may over constrain the problem such that specialized procedures are required to enrich the grid where otherwise a full 3D approach would naturally extract topologic equivalence from the grid. However, it is clear that mesh enrichment strategies will still need to be employed, particularly for features significantly smaller than the grid cell size and for vertex valence greater than 6. We contend that the use of such procedures may be reduced by limiting specialized mesh enrichment procedures that impose high valent or marginal transition elements, ultimately improving overall mesh quality.

3   Embedding Procedures

Beginning from the solid model boundary representation of the model, the current implementation may first employ defeaturing as described by Quadros et. al. [7] and then extract the facets from the model to use as an approximate geometry representation. Using the faceted form of the model allows for easy integration of this procedure with the discrete feature suppression procedures also described in [7]. Also beneficial is that geometry evaluation during the embedding procedures can be limited to evaluation of planar triangles and linear edges rather than evaluations using a full B-spline library. A Cartesian grid that completely encloses the geometry with a user-defined resolution is then established. While there is no explicit requirement on grid cell size, for practical purposes the cells should be approximately smaller than the smallest feature size in the geometry. The grid may also be optionally oriented so that a tight fitting bounding box is established to help align curves and surfaces with the principal axes of the grid. To maintain the advantages of a Cartesian grid, rather than transforming the grid itself, the geometry can be transformed into the Cartesian space during the embedding procedures and transformed back when complete.

The approach taken for embedding features follows roughly the bottom-up method of mesh generation, where successive dimensions are embedded starting from vertices and continuing through curves, surfaces and volumes. If we define a Cartesian grid $\Omega_M = \{M_i^r | r = 0, 1, 2, 3\}$ and a B-Rep $\Omega_G = \{G_i^r | r = 0, 1, 2, 3\}$, our objective is to find groups of grid entities, M_i^r of dimension r that will be owned by corresponding B-Rep entities, G_i^r of the same dimension. Corresponding grid and B-Rep entities are enumerated in table 1. The embedding $E_{G \to M} \subseteq \Omega_M$ is defined as $E_{G \to M} = \{M_i^r : M_i^r \mapsto G_i^r\}$, where the non-unique mapping, $M_i^r \mapsto G_i^r$, is a set of procedures for each dimension that assigns grid entities to individual B-Rep entities. This

Table 1. Corresponding B-Rep and Cartesian grid entities

B-Rep Entity	Symbol	Cartesian Grid Entity	Symbol
Vertex	G^0	Node	M^0
Curve	G^1	Edge	M^1
Surface	G^2	Face	M^2
Volume	G^3	Cell	M^3

mapping will necessarily be *collision-free*; that is, a grid entity, M_i^r may be mapped to one and only one B-Rep entity, G_i^r.

The embedding procedures $M_i^r \mapsto G_i^r$ for each dimension have similar characteristics. Each uses a combination of local geometric and topologic information to determine grid entities that will map to a given B-Rep entity. The procedures generally proceed by looping through each B-Rep entity G_i^r for dimension r and capturing grid entities M_i^s for dimensions $s \geq r$. For example, the vertex embedding procedure, while incorporating node capture, also includes methods to capture nearby edge, face and cell entities. Similarly, the curve embedding procedure includes the capture of faces and cells. This has proven effective in avoiding or controlling collision conditions as the algorithm proceeds. The overall procedure is illustrated in algorithm 1, where the function $\{M_j^s\} \leftarrow \texttt{Capture}(s, G_i^r)$ defines procedures for mapping grid entities of dimension s to the geometry entities of dimension s immediately at or adjacent to the geometry entity G_i^r. The following sections detail the $\texttt{Capture}(s, G_i^r)$ procedures for each dimension.

Input: B-Rep Ω_G, Cartesian Grid Ω_M
Output: Embedding $E_{G \to M} \subseteq \Omega_M$

```
1  for dimension r ← 0 to 3 do
2  |    for dimension s ← r to 3 do
3  |    |    foreach geometry entity G_i^r ∈ Ω_G do
4  |    |    |    {M_j^s} ← Capture(s, G_i^r);
5  |    |    |    E_{G→M}+ = {M_j^s};
6  |    |    end
7  |    end
8  end
```

Algorithm 1. Algorithm for computing embedding set $E_{G \to M} \subseteq \Omega_M$

3.1 Embedding Vertices

Capturing nodes at vertices

Vertices can be embedded in most cases by simply finding the closest node in the grid to each vertex in the B-Rep. Collisions may occur where multiple vertices may claim the same grid node. Collision resolution in this case is handled by selecting the closest vertex to the node in question and successively assigning ownership to nearby grid nodes based on their relative distance.

Capturing edges at vertices

Once vertex locations are assigned, grid edges at the selected grid node must be matched with appropriate curves. Figure 5 shows an example where the vertex G_0^0 at the apex of a pyramid maps grid node M_0^0. For this vertex, four grid edges must be selected that match the four curves that share the vertex. At this location, any of the curves $G_{i=0,1,2,3}^1$ may choose any one of the six grid edges $M_{j=0,1,...5}^1$ sharing grid node M_0^0. Clearly vertices with valence greater than six would not be permitted when utilizing a Cartesian grid. To facilitate edge selection at a selected grid node, the metric μ_{ij} is computed as follows:

$$\mu_{ij} = \frac{1 + (\mathbf{V}_{M_j} \cdot \mathbf{V}_{G_i})}{2} \tag{1}$$

where μ_{ij} is a value $0 \le \mu_{ij} \le 1$ that represents how well curve G_i^1 matches geometrically with edge M_j^1 where $\mu_{ij} = 1$ is a perfect match and $\mu_{ij} = 0$ is very poor. In equation 1 the variables $\mathbf{V}_{M_j}$ and $\mathbf{V}_{G_i}$ are the normalized

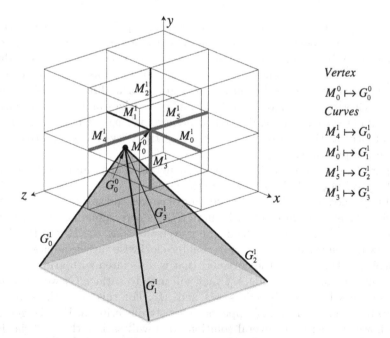

Fig. 5. A representation of the local grid near the apex of a pyramid. Four of the grid edges selected from $M_{j=0,1,...5}^1$ that are connected to the node M_0^0 must be paired to the four curves in the B-Rep $G_{i=0,1,2,3}^1$.

outward pointing vectors of the grid edges M_j^1 and curves G_i^1 at the node M_0^0 and vertex G_0^0 respectively. This equation simply incorporates the dot product of edge and curve vectors to indicate how well each grid edge is oriented with respect to a particular curve. In practice, since the B-Rep is comprised of facets, the curve vector $\mathbf{V}_{G_i}$ can be approximated from the first facet edge that shares the vertex G_0^0 and curve G_i^j.

Once the metric μ_{ij} has been computed for each edge with respect to each curve at the vertex, all permutations of pairings between edges and curves are computed and the sum of μ_{ij} for each permutation is determined. The pairings can then be ranked from best to worst based on their μ_{ij} sums. For example, in figure 5 a ranking of all possible curve-edge pairings are compiled. Given 6 possible edges assigned to 4 curves, the total number of permutations of edge pairings would be $6 \cdot 5 \cdot 4 \cdot 3 = 360$ permutations. For the worst case of a 5 or 6 valent vertex, the total number of permutations would be 6! or 720. The curve-edge pairing with the highest μ_{ij} sum is used as the candidate set of edges to be used to represent the curves at the vertex. For the example in figure 5 the selected candidate edge-curve pairings are shown at right. Note that multiple edge-curve permutations may yield identical μ_{ij} sums as in this simple example.

Capturing faces at vertices

Even after selecting the best candidate edge-curve pairing for a given vertex, there is no guarantee that a valid topology that matches the local surface configuration can be established. For this reason a face-surface pairing procedure is also used. The best candidate edge-curve pairing is used as the starting point for this procedure. For any B-Rep graph of a three-dimensional domain, each vertex will have an equal number of edges and surfaces sharing the common vertex. Having defined edge-curve pairings, it remains to find a path of faces at the vertex between existing edge-curve pairs such that each surface is represented by at least one face.

Figure 6 shows the same example of a pyramid apex and the local grid topology at node M_0^0. Also shown are the faces $M_{j=0,1,\ldots11}^2$ sharing the node. For this problem we use the cyclic ordering of curves and surfaces at the vertex to guide a traversal. Staring with curve G_0^1, its corresponding grid edge (in this case M_4^1) is selected to begin the traversal. Using the B-Rep topology, we know that surface G_1^2 is the next counter-clockwise surface adjacent curve G_0^1; which in turn shares the curve G_1^1 that is associated with grid edge M_0^1.

To find the face(s) at node M_0^0 that will map to surface G_1^2 we must find a set of faces bounded by edges M_4^1 and M_0^1. For this case, the choice is relatively trivial, as face M_5^2 appears to satisfy our criteria. For the general case, however there are several solutions that will satisfy this criteria. For example the set of faces $\{M_7^2, M_6^2, M_4^2\}$, $\{M_{11}^2, M_1^2\}$ and $\{M_{10}^2, M_3^2\}$ are each bounded by edges M_4^1 and M_0^1. Figure 7 is a representation of the edge-face graph at any node in a Cartesian grid. Using this connectivity information,

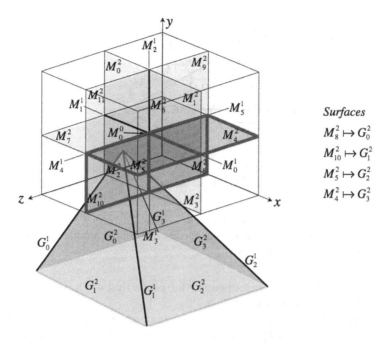

Fig. 6. A representation of the local grid near the apex of a pyramid. Four of the grid faces selected from $M^2_{j=0,1,\ldots11}$ that are connected to the node M^0_0 must be paired to the four surfaces in the B-Rep $G^2_{i=0,1,2,3}$.

a complete set of unique paths can be derived between any two edges. A metric m_{ij}, where $0 \leq m_{ij} \leq 1$, can be computed for each possible path $P^2_{ij} \subseteq M^2_{j=0,1,\ldots11}$ based on the following criteria:

$$m_{ij} = \frac{1 + 0.5(\mathbf{T}_{G^2_i} \cdot \mathbf{T}_{P^2_{ij}})}{1 + \left| J_i - N_{P^2_{ij}} \right|} \qquad (2)$$

where $\mathbf{T}_{G^2_i}$ is the normalized outward pointing tangent vector at the common vertex G^0_i and in the plane of the surface G^2_i that bisects curves G^1_i and G^1_{i+1}; $\mathbf{T}_{P^2_{ij}}$ is the average normalized outward pointing tangent vector at node M^0_i to the faces in P^2_{ij}; $N_{P^2_{ij}}$ is the number of faces in P^2_{ij} and J_i is defined as:

$$J_i = \begin{cases} 1, & \theta_i \leq \frac{3\pi}{4} \\ 2, & \frac{3\pi}{4} \leq \theta_i \leq \frac{5\pi}{4} \\ 3, & \theta_i > \frac{5\pi}{4} \end{cases} \qquad (3)$$

where J_i is an integer that represents the ideal number of faces that should represent surface G^2_i given the angle θ_i between curves G^1_i and G^1_{i+1} on the surface. Equation 2 computes the relative alignment between the path P^2_{ij}

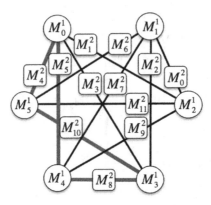

Fig. 7. The graph of local faces and edges at a node illustrating all possible paths of faces that can be selected representing the set of surfaces around a vertex. The path used in figure 6 is highlighted.

and the surface G_i^2 and assigns a penalty for paths that are longer or shorter than the ideal path defined by J_i.

A suitable path P_i^2 can now be selected that effectively matches surface G_i^2 by selecting the path P_{ij}^2 with the maximum m_{ij}. Subsequent paths can also be computed in a similar manner. The fact that edges and faces can only be used once, may block an optimal path between any given pair of edges, however alternate paths can generally be found. The final set of paths is represented in figure 7 as thicker lines and are illustrated as faces in figure 6.

Once a valid set of faces and edges have been determined using the preceding algorithm, there is still no guarantee that an optimal solution has been found. Given the fact that equation 1 may yield identical or similar metrics μ_{ij} for different permutations of edge-curve pairings, it may be necessary to check multiple edge-curve pairings by determining their associated face-surface pairings. This can be done by keeping track of the sum of m_{ij} for the face-surface pairings associated with each edge-curve pairing. For most cases, a limited number of edge-curve pairings will need to be tested before an optimal is determined.

Capturing cells at vertices

The final step in matching grid topology at a vertex is to capture the grid cells that will be inside the volume. In practice this is done by finding the face that is well-aligned with its associated surface. We can determine alignment based on the dot product of the face with the surface normal. Using the surface normal we can then determine which side of the face is defined as inside and which is defined as outside. By selecting the inside cell and traversing to neighboring cells sharing the node M_i^0 that are on the same side as the inside cell, all cells at the node inside the volume can be selected.

An example of topology captured at the vertices of the simple pyramid problem is shown in figure 8.

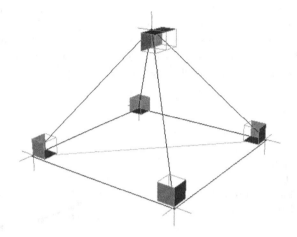

Fig. 8. The grid topology captured for vertices of a simple pyramid model is illustrated. The procedure described here has been used to select the appropriate grid topology.

3.2   Embedding Curves

Capturing edges at curves

The next step of the procedure involves embedding curves into the grid topology, or specifically the set $E_{G \to M} = \{M_i^1 : M_i^1 \mapsto G_i^1\}$. We can gather a set of grid edges for each curve by starting from the grid edge associated with the curve at each vertex that we determined in the previous step and finding a collision free path between the start and ending grid edges on the curve. For this procedure, knowing the edge at position k on the curve, we can determine the next grid edge at position $k + 1$ by examining the 5 connected edges at its end. Equation 6 can then be used to select the best next edge in the path.

$$T_{ij} = \frac{1 + (\mathbf{V}_{M_j} \cdot \mathbf{V}_{G_i})}{2} \tag{4}$$

$$D_{ij} = \frac{\left|\mathbf{P}_{M_j} - \mathbf{P}_{G_i}\right|^{-2}}{\sum_{j=1}^{5} \left|\mathbf{P}_{M_j} - \mathbf{P}_{G_i}\right|^{-2}} \tag{5}$$

$$\mu_{ij} = T_{ij}\left(1 - \frac{d_i}{2s}\right) + D_{ij}\frac{d_i}{2s} \tag{6}$$

The variables in equations 4 to 6 are illustrated in figure 9. Equation 6 incorporates both a tangent component, T_{ij} in equation 4 and a distance component, D_{ij} in equation 5. The tangent component is the same as that used in equation 1 to determine alignment of edges at a vertex. The distance component is used to ensure the selected grid edges do not deviate too far from its parent curve. Equation 5 incorporates proximity information, where a normalized inverse distance weighted value between 0 and 1 is computed so that edges that are close are given a larger weight than those farther away. The vector $\mathbf{P}_{M_j}$ in this case is defined as the midpoint of edge M_j^1 and the vector $\mathbf{P}_{G_i}$ is a point on curve G_i^0 a distance $\frac{1}{2}length(M_j^1)$ from a projection of the end point of edge M_k^1 on the curve G_i^1.

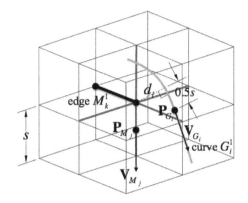

Fig. 9. The selection of the next edge in a sequence following edge M_k^1 is illustrated. The five edges attached to this edge are the only candidates. Equations 4 to 6 are used to select the edge.

Equation 6 incorporates both the tangent and distance components from equations 4 and 5 by using a simple linear blending function pair $1 - d_i/2s$ and $d_i/2s$ where d_i is the projected distance from the end point of edge M_k^1 and the parent curve G_i^1 and s is the constant grid spacing size as illustrated in figure 9. This ensures that when the curve is close to the grid edges, that the tangent component will control, while for distances beyond the grid spacing, s, the distance component will control.

Some of the possible edge selections may be eliminated by examining the end node of each of the edges to see if they are already in use. While this will avoid collisions, in some cases a non-optimal path may be selected. Possible alternative paths may be generated by starting from opposite ends of the curve or by changing the order in which curves are processed. Optimal curve paths may be generated by minimizing the direction changes between grid edges representing the curve, and attempting to modify the edge-curve pairing of those curves where μ_{ij} is small while attempting to maximize the average μ_{ij}. This may require multiple iterations to define an optimal curve path.

Figure 10 left shows edge paths representing the curves for the pyramid problem. Where curves are not naturally aligned with one of the coordinate axes, multiple direction changes between grid edges are unavoidable, which will ultimately reduce final mesh quality.

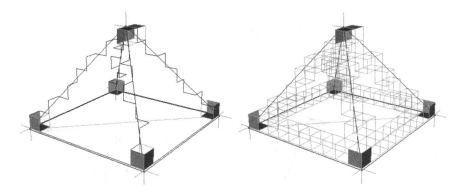

Fig. 10. Continuing with the embedding procedures, the figure on the left illustrates the edges selected to represent each of the curves in the model. On the right, the cells adjacent each of these edges has been selected and illustrated here.

Capturing faces and cells at curves

Another important aspect of the curve capture problem is selecting the cells at the curve that will be interior to the volume. Knowing the orientation and angle between surfaces at the curve, we can attempt to control the number and position of each grid cell so that final mesh topology will match as close as possible to the B-Rep. Figure 11(a) illustrates two edges M_k^1 and M_{k+1}^1 that have been selected to represent a segment on a curve G_i^1. The four shaded cells adjacent edge M_k^1 represent the possible choices for cells where at least one and no more than three cells must be selected to represent the interior of the volume at the curve. Figure 11(b) shows the 4 cells adjacent edge M_k^1. J_i from equation 3 may be used to determine the number of cells to select where θ_i is the interior angle formed by adjacent surfaces G_i^2 and G_{i+1}^2 at the curve. If we compute vector $\mathbf{B}_i$ as the bisecting vector at edge G_i^1 projected onto the plane normal to edge M_k^1, then the quadrant in which $\mathbf{B}_i$ falls with respect to the four cells adjacent edge M_k^1 will determine which initial cell M_i^3 to select. Where $J_i > 1$, then one or two additional cells adjacent the cell M_i^3 at edge M_k^1 may be selected based on the relative orientation of $\mathbf{B}_i$ within the cell. Once cells at the edge have been selected, then faces M_i^2, also shown in figure 11(b), can be selected and associated with their respective surfaces G_i^2 and G_{i+1}^2

Another issue which must be resolved is potential collisions between cells as the algorithm proceeds. This is common especially at adjacent edges where a 90 degree turn is required to better capture geometry. This is illustrated in figure 11(a), where edge M_k^1 is oriented 90 degrees to M_{k+1}^1. In this case it is not uncommon for $\mathbf{B}_i$ and J_i to be computed such that the cells selected adjacent M_{k+1}^1, conflict with those selected adjacent M_k^1. Where conflicts arise, they are resolved by using the cell selection from the edge whose tangent most closely aligns with that of the curve at its closest point.

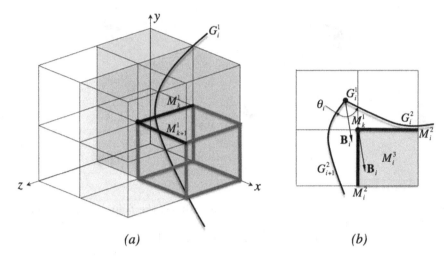

Fig. 11. (a) Representation of the curve G_k^1 in the geometry and two edges M_k^1 and M_{k+1}^1 that have been associated with the curve. The cell selected at the edges representing the interior of the volume is highlighted. (b) The side view of the four cells adjacent the same edge M_k^1 along with the curve G_i^1 and its adjacent surfaces. The angle θ_i between the surfaces and its bisecting vector $\mathbf{B}_i$ are used to determine which of the four cells at the edge will be captured.

3.3   Embedding Surfaces

Capturing faces at surfaces

The next stage of the algorithm is to ensure that a continuous set of faces M_i^2 is associated with their respective surfaces G_i^2. For this we seek the set of faces $E_{G \rightarrow M} = \{M_i^2 : M_i^2 \mapsto G_i^2\}$. Since both vertex and curve embedding procedures also included selection of adjacent faces, we can begin with the assumption that at least one continuous layer of faces has been captured at each surface. This is illustrated in figure 12 where the image on the left shows one of the surfaces with only the faces near the curves that have been captured.

This procedure can be compared to an advancing front algorithm where boundary faces are first captured and interior faces to the surfaces are progressively discovered and added to the surface until a continuous set of grid faces have been established that represent the surface. It is advantageous to order the procedure such that loops of faces progressively advance towards the interior of the surface until they close on themselves. Figure 13 illustrates the procedure for advancing a single front where a current front edge defined by M_i^1 has adjacent face M_i^2 which has already been associated with its parent surface G_i^2. The objective here is to determine which of the three faces, illustrated as $M_{i \rightarrow up}^2$, $M_{i \rightarrow forward}^2$, or $M_{i \rightarrow down}^2$, should be selected as the next face onto which the front will advance. The candidate face can be

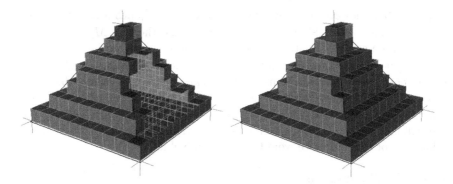

Fig. 12. Continuing with the same pyramid example, the figure at left illustrates the faces that have been captured near the curves of one of the surfaces. The figure on the right illustrates the final grid topology with all features embedded.

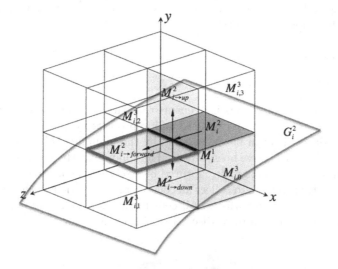

Fig. 13. The cells adjacent a grid edge and face M_i^2 that has been associated with the surface G_i^2 is illustrated to represent the current front of the advancing front procedure. The next face in the surface will be selected from the three faces $M_{i\rightarrow up}^2$, $M_{i\rightarrow forward}^2$ or $M_{i\rightarrow down}^2$.

selected by computing μ_{ij} using equation 6 as we did for edges. In this case the vectors $\mathbf{V}_{M_j}$ and $\mathbf{V}_{G_i}$ are the normal vectors of the grid face and surface, and the points $\mathbf{P}_{M_j}$ and $\mathbf{P}_{G_j}$ is the midpoint of the face and its projection to the surface G_i^2 respectively.

The selection of the next front is also effected by the placement of the existing cells associated with the volume. From our definition of the cells captured at curves, we will always have a volume cell $M_{i,0}^3$ immediately adjacent the face M_i^2. It is however conceivable, that the cell $M_{i,1}^3$ has already been selected for the volume. Should this case arise, then the face $M_{i\rightarrow down}^2$ would be eliminated as a candidate for the next advancement. Likewise should $M_{i,2}^3$ and $M_{i,1}^3$ be in use, the only candidate for advancement would be $M_{i\rightarrow up}^2$. To ensure non-manifold connections are not created, we would ensure that cases where only $M_{i,0}^3$ and $M_{i,1}^3$ are selected at any grid edge are not permitted. Similarly, we ensure that the case where cells $M_{i,0}^3$ and $M_{i,1}^4$ are both used by the volume would not be permitted, as this would create a condition where a hanging or dangling face would exist in the volume.

Capturing cells at surfaces

Also part of this procedure is the selection of the cells adjacent new faces as the algorithm progresses. Cells can be selected by choosing from the two adjacent cells at the new face. Depending on which of the three candidate faces is selected, zero, one or two new cells may be added to the volume with each new face added to a surface.

3.4   Embedding Volumes

Capturing cells in a volume

The final step of the algorithm is to ensure that all remaining cells M_i^3 interior to the volume G_i^3 have been captured and appropriately assigned. Similar to other procedures this can be represented by the mapping $M_i^3 \mapsto G_i^3$. At this stage, all surfaces will have been captured and have exactly one adjacent cell. The remaining cells can now be captured by selecting one known interior cell and recursively traversing to collect all cells that are bounded by the grid faces that are associated with surfaces. All cells in the Cartesian grid not selected as part of this procedure are discarded. An example of a base grid with

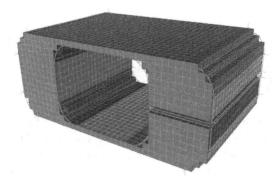

Fig. 14. A completed embedding procedure is illustrated using the C-geometry model shown in figure 3. Note that topological equivalence is maintained through the thin sections and narrow gap.

embedding procedures completed is shown in figure 14 where the different colors (shades) represent the separate captured surfaces in the Cartesian grid.

4 Completing the Mesh

The embedding procedures discussed in this work focus specifically on providing a base mesh that is topologically equivalent to a given boundary representation. Once this is achieved, subsequent sheet insertion and mesh optimization steps would be employed to build a final mesh as described by Shepherd [10]. Up to this point in the procedure, a Cartesian grid has been used because of its efficiency and low memory requirements. Sheet insertion procedures, however, require a full unstructured mesh data representation to insert appropriate layers of hexes. As a result, the base mesh is transferred to an unstructured data representation before sheet insertion and mesh optimization algorithms are executed. A simple implementation of a completed mesh on the example C-Geometry model illustrated in figure 4 is shown in figure 15. On the left is the base mesh shown in figure 14 with its associated grid entities projected to its parent geometry. The curves in the original geometry are also shown demonstrating the ability of the procedure to embed arbitrary B-Rep topology within a grid. This figure also illustrates that poor quality elements arise as a result of multiple edges and faces from a single cell being projected to the same geometric parent. Although not specifically part of this study, the figure on the right illustrates the insertion of buffer layers or sheets to improve element quality near the boundary using the *pillow* operation in the CUBIT [2] Meshing Toolkit. Future work will focus on the automatic insertion of boundary sheets.

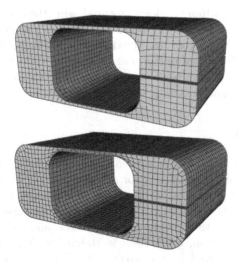

Fig. 15. The mesh is completed by projecting grid nodes to their associated parent geometry, inserting sheets or layers at the boundaries and then smoothing.

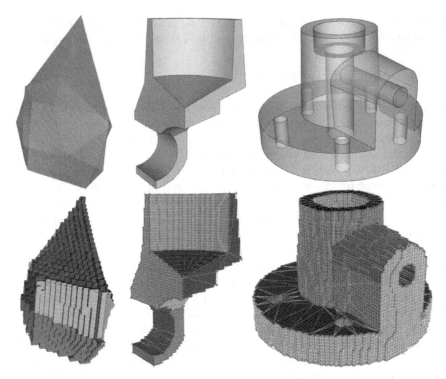

Fig. 16. Examples of embedding results on three solid models. Solid model is shown top with its resulting topologically equivalent base mesh on the bottom. Projection, smoothing and sheet insertion operations have not yet been performed.

A few examples of B-Rep topology captured using the proposed embedding procedures are also illustrated in figure 16. The top row shows the original CAD model, while the bottom row illustrates the topologically equivalent set of grid cells that will be used in the final meshing procedures. Different colors (shades) in the grids illustrate the distinct surfaces that have been captured from the grid and associated with its parent geometry.

5   Conclusion

Hexahedral methods that begin with a Cartesian grid as a base mesh are attractive because of their potential for automation. Their primary use, however has been for bio-medical applications that do not include significant topology. These methods have not been as effective for topology rich models such as those common to three-dimensional CAD solid models described by a boundary representation. While some work has been done to ensure curves and surfaces are captured in a Cartesian-based mesh, this work proposes a new systematic approach to ensure that features of a B-Rep model are adequately represented in a final hexahedral mesh.

This work introduces an approach to embedding features in a Cartesian grid. Without such procedures there is no guarantee that important features defined in a CAD model will be captured in the simulation model. Using traditional in-out procedures can often neglect topological equivalence, resulting in non-manifold or disjoint cells, as well as insufficient local node valence for a given B-Rep topology. Current methods generally resolve these issues by using mesh enrichment strategies that can introduce highly refined and poorly shaped elements. The embedding procedures described here build a base mesh from a Cartesian grid that is intended to meet topological equivalence requirements of a B-Rep while reducing the need for mesh enrichment.

While a Cartesian grid provides efficiency and ease-of-use, for complete generality, it is clear that mesh enrichment strategies will need to be applied. However it is expected that the proposed embedding procedures will limit the use of such strategies. Future work will need to extend these procedures to combine mesh enrichment with topology embedding to ensure that topological equivalence is maintained for an arbitrary B-Rep configuration.

In contrast to traditional in-out procedures for generating a base mesh, using the proposed embedding algorithms also provides more control over mesh topology, such that numbers of hexes placed at curves and vertices can better account for local geometry. Future work will study how sheet insertion procedures can work together with embedding algorithms to better control placement of sheets to maximize element quality.

The topology embedding problem, while a key component of automatic hexahedral meshing, is clearly only one part of a fully automatic procedure. Careful insertion of boundary sheets as well as projection and mesh optimization methods are necessary to provide a robust, quality hexahedral mesh. Ongoing work to couple the proposed procedures into a fully automatic hexahedral meshing algorithm for topology rich models is still necessary and under development along with its extension to muti-volume assemblies.

Acknowledgements

We express appreciation to the Computer Science Research Foundation at Sandia National Laboratories for funding this work. Also to the CUBIT research and development team at Sandia for their input and feedback to the algorithms proposed and to Jenna Kallaher and Brett Clark for their insightful review of the manuscript.

References

1. Blacker, T.D., Meyers, R.J.: Seams and Wedges in Plastering: A 3D Hexahedral Mesh Generation Algorithm. Engineering with Computers 2(9), 83–93 (1993)
2. CUBIT Meshing and Geometry Toolkit, Version 11.1, Sandia National Laboratories (2009), http://cubit.sandia.gov

3. Ito, Y., Shih, A.M., Soni, B.K.: Octree-based reasonable-quality hexahedral mesh generation using a new set of refinement templates. International Journal for Numerical Methods in Engineering 77(13), 1809–1833 (2009)
4. Kwak, D.Y., Im, Y.T.: Remeshing for metal forming simulations - Part II: Three-dimensional hexahedral mesh generation. International Journal for Numerical Methods in Engineering 53, 2501–2528 (2002)
5. Ledoux, F., Shepherd, J.F.: Topological and geometrical properties of hexahedral meshes. Submitted to Engineering with Computers (2009)
6. Ledoux, F., Shepherd, J.F.: Topological modifications of hexahedral meshes via sheet operations: A theoretical study. Submitted to Engineering with Computers (2009)
7. Quadros, W.R., Owen, S.J.: Geometry Tolerant Mesh Generation. Accepted to 18th International Meshing Roundtable (2009)
8. Schneiders, R., Schindler, F., Weiler, F.: Octree-based Generation of Hexahedral Element Meshes. In: Proceedings of the 5th International Meshing Roundtable, pp. 205–216 (1996)
9. Shepherd, J.F.: Topologic and Geometric Constraint-based Hexahedral Mesh Generation, PhD Thesis, University of Utah, Utah (2007)
10. Shepherd, J.F.: Conforming Multi-Volume Hexahedral Mesh Generation via Geometric Capture Methods. Accepted to 18th International Meshing Roundtable (2009)
11. Staten, M.L., Kerr, R.A., Kerr, O.S., Blacker, T.D.: Unconstrained Paving and Plastering: Progress Update. In: Proceedings, 15th International Meshing Roundtable, pp. 469–486 (2006)
12. Tautges, T.J., Blacker, T.D., Mitchell, S.A.: The Whisker Weaving Algorithm: A Connectivity-Based Method for Constructing All-Hexahedral Finite Element Meshes. International Journal for Numerical Methods in Engineering 39, 3327–3349 (1996)
13. Tchon, K.F., Hirsch, C., Schneiders, R.: Octree-based Hexahedral Mesh Generation for Viscous Flow Simulations. American Institute of Aeronautics and Astronautics A97-32470, 781–789 (1997)
14. Thakur, A., Banerjee, A.G., Gupta, S.K.: A survey of CAD model simplification techniques for physics-based simulation applications. Computer Aided Design 41(2), 65–80 (2009)
15. Yin, J., Teodosiu, C.: Constrained mesh optimization on boundary. Engineering with Computers 24, 231–240 (2008)
16. Zhang, H., Zhao, G.: Adaptive hexahedral mesh generation based on local domain curvature and thickness using a modified grid-based method. Finite Elements in Analysis and Design 43, 691–704 (2007)
17. Zhang, Y., Bajaj, C.L.: Adaptive and Quality Quadrilateral/Hexahedral Meshing from Volumetric Data. Computer Methods in Applied Mechanics and Engineering 195, 942–960 (2006)
18. Zhang, Y., Hughes, T.J.R., Bajaj, C.L.: Automatic 3D Mesh Generation for a Domain with Multiple Materials. In: Proceedings of the 16th International Meshing Roundtable, pp. 367–386 (2007)

Label-Invariant Mesh Quality Metrics

Patrick Knupp

Sandia National Laboratories
pknupp@sandia.gov

Abstract. Mappings from a master element to the physical mesh element, in conjunction with local metrics such as those appearing in the Target-matrix paradigm, are used to measure quality at points within an element. The approach is applied to both linear and quadratic triangular elements; this enables, for example, one to measure quality within a quadratic finite element. Quality within an element may also be measured on a set of *symmetry points*, leading to so-called *symmetry metrics*. An important issue having to do with the labeling of the element vertices is relevant to mesh quality tools such as Verdict and Mesquite. Certain quality measures like area, volume, and shape should be *label-invariant*, while others such as aspect ratio and orientation should not. It is shown that local metrics whose Jacobian matrix is non-constant are label-invariant only at the center of the element, while symmetry metrics can be label-invariant anywhere within the element, provided the reference element is properly restricted.

1 Measuring Quality *Within* Mesh Elements

Mesh quality is important for maintaining accuracy and efficiency of numerical simulations based on the solution of partial differential equations [6]. Mesh quality metrics are used to measure mesh quality and there is an extensive literature on the subject, particularly for finite element meshes [8], [14], [16], [17], [19]. For simplicial elements, 'shape' is an important measure [3], [15]. A shape measure based on condition number was proposed in [4], [5]. In [2] and [7] the notion of shape for simplicial elements was formalized. Few works discuss quality measures for quadratic elements [1], [18]; the latter reference being the only example that goes beyond detecting singular points. Significantly, the latter is limited to triangle elements.

Engineers usually measure mesh quality by one of two basic approaches, depending on whether they are working with unstructured or structured meshes. The quality of an unstructured mesh is most often studied in terms of the individual elements within the mesh. Elements are most often polygons or polyhedra, with triangles, tetrahedra, quadrilaterals, hexahedra, prisms, and pyramids being the most commonly used types. A mesh element contains vertices and/or nodes, usually given in some canonical ordering. The vertices/nodes have coordinates $x_m \in R^d$, with $d = 2, 3$ and $m = 0, 1, 2, ..., M$, with M depending on the element type and order. The quality q_ε of an element is most often defined as some continuous function of the element coordinates.

Triangular element *aspect ratio*, given by the formula $q_\varepsilon = \frac{L_{max}}{2\sqrt{3}\,r}$ is an example of the first approach to measuring quality. Because the lengths in the formula depend on the coordinates of the vertices in the triangle, the element metric is a function of the vertex coordinates. The formula only applies to straight-sided (low-order) triangles.

The second approach to measuring mesh quality arises in the structured meshing community. A global mapping from a logical block U to a physical block $\Omega \subset R^d$ is found and serves to define a discrete grid. When $d = 3$, the map takes the form $x = x(\Xi)$, with $\Xi = (\xi_1, \xi_2, \xi_3) \in U$ and $x = (x_1, x_2, x_3) \in \Omega$. The tangents to the map, $dx_i/d\xi_j$, $i, j = 1, 2, 3$, are used to define local mesh quality at a point within the domain. For example, for $d = 2$, one measures *orthogonality* at a point in U via the *local* metric $x_{\xi_1} \cdot x_{\xi_2}$.

Over the past decade, the author has used a third approach to measuring quality that is a hybrid of the two basic approaches [8], [9]. For each *element* of a mesh, let there be a map from a logical (or master) element to the physical element. Then one can measure local quality *within* the element using formulas based on the local tangents of the map, just as is done in the structured meshing community. Because the element map depends on the coordinates of the vertices/nodes within the element, the local quality at a point within the element also depends on these coordinates. Although the third approach uses the master element concept from the finite element method, it can be used to measure quality whether or not the mesh is intended to be used in a finite element simulation. That is, measuring quality by the third approach applies equally well to finite element, finite volume, finite difference, or even spectral element simulations, as is the case with the first approach.[1]

The third approach does not preclude the measurement of *element* quality, if desired. Let μ be a local quality metric and $\mu(\Xi_n)$, $n = 1, ..., N$, be the local qualities measured at N points Ξ_n within the master element. Then element quality may be defined to be, for example, $q_\varepsilon = \max_n\{\mu(\Xi_n)\}$, $q_\varepsilon = \min_n\{\mu(\Xi_n)\}$, or the pth power-mean, $p \neq 0$, of the local qualities:

$$q_\varepsilon = \left(\frac{1}{N} \sum_{n=1}^{N} [\mu(\Xi_n)]^p \right)^{1/p} \tag{1}$$

with $\mu > 0$. The power-mean, minimum, and maximum are attractive as a means to combine the local metrics because the range of the resulting element metric is the same as the range of the local metric.

---

[1] For the sake of clarity, we propose to call the first approach to measuring quality the 'element' quality method, the second approach the 'pointwise' quality method, and the third approach the 'hybrid' quality method.

Three examples are given to show why this third approach may be attractive. First, consider a planar quadrilateral element, with *area* as the quantity of interest. Let the four vertices be labeled x_m, $m = 0, 1, 2, 3$, in counterclockwise order. The linear map is $x(\xi_1, \xi_2)$ and the Jacobian matrix $A(\xi_1, \xi_2)$

$$A = [\, (x_1 - x_0) + (x_0 - x_1 + x_2 - x_3)\, \xi_2, \ (x_3 - x_0) + (x_0 - x_1 + x_2 - x_3)\, \xi_1 \,]$$

The signed area at any given point within the element is $\alpha = det(A)$.[2] In the 'element' quality method, a standard quadrilateral area measure is

$$A_1 = \frac{1}{2}\, det([x_1 - x_0, x_3 - x_0]) + \frac{1}{2}\, det([x_3 - x_2, x_1 - x_2]) \qquad (2)$$

In the 'pointwise' quality method, an area measure based directly on the local metric $\alpha(\xi_1, \xi_2)$ is

$$A_2 = \min\{\, \alpha(0, 0), \alpha(1, 0), \alpha(1, 1), \alpha(0, 1) \,\} \qquad (3)$$

To compare these two area measures, consider the quadrilateral with vertex coordinates $x_0 = (0, 0)$, $x_1 = (1, 0)$, $x_2 = (\frac{1}{4}, \frac{1}{4})$, and $x_3 = (0, 1)$[3] Then formula (2) yields $A_1 = \frac{1}{4}$, while formula (3) gives $A_2 = -\frac{1}{2}$. Therefore, the latter formula detects the negative Jacobian, while the former does not.[4]

In the second example, consider the quality of a high-order finite element such as a quadratic triangle having three mid-side nodes. With the exception of [18], there are no examples in the literature that measure the quality of a quadratic finite element by the 'element' quality method and this reference is limited to triangular elements. The quality of high-order finite elements such as tetrahedra and hexahedra can be assessed using the 'hybrid' quality method. In fact, the method is exactly the same as it is for linear elements: evaluate the local quality metric at a point by computing the Jacobian of the relevant map from the master to the physical element and combine the local qualities via the formulas for maximum or minimum quality or the power-mean (1). *Although this method often does not bound the worst quality within the element, a judicious choice of sample points within the element can provide a lot of useful information.* For a quadratic triangle, for example, it is reasonable to measure the local quality at the three corner vertices and at the three mid-side nodes. In this manner one can measure local shape, size, and orientation within an element using, for example, metrics from the Target-matrix paradigm [10], [11], [12].

In the third example, suppose one wanted to optimize the quality of a locally-refined mesh containing, as a submesh, the two linear quadrilaterals

---

[2] The notation $A = [x_{\xi_1}, x_{\xi_2}]$ signifies that the first column of A is the $1 \times d$ vector x_{ξ_1} and that the second column of A is the $1 \times d$ vector x_{ξ_2}. Similar notation is used throughout. Also, $det(A)$ signifies the determinant of A.

[3] This poor quality quadrilateral is called an *arrow* due to the re-entrant corner.

[4] Formula (2) can also be written in terms of local metrics. In this example, the ability to detect negative Jacobians is a matter of choosing the minimum instead of the linear average.

on the right in Figure 1 and one quadratic quadrilateral on the left. An objective function could be based on (1), for example, in which μ is any desired local quality metric. The points Ξ_n would include the logical corners of the three quadrilaterals, along with the logical points corresponding to the mid-edge nodes in the quadratic 'quadrilateral'. If the mid-side node is allowed to be 'free' in the optimization, then the quadratic map is required for the left quadrilateral, while if the mid-side node is constrained to the mid-point of the straight edge, then only a linear map is needed. Note that in the hybrid quality method one can have more than one quality measurement per vertex within the mesh.

Fig. 1. Three non-conformal quadrilateral elements

The hybrid method is clearly more flexible than the element quality method. It becomes yet more powerful when combined with concepts from the Target-matrix paradigm which provides numerous referenced local metrics. Figure 2 shows the basic idea: let there be maps from the logical element to the physical element, and from the logical element to a reference element which gives the desired element configuration. Let the Jacobian matrix of the first map be denoted by $A(\Xi)$ and the Jacobian matrix of the second map be $W(\Xi)$. It is reasonable to assume that the reference element is non-degenerate; in that case, W is non-singular. To compare the two matrices, form $T = AW^{-1}$ so that when $A = W$, $T = I$. The quality at a point Ξ within the element is given by a quality metric $\tilde{\mu}(\Xi) = \mu[T(\Xi)]$. A variety of useful local quality metrics $\mu(T)$ are studied in [11]. Most of the quality metrics are combinations of the fundamental quantities $\tau = det(T)$, $|T|$, $|T^tT|$, $tr(T)$, and $|T^{-1}|$, so the analysis to follow is focused on these.

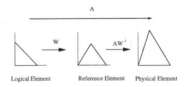

Logical Element   Reference Element   Physical Element

Fig. 2. Relation between the Logical, Reference, and Physical Elements

2   Label-Invariance of Quality Metrics

An important practical issue arises in the measurement of a priori unstructured mesh quality that has to do with the labeling of the vertices within an element. Consider the logical (left) and physical (right) triangles in Figure 3. The vertices in the logical triangle are labeled 0,1,2 in counter-clockwise order, while the vertices in the physical triangle are labeled m, $m + 1$, $m + 2$, again in counter-clockwise order. If $m = 0$, then physical vertex m corresponds to logical vertex 0, physical vertex $m + 1$ to logical vertex 1, and so on. However, if $m = 1$, then physical vertex m corresponds to logical vertex 1, physical vertex $m + 1$ to logical vertex 2, and so on. In most unstructured mesh generation software, the value of m is determined by the order of the nodes in the list of nodes for the given element. As an example, in the Verdict mesh quality assessment code [19], one step in calculating the quality of an element is to obtain the list of vertices that are contained by the element. No sorting of this list is done and so the first vertex in the list automatically becomes the image vertex 0, and the second vertex in the list becomes image vertex 1, etc. The impact of this can be seen in the two examples to follow.

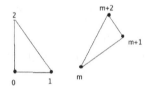

Fig. 3. Vertex Labeling Choices

First, consider the metric A_1 in (2). If the indices are cyclically permuted by 1 the formula becomes

$$A_1' = \frac{1}{2} \, det(x_2 - x_1, x_0 - x_1) + \frac{1}{2} \, det(x_0 - x_3, x_2 - x_3) \qquad (4)$$

One can show that $A_1' = A_1$, that is, the element area is independent of the choice of labeling of the vertices. This is an example of what we will call a label-invariant metric. Not all element metrics enjoy this property. For example, let

$$L_h = \tfrac{1}{2} \left\{ |x_1 - x_0| + |x_2 - x_3| \right\}, L_v = \tfrac{1}{2} \left\{ |x_3 - x_0| + |x_2 - x_1| \right\}$$

and define the quadrilateral aspect ratio metric to be $AR = L_v/L_h$. Then the cyclically permuted formula is $(AR)' = 1/(AR)$.

Definition 1. Label-invariance for Element Quality Metrics
An element quality metric is *label-invariant* if, for an arbitrary physical element, its value is the same no matter which corner vertex of the physical element is labeled zero.

The question arises as to whether or not quality metrics should be label-invariant. In general, the answer is no because while metrics such as area, volume, and shape should probably be label-invariant, metrics like aspect ratio may be more informative if they are not label-invariant.

The labeling issue above was described within the context of the first approach to the measurement of element quality. It also exists within the context of the third approach, with a few additional twists, due to the fact that the mapping from the logical to the physical element depends on the choice of labeling and the Jacobian matrix thus depends on m. The first twist in the third approach is that even the local metric at a point may or may not be label-invariant, so that one can speak of label-invariant local metrics in addition to label-invariant element metrics. Second, the label-invariance may depend on the choice of reference element. For example, if an isotropic reference element is selected, it is more likely that the local metric is label-invariant. Third, label-invariance of a local metric may depend on the point within the element at which it is evaluated. This necessitates a modification of the previous definition for the case of measuring quality within elements.

Definition 2. Label-invariance for Local Quality Metrics
Let $\mu_m(\Xi) = \mu(T_m(\Xi))$ be a local (target-matrix) quality metric. Let the reference element be a particular type and have a particular configuration within that type. Then the local quality metric is label-invariant at Ξ (with particular reference) if, for an arbitrary physical element (whose type is the same as the reference element), $\mu_m(\Xi)$ is a constant for all m.

In additional to the above definition, there is another concept of importance that arises in the third approach to measuring quality. Let $\{\Xi^{(0)}, \ldots, \Xi^{(N)}\}$ be a collection of *symmetry points* within the master element.[5] The symmetry points are each functions of Ξ. Define a non-local *symmetry metric* $\sigma_m(\Xi)$, similar to (1), based on an associated local metric μ. For example, in terms of the power-mean

$$\sigma_m(\Xi) = \left(\frac{1}{N} \sum_{n=0}^{N} [\mu_m(\Xi^{(n)})]^p \right)^{1/p} \tag{5}$$

and for the minimum and maximum

$$\sigma_m(\Xi) = \min_n \{\mu_m(\Xi^{(n)})\} \tag{6}$$

$$\sigma_m(\Xi) = \max_n \{\mu_m(\Xi^{(n)})\} \tag{7}$$

Definition 3. Label-invariance for Symmetry Quality Metrics
Let σ_m be a symmetry metric derived from the metric μ_m. Let the reference element be a particular type and have a particular configuration within that

---

[5] It will become apparent later what is meant by a symmetry point.

type. Then the symmetry metric is label-invariant at Ξ (with particular reference) if, for any arbitrary physical element (whose type is the same as the reference element), σ_m is a constant for all m.

It is noted that the concept of label-invariance is not the same as the concept of orientation-invariance. As a example, the aspect ratio metric AR is orientation-invariant because, if the element is rigidly rotated, the value of the metric does not change; in contrast, the metric is not label-invariant.

The comments and definitions presented in this section should become clearer in the sections to follow, where the hybrid quality method is studied on triangles with linear and quadratic maps.

3 Linear Planar Triangles

3.1 The Linear Map

Let $\Xi = (\xi, \eta)$ and $U = \{\Xi \mid \xi \geq 0, \eta \geq 0, \xi + \eta \leq 1\}$ be a logical triangle. Let x_0, x_1 and x_2 be the three (ordered) vertices of a physical triangle in R^2. For linear triangles in the xy-plane, the mapping from U to the physical triangle is

$$x(\Xi) = x_0 + (x_1 - x_0)\xi + (x_2 - x_0)\eta \tag{8}$$

Then, $x_\xi = x_1 - x_0$ and $x_\eta = x_2 - x_0$, so the Jacobian matrix is $A = [x_\xi, x_\eta]$. For the linear triangle map, the Jacobian matrix and its determinant, $det(A)$, are independent of ξ and η and are thus constant over the element (i.e., the same at every point in U).

3.2 The Reference Element

Let a reference triangle with vertex coordinates w_0, w_1, and w_2 be given. The Jacobian matrix W of the reference triangle is obtained from the previous relations by replacing x_0 with w_0, x_1 with w_1, and x_2 with w_2, yielding $W = [w_1 - w_0, w_2 - w_0]$. The reference element is assumed to be non-degenerate, i.e., $det(W) \neq 0$; therefore W^{-1} exists. Let $T = AW^{-1}$ be the weighted Jacobian matrix. Both W and T are constant over the linear triangle.

Let $\rho > 0$, R be any 2×2 rotation matrix, and

$$V = \begin{pmatrix} 1 & 1/2 \\ 0 & \sqrt{3}/2 \end{pmatrix} \tag{9}$$

Then if the reference triangle is equilateral, the matrix W representing the reference Jacobian belongs to the set of 2×2 matrices $\mathcal{M}$ of the form $\rho R V$.

3.3 Label-Invariance

The map (8) assigns the vertex $(0,0) \in U$ to the image vertex 0 in the physical triangle. In general, the vertex $(0,0)$ could have been assigned to any of the image vertices 0, 1, or 2 (see Figure 3). To preserve orientation, it is assumed that if reference vertex $(0,0)$ is assigned to image vertex m, with $m = 0, 1, 2$, then reference vertex $(1,0)$ is automatically assigned to image vertex $m + 1$, and $(0,1)$ to image vertex $m + 2$. There are thus three ways one can define a properly oriented mapping for a linear triangle, depending on which physical vertex, m is selected to be the image of $\Xi = 0$. The previous map (8) is modified to emphasize this dependence. Define the map

$$x(\Xi, m) = x_m + (x_{m+1} - x_m)\,\xi + (x_{m+2} - x_m)\,\eta \tag{10}$$

where $m = 0, 1$, or 2.[6] Quantities derived from the map, such as the Jacobian matrices and quality metrics, will also depend on m in general.

Let the Jacobian matrices of the map (10) be A_m. From the map it is clear that $A_m = [x_{m+1} - x_m, \, x_{m+2} - x_m]$. The three Jacobian matrices are, in general, not equal to one another, and thus A_m is not a label-invariant quantity. It is straightforward to show that the Jacobian matrices obey the relation

$$A_{m+1} = A_m\, P \tag{11}$$

where P is the constant matrix

$$P = \begin{pmatrix} -1 & -1 \\ +1 & 0 \end{pmatrix} \tag{12}$$

Accordingly, $A_1 = A_0\, P$ and $A_2 = A_1\, P = A_0\, P^2$. Because $det(P) = 1$, $det(A_0) = det(A_1) = det(A_2)$. Thus $det(A_m)$ is a label-invariant quantity.

For this map, there are three weighted Jacobian matrices $T_m = A_m W^{-1}$, for which $det(T_0) = det(T_1) = det(T_2)$. Thus the local metric $\tau_m = det(T_m)$ for the linear planar triangle is label-invariant *for any choice of Ξ or W*.

Proposition 1. The local quantities $|T_m|$ and $|(T_m)^t (T_m)|$ are label-invariant for arbitrary Ξ if and only if $W \in \mathcal{M}$. The quantity $tr(T_m)$ is not label-invariant for any choice of W.

Proof. Suppose that $W \in \mathcal{M}$. Then let $W = \rho R V$. One can show by direct calculation that the matrix $V\, P^m\, V^{-1}$ is a rotation. Therefore, $W\, P^m\, W^{-1} = R\,(V\, P^m\, V^{-1})\, R^t$ is also a rotation. Let $Q_m = W\, P^m\, W^{-1}$. Then, $T_m = A_m W^{-1} = A_0\, P^m\, W^{-1} = A_0\, W^{-1} Q_m$. Because the Frobenius norm is invariant to orthogonal matrices,

$$|T_m| = |A_0 W^{-1} Q_m| = |A_0 W^{-1}| = |T_0|$$

The steps in this proof are reversible, so if $|T_m|$ is label-invariant, then $W \in \mathcal{M}$. The proof for $|(T_m)^t (T_m)|$ is similar. Since the trace is not invariant to a rotation, it is not label-invariant.   §

---

[6] All vertex subscripts in this section are modulo 3 so, for example, $x_4 = x_1$.

Corollary. Any local metric $\mu_m = \mu(T_m)$, which is a function of the quantity τ_m only (e.g. $\mu(T) = \tau^2$), is label-invariant for any $\varXi$ and W. Any local metric which is a combination of only τ_m, $|T_m|$, and/or $|(T_m)^t(T_m)|$ is label-invariant provided $W \in \mathcal{M}$. For example, the local inverse mean ratio metric $\mu(T) = \frac{|T|^2}{2\tau}$ (which when $d = 2$ is the same as the condition number metric) is label-invariant when $W \in \mathcal{M}$. Invariance is desirable in this case since mean ratio is intended to measure the local shape within a triangle relative to the reference shape.[7] Any local metric containing $tr(T_m)$ is not label-invariant. For example, the metric $\mu(T) = |T-I|^2$, which is the same as $|T|^2 - 2\,tr(T) + 2$, is not label-invariant. Lack of invariance is acceptable for this metric since it is intended to control the orientation within mesh elements [11].   §

These results, of course, apply to the linear map (10) and must be re-examined when the map is different.

4   Quadratic Planar Triangles

4.1   The Quadratic Map

There are three ways one can define the mapping for a quadratic planar triangle, depending on which vertex, m ($m = 0, 1, 2$) is selected to be the image of $\varXi = 0$. Write the quadratic map on U as

$$
\begin{aligned}
x(\varXi, m) = c_{0,m} &+ c_{1,m}\,\xi + c_{2,m}\,\eta \\
&+ c_{3,m}\,\xi^2 + c_{4,m}\,\xi\eta + c_{5,m}\,\eta^2
\end{aligned} \tag{13}
$$

The tangent vectors of the map are[8]

$$
x_\xi(\varXi, m) = c_{1,m} + 2\,c_{3,m}\,\xi + c_{4,m}\,\eta \tag{14}
$$
$$
x_\eta(\varXi, m) = c_{2,m} + c_{4,m}\,\xi + 2\,c_{5,m}\,\eta \tag{15}
$$

In contrast to the linear map, the tangent vectors for the quadratic map depend on $\varXi$, thus the Jacobian matrix $A_m = A_m(\varXi)$ depends on $\varXi$. It is easy to show that the Jacobian matrix A_m for the quadratic map is constant (i.e., independent of $\varXi$) if and only if the physical triangle has straight sides.

Recall the relations between the logical, reference, and physical elements shown in Figure (2). When the map from the logical to the physical element is quadratic, there is no reason why the map from the logical to the reference element cannot also be quadratic. In that case, $W = W(\varXi)$, i.e, the reference Jacobian matrix can vary from one position to another. The discussion that follows does not assume W is constant so that reference elements having

---

[7] Because the Jacobian of the linear triangle map is constant, mean ratio also measures the shape of the triangle itself.

[8] To save space, the formulas for the coefficients in terms of the element nodes is not given since they are well-known.

curved sides are allowed.[9] Define $T_m(\Xi) = A_m(\Xi)[W(\Xi)]^{-1}$ and $\tau_m = det(T_m)$.

4.2   Symmetry Points for Maps to Triangular Elements

Recall that for both the linear and quadratic triangular elements there is a map x of the form $x(\Xi, m)$, where $m = 0, 1, 2$ denotes the vertex of the triangle that serves as the base in the construction the map. Let $X^{(0)}$ be an arbitrary point in the physical triangular element. Setting $X^{(0)} = x(\Xi^{(0)}, m)$ for some fixed choice of m, we have $\Xi^{(0)}$ as the pre-image of $X^{(0)}$. For each such point $X^{(0)}$ in the triangle, there are two additional points $X^{(1)} = x(\Xi^{(1)}, m)$ and $X^{(2)} = x(\Xi^{(2)}, m)$ with pre-images $\Xi^{(1)}$ and $\Xi^{(2)}$ which we define by the relations[10]

$$x(\Xi^{(1)}, m) = x(\Xi^{(0)}, m+1) \tag{16}$$
$$x(\Xi^{(2)}, m) = x(\Xi^{(0)}, m+2) \tag{17}$$

The points $X^{(k)}$ and their pre-images $\Xi^{(k)}$ ($k = 0, 1, 2$) are points of symmetry because point $X^{(k+1)}$ can be obtained either from the map based at vertex m evaluated at $\Xi^{(k+1)}$ or from the map based at vertex $m+1$ evaluated at $\Xi^{(k)}$. The symmetry points are defined by the relations above and hold on any triangle of any shape and includes both linear and quadratic maps.[11]

The relations above can be used to find the pre-image points $\Xi^{(k)}$ in terms of $\Xi^{(0)}$. Solving for $\Xi^{(1)} = (\xi^{(1)}, \eta^{(1)})$ and $\Xi^{(2)} = (\xi^{(2)}, \eta^{(2)})$, one obtains

$$\Xi^{(1)} = (1, 0) + P\,\Xi^{(0)} \tag{18}$$
$$\Xi^{(2)} = (0, 1) + P^2\,\Xi^{(0)} \tag{19}$$

where P is the matrix given in (12). Explicitly, $\Xi^{(0)} = (\xi^{(0)}, \eta^{(0)})$, $\Xi^{(1)} = (1 - \xi^{(0)} - \eta^{(0)}, \xi^{(0)})$, and $\Xi^{(2)} = (\eta^{(0)}, 1 - \xi^{(0)} - \eta^{(0)})$. Notably, the logical symmetry points do not depend on the vertices x_m, x_{m+1}, and x_{m+2} of the triangle.

4.3   Symmetry Relation for Jacobian of the Quadratic Map

Proposition 2. Let the Jacobian of the quadratic map (13) at the point Ξ be given by $A_m(\Xi) = [x_\xi(\Xi, m), x_\eta(\Xi, m)]$, with the latter given in (14)-(15). Then, for $k, m = 0, 1, 2$, the following relations hold

$$A_m(\Xi^{(k)}) = A_0(\Xi^{(k+m)})P^m \tag{20}$$

with P defined as in (12).

---

[9] Note, however, that W does not depend on m since there is no labeling issue with the reference element.

[10] More generally, one can write $x(\Xi^{(r+s)}, m) = x(\Xi(s), m+r)$ with $s = 0, 1, 2$, which leads to the same symmetry points.

[11] The indices k are cyclic with period 3.

Proof. The logical symmetry points derived in the previous section can be regarded as functions of $\xi^{(0)}$ and $\eta^{(0)}$. Differentiation of the pre-image formulas with respect to these variables, one finds

$$
\begin{pmatrix}
\frac{\partial \xi^{(k)}}{\partial \xi^{(0)}} & \frac{\partial \xi^{(k)}}{\partial \eta^{(0)}} \\
\frac{\partial \eta^{(k)}}{\partial \xi^{(0)}} & \frac{\partial \eta^{(k)}}{\partial \eta^{(0)}}
\end{pmatrix} = P^k
\tag{21}
$$

From the relations (16)-(17) that define the symmetry points of the map, one can deduce the general statement

$$
x(\Xi^{(k)}, m) = x(\Xi^{(k-1)}, m+1) = x(\Xi^{(k-2)}, m+2)
$$

for $k, m = 0, 1, 2$. Differentiation of these relationships with respect to $\xi^{(0)}$ and $\eta^{(0)}$ and applying (21) yields

$$
A_m(\Xi^{(k)}) P^k = A_{m+r}(\Xi^{(k-r)}) P^{k-r}
$$

Simplifying,

$$
A_m(\Xi^{(k)}) = A_{m+r}(\Xi^{(k-r)}) P^{-r}
\tag{22}
$$

Now let $r = -m$ to obtain the result.   §

4.4   Label-Invariance

Let $\alpha_m = det(A_m)$. From Proposition 2, it is immediate that

$$
\alpha_m\left(\Xi^{(r)}\right) = \alpha_0\left(\Xi^{(r+m)}\right)
\tag{23}
$$

$$
T_m\left(\Xi^{(r)}\right) W\left(\Xi^{(r)}\right) = T_0\left(\Xi^{(r+m)}\right) W\left(\Xi^{(r+m)}\right) P^m
\tag{24}
$$

$$
\tau_m\left(\Xi^{(r)}\right) \omega\left(\Xi^{(r)}\right) = \tau_0\left(\Xi^{(r+m)}\right) \omega\left(\Xi^{(r+m)}\right)
\tag{25}
$$

for any W. The relation $A_{m+1}(\Xi) = A_m(\Xi) P$ that held for the linear map does not hold for the quadratic map at arbitrary Ξ. As a consequence, the local metrics of interest are not label-invariant for arbitrary Ξ as they were in the linear case.

Proposition 3. Let $\Xi^c = \left(\frac{1}{3}, \frac{1}{3}\right)$. Then the metrics $\mu(T) = \tau$ and $\mu(T) = |T|$ are label-invariant at Ξ^c, the first for arbitrary reference element and the second for an equilateral element.

Proof. When $\Xi = \Xi^c$, the three symmetry points are all equal to Ξ^c. Then (25) becomes

$$
\tau_m\left(\Xi^{(c)}\right) = \tau_0\left(\Xi^{(c)}\right)
\tag{26}
$$

Therefore, the metric $\mu(T) = \tau$ is label-invariant. Similarly, (24) becomes

$$T_m\,(\Xi^{(c)})\,W\,(\Xi^{(c)}) = T_0\,(\Xi^{(c)})\,W\,(\Xi^{(c)})\,P^m \tag{27}$$

Therefore,

$$|\,T_m\,(\Xi^{(c)})\,| = |\,T_0\,(\Xi^{(c)})\,W\,(\Xi^{(c)})\,P^m\,[\,W\,(\Xi^{(c)})\,]^{-1}\,| \tag{28}$$

But one can show that when the reference element is equilateral, $W(\Xi^c) \in \mathcal{M}$, so, as was shown in Proposition 1, $W\,P^m\,W^{-1}$ is a rotation. Using the Frobenius invariance property, we have

$$|\,T_m\,(\Xi^{(c)})\,| = |\,T_0\,(\Xi^{(c)})\,| \tag{29}$$

and thus the local metric $|T|$ is label-invariant.   §

A similar proof can be constructed to show that $\mu(T) = |T^t T|$ is also label-invariant at Ξ^c, provided the reference element is equilateral.

Corollary. Metrics $\mu(T)$ that involve combinations of τ, $|T|$, and/or $T^t T$ are label-invariant at $\Xi = \Xi^c$, provided $W(\Xi^c) \in \mathcal{M}$. For example, the Mean Ratio metric.

For the linear triangle map, the local metrics τ, $|T|$, and $|T^t T|$ were label-invariant for any Ξ because the Jacobian matrices were constant. Therefore, there was no need to consider symmetry metrics. Since the Jacobians vary with Ξ in the quadratic case, it is necessary to investigate symmetry metrics.

Proposition 4. If the local metric satisfies $\mu_{m+r}\,(\Xi^{(s)}) = \mu_m\,(\Xi^{(r+s)})$ for a particular reference element, then σ_m in (7) with $N = 3$ is label-invariant.

Proof

$$\sigma_{m+r} = \max\{\,\mu_{m+r}(\Xi^{(0)}),\ \mu_{m+r}(\Xi^{(1)}),\ \mu_{m+r}(\Xi^{(2)})\,\} \tag{30}$$
$$= \max\{\,\mu_m(\Xi^{(r)}),\ \mu_m(\Xi^{(r+1)}),\ \mu_m(\Xi^{(r+2)})\,\} \tag{31}$$

For any choice of r, we have $\sigma_{m+r} = \sigma_m$, and thus this symmetry metric is label-invariant.   §

Corollary. When the local metric is $\mu(T) = \tau$ then σ_m is label-invariant, provided the reference triangle has straight sides.

Proof. When $k - r = s$, the relation (22) becomes

$$A^{(s)}_{m+r} = A^{(r+s)}_m\,P^r$$

from which one obtains

$$T^{(s)}_{m+r}\,W^{(s)} = T^{(r+s)}_m\,W^{(r+s)}\,P^r$$

and so

$$\tau_{m+r}^{(s)}\,\omega^{(s)} = \tau_m^{(r+s)}\,\omega^{(r+s)}$$

When the reference element has straight sides, $det(W)$ is constant, the previous becomes $\tau_{m+r}^{(s)} = \tau_m^{(r+s)}$ and therefore $\mu_{m+r}^{(s)} = \mu_m^{(r+s)}$. Thus the assumption of Proposition 4 is satisfied.   §

Corollary. When the local metric is $\mu(T) = |T|$ then σ_m is label-invariant, provided the reference triangle is equilateral (with straight sides).

Proof. From the previous corollary

$$T_{m+r}\left(\Xi^{(s)}\right) = T_m\left(\Xi^{(r+s)}\right) W^{(r+w)}\, P^r\,[\,W^{(s)}\,]^{-1} \tag{32}$$

When the reference element is equilateral, it has straight sides, and then W is independent of Ξ. Moreover, $W\,P^r\,W^{-1}$ is a rotation. Taking the norm of both sides of the above relation and using the rotation-invariance property of the Frobenius norm shows that the assumption of Proposition 4 is satisfied. §

One can similarly show that the symmetry metric σ_m based on the local metric $\mu(T) = |T^t T|$ is label-invariant provided the reference element is equilateral.

Corollary. When the local metric is $\mu(T) = |T|^2/2\tau$ then σ_m is label-invariant, provided the reference triangle is equilateral.

Proof. The previous corollaries showed $\tau_{m+r}^{(s)} = \tau_m^{(r+s)}$ and $|T_{m+r}(\Xi^{(s)})| = |T_m(\Xi^{(r+s)})|$. From this, one can readily see that the assumption of Proposition 4 is satisfied.   §

More generally, any local metric that is a combination of τ and $|T|$ can be used to form a label-invariant symmetry metric provided the reference triangle is equilateral.

The local metric $\mu(T) = tr(T)$ is never label-invariant, nor is its associated symmetry metric label-invariant.

The results of Proposition 4 and its corollaries apply equally well to symmetry metrics based on the minimum (6) or the power mean (5) instead of the maximum (7).

If $\mu_{m+r}\left(\Xi^{(s)}\right) = \mu_m\left(\Xi^{(r+s)}\right)$ for a particular choice of reference element, then for $s = 0$ we have $\mu_{m+r}\left(\Xi^{(0)}\right) = \mu_m\left(\Xi^{(r)}\right)$. Then one can show that, for example, (7) becomes

$$\sigma_m = \max\{\,\mu_0\left(\Xi^{(0)}\right), \mu_1\left(\Xi^{(0)}\right), \mu_2\left(\Xi^{(0)}\right)\} \tag{33}$$

This directly shows that the maximum-symmetry metric is label-invariant for a particular choice of reference element, and that one can evaluate it either by fixing the map and evaluating the local metrics at the three symmetry points, or, by varying the map and evaluating the local metric at the first symmetry point. The same is true for the minimum-symmetry and power-symmetry metrics.

4.5 The Shape Quality of Quadratic Triangles

As noted previously, [18] is the only reference which proposes quality metrics for quadratic elements. In theory one might also create an *element* metric based on some local metric μ as in the following example:

$$q_\varepsilon = \max_{\Xi \in U} \{\, \mu(\Xi) \,\}$$

Unfortunately, the definition is impractical to compute efficiently due to the infinite number of points at which μ must be evaluated. As a practical alternative, consider using one or more sets of symmetry metrics. For example, in the quadratic triangle case let $\mathcal{S}_v$ consist of the symmetry points $\{\, \Xi^{(0)},\ \Xi^{(1)},\ \Xi^{(2)} \,\}$ when $\Xi = (0,0)$, i.e., $\mathcal{S}_v = \{\, (0,0),\ (1,0),\ (0,1) \,\}$. Likewise, let $\mathcal{S}_n = \{\, (\frac{1}{2},\frac{1}{2}),\ (0,\frac{1}{2})\ (\frac{1}{2},0) \,\}$, obtained when $\Xi = (\frac{1}{2},\frac{1}{2})$. Now define three symmetry metrics

$$\sigma_v = \max_{\Xi \in \mathcal{S}_v} \{\, \mu(\Xi) \,\} \tag{34}$$

$$\sigma_n = \max_{\Xi \in \mathcal{S}_n} \{\, \mu(\Xi) \,\} \tag{35}$$

$$\sigma_{v+n} = \max_{\Xi \in (\mathcal{S}_v \cup \mathcal{S}_n)} \{\, \mu(\Xi) \,\} \tag{36}$$

In words, the first symmetry metric evaluates the local metric at only the three corner vertices of the element, the second at only the three mid-side nodes, and the third on both sets.

To illustrate, the three metrics σ_v, σ_n, and σ_{v+n} were computed for ten thousand randomly generated quadratic triangles. The local metric μ was chosen to be the inverse mean ratio (shape) metric with an equilateral reference element. Each of these values was compared to the value q_ε, which was approximated by evaluating μ on 1250 points uniformly distributed over the logical triangle. Figure 4 compares the pairs $(\sigma_v, q_\varepsilon)$, $(\sigma_n, q_\varepsilon)$, and $(\sigma_{v+n}, q_\varepsilon)$ in three scatter plots whose range on both the x- and y-axes is -1 to 1 (1 being the best quality). Since for any triangle, $\sigma_v \leq q_\varepsilon$ (and likewise for the other cases), the upper left side of each plot is empty. Sampling at only the mid-side nodes is the least effective of the three cases. As one can see, for some triangles even σ_{v+n} can be a poor approximation to q_ε.

On the positive side, it appears from the σ_{v+n} plot that the approximation to q_ε improves as the quality of the triangle improves. For example, there are relatively few points found below the $\sigma_{v+n} = q_\varepsilon$ line when the quality is better than 0.0, whereas there are a lot of points below the line when the quality is less than 0.0. That means for non-inverted elements, the approximation to element shape quality is not too bad in most instances. Further investigation is required in order to determine whether or this observation holds for other metrics and/or element types, but it is encouraging, at least. In any case, this third approach to measuring the quality of non-triangular quadratic elements is the only known approach.

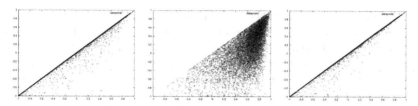

Fig. 4. Quadratic Planar Tri: σ_v, σ_n, σ_{v+n} vs. q_ε. (Left) Three Corner Vertices, (Middle) Three Mid-side Nodes, (Right) Six Vertices

Finally, we close this section with a proposition that applies to symmetry metrics for any map and further, makes it clear that the symmetry metric σ_{v+n} discussed in this section is label-invariant.

Proposition 5. Let $\mathcal{S}(\Xi) = \{\, \Xi^{(0)}, \Xi^{(1)}, \ldots, \Xi^{(N)} \,\}$ be a set of symmetry points. Let $\Xi_1 \in U$ and $\Xi_2 \in U$ be two values of Ξ, and further let $\mathcal{S}_1 = \mathcal{S}(\Xi_1)$ and $\mathcal{S}_2 = \mathcal{S}(\Xi_2)$. Let $\sigma_m(\Xi)$ be defined as one of (5)-(7). Suppose both $\sigma_m(\Xi_1)$ and $\sigma_m(\Xi_2)$ are label-invariant. Then σ_m evaluated on $\mathcal{S}_1 \cup \mathcal{S}_2$ is label-invariant.

Proof. The proof is constructed for the case where the symmetry metric is based on the maximum function. Similar proofs can be given for the other cases. Then

$$\sigma_m(\Xi_1) = \max_{\Xi^{(s)} \in \mathcal{S}_1} \{\, \mu_m(\Xi^{(s)}) \,\} \tag{37}$$

$$\sigma_m(\Xi_2) = \max_{\Xi^{(s)} \in \mathcal{S}_2} \{\, \mu_m(\Xi^{(s)}) \,\} \tag{38}$$

are label-invariant. Also define

$$\sigma_m(\Xi_1, \Xi_2) = \max_{\Xi^{(s)} \in (\mathcal{S}_1 \cup \mathcal{S}_2)} \{\, \mu_m(\Xi^{(s)}) \,\} \tag{39}$$

Therefore

$$\sigma_m(\Xi_1, \Xi_2) = \max\{\, \sigma_m(\Xi_1), \sigma_m(\Xi_2) \,\} \tag{40}$$

and

$$\begin{aligned} \sigma_{m+r}(\Xi_1, \Xi_2) &= \max\{\, \sigma_{m+r}(\Xi_1), \sigma_{m+r}(\Xi_2) \,\} \\ &= \max\{\, \sigma_m(\Xi_1), \sigma_m(\Xi_2) \,\} \\ &= \sigma_m(\Xi_1, \Xi_2) \end{aligned} \tag{41}$$

§

5   Summary

Quality measurement within mesh elements can be achieved using local metrics such as those given in the Target-matrix paradigm, along with a map from the logical to the reference and physical elements. Local area, volume, shape, and orientation can thus be measured with respect to the same local quantities within the reference element. This provides a method for assessing the quality of elements having curved sides, such as those associated with the quadratic map. The minimum or maximum value of these quantities over all points in the element can be approximated by taking local measurements on a small, finite, set of points. For elements whose quality is not too bad (e.g., non-inverted), the approximations appear reasonably good, as seen in the quadratic triangle example. In any case, this 'hybrid' quality method is the only known approach to measuring the quality of non-triangular high-order elements.

Label-invariance is a desirable property for some quality metrics. Local metrics can be made label invariant by evaluating them at the center of the element and using a particular reference element. For shape metrics, the appropriate reference element for label-invariance was the regular shape corresponding to the given element type. For size metrics, the reference element was arbitrary. Metrics that are sensitive to orientation, such as those involving the trace, are not label-invariant for any choice of reference element.

Another type of label-invariance can be obtained by defining a symmetry metric, based on an associated local metric and a set of symmetry points that differs from one element type to another. As in the local metric case, the symmetry metrics can be made label invariant provided the reference element is regular. An advantage of the symmetry metrics is that one does not have to evaluate the local metric at the center of the element in order to obtain label-invariance. This is important because, for example, quality at the corners can often provide a more discerning criterion than quality at the center.

Similar results for the linear and quadratic tetrahedron, and the linear planar quadrilateral, are given in [13]. It is expected the similar results would hold for the quadratic quadrilateral and for the linear and quadratic hexahedral elements. Pyramid and prismatic elements are not naturally isotropic, so probably local metrics on these cannot be made label-invariant. Non-planar quadrilaterals and non-planar quadratic triangles have not been investigated for label-invariance.

References

1. Branets, L., Carey, G.: Extension of a Mesh Quality Metric for Elements With a Curved Boundary Edge or Surface. J. Comput. Inf. Sci. Eng. 5(4), 302–309 (2005)
2. Dompierre, J., Labbe, P., Guibault, F., Camerero, R.: Proposal of Benchmarks for 3D Unstructured Tetrahedral Mesh Optimization. In: Proceedings of the 7-th International Meshing Roundtable 1998, Dearborn MI, pp. 459–478 (1998)
3. Field, D.A.: Qualitative measures for initial meshes. Intl. J. for Num. Meth. Engr. 47, 887–906 (2000)

4. Freitag, L., Knupp, P.: Tetrahedral Element Shape Optimization via the Jacobian Determinant and Condition Number. In: Proceedings of the 8th International Meshing Roundtable, Lake Tahoe CA, pp. 247–258 (1999)
5. Freitag, L., Knupp, P.: Tetrahedral mesh improvement via optimization of the element condition number. Intl. J. Numer. Meth. Engr. 53, 1377–1391 (2002)
6. Knupp, P.: Remarks on Mesh Quality, 45th AIAA Sciences Meeting and Exhibit, Reno NV (January 7-10, 2007)
7. Knupp, P.: Algebraic Mesh Quality Measures. SIAM J. Sci. Comput. 23(1), 193–218 (2001)
8. Knupp, P.: Algebraic Mesh Quality Measures for Unstructured Initial Meshes. Finite Elements in Design and Analysis 39, 217–241 (2003)
9. Knupp, P.: Hexahedral and Tetrahedral Mesh Shape Optimization. Intl. J. Numer. Meth. Engr. 58, 319–332 (2003)
10. Knupp, P.: Formulation of a Target-Matrix Paradigm for Mesh Optimization, SAND2006-2730J, Sandia National Laboratories (2006)
11. Knupp, P., Hetmaniuk, U.: Local 2D Metrics for Mesh Optimization in the Target-Matrix Paradigm, SAND2006-7382J, Sandia Nat'l. Laboratories (2006)
12. Knupp, P.: Analysis of 2D, Rotation-invariant, Non-Barrier Metrics in the Target-matrix Paradigm, SAND2008-8219P, Sandia Nat'l. Laboratories (2008)
13. Knupp, P.: Measuring Quality within Mesh Elements, SAND2009-3081J, Sandia National Laboratories (2009)
14. Kwok, W., Chen, Z.: A simple and effective mesh quality metric for hexahedral and wedge elements. In: Proceedings of the 9th International Meshing Roundtable, New Orleans LA, pp. 325–333 (2000)
15. Parasarathy, V.N., et al.: A Comparison of tetrahedron quality measures. Finite Elem. Anal. Des. 15, 255–261 (1993)
16. Pebay, P.: Planar quadrangle quality measures. Engr. w/Comp. 20(2), 157–173 (2004)
17. Robinson, J.: CRE method of element testing and the Jacobian shape parameters. Eng. Comput. 4, 113–118 (1987)
18. Salem, A., Canann, S., Saigal, S.: Robust Distortion Metric for Quadratic Triangular 2D Finite Elements. Trends in Unstructured Mesh Generation, vol. AMD-220, pp. 73–80. ASME (1997)
19. Stimpson, C., Ernst, C., Knupp, P., Pebay, P., Thompson, D.: The Verdict Geometric Quality Library, SAND2007-1751. Sandia National Laboratories (2007)

Perturbing Slivers in 3D Delaunay Meshes

Jane Tournois[1], Rahul Srinivasan[2], and Pierre Alliez[1]

INRIA Sophia Antipolis - Méditerranée, France
[1]firstname.lastname@sophia.inria.fr
[2]firstname.lastname@iitb.ac.in

Abstract. Isotropic tetrahedron meshes generated by Delaunay refinement algorithms are known to contain a majority of well-shaped tetrahedra, as well as spurious sliver tetrahedra. As the slivers hamper stability of numerical simulations we aim at removing them while keeping the triangulation Delaunay for simplicity. The solution which explicitly perturbs the slivers through random vertex relocation and Delaunay connectivity update is very effective but slow. In this paper we present a perturbation algorithm which favors deterministic over random perturbation. The added value is an improved efficiency and effectiveness. Our experimental study applies the proposed algorithm to meshes obtained by Delaunay refinement as well as to carefully optimized meshes.

1 Introduction

Delaunay refinement algorithms [9, 26, 23, 25] have been extensively studied in the literature. They are amenable to analysis, and hence are reliable algorithms. In addition, the robust implementations of Delaunay triangulations which are now available greatly facilitate the implementation of Delaunay-based mesh refinement algorithms. However, most Delaunay refinement algorithms fail at removing all badly-shaped tetrahedra, and a special class of almost-flat tetrahedra (so-called slivers) may remain in the triangulation. These slivers, with dihedral angles close to 0 and to π, are problematic for many numerical simulations.

1.1 Slivers

Many finite element methods require discretizing a domain into a set of tetrahedra. These applications require more than just a triangulation of the domain for simulation and rendering. The accuracy and the convergence of these methods depend on the size and shape of the elements apart from the fact that the mesh should conform to the domain boundary [28]. Both the bad quality and the large number of the mesh elements can negatively affect the execution of a simulation. It is required that all elements of the mesh are well-shaped as the accuracy of the simulations and computations can be compromised by the presence of even a single badly shaped element. In general it is desirable to bound the smallest dihedral angle in the mesh, from below. The Delaunay refinement technique guarantees a bound on the radius-edge ratio of all mesh elements, which is the ratio of the circumradius to the shortest edge length of a tetrahedron. Although in 2D this translates into a lower bound on the minimum angle in the mesh, in 3D it does not: a bound on the radius-edge ratio is not equivalent to a bound on the smallest dihedral angle.

The only bad elements that remain after Delaunay refinement are slivers. A sliver tetrahedron is formed by almost evenly placing its 4 vertices near the equator of its circumsphere (see Figure 1), and has a bounded radius-edge ratio. In such a sliver the smallest dihedral angle can be very close to $0°$, and a numerical simulation may be far from accurate in the presence of slivers.

1.2 Tetrahedron Quality

Several tetrahedron quality criteria have been defined and used in the literature depending on the application. The radius edge ratio ρ of a simplex is defined as the ratio of its circumradius to the length of the shortest edge. This measure, which is minimal for the regular tetrahedron, unfortunately cannot detect slivers, though it is used in Delaunay refinement algorithms to define bad simplices. The radius ratio, defined as the ratio of the inradius (insphere radius) to the circumradius (circumsphere radius), is another popular measure of tetrahedron quality. It is desired to ensure that radius ratio of all tetrahedra are bounded from below by a constant.

Another criterion for mesh generation is the minimum dihedral angle θ_{min}. It can be shown that a lower bound on the radius ratio is equivalent to a lower bound on the minimum dihedral angle. In the sequel we choose this measure to evaluate the mesh quality as it is more intuitive and geometrically meaningful than, e.g., the radius ratio, which combines the six dihedral angles of a tetrahedron.

Consider an arbitrary tetrahedron τ with triangular faces T_1, T_2, T_3, T_4. Let the areas of these triangles be denoted by S_1, S_2, S_3, S_4 respectively, the dihedral angle between T_i and T_j by θ_{ij} and the length of the edge shared by T_i and T_j by l_{ij}. The volume V of τ is given by

$$V = \frac{2}{3l_{ij}} S_i S_j sin\theta_{ij} \quad \text{for } i \neq j \text{ in } \{1, 2, 3, 4\}. \tag{1}$$

Let r_C, r_I be the circumradius and inradius of τ and r_i be the circumradius of T_i for i in $\{1, 2, 3, 4\}$. We know that for any tetrahedron, $r_i \leq r_C$. This gives $S_i \leq \pi r_i^2 \leq \pi r_c^2$, and we also have a bound on the volume $V \geq \frac{4}{3} \pi r_I^3$.

Using Equation 1, for $i \neq j$ we have

$$\frac{4}{3}\pi r_I^3 \leq \frac{2}{3l_{ij}}S_i S_j sin\theta_{ij} \leq \frac{2}{3l_{ij}}\pi^2 r_c^4 sin\theta_{ij},$$

and we get

$$sin\theta_{ij} \geq \frac{2}{\pi} \cdot \frac{r_I^3}{r_c^3} \frac{l_{ij}}{r_C} \geq \frac{2}{\pi} \cdot \frac{a_0^3}{\rho_0},$$

where a_0 is the radius-radius ratio and ρ_0 is the radius-edge ratio. Finally,

$$\theta_{ij} \geq sin^{-1}\left(\frac{2}{\pi} \cdot \frac{a_0^3}{\rho_0}\right).$$

Li [21] uses a different parameter of tetrahedron quality to define a sliver. Denote the volume of tetrahedron $pqrs$ by V and its shortest edge length by l. The volume per cube of shortest edge length ($\sigma = \frac{V}{l^3}$) is used as a measure of the shape quality along with the radius-edge ratio ρ, or on its own [7]. According to Li, a tetrahedron $pqrs$ is called sliver if $\rho(pqrs) \leq \rho_0$ and $\sigma(pqrs) \leq \sigma_0$, where ρ_0 and σ_0 are constant.

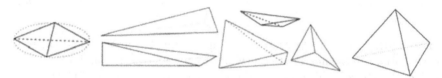

Fig. 1. Tetrahedron shapes. A sliver (left) has its four vertices close to a circle, four very small dihedral angles (close to 0°), and two very large (close to 180°). A regular tetrahedron (right) is well shaped and has its dihedral angles close to 70.5°. Each of the other tetrahedra (middle) present a different type of degeneracy.

1.3   Previous Work

The problem of removing slivers from a 3D Delaunay mesh has received some attention over the last decade. Delaunay refinement gets so close to providing a perfect output that removing the leftover slivers is generally performed as a post-processing step that is worth it. Previous work on removing and avoiding the creation of slivers can be classified into three parts: The Delaunay-based methods, the weighted Delaunay-based methods, and the non-Delaunay methods. For each part, post-processing steps and complete mesh generation algorithms can be studied. This paper focuses on a post-processing step, devised to take as input a Delaunay mesh and to improve its quality in terms of dihedral angles.

Delaunay-based

Vertex Perturbation

Li [21, 14] proposes to explicitly perturb the vertices incident to a sliver in an almost-good mesh, by locally relocating them so as to remove the incident slivers. The idea is based on the fact that, for any triangle qrs, the region of locations of the vertex p such that the tetrahedron $pqrs$ is a sliver, is very small. Moving the point p out of this region ensures that the tetrahedron is not a sliver anymore, or has disappeared once the Delaunay connectivity is updated. This is achieved by moving the point p to a new location inside a small ball centered at p, whose radius is proportional to the distance from p to its the nearest neighbor. The author shows that for certain values of the involved parameters, there always exists some points in this ball which are outside all regions that form slivers with nearby triangles. Li uses the union graph concept to avoid circular dependencies on vertex perturbations. The following theorem [21] proves the existence of such a point that makes the mesh locally sliver-free.

Theorem 1 (Sliver theorem). *If every simplex in a Delaunay triangulation has radius-edge ratio of at least ρ_0, then there is a constant $\sigma_0 > 0$ and a very mild perturbation S' with $\sigma(\tau) \geq \sigma_0$ for each tetrahedron τ in the perturbed triangulation.*

Based on this theorem, Li proposes an algorithm that applies mild random perturbations to the mesh until one which removes slivers is found. One drawback of the above result is the pessimistic theoretical estimate of the bounds on the involved parameters. These bounds are either too small or too large to have any significance. In practice, though this technique is very effective, when targeting a large bound on the minimum dihedral angle (e.g. $15°$), the average number of trials of random perturbations required is very large. In our experiments, it is not rare to apply hundreds of random perturbation trials on a single vertex before succeeding in removing a sliver. This number is not surprising when seeking a high minimum dihedral angle such as $15°$ since the corresponding tetrahedron is not a real sliver anymore. However the fact that the perturbation succeeds even for a high minimum dihedral angle is at the core of our motivation. Finally, the fact that this method always maintains the mesh as a true Delaunay triangulation makes it both robust and practical.

Sliver-free mesh generation

Some mesh generation algorithms are designed to avoid creating slivers. For example, Delaunay refinement can be modified by choosing a new type of Steiner point which does not create any sliver [22, 23, 24]. As an example, Chew's algorithm [9] inserts Steiner points in a randomized manner, to avoid the creation of slivers. This method has a theoretical lower bound of $\arcsin 1/4 \approx 14.5°$ on the angles of the triangular faces of the mesh.

Weighted Delaunay-based

Sliver exudation

First described by Cheng et al. [7], sliver exudation is a technique based on turning a Delaunay triangulation into a weighted Delaunay triangulation [3], devised to trigger flips so as to increase the minimal angle. Edelsbrunner and Guoy [13] provide an experimental study of sliver exudation, and show that it works pretty well in practice as a post-treatment applied to a triangulation obtained by Delaunay refinement [25]. The main strategy of the algorithm consists of assigning a weight to each vertex so that the weighted Delaunay triangulation is free of any slivers after connectivity updates, without any changes over the vertex locations. This method successfully increases all dihedral angles above 5° in the best configuration (see Section 3), but as admitted in [13], the theoretical bound on the dihedral angle is too small to be of any practical significance.

Beside being not strictly Delaunay anymore, the main disadvantage of sliver exudation is that the process often ends with leftover slivers near the boundary [13]. This is mainly due to the fact that sliver exudation is not allowed to modify the topology of the boundary of the mesh. Hence, weight assignments close to the boundary are constrained and do not always manage to remove the slivers.

Complete algorithm

Cheng and Dey [6] propose a complete Delaunay refinement algorithm, combined with the sliver exudation technique. This type of weighted-Delaunay algorithm is also used to handle input domains containing sharp creases subtending small angles [8].

Non-Delaunay

Local combinatorial operations

Though a Delaunay-refined triangulation is known to have nice properties on its angles in 2D [12], there is no theoretical guarantee on the dihedral angles in 3D. One valid choice consists of leaving the Delaunay framework by flipping some well-chosen simplices [27, 18], either as a post-processing step to the meshing process [19], or during the whole process [10]. As long as the triangulation remains valid, flips can be performed on its edges and facets. Joe gives a description of all possible flips [17] that can be made in a triangulation, and a triangulation improvement algorithm through these flips. Although each improvement in this algorithm is local, the complete algorithm succeeds in improving the overall quality of the mesh.

Dealing with non-Delaunay meshes can also be combined with optimization steps, such as Laplacian smoothing [15], which relocates each vertex to a new location computed as an average of the incident vertex positions. Laplacian smoothing can be applied to any valid triangulation.

Complete algorithm

Some other types of triangulations, such as for example max-min solid angle triangulations [16] can be computed to improve the solid angles as compared to that in a Delaunay triangulation. This method generates a set of well-distributed points in the input polyhedral domain and first computes a Delaunay triangulation of these vertices. Then, local combinatorial transformations are applied to satisfy the local max-min angle criterion. These local transformations can in fact be applied to any triangulation as a postprocessing step.

Instead of performing local improvements through flips in a Delaunay mesh, Labelle and Shewchuck [20] propose a fast lattice refinement technique which constructs a triangulation based on two nested regular or adapted grids. In its graded version this algorithm provides a theoretical bound on the dihedral angles which is much more practical than provided by other algorithms.

1.4   Contribution

We present a sliver removal algorithm inspired by Li's random perturbation algorithm [21]. Our algorithm is made more deterministic by choosing a favored perturbation direction for each vertex incident to one or more slivers,

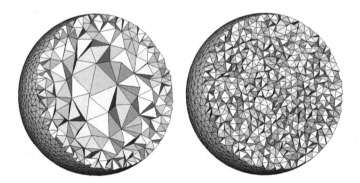

Fig. 2. Sphere. (Left) Graded mesh with 3195 vertices, output angles are in [23.5; 142.5]. (Right) Uniform mesh with 7041 vertices, output angles are in [30.02; 138.03].

before resorting to Li's random perturbation if the favored perturbation fails at removing the incident slivers. Our experiments show that the chosen deterministic directions are sufficient to remove more than 80% of the slivers of a mesh, leading to shorter computational times. In addition, our approach reaches higher minimum dihedral angles in practice.

2 Algorithm

We describe a sliver perturbation algorithm which improves in a hill-climbing manner the dihedral angles of an input isotropic Delaunay mesh. This algorithm can be used as a post-processing step after refinement or optimization.

To improve the dihedral angles of the mesh tetrahedra, the rationale behind our approach is as follows: each vertex v incident to at least one sliver is repeatedly relocated through a perturbation vector $\mathbf{p_v}$ such that when v moves to $v + \mathbf{p_v}$, the incident slivers get flipped. More specifically, the chosen direction for $\mathbf{p_v}$ is not devised to improve the shape of the slivers, but rather to worsen them instead, so that they get flipped. Two directions are favored by the algorithm: the incident squared circumradius gradient ascent (see Section 2.1) and the sliver volume gradient descent (see Section 2.2). The length of the perturbation vector is heuristically chosen as a fraction (usually between 0.05 and 0.2) of the minimum incident edge length. If neither of these two perturbation vectors succeed in flipping a sliver we resort to random perturbations (see Section 2.3). If the whole sequence does not improve the local minimum dihedral angle then we restore the vertex to its original location before perturbation.

By construction, our combined perturbation algorithm is hill-climbing in the sense that the dihedral angles in the output mesh must be higher than the ones in the input mesh. The theoretical proofs of Li's method [21] concerning random perturbation apply to this combined perturbation method as we resort to it in case of failure of the deterministic perturbation.

When more than one sliver is incident to a vertex v, all perturbation vectors must be compatible (i.e., pushing in a similar direction) to be effective. In our current algorithm, a set of perturbation vectors are said to be compatible if all their pairwise dot products are positive. The perturbation vector $\mathbf{p_v}$ is then set to be the average of these vectors. When not compatible, v is perturbed only using random perturbations. The algorithm relies on a modifiable priority queue, built in a way such that vertices incident to fewer slivers are processed first. Hence, any "chain" of slivers (set of slivers sharing at least one vertex) is treated starting from its endpoints thereby minimizing the need to process vertices incident to more than one sliver.

Algorithm 1. Sliver perturbation

Input: $\mathcal{T}$: *a Delaunay triangulation,*
   α: *the angle bound defining slivers, and*
   N_{max}: *the maximum number of random trials, or gradient steps.*

Let $\mathcal{P}$ be a priority queue of Delaunay vertices.
Fill $\mathcal{P}$ with vertices incident to slivers,
Compute perturbation vector $\mathbf{p_v}$ for each vertex v in $\mathcal{P}$,

while $\mathcal{P}$ non-empty **do**
   Pop v from $\mathcal{P}$,
   $v' \leftarrow v$,
   while relocating v to v' would not trigger a combinatorial change,
   and #*loops* $< N_{max}$ **do**
      $v' \leftarrow v' + \mathbf{p_v}$,
      if $\mathbf{p_v}$ is random, **then**
         compute a new $\mathbf{p_v}$,
         $v' \leftarrow v$.
      end if
   end while

   Conditionally relocate v to v'.

   if v is still incident to slivers and $\mathbf{p_v}$ is not random, **then**
      Compute a new perturbation vector (another type, if possible),
      and re-insert v into $\mathcal{P}$.
   end if

   Insert all vertices affected by relocation into $\mathcal{P}$,
   with their new perturbation vector.
end while

Note that each vertex relocation is conditional, as we want our algorithm to be hill-climbing in terms of dihedral angles. We need to check that the minimum dihedral angle of the triangulation does not decrease, and that the topology of the boundary is not affected. Otherwise, the relocation is canceled.

Each time a vertex is effectively relocated, the priority queue is updated. Moving v to v' in a Delaunay mesh makes combinatorial changes (and, hence, changes on incident dihedral angles) on the vertices incident to v before its removal, and the ones incident to v' after its insertion. We first compute the perturbations associated with all these vertices, and insert them into the priority queue.

The order in which the vertices are processed in the priority queue is related to the vertex type. Interior vertices are processed first, since they are more likely to be easily perturbable than boundary vertices. The boundary vertices are constrained to be located on the boundary, and their move must not break the topology of the mesh. These constraints make them more difficult to perturb. The other ordering criteria are discussed in Section 3.

2.1 Circumsphere Radius

In an almost-good isotropic tetrahedron mesh, the distribution of the mesh vertices is locally uniform. Hence, perturbing the vertex locations so as to make the radius of the sliver's circumsphere explode triggers many flips as the empty circumsphere property must hold after Delaunay connectivity update.

Let τ be the sliver, and $\{p_i\}_{i=0,1,2,3}$ its vertices. Without loss of generality, and since the sequel remains true by translation, we can assume that $p_0 = 0_{\mathbb{R}^3}$. We also assume that this vertex is fixed. Let c be τ's circumcenter. We have $||c|| = R$ the radius of τ's circumsphere. Then, $\nabla R^2 = \nabla ||c||^2$. We aim at computing ∇R^2.

Let $p_i = (x_i, y_i, z_i)$ for i in $\{1, 2, 3\}$ be τ's vertices, with $p_0 = 0_{\mathbb{R}^3}$. Also, let p_i^2 be $(x_i^2 + y_i^2 + z_i^2)$. The center c of the circumsphere of τ is given by

$$c = \begin{pmatrix} x_c \\ y_c \\ z_c \end{pmatrix} = \begin{pmatrix} \frac{D_x}{2a} \\ \frac{D_y}{2a} \\ \frac{D_z}{2a} \end{pmatrix}, \text{ where}$$

$$a = \begin{vmatrix} x_1 & y_1 & z_1 \\ x_2 & y_2 & z_2 \\ x_3 & y_3 & z_3 \end{vmatrix}, D_x = -\begin{vmatrix} p_1^2 & y_1 & z_1 \\ p_2^2 & y_2 & z_2 \\ p_3^2 & y_3 & z_3 \end{vmatrix}, D_y = +\begin{vmatrix} p_1^2 & x_1 & z_1 \\ p_2^2 & x_2 & z_2 \\ p_3^2 & x_3 & z_3 \end{vmatrix}, \text{ and } D_z = -\begin{vmatrix} p_1^2 & x_1 & y_1 \\ p_2^2 & x_2 & y_2 \\ p_3^2 & x_3 & y_3 \end{vmatrix}.$$

Thus we have, $\nabla_{p_1} ||c||^2 = \begin{pmatrix} \frac{\partial ||c||^2}{\partial x_1} \\ \frac{\partial ||c||^2}{\partial y_1} \\ \frac{\partial ||c||^2}{\partial z_1} \end{pmatrix}$ with

$$\frac{\partial ||c||^2}{\partial x_1} = \frac{\partial}{\partial x_1}\left(\frac{D_x^2 + D_y^2 + D_z^2}{4a^2}\right)$$

$$= \frac{1}{2a^3}\cdot\left(a\cdot(D_x\cdot\frac{\partial D_x}{\partial x_1} + D_y\cdot\frac{\partial D_y}{\partial x_1} + D_z\cdot\frac{\partial D_z}{\partial x_1}) - \frac{\partial a}{\partial x_1}\cdot(D_x^2 + D_y^2 + D_z^2)\right),$$

$$\frac{\partial ||c||^2}{\partial y_1} = \frac{1}{2a^3}\cdot\left(a\cdot(D_x\cdot\frac{\partial D_x}{\partial y_1} + D_y\cdot\frac{\partial D_y}{\partial y_1} + D_z\cdot\frac{\partial D_z}{\partial y_1}) - \frac{\partial a}{\partial y_1}\cdot(D_x^2 + D_y^2 + D_z^2)\right),$$

$$\frac{\partial ||c||^2}{\partial z_1} = \frac{1}{2a^3}\cdot\left(a\cdot(D_x\cdot\frac{\partial D_x}{\partial z_1} + D_y\cdot\frac{\partial D_y}{\partial z_1} + D_z\cdot\frac{\partial D_z}{\partial z_1}) - \frac{\partial a}{\partial z_1}\cdot(D_x^2 + D_y^2 + D_z^2)\right).$$

and

$$\nabla_{p_1} a = \begin{pmatrix} \frac{\partial a}{\partial x_1} \\ \frac{\partial a}{\partial y_1} \\ \frac{\partial a}{\partial z_1} \end{pmatrix} = \begin{pmatrix} y_2 z_3 - y_3 z_2 \\ -(x_2 z_3 - x_3 z_2) \\ x_2 y_3 - x_3 y_2 \end{pmatrix},$$

$$\nabla_{p_1} D_x = \begin{pmatrix} \frac{\partial D_x}{\partial x_1} \\ \frac{\partial D_x}{\partial y_1} \\ \frac{\partial D_x}{\partial z_1} \end{pmatrix} = \begin{pmatrix} -2x_1\frac{\partial a}{\partial x_1} \\ -2y_1\frac{\partial a}{\partial x_1} + p_2^2 z_3 - p_3^2 z_2 \\ -2z_1\frac{\partial a}{\partial x_1} - p_2^2 y_3 + p_3^2 y_2 \end{pmatrix},$$

$$\nabla_{p_1} D_y = \begin{pmatrix} \frac{\partial D_y}{\partial x_1} \\ \frac{\partial D_y}{\partial y_1} \\ \frac{\partial D_y}{\partial z_1} \end{pmatrix} = \begin{pmatrix} -2x_1\frac{\partial a}{\partial y_1} - p_2^2 z_3 + p_3^2 z_2 \\ -2y_1\frac{\partial a}{\partial y_1} \\ -2z_1\frac{\partial a}{\partial y_1} + p_2^2 x_3 - p_3^2 x_2 \end{pmatrix},$$

$$\nabla_{p_1} D_z = \begin{pmatrix} \frac{\partial D_z}{\partial x_1} \\ \frac{\partial D_z}{\partial y_1} \\ \frac{\partial D_z}{\partial z_1} \end{pmatrix} = \begin{pmatrix} -2x_1\frac{\partial a}{\partial z_1} + p_2^2 y_3 - p_3^2 y_2 \\ -2y_1\frac{\partial a}{\partial z_1} - p_2^2 x_3 + p_3^2 x_2 \\ -2z_1\frac{\partial a}{\partial z_1} \end{pmatrix}.$$

Following a gradient ascent scheme, the vertex position p_i evolves this way:

$p_i^{next} = p_i + \epsilon \nabla p_i R_\tau^2 / ||\nabla p_i R_\tau^2||$, where the step length ϵ is taken as a fraction of the minimum incident edge length to p_i. A relocation is performed only if the new minimal dihedral angle in the tetrahedra impacted by the relocation is not smaller than it was before relocation. As shown by Figure 3, the squared radius of τ's circumsphere increases very fast for a small perturbation of one of its vertices' positions. The circumsphere, now huge, most probably includes other mesh vertices, which triggers a flip to maintain the empty sphere Delaunay property.

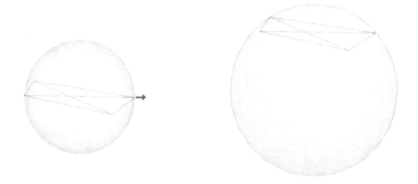

Fig. 3. Circumsphere of a sliver. Before perturbation (left), the sliver is close to the equatorial plane of its circumsphere. A very mild perturbation of one of the sliver vertices (right) makes its circumradius increase considerably.

2.2 Volume

One of the main characteristics of a sliver is that its volume is strictly positive albeit small with respect to its smallest edge length, and possibly arbitrarily small. This property can be exploited in order to apply a perturbation devised to generate a sliver with negative volume and hence to trigger a combinatorial change.

Let $\{p_i\}_{i=1,2,3}$ be the three fixed points of τ, and p_0 the vertex to be perturbed. The volume of τ is

$$V_\tau = \frac{1}{6} \begin{vmatrix} x_0 & y_0 & z_0 & 1 \\ x_1 & y_1 & z_1 & 1 \\ x_2 & y_2 & z_2 & 1 \\ x_3 & y_3 & z_3 & 1 \end{vmatrix}.$$

Then, we get the volume gradient:

$$\nabla_{p_0} V_\tau = \frac{1}{6} \begin{pmatrix} y_2 z_3 + y_1 (z_2 - z_3) - y_3 z_2 - z_1 (y_2 - y_3) \\ -x_2 z_3 - x_1 (z_2 - z_3) + x_3 z_2 + z_1 (x_2 - x_3) \\ x_2 y_3 + x_1 (y_2 - y_3) - x_3 y_2 - y_1 (x_2 - x_3) \end{pmatrix}.$$

Following a gradient descent scheme, the vertex position p_i evolves this way: $p_i^{next} = p_i - \epsilon \nabla_{p_i} V_\tau / \|\nabla_{p_i} V_\tau\|$, where the step length ϵ is taken as a fraction of the minimum incident edge length to p_i. A relocation is performed only if the new minimal dihedral angle of the tetrahedra impacted by the relocation is not smaller than it was before relocation. A negative tetrahedron volume triggers a flip to maintain a valid Delaunay triangulation.

2.3   Random Perturbation

When both ∇V and ∇R^2 fail at flipping the considered slivers by vertex perturbation, we use a random perturbation based on Li's approach [21]. A perturbation satisfying three conditions (flip sliver, improve minimum dihedral angles, preserve restricted Delaunay triangulation) is searched for randomly inside a sphere centered at v. In accordance with Li's algorithm, the magnitude of the perturbation vector is set to fraction of the minimum incident edge length.

3   Experiments and Results

The algorithm presented has been implemented with the 3D Delaunay triangulation of the *Computational Geometry Algorithms Library* [1]. Our implementation of Li's random perturbation algorithm is based upon Algorithm 1, with one single perturbation type: the random one, described in Section 2.3. For each of the following experiments we set 100 trials of random perturbations (in our combined version as well as in the purely random algorithm).

The order in which the vertices are processed in the priority queue has been chosen empirically as a result of many experiments. Interior vertices are processed first, with priority over boundary vertices. Boundary vertices are constrained so as to remain on the domain boundary and their relocation is invalid if they modify the local restricted triangulation. This makes boundary vertices more difficult to perturb than interior vertices. The second order criterion is the number of incident slivers to the processed vertex. The idea behind this choice is that a *chain* of slivers (several incident slivers) is more difficult to perturb than an isolated sliver as the directions of gradients may not be compatible. However, if the endpoints of the chain are successfully perturbed, we ideally would not have to deal with vertices incident to more than one sliver. Thirdly, the vertex incident to a smaller dihedral angle is processed first, as our first goal is to remove the worst tetrahedra.

In our experiments, the ∇R^2 direction turns out to be more effective than ∇V at perturbing a sliver. On average, this perturbation is responsible for about 80% of all sliver flips. The ∇V perturbation accounts for about 15% of the flips while the random perturbation counts for the remaining 5%. The priority given to ∇R^2 over ∇V and random while picking the perturbation vector can be blamed for distorting these statistics, but we have chosen this order because it turns out to be the most effective. Giving priority to ∇V results in an overall slowdown. Random perturbation always remains the last resort in the combined perturbation algorithm as the deterministic directions are favored.

The following experiments show what our combined algorithm can achieve on meshes generated by Delaunay refinement alone and on some meshes which have been optimized after refinement. A mesh optimization algorithm is in general devised to improve the mesh quality [2] while simpler algorithms aim at evenly distributing the vertices in accordance to a given mesh sizing function. Note that a mesh with well-spaced vertices does not mean an absence of slivers inside the mesh [29], and hence sliver removal is still required. The mesh optimization schemes used in our experiments are the centroidal Voronoi tessellation [11] using the Lloyd iteration, and the Optimal Delaunay triangulation (ODT for short) [5]. Both of these optimization methods have been implemented in a way that respects the local density of the mesh. It is important to not modify the density of a graded mesh, and to not decrease its quality.

Figures 4 and 5 provide the computation times and the best minimum dihedral angles obtained in our experiments. The same experiment has been carried out on many other models (not shown), giving similar results. Figures 4 and 5 emphasize that, for the same definition of a sliver (in terms of smallest dihedral angle), the combined algorithm is faster in removing all slivers by explicit perturbation compared to using Li's random perturbation alone. Moreover the combined algorithm reaches higher minimum dihedral angles.

The algorithm obtains fairly high minimum dihedral angles when the input is a mesh obtained by Delaunay refinement. Figures 4 and 5 illustrate that when the mesh is optimized prior to perturbation, the time taken for the algorithm to succeed in removing all slivers is shorter and that it can reach a higher minimum dihedral angle. As shown by histograms of Figure 4, the algorithm takes 611 seconds to perturb the mesh obtained after Delaunay refinement so that no dihedral angle is below 17°. If the same mesh is optimized prior to perturbations, the time taken goes down to 76 seconds for Lloyd and even further down to 11 seconds for ODT. Overall the same histograms show that a mesh optimized by ODT is easier to perturb and can reach a higher minimum angle (25°) than a mesh optimized by Lloyd (21°). However, optimization can be costly. The optimizations performed on Figure 4 meshes before applying perturbation took about 200 seconds. In spite of this additional cost, the combined perturbation algorithm remains more efficient than the random one. The same comments apply to Figure 5. The gradation of the mesh in Figure 5, along with the numerous high curvature

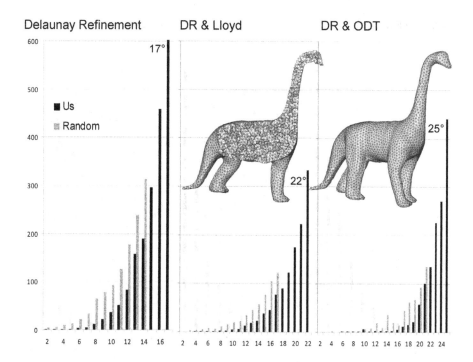

Fig. 4. Dinosaur. Comparison of the timings for our perturbation and random perturbation (in seconds) w.r.t. the sliver angle bound α on the Dinosaur model meshes obtained by Delaunay refinement (left), followed by Lloyd optimization (middle) and ODT optimization (right).

regions, make it more difficult to perturb in a way that still preserves the gradation, even after optimization. Even in this case, ODT reaches a higher minimum angle.

For comparison we have also performed sliver exudation on meshes generated by Delaunay refinement and on meshes optimized after refinement. As expected sliver exudation performs better on the optimized meshes.

We performed two other experiments that were abandoned since they rarely succeeded in improving the mesh quality. While computing the perturbation of a vertex incident to more than one sliver, we tried combining ∇V vector of one of the slivers and ∇R^2 of the other by using their average as perturbation direction if they were compatible. In practice such a combination was almost never successful removing the slivers. The other aborted experiment consisted of removing from the mesh the vertices that every explicit perturbation failed to perturb. In practice this never resulted in improving the minimum dihedral angle.

Moreover, our experiments show that successively applying our combined algorithm to the mesh several times while progressively increasing the angle

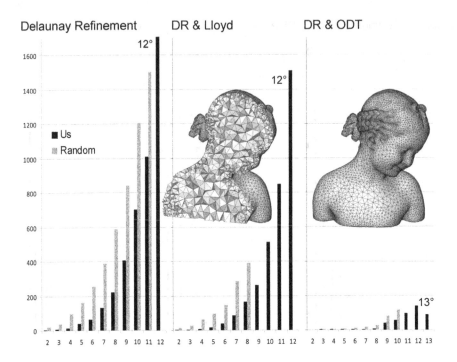

Fig. 5. Bimba. Comparison of the timings for our perturbation and random perturbation (in seconds) w.r.t. the sliver angle bound α on the Bimba model meshes obtained by Delaunay refinement (left), followed by Lloyd optimization (middle) and ODT optimization (right).

Table 1. Angles. Minimum dihedral angles obtained by the different perturbation algorithms (combined perturbation, random perturbation, and sliver exudation). To achieve these maxima, combined perturbation takes about twice the exudation time, and random perturbation takes about six times the exudation time.

Mesh	input	combined	random	exudation
Dinosaur (DR)	0.65	25.0	24.2	2.62
Dinosaur (DR & Lloyd)	0.24	26.15	23.5	4.47
Dinosaur (DR & ODT)	2.26	28.55	22.0	4.55
Bimba (DR)	0.16	15.51	15.64	1.11
Bimba (DR & Lloyd)	0.11	16.02	15.63	3.84
Bimba (DR & ODT)	0.84	19.8	18.85	4.47

bound that defines a sliver provides higher minimal dihedral angles at the price of higher computation times. This amounts to giving priority to vertices incident to the worst slivers, cluster by cluster of minimum dihedral angles.

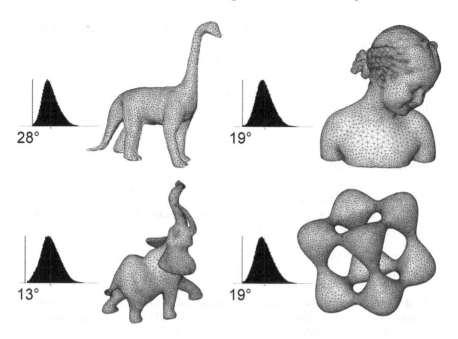

Fig. 6. Delaunay meshes perturbed with combined perturbation algorithm after ODT optimization.

Table 1 summarizes the best angles obtained in this way using combined perturbation, random perturbation and sliver exudation. In this labor-intensive experiment we only measure how far we can go in terms of dihedral angles and do not consider timing. Finally, Figure 6 shows some Delaunay meshes obtained by Delaunay refinement followed by ODT optimization and perturbed with the combined algorithm along with their dihedral angle histograms.

4 Conclusion and Discussion

We have presented a practical vertex perturbation algorithm for improving the dihedral angles of a 3D isotropic Delaunay triangulation. The key idea consists of performing a gradient ascent over the sliver circumsphere radius as well as a gradient descent over the sliver volume. All vertices incident to slivers are processed, in an order devised to improve effectiveness and computation times. We compare our approach with pure random perturbation and sliver exudation.

Our experiments show that we are both faster and able to reach higher minimum dihedral angles. Our scheme is particularly well suited as a post-processing step after mesh optimization [30]. We also plan to use it in the context of mesh generation from multi-material voxel images [4].

In the cases where all vertices of a sliver are on the domain boundary, the perturbation can fail in removing a sliver as the boundary vertices are too constrained. One way to extend our approach would be to also perturb the vertices of the sliver's adjacent tetrahedra whose relocation can impact the sliver. Future work will focus on obtaining a proof of termination of our combined perturbation algorithm, and some tighter lower bounds on output dihedral angles.

Acknowledgments

The authors wish to thank Mathieu Desbrun for helpful discussions and suggestions. We also thank Mariette Yvinec and Laurent Rineau for providing us with their implementation of sliver exudation.

References

1. Cgal, Computational Geometry Algorithms Library, http://www.cgal.org
2. Amenta, N., Bern, M., Eppstein, D.: Optimal point placement for mesh smoothing. In: Proc. of the 8th ACM-SIAM Symposium on Discrete Algorithms, pp. 528–537. SIAM, Philadelphia (1997)
3. Boissonnat, J., Wormser, C., Yvinec, M.: Curved Voronoi diagrams. Effective Computational Geometry for Curves and Surfaces, 67–116 (2006)
4. Boltcheva, D., Yvinec, M., Boissonnat, J.-D.: Mesh generation from 3d multi-material images. In: MICCAI 2009. LNCS, Springer, Heidelberg (2009)
5. Chen, L., Xu, J.: Optimal Delaunay triangulation. Journal of Computational Mathematics 22, 299–308 (2004)
6. Cheng, S., Dey, T.: Quality meshing with weighted Delaunay refinement. In: Proc. of the 13th ACM-SIAM Symposium on Discrete Algorithm, pp. 137–146. SIAM, Philadelphia (2002)
7. Cheng, S.W., Dey, T.K., Edelsbrunner, H., Facello, M.A., Teng, S.H.: Sliver exudation. Journal of the ACM 47(5), 883–904 (2000)
8. Cheng, S., Dey, T., Ray, T.: Weighted Delaunay refinement for polyhedra with small angles. In: Proc. of the 14th Int. Meshing Roundtable (2005)
9. Chew, L.: Guaranteed-quality Delaunay meshing in 3D (short version). In: Proc. of the 13th Symposium on Computational Geometry, pp. 391–393. ACM Press, New York (1997)
10. Cutler, B., Dorsey, J., McMillan, L.: Simplification and improvement of tetrahedral models for simulation. In: Proc. of the Eurographics/ACM SIGGRAPH Symposium on Geometry Processing, pp. 93–102. ACM, New York (2004)
11. Du, Q., Faber, V., Gunzburger, M.: Centroidal Voronoi tessellations: applications and algorithms. SIAM review 41(4), 637–676 (1999)
12. Edelsbrunner, H.: Algorithms in Combinatorial Geometry. Springer, Heidelberg (1987)
13. Edelsbrunner, H., Guoy, D.: An experimental study of sliver exudation. Engineering with computers 18(3), 229–240 (2002)

14. Edelsbrunner, H., Li, X., Miller, G., Stathopoulos, A., Talmor, D., Teng, S., Üngör, A., Walkington, N.: Smoothing and cleaning up slivers. In: Proc. of the 22nd ACM symposium on Theory of computing, pp. 273–277. ACM, New York (2000)

15. Forsman, K., Kettunen, L.: Tetrahedral mesh generation in convex primitives by maximizing solid angles. IEEE Transactions on Magnetics 30(5) (Part 2), 3535–3538 (1994)

16. Joe, B.: Delaunay versus max-min solid angle triangulations for 3-dimensional mesh generation. Int. Journal for Numerical Methods in Engineering 31(5) (1991)

17. Joe, B.: Construction of 3-dimensional improved-quality triangulations using local transformations. SIAM Journal on Scientific Computing 16(6), 1292–1307 (1995)

18. Klingner, B., Shewchuck, J.: Aggressive Tetrahedral Mesh Improvement. In: Proc. of 16th Int. Meshing Roundtable, pp. 3–23 (2007)

19. Krysl, P., Ortiz, M.: Variational Delaunay approach to the generation of tetrahedral finite element meshes. Int. Journal for Numerical Methods in Engineering 50(7), 1681–1700 (2001)

20. Labelle, F., Shewchuk, J.: Isosurface stuffing: fast tetrahedral meshes with good dihedral angles. In: Int. Conference on Computer Graphics and Interactive Techniques. ACM Press, New York (2007)

21. Li, X.: Sliver-free 3-Dimensional Delaunay mesh generation. PhD thesis, University of Illinois (2000)

22. Li, X.: Spacing control and sliver-free Delaunay mesh. In: Proc. of the 9th Int. Meshing Roundtable, pp. 295–306 (2000)

23. Li, X., Teng, S.: Generating well-shaped Delaunay meshes in 3D. In: Proc. of the 12th ACM-SIAM Symposium on Discrete Algorithms, pp. 28–37. SIAM, Philadelphia (2001)

24. Liu, C.Y., Hwang, C.J.: New strategy for unstructured mesh generation. AIAA journal 39(6), 1078–1085 (2001)

25. Rineau, L., Yvinec, M.: Meshing 3D domains bounded by piecewise smooth surfaces. In: Proc. of the 16th Int. Meshing Roundtable, pp. 442–460 (2007)

26. Shewchuk, J.: Tetrahedral mesh generation by Delaunay refinement. In: Proc. of the 14th Symposium on Computational Geometry, pp. 86–95. ACM Press, New York (1998)

27. Shewchuk, J.: Two discrete optimization algorithms for the topological improvement of tetrahedral meshes (Unpublished manuscript) (2002)

28. Shewchuk, J.: What is a good linear element? interpolation, conditioning, and quality measures. In: Proc. of the 11th Int. Meshing Roundtable, pp. 115–126 (2002)

29. Talmor, D. Well-spaced points for numerical methods. PhD thesis, University of Minnesota (1997)

30. Tournois, J., Wormser, C., Alliez, P., and Desbrun, M. Interleaving Delaunay refinement and optimization for practical isotropic tetrahedron mesh generation. ACM Transactions on Graphics 28(3) (2009)

Mesh Smoothing Algorithms for Complex Geometric Domains

Hale Erten, Alper Üngör*, and Chunchun Zhao

University of Florida, Dept. of Computer & Info. Sci. & Eng.
{herten,ungor,czhao}@cise.ufl.edu

Abstract. Whenever a new mesh smoothing algorithm is introduced in the literature, initial experimental analysis is often performed on relatively simple geometric domains where the meshes need little or no element size grading. Here, we present a comparative study of a large number of well-known smoothing algorithms on triangulations of complex geometric domains. Our study reveals the limitations of some well-known smoothing methods. Specifically, the optimal Delaunay triangulation smoothing and weighted centroid of circumcenter smoothing methods are shown to have difficulty achieving smooth grading and adapting to complex domain boundary. We propose modifications and report significant improvements and behavior change in the performance of these algorithms. More importantly, we propose three new smoothing strategies and show their effectiveness in computing premium quality triangulations for complex geometric domains. While the proposed algorithms give the practitioners additional tools to chose from, our comparative study of over a dozen algorithms should guide them selecting the best smoothing method for their particular application.

1 Introduction

Mesh smoothing is an important research problem for scientific simulation applications where high quality elements are desired for accurate numerical calculations. While mesh generation [5, 6, 7, 22] focuses on computing a subdivision (e.g. triangulation) of a given input domain from scratch, mesh smoothing [1, 2, 4, 8, 9, 10, 15, 19, 18, 21, 23, 26, 29, 28, 27] aims to improve an existing subdivision (e.g., for instance by relocating its vertices). Effectiveness of proposed smoothing methods are often shown on meshes of relatively simple domains, e.g., points distributed inside a square box. When the input mesh models a complex geometric domain, the existing smoothing strategies may struggle. For instance, Chen [10] proposed a smoothing method called optimal Delaunay triangulation (ODT) and showed that it is effective for smoothing meshes of simple geometric domains. Figure 1 illustrates

---

* This research is partially supported by NSF CAREER Award CCF-0846872.

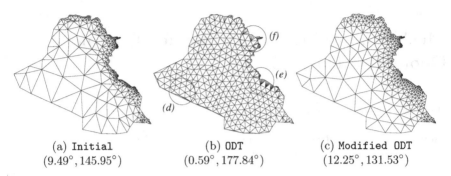

(a) Initial
$(9.49°, 145.95°)$

(b) ODT
$(0.59°, 177.84°)$

(c) Modified ODT
$(12.25°, 131.53°)$

Fig. 1. Smoothing on complex geometry. (a) Initial Iraq mesh; (b) ODT smoothing output with regions depicting suppressed (d) and stretched elements ((e) and (f)); (c) Output of our modified ODT smoothing method. Smallest and largest angles in each mesh are also shown.

the poor performance ODT smoothing when the input geometry is complex. The smoothed mesh contains very many bad quality triangles (suppressed or stretched) near the domain boundary. We later present a modification of ODT smoothing that can handle complex geometric domains (Figure 1 (c)) illustrates the output of this improved version).

In this paper, we compare the performance of a number of smoothing methods when the input is a complex geometric domain. We propose modifications for some of these techniques to handle complex geometry. In addition, we introduce three new smoothing methods, which are shown to be successful to obtain high quality triangulations on complex domains.

2   Review of Smoothing Algorithms

In this section, we examine a number of smoothing methods, each of which has been shown to be easy to implement, fast and/or effective. Here, we review each approach to better understand their similarities, differences, strengths and drawbacks regarding complex input domains. Note that, in this study we mainly focus on smoothing methods which are based on geometric concepts.

First, we define a vertex $\mathbf{x}_i$ to be *free* if it is allowed to be relocated. The *star* of $\mathbf{x}_i$ is denoted by Ω_i which consists of all triangles incident to $\mathbf{x}_i$. Similarly, the *link* of $\mathbf{x}_i$ consists of all surrounding edges of triangles in the star that are not incident to $\mathbf{x}_i$.

In general, smoothing algorithms iteratively relocate free vertices of a triangulation until a threshold value is satisfied or a specified number of iterations are completed. This is also incorporated with potential edge flipping operations to obtain higher quality triangulations. Now, we describe how to compute a new location for a given vertex based on given algorithms.

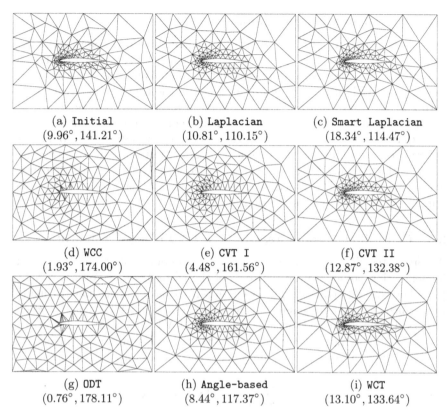

Fig. 2. (a) Initial airfoil mesh. (b)-(i) Output of existing smoothing algorithms. Smallest and largest angles in each mesh are listed.

2.1 Laplacian Smoothing

In Laplacian smoothing [18], a free vertex is simply relocated to the centroid of the vertices connected to that vertex (see Equation 1). This technique is widely used due to its simplicity and effectiveness. However, it does not guarantee quality improvement where inverted elements might be generated. Therefore, many studies introduced different versions of Laplacian smoothing or combined with other methods. An overview of those can be found in [9].

$$\mathbf{x}^* = \frac{1}{k} \sum_{\mathbf{x}_j \in \Omega_i, \mathbf{x}_j \neq \mathbf{x}_i} \mathbf{x}_j, \tag{1}$$

where Ω_i is the star of the vertex $\mathbf{x}_i$ having k points and $\mathbf{x}^*$ is the new location. Notice that, this formulation can also be interpreted as a torsion spring system [29], where a central node in a star polygon is located at the centroid of the polygon balancing out the system to stay in equilibrium (see also Equation 10).

2.2 Smart Laplacian Smoothing

One of the variations of the standard Laplacian smoothing algorithm is the smart Laplacian smoothing [9, 19] which is also called constrained Laplacian smoothing. The main difference between the original and this variant is relocating points only if there is an improvement in the neighborhood. Briefly, a vertex is relocated when the new candidate location improves the quality of the elements in the star. Otherwise, the node remains in its current location. Note that, the candidate location is the centroid of the surrounding points as before. Cost of evaluating the quality is also added to overall computation, however this method is still quite inexpensive and avoids inverted elements. The quality metric can be chosen based on the application, where we use minimum angle as our quality criterion.

2.3 Centroidal Voronoi Tessellation (CVT)-Based Smoothing

A Voronoi tessellation is called a centroidal Voronoi tessellation (CVT) when its generating points are the centroids of the corresponding Voronoi regions [13]. This special structure has been shown to have applications in many fields, such as image processing, clustering, cell division, and others. Du *et al.* discussed many properties and applications of CVT in [13, 14, 15, 27]. They also showed that the Lloyd iteration [20] which is a minimizer for the energy function given in Equation 2, converges to centroidal Voronoi tessellation.

$$\Psi_{CVT} = \sum_{i=1}^{N} \int_{V_i} \rho(\mathbf{x})\|\mathbf{x} - \mathbf{x}_i\|^2 d\mathbf{x}, \qquad (2)$$

where V_i is the Voronoi region generated by the vertex $\mathbf{x}_i$ and ρ is the density function.

Hence, an immediate algorithm is to apply the Lloyd iteration by iteratively computing the Voronoi regions for each point and updating their locations to the centroid of each region until convergence. In general, the new location can be defined as follows:

$$\mathbf{x}^* = \frac{\int_{V_i} \mathbf{x}\rho(\mathbf{x})d\mathbf{x}}{\int_{V_i} \mathbf{x}d\mathbf{x}} \qquad (3)$$

Note that, the Lloyd iteration is not the only technique to determine CVTs. A discussion on other deterministic and probabilistic methods can be found in [13]. However, this method is naturally suitable for mesh smoothing, although the Lloyd iteration converges slow and hard to analyze [12]. Thus, researchers proposed different smoothing techniques based on centroidal Voronoi tessellation concept [15, 27, 1, 10]. Now, we overview a few which mainly differ in density function and regions leading to less computation time.

Weighted Centroid of Circumcenters (WCC). In this method, Alliez *et al.* [1] proposed to relocate each vertex to a weighted average of corresponding circumcenters, where the weights are based on the physical size of each simplex. This is same as computing a weighted centroid of the corresponding Voronoi region, where density function is represented by the weights. Then, the new location is represented as follows:

$$\mathbf{x}^* = \frac{1}{|\Omega_i|} \sum_{\tau_j \in \Omega_i} |\tau_j| \mathbf{c}_j, \tag{4}$$

where Ω_i is the star of the point of interest, $\mathbf{c}_j$ is the circumcenter of the simplex τ_j and $|.|$ denotes area in two-dimensions. The circumcenter can be defined as the intersection point of the bisectors of a triangle, which is also the center of the circle that passes through all three vertices of the triangle.

CVT I. In this variant, Chen [10] used the same density function as above, but instead of circumcenters, he proposed to use the centroid of each simplex. This allows to compute the surrounding points, i.e., the region, faster. This smoother can be represented as the following expression:

$$\mathbf{x}^* = \frac{1}{|\Omega_i|} \sum_{\tau_j \in \Omega_i} |\tau_j| \mathbf{x}_{\tau_j}, \tag{5}$$

where $\mathbf{x}_{\tau_j}$ denotes the centroid of the simplex τ_j, which is equal to $\frac{\sum_{\mathbf{x}_k \in \tau_j} \mathbf{x}_k}{n+1}$, while n is the dimension.

CVT II. For non-uniform domains, Chen [10] proposed to incorporate the mesh density function to the calculations. Hence, the smoother can generate an appropriate grading for modulation regions. So, in general by the addition of a density function ρ, Equation 5 becomes

$$\mathbf{x}^* = \frac{\sum_{\tau_j \in \Omega_i} |\tau_j| \mathbf{x}_{\tau_j} \rho_{\tau_j}}{\sum_{\tau_j \in \Omega_i} |\tau_j| \rho_{\tau_j}}. \tag{6}$$

In two-dimensions, when the density function is chosen as $\rho_{\tau_j} = |\tau_j|^{-\frac{n}{2}}$, the above expression is equal to

$$\mathbf{x}^* = \frac{2}{3} \frac{\sum \mathbf{x}_j}{k} + \frac{1}{3} \mathbf{x}_i. \tag{7}$$

Chen described this as a "lumped Laplacian Smoothing", which partially explains the reason behind the effectiveness of the Laplacian smoothing [10].

2.4   Optimal Delaunay Triangulation (ODT)-Based Smoothing

Chen and Xu [11] studied Delaunay triangulations in terms of linear interpolation error for a given function. They showed that when $f(\mathbf{x}) = \|\mathbf{x}\|^2$, the Delaunay triangulation can be referred as optimal in terms of minimizing the interpolation error among all the triangulations for a given number of vertices. This helped Chen to design a new mesh smoothing strategy based on the optimal Delaunay triangulation (ODT) concept [10], which aims to equally distribute the edge lengths based on the function to be approximated. Basically, the energy function from the interpolation error is exactly solvable, which results in the new location for each query point (see Equation 8).

$$\Psi_{ODT} = \frac{1}{1+n} \sum_{i=1}^{N} \int_{\Omega_i} \|\mathbf{x} - \mathbf{x}_i\|^2 d\mathbf{x}, \tag{8}$$

where Ω_i is the star of the point of interest $\mathbf{x}_i$, while n represents the dimension. For a regular-shaped uniform mesh generation, the function to be interpolated can be chosen as $f(\mathbf{x}) = \|\mathbf{x}\|^2$. Then, the new location can be computed as in Equation 9. Note that, convergence rate is slower while the time for each relocation calculation is increased compared to the others.

$$\mathbf{x}^* = -\frac{1}{2|\Omega_i|} \sum_{\tau_j \in \Omega_i} (\nabla|\tau_j(\mathbf{x})| \sum_{\mathbf{x}_k \in \tau_j, \mathbf{x}_k \neq \mathbf{x}_i} \|\mathbf{x}\|^2). \tag{9}$$

2.5 Angle-Based Smoothing (AB)

Zhou and Shimada [29] proposed a smoothing method which aims to improve the geometric quality of a mesh in a computationally easy way. Their method has been shown to provide better results than Laplacian smoothing in quality and to avoid creating inverted elements. They formulated local neighborhood of a vertex as a torsion spring system similar to Laplacian smoothing. (See Equations 10 and 11.)

$$\Psi_{LAPLACIAN} = \sum_{j=1}^{k} \frac{1}{2} K \|\mathbf{v}_j\|^2, \tag{10}$$

$$\Psi_{ANGLE} = \sum_{j=1}^{2k} \frac{1}{2} K \theta_j{}^2, \tag{11}$$

where k is number of nodes in the star of the vertex $\mathbf{x}_i$, K is the spring constant, $\mathbf{v}_j$ is the vector from $\mathbf{x}_i$ to every surrounding node and θ_j is the angle formed by each side of the polygon and the central point $\mathbf{x}_i$.

However, their spring system is based on angles instead of distances. The system has its minimum potential energy when the new location for the central node yields to an optimum angle distribution on the angles formed by the each side of the star. In order to find the optimum location, following two approaches have been introduced.

Original Version. Zhou and Shimada [29] introduced a heuristic method which gives an approximate position for the optimum location. A four-step procedure, summarized below, is followed to find the new location for each free vertex. Note that, this method is more expensive than Laplacian smoothing, but easier and faster than optimization techniques.

Given a vertex $\mathbf{x}_i$ and its corresponding neighborhood (star) and link, the vertices of the link is denoted by $\mathbf{y}_j$, $j = \{1, 2, ..., k\}$ in counterclockwise order.

1. Compute the side angles $\alpha_{j_1} = \angle \mathbf{x}_i \mathbf{y}_j \mathbf{y}_{j+1}$ and $\alpha_{j_2} = \angle \mathbf{y}_{j-1} \mathbf{y}_j \mathbf{x}_i$.
2. Compute the difference between each adjacent angles: $\beta_j = (\alpha_{j_2} - \alpha_{j_1})/2$.
3. Compute the new location $\mathbf{x}_i^j$ for the central node $\mathbf{x}_i$ by a vector rotation based on β_j, $\mathbf{y}_j$ and $\mathbf{x}_i$, while keeping the size of the vector from $\mathbf{y}_j$ to $\mathbf{x}_i$.
4. Compute the average of the new location suggestions for the final location: $\mathbf{x}^* = \frac{1}{k} \sum_{j=1}^{k} \mathbf{x}_i^j$

Optimization Version. Xu and Newman [28] approached to the same idea of angle-based smoothing from an optimization point of view. They sacrificed computation time to have better quality results. Instead of heuristically calculating an approximate new location, they used optimization techniques to find a close optimum. The same torsion spring system is formulated as an optimization problem and the Gauss-Newton approximation has been used to solve the least-squares formulation of the following objective function:

$$s = \sum_{j=1}^{k} [distance(\mathbf{x}^*, l_j)], \tag{12}$$

where $\mathbf{x}^*$ is the optimized location of the query point $\mathbf{x}_i$, and l_j is the angle bisector line for each side angle of the star Ω_i. This method has been shown to produce slightly better quality meshes than original angle-based method and also shown to converge faster. However, overall computation can take more than the heuristic version, although the number of iterations is less. Therefore, we will use the original version of this method in our experiments due to its simplicity, efficiency and availability.

2.6 Well-Centered Triangulation (WCT) Smoothing

Another iterative method has been proposed by Vanderzee *et al.* in [26], mainly focusing on providing well-centered triangulations (WCT), i.e., acute triangulations in the plane. They introduced a global energy function, which aims to minimize the maximum angle of a mesh M. They also claim that their energy function penalizes small angles as well.

$$\Psi_{WCT} = \sum_{\theta \in M} |cos(\theta) - 1/2|^p, \tag{13}$$

where p is a finite number (it is sufficient to set $p \in \{4, 6, 8\}$) and θ represents each angle in the triangulation M. This algorithm is computationally more expensive, however considering this approach concentrates on maximum angle, it would be interesting to observe its behavior among others. We should also note that, this algorithm preserves the topological connections unlike the above mentioned methods where they incorporate edge flip operations into smoothing.

3   New Smoothing Algorithms

We first give modifications to two of the methods described in the previous section, so that they handle complex geometric input domains. Then, we propose three new smoothing algorithms in Sections 3.2, 3.3 and 3.4.

3.1   Modifying Existing Algorithms for Complex Geometry

As we mentioned, when the input domain has a complex geometry, most of the known smoothing methods struggle and generate bad quality elements (see Figure 1 and 2). In particular, although the output triangulation overall contains good quality elements, the elements which are close to the boundary, are problematic. We observed this boundary problem especially in ODT and WCC methods. Here, we propose simple modifications for each algorithm to handle boundary issues to an extend.

ODT with boundary fix via incenters. ODT smoothing approach [10] tends to create bad quality elements close to input features (see Figures 1, 2 (g) and 3). The original ODT algorithm relocates a vertex x_i, by exactly solving an optimization problem within the star of x_i and trying to achieve uniformity in the edge lengths of the triangles. Since we consider the boundary vertices as input features representing the complex domain and chosen not to be relocated, this approach causes some triangles near the boundary to be stretched or suppressed (see Figure 3). As a modification, we now perform the same relocation strategy within the star-shaped polygon formed by the incenters of the triangles in the star of x_i. This way, we aim to make the distances of the relocated vertex to the star vertices proportional to the length of the corresponding edges on the star. Experiments indicate that this idea works fairly well in producing graded and quality triangulations. Therefore, we suggest to use a patch formed by the incenters of the connected elements instead of elements themselves (see Figures 1, 3, 5 and 7).

WCC with boundary push. WCC smoothing approach [1] has the similar boundary issues (see Figure 2 (g)). This is partly because the algorithm tries to iteratively reach a regular-shaped centroidal Voronoi tessellation, where in case of complex geometric domains it fails to consider the input features. In order to alleviate this problem, while calculating the new location, we propose to use incenters of the triangles whose circumcenters are located outside the boundary. Note that, by definition an incenter is always located inside of the triangle. This idea reasonably eliminates suppressed elements and increase the quality, however stretched elements can remain near the boundary (see Figures 5 and 7). Therefore, we explore other smoothing ideas as we propose next.

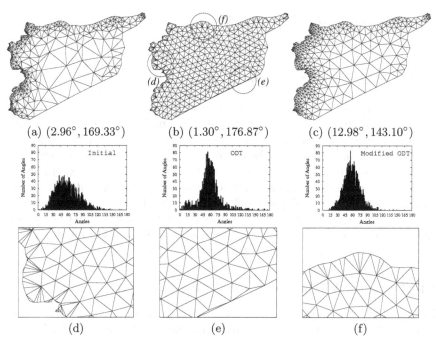

Fig. 3. (a) Initial Syria mesh and histogram of angles. (b) Output and histogram of angles after 50 iterations of ODT applied on (a). (c) Output and histogram of angles after applying 10 iterations of modified ODT on (a). ODT output is zoomed in for stretched ((d) and (f)) and suppressed (e) elements.

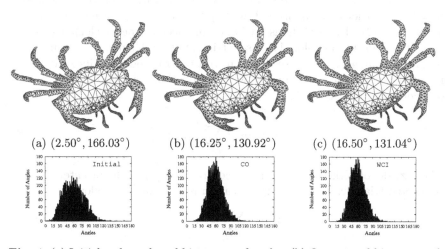

Fig. 4. (a) Initial crab mesh and histogram of angles. (b) Output and histogram of angles after 10 iterations of CO smoothing applied on (a). (c) Output and histogram of angles after applying 10 iterations of WCI on (a).

3.2   Centroid of Off-Centers (CO)

Off-centers [25] have been introduced to improve Delaunay refinement algorithms as another type of Steiner point alternative to circumcenters. Given a triangle with the shortest edge pq, the off-center is a point o on the bisector of pq, which is furthest from p (or q) such that the angle $\angle poq$ is α.

Here, we introduce a new method to benefit from off-centers for smoothing. Given a point to be relocated, similar to WCC method, with an additional quality constraint, we compute the off-center of each simplex connected to that point. Then, we calculate the centroid of this region formed by the computed centers and relocate our point to that location. In our experiments, we chose a reasonably large α value for computing the off-centers and obtained good results (see Figures 4, 5 and 7). Note that, here the uniformity constraint is lifted, however a weighted version of this method (WCO) is also explored and observed to have a similar behavior to WCC. WCO is also prone to boundary problems, but off-centers help to reduce.

3.3   Distance Weighted Centroid of Incenters (WCI)

The incenter of a triangle is the intersection point of the angle bisectors, which is the center of the largest circle that fits inside the triangle. Hence, the location of the incenter of a triangle is determined by the angles and edge lengths of that element, while remaining always inside of the element. Here, we relocate free vertices to the centroid of the incenters of the elements in their stars. A weighted version of this idea have been introduced as part of an hybrid smoothing approach for mixed meshes which classifies elements based on their local neighborhoods and apply different smoothing algorithms accordingly [21]. Hence, WCI is used only for specific types of elements. However, it has not been explored as a sole triangular mesh smoother. As in some other smoothing methods (see Section 2) a central node and its star is formulated as a torsion spring model, where relocation point is computed based on energy minimization.

$$\Psi_{INCENTER} = \sum_{j=1}^{k} \frac{1}{2} K \|\mathbf{z}_j\|^2, \qquad (14)$$

where $\mathbf{z}_j$ denotes the vector from the central node $\mathbf{x}_i$ to the incenter of each connected triangle. Then, the new location is calculated as follows:

$$\mathbf{x}^* = \sum_{\tau_j \in \Omega_i} w_{\tau_j} \mathbf{p}_{\tau_j}, \qquad (15)$$

where $\mathbf{p}_{\tau_j}$ is the incenter of the simplex τ_j and weight $w_{\tau_j} = \frac{\|\mathbf{z}_j\|}{\sum_{t=1}^{k} \|\mathbf{z}_j\|}$. Note that, here the weight function is different than the previously mentioned methods. Our experiments show that this smoothing method achieves good quality bounds while generating proper grading (see Figure 4, 5 and 7).

3.4   Sliced-Petal Smoothing

Among many mesh quality criteria [24], minimum angle quality is widely popular due to its direct influence on applications [3, 24]. Here, we designed a new smoothing method to obtain large minimum angles for given triangulations. In our method, we guarantee to preserve and improve (if possible) the initial minimum angle while smoothing the mesh elements. As can be seen from Figure 2, this is not always true for other smoothing methods. They potentially reduce the minimum angle, in particular, for complex geometric domains.

Our method relies on the geometric concepts that have been introduced in [17, 16], where a similar relocation strategy has been used as part of the Delaunay refinement process.

Consider a bad quality triangle pqr (where its minimum angle does not satisfy the angle constraint α) in a triangulation with its shortest edge pq. Let α-$petal(pq)$ be the disk bounded by the circle that goes through p, q and a third point y such that y and r are on the same side of pq and $\angle pyq = \alpha$ [17, 16]. Then, α-$slice(pq)$ is defined as the intersection of the α-$petal(pq)$ and the region between the two lines one going through p, the other q where both making an angle α to line segment pq. Note that, the α-slice is the feasible region, where r needs to be relocated in order to make all the angles of pqr greater than or equal to α. Here, due to space restriction we omit descriptive figures, however the reader can refer to the figures in [17, 16].

Given a vertex to be relocated x_i and its star, in order to have minimum angles of all elements inside the star to be at least α, the intersection of the α-slices of the edges on the link must be non-empty. Based on this observation, first we search for approximately the best possible α value. We start our search with an initial α value which is determined according to the current minimum angle of the star. After the search operation, we construct the feasible relocation region. In order to find the best location inside this region for x_i which strictly improves the local quality, we check sufficiently large number of sample points. If such a location could not be found, simply the point location is kept the same. Thus, every local relocation definitely improves the overall quality of the mesh.

Compared to the other smoothing methods, our results indicate a considerable amount of improvement in quality (see Figures 5, 6, 7 and 8). Clearly, this method requires more computation time than many of the algorithms mentioned, however it successfully handles complex geometric domains with proper grading and has high convergence rate. Although our method obtains small maximum angle values, our ultimate goal is to incorporate the maximum angle constraint (γ) to our relocation strategy as in [16].

4   Results and Discussions

Results of our experiments have been given throughout the paper, however in this section we compare all smoothing algorithms that are reviewed and introduced. We generated several experimental set-ups which include various input domains in different complexities. We discuss the results mostly based on the distribution of angles and edge lengths, uniformity, minimum and maximum angle and time.

4.1 Implementation and Data Sets

The algorithms discussed in this paper are all implemented in the same platform and using the C programming language. The examples and data sets presented in this limited space are only a small but representative subset of our extensive experimental study. The output triangulations (Figures 5 and 7) are accompanied by the histogram of angles in each case (Figures 6 and 8) and the time performance is summarized in Table 2.

4.2 Experiments

Our experimental study is conducted on a large number of data sets including the representative China and Hawaii maps, which have 1158 and 3216 points and 1876 and 5024 triangles, respectively.

In Table 1, we give an overall qualitative comparison of smoothing methods experimented in this study. These classifications have been obtained based on the results of our qualitative visual experiments. In particular, ease of implementation and speed incorporate our experimental observations and the main idea of each approach, where the speed levels are determined according to have almost an order of magnitude difference. Minimum and maximum

Table 1. A qualitative comparison of the known and the proposed smoothing methods, where we label the performance of each method on a scale of 3, as 1. Poor, 2. Medium, or 3. Good.

Smoothing Method	Ease of Implementation	Speed	Quality α	Quality γ	Grading
Laplacian [18]	Good	Good	Medium	Medium	Good
Smart Lap. [9, 19]	Good	Good	Medium	Good	Good
WCC [1]	Medium	Good	Poor	Poor	Medium
CVT I [10]	Good	Good	Medium	Medium	Good
CVT II [10]	Good	Good	Medium	Medium	Good
ODT [10]	Medium	Medium	Poor	Poor	Poor
AB [29]	Medium	Good	Medium	Good	Medium
WCT [26]	Poor	Poor	Poor	Medium	Good
M. ODT	Medium	Good	Medium	Medium	Good
M. WCC	Medium	Good	Medium	Medium	Medium
WCO	Medium	Good	Medium	Medium	Medium
CO	Medium	Good	Medium	Medium	Good
WCI [21]	Medium	Good	Medium	Medium	Good
Sliced-petal	Medium	Medium	Good	Good	Good

Table 2. Time comparison between the known and the proposed methods.

Smoothing Method	China Quality (α, γ)	China Total Time (msec)	Hawaii Quality (α, γ)	Hawaii Total Time (msec)
Initial	$(8.82°, 161.47°)$	N/A	$(5.80°, 164.76°)$	N/A
Laplacian [18]	$(22.02°, 117.83°)$	377	$(6.88°, 151.59°)$	1447
Smart Lap. [9, 19]	$(23.60°, 121.84°)$	420	$(13.92°, 148.32°)$	1988
WCC [1]	$(10.54°, 138.30°)$	382	$(0.14°, 179.72°)$	1866
CVT I [10]	$(12.88°, 138.30°)$	449	$(0.70°, 166.35°)$	1872
CVT II [10]	$(21.13°, 115.58°)$	415	$(8.27°, 145.28°)$	1849
ODT [10]	$(0.72°, 177.96°)$	2073	$(0.01°, 179.97°)$	18669
AB [29]	$(18.54°, 114.92°)$	407	$(5.44°, 154.95°)$	2053
WCT [26]	$(15.24°, 118.58°)$	8129	$(1.96°, 168.18°)$	21534
M. ODT	$(17.15°, 117.00°)$	415	$(2.12°, 161.30°)$	1984
M. WCC	$(10.54°, 138.30°)$	996	$(0.58°, 168.71°)$	13112
WCO	$(10.14°, 144.60°)$	483	$(0.40°, 171.51°)$	2225
CO	$(22.02°, 117.83°)$	443	$(6.88°, 151.59°)$	1938
WCI [21]	$(21.88°, 118.09°)$	421	$(7.50°, 148.32°)$	1839
Sliced-petal	$(30.48°, 113.42°)$	1550	$(18.98°, 133.73°)$	5347

angle (α and γ respectively) quality is categorized based on the consistency of the methods. Average performance has been chosen as medium level while the others indicate notably better or worse performance. Grading values reflect the overall appearence of the output triangulations, where three different classes naturally emerge.

We observed that the complexity of geometric domains plays a significant role in the performance of the smoothing methods. For complex geometric domains, methods incorporating input features into their relocation strategy perform well. On the other hand, algorithms that aim to converge to lattice structures have trouble computing graded meshes. Our proposed sliced-petal method is shown to be performing considerably better than others in quality, while smart Laplacian method obtains good results as well. However, its behavior is not as stable as our sliced-petal method.

Quality. Figures 5 and 7 together with the histograms in Figures 6 and 8 show the performance of smoothing algorithms on complex geometric domains. Here, Laplacian and smart Laplacian methods are shown to be performing well on complex geometric domains, however ODT and WCC methods have clear boundary problems. Our modified versions alleviate the problem considerably for ODT and reduce it for WCC. The CO and WCI methods behave similar to the angle-based smoothing method, whereas our sliced-petal smoothing method outperforms them all. Both the smallest and the largest angle values observed in the output of the sliced-petal smoothing method is significantly better than those of the other methods. Histograms also support our observations by showing the angle quality of each element.

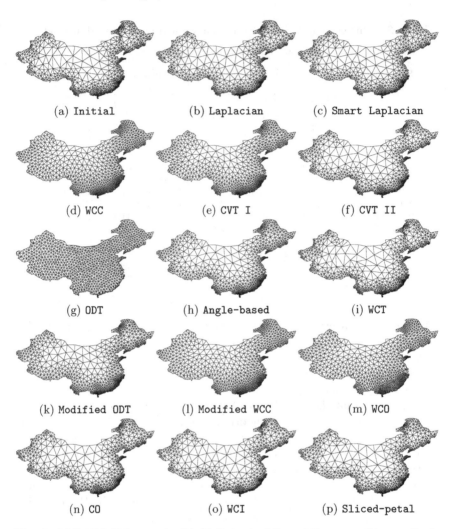

Fig. 5. (a) Initial China mesh. (b)-(j) Output of the existing smoothing methods. (k)-(p) Output of the proposed smoothing methods.

Time. In Table 2, we report the total execution time for each smoothing method. As can be seen, WCT and ODT take significantly more time than the others. In case of WCT, computing the energy function is costly, whereas for ODT convergence takes time. Mainly, due to the current implementation of the feasible region calculation, our sliced-petal method is slower than the remaining. As we expected, the Laplacian smoothing is the fastest, whereas others perform similar to each other.

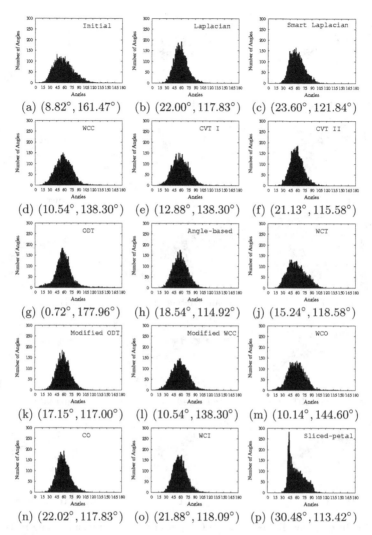

Fig. 6. (a) Initial China mesh histogram. (b)-(j) Output histograms of the existing smoothing methods. (k)-(p) Output histograms of the proposed smoothing methods.

Grading. Experiments show that smoothing on complex geometric domains mostly requires proper grading in order to achieve high quality triangulations. As we pointed out, smoothing methods designed to reach regular meshes fails to conform to the boundary (see Figures 5 and 7). While others achieve proper grading as well, our sliced-petal method converges to significantly higher quality meshes than the existing methods.

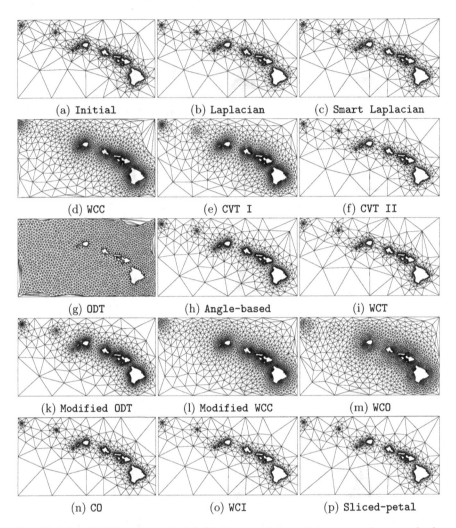

Fig. 7. (a) Initial Hawaii mesh. (b)-(j) Output of the existing smoothing methods. (k)-(p) Output of the proposed smoothing methods.

4.3   Future Work

Combining smoothing methods to benefit from their properties remains as future work. A thorough convergence study on these algorithms and a similar study on quadrilateral/three-dimensional meshes would be helpful for future studies. Our sliced-petal smoothing performance can easily be improved by careful implementation and search strategies. We believe this study will be useful for applications involving complex geometric domains, especially in choosing the appropriate smoothing method.

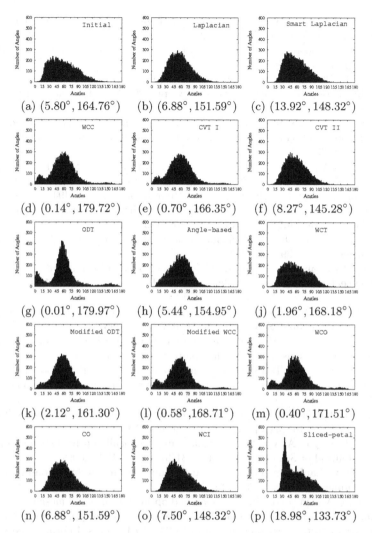

Fig. 8. (a) Initial Hawaii mesh histogram. (b)-(j) Output histogram of the existing smoothing methods. (k)-(p) Output histogram of the proposed smoothing methods.

References

1. Alliez, P., Cohen-Steiner, D., Yvinec, M., Desbrun, M.: Variational tetrahedral meshing. ACM Transactions on Graphics 24, 617–625 (2005)
2. Amenta, N., Bern, M.W., Eppstein, D.: Optimal point placement for mesh smoothing. In: ACM-SIAM Symp. Discrete Algorithms, pp. 528–537 (1997)
3. Babuška, I., Aziz, A.: On the angle condition in the finite element method. SIAM J. Numer. Analysis 13, 214–227 (1976)
4. Bank, R.E., Smith, R.K.: Mesh smoothing using A posteriori error estimates. SIAM Journal on Numerical Analysis 34(3), 979–997 (1997)

5. Bern, M., Eppstein, D.: Mesh generation and optimal triangulation. Computing in Euclidean Geometry, 23–90 (1992)
6. Bern, M., Eppstein, D., Gilbert, J.R.: Provably good mesh generation. J. Comp. System Sciences 48, 384–409 (1994)
7. Bern, M., Plassmann, P.: Mesh generation. In: Sack, J.-R., Urrutia, J. (eds.) Handbook of Computational Geometry, pp. 291–332. Elsevier, Amsterdam (1998)
8. Canann, S.A., Tristano, J.R., Blacker, T.: Opti-smoothing: An optimization-driven approach to mesh smoothing. Finite Elements in Analysis and Design 13, 185–190 (1993)
9. Canann, S.A., Tristano, J.R., Staten, M.L.: An approach to combined laplacian and optimization-based smoothing for triangular, quadrilateral, and quad-dominant meshes. In: 7th Int. Meshing Roundtable, pp. 479–494 (1998)
10. Chen, L.: Mesh smoothing schemes based on optimal Delaunay triangulations. In: 13th Int. Meshing Roundtable, pp. 109–120 (2004)
11. Chen, L., Xu, J.: Optimal Delaunay triangulation, Journal of Computational Mathematics 22, 299–308 (2004)
12. Du, Q., Emelianenko, M., Ju, L.: Convergence of the Lloyd algorithm for computing centroidal Voronoi tessellations. SIAM. J. Numer. Anal. 44(1), 102–119 (2006)
13. Du, Q., Faber, V., Gunzburger, M.: Centroidal Voronoi tessellations: Applications and algorithms. SIAM Review 41, 637–676 (1999)
14. Du, Q., Gunzburger, M.: Grid generation and optimization based on centroidal Voronoi tessellations. Appl. Math. Comp. 133, 591–607 (2002)
15. Du, Q., Wang, D.: Tetrahedral mesh generation and optimization based on centroidal Voronoi tessellations. Inter. J. Numer. Meth. Eng. 56(9), 1355–1373 (2002)
16. Erten, H., Üngör, A.: Computing no small no large angle triangulations, To be appear in Proc. Int. Symp. Voronoi Diagrams (2009)
17. Erten, H., Üngör, A.: Quality triangulations with locally optimal steiner points. SIAM Journal of Scientfic Computing 31(3), 2103–2130 (2009)
18. Field, D.A.: Laplacian smoothing and Delaunay triangulations. Comm. in Applied Numer. Analysis 4, 709–712 (1988)
19. Freitag, L.A.: On combining laplacian and optimization-based mesh smoothing techniques. In: Trends in Unstructured Mesh Generation, pp. 37–43 (1997)
20. Lloyd, S.: Least square quatization in PCM. IEEE Trans. Inform. Theory 28, 129–137 (1982)
21. Mukherjee, N.: A hybrid, variational 3D smoother for orphaned shell meshes. In: 11th Int. Meshing Roundtable, pp. 379–390 (2002)
22. Owen, S.: A survey of unstructured mesh generation technology. In: 7th Int. Meshing Roundtable, pp. 239–267 (1998)
23. Parthasarathy, V.N., Kodiyalam, S.: A constrained optimization approach to finite element mesh smoothing. Finite Elements in Analysis and Design 9, 309–320 (1991)
24. Shewchuk, J.R.: What is a good linear element? interpolation, conditioning, and quality measures. In: 11th Int. Meshing Roundtable, pp. 115–126 (2002)
25. Üngör, A.: Off-centers: A new type of steiner points for computing size-optimal quality-guaranteed Delaunay triangulations. Comput. Geom. Theory Appl. 42(2), 109–118 (2009)

26. VanderZee, E., Hirani, A.N., Guoy, D., Ramos, E.: Well-centered planar triangulation–an iterative approach. In: 16th Int. Meshing Roundtable, pp. 121–138 (2007)
27. Wang, D., Du, Q.: Mesh optimization based on the centroidal Voronoi tessellation. Int. J. Numer. Anal. Mod. 2, 100–113 (2005)
28. Xu, H., Newman, T.S.: An angle-based optimization approach for 2d finite element mesh smoothing. Finite Elem. Anal. Des. 42(13), 1150–1164 (2006)
29. Zhou, T., Shimada, K.: An angle-based approach to two-dimensional mesh smoothing. In: 9th Int. Meshing Roundtable, pp. 373–384 (2000)

A Novel Method for Surface Mesh Smoothing: Applications in Biomedical Modeling

Jun Wang and Zeyun Yu[*]

Department of Computer Science, University of Wisconsin-Milwaukee
{wang43,yuz}@uwm.edu

Abstract. In this paper, we present a surface-fitting based smoothing algorithm for discrete, general-purpose mesh models. The surface patch around a mesh vertex is defined in a local coordinate system and fitted with a quadratic polynomial function. An initial mesh smoothing is achieved by projecting each vertex onto the fitted surface. At each vertex of the initial mesh, the curvature is estimated and used to label the vertex as one of four types. The curvature-based vertex labeling, together with the curvature variation within a local region of a vertex, is utilized to adaptively smooth the mesh with fine features well preserved. Finally, three post-processing methods are adopted for mesh quality improvement. A number of realworld mesh models are tested to demonstrate the effectiveness and robustness of our approach.

Keywords: Surface mesh smoothing, Quadric surface fitting, Curvature labeling, Mesh quality improvement.

1 Introduction

The surface reconstruction and visualization from 3D imaging data have found wide applications in biomedical fields, such as computer-aided diagnosis, intervention planning, and realistic pathological analysis and prediction. The models used for such purposes are typically represented by surface meshes in triangular or quadrilateral forms. Going from imaging data to surface meshes involves a number of computational approaches. Two of the critical steps are image segmentation and surface mesh generation and processing. In digital images, image segmentation normally produces a vast number of discrete voxels (small cubes) that approximate the boundary of an object of interest. These voxels can certainly be converted into surface manifolds represented by quadrilateral meshes by extracting the faces of the cubes that face towards either outside or inside of the boundaries. But the resulting meshes suffer from an extremely bumpiness on the surface, as shown in the examples in Fig. 1. One of the goals in the present paper is hence to smooth a given mesh to reduce the bumpiness on the surface.

---

[*] Corresponding author.

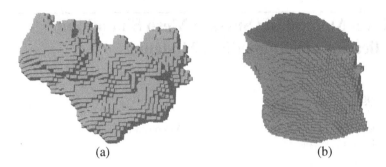

(a)   (b)

Fig. 1. Examples of surface meshes generated from image segmentation. The meshes appear very bumpy without further mesh processing.

Surface mesh smoothing has been studied for decades. Laplace iterative smoothing is one of the most common and simplest techniques for mesh smoothing [1-5]. During each iteration, each vertex of a mesh is adjusted to the barycenter of its neighboring region. It is very fast but often results in volume shrinkage. Taubin [2] used a signal processing method, the Laplacian operator, to fair surface design and reduce the shrinkage. Desbrun et al. [3] extended Taubin's work to smooth irregular meshes by using geometric flows. Their approach provides a better way to prevent the volume shrinkage. Peng [6] presented a denoising approach for geometric data represented as a semiregular mesh on the basis of locally adaptive Wiener filtering.

Other popular smoothing approaches include the energy minimization techniques. Kobblet [7, 8] proposed a general algorithm to fair a triangular mesh with arbitrary topology in R^3 by estimating the curvature for the discrete mesh model. Welch and Witkin [9] described an approach to designing fair, freeform shape by using triangulated surfaces. These methods are time-consuming due to the complicated energy functions that need to be solved.

Recently, feature-preserving smoothing methods [10-21] have drawn more and more attentions. Jones et al. [19] developed a feature-preserving smoothing algorithm by adopting statistics and local first-order predictors of triangulated surface meshes. Bajaj et al. [20] proposed a PDE-based anisotropic diffusion approach for processing noisy geometric surfaces and functions defined on surfaces. Li et al. [21] adopted the weighted bi-quadratic Bezier surface fitting and uniform principal curvature techniques to smooth surface meshes.

In the present paper, we describe a novel approach, combining the surface fitting, curvature estimation, and vertex labeling, to smooth meshes. Beside the surface smoothness that we shall demonstrate below, Antoher main goal in this paper is to preserve or improve the mesh quality so that the processed meshes have no sharp angles. The mesh quality is extremely critical in some applications such as the numerical simulation using finite or boundary element methods. However, not all meshes that appear smooth have high quality in terms of angles. For example, the marching cube algorithm can be used to generate a very smooth surface triangulation of a volumetric function but the resulting mesh usually suffers from too many sharp or skinny angles, which often cause poor accuracy in numerical simulation. To this end, improving mesh quality, sometimes also referred to as *mesh smoothing*, is taken into serious consideration in our algorithms presented here.

While triangular surface meshes are perhaps the most common form in representing a 3D model and a majority of previous work has been focused on this type of meshes, we will consider in this paper quadrilateral meshes, especially those extracted from 3D imaging data and represented by cube-like bumpy meshes. The resulting meshes will be smoothed quadrilateral meshes with improved quality, which can be readily converted into triangular meshes if needed. In addition, the data structures we used can be easily modified to read triangular meshes as inputs. In other words, the approach we propose here will provide a general-purpose tool for mesh processing in a variety of applications.

2   Surface Mesh Smoothing Algorithm

Our algorithm includes four steps. At first, the geometric and topological relationship is constructed from the input surface mesh model. For each vertex, we construct a local coordinate system and then fit the neighboring vertices in the local coordinate system with an analytical quadric surface function. The vertex being considered is projected onto this quadric surface. The initial mesh smoothing is achieved by updating each vertex with its projection. Then, the curvature of each vertex is estimated based on the first and second fundamental forms of its quadric surface obtained and then the vertex is labeled with the corresponding type. Additionally, for each vertex, the size of its neighborhood is dynamically determined according to the curvature and vertex type so that the mesh is adaptively smoothed by similar approaches as in the first step. Finally, the mesh is further improved with a few post-processing algorithms.

2.1   Initial Mesh Smoothing

The local surface patch around a point can be approximated with a quadric surface [22]: $S(u, v) = (u, v, h(u, v))$, a parametric representation in a local coordinate system as shown in Fig 2, where p is the origin; h-axis directs along the normal vector n of p on S; and u-, v- axes are orthogonal vectors in the tangent plane of p on S. Obviously, the local coordinate system (p-uvh) can be transformed from the global coordinate system (o-xyz). According to the surface theory [23], the local shape of surface around p can be represented with Darboux system $D(p) = (p, T_1, T_2, n, k_1, k_2)$, where (T_1, T_2), (k_1, k_2) are principal directions and curvatures, respectively.

We extend this principle to discrete mesh models and construct the local coordinated system for each vertex on the meshes. Since the construction of a local system mainly relies on the h-axis, i.e., the normal vector of the vertex, we first estimate the normal vector for each vertex with the area-weighted averaging method.

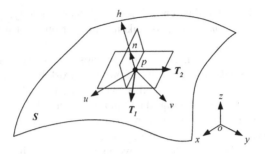

Fig. 2. An illustration of the local coordinate system on a surface

Let $NV\_i(k)$ be the set of neighboring vertices of vertex v_i within the **k-ring** (that is, the minimum number of edges from v_i to the neighboring vertices is equal to or less than k), $E\_i = \{<v_i, v_j>|v_j \in NV\_i(1)\}$ the set of edges incident to v_i, and F_i the set of faces incident to v_i. Then the normal vector n_i of v_i is calculated as:

$$n_i = \frac{1}{N} \sum_{j=1}^{N} \left(area_{f_j} \cdot n_{f_j} \right), \quad f_j \in F\_i. \tag{1}$$

where $area_f$, n_f are respectively the area and normal vector of face f in F_i. After the normal vector n_i is calculated, we consider v_i, n_i as the origin and z-axis of the local coordinate system respectively. Then the x- and y- axes of the local coordinate system are arbitrarily chosen in the plane locating at v_i that is orthogonal to the normal n_i.

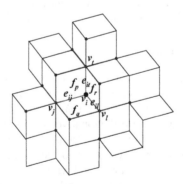

Fig. 3. The 2-ring neighboring vertices at a vertex. The minimal number of edges from these vertices to the center vertex is equal to or less than 2.

After constructing the local coordinate system, we find all the neighboring vertices for each vertex with a ring-by-ring scheme, referred to as the *k-ring neighborhood*. At the initial mesh smoothing phase, k is fixed as 2 in our algorithm. Fig. 3 shows the 2-ring neighboring vertices of v_i, in which the red and blue vertices are the first and second ring neighboring vertices respectively, and all these vertices constitute the 2-ring neighbors of v_i.

For each vertex, the global coordinates of all the neighboring vertices are transformed to the local coordinates by homogeneous transformation and then the local coordinates are used to fit an analytical quadric surface:

$$h(u, v) = au^2 + buv + cv^2 + eu + fv + g \qquad (2)$$

In our algorithm, the least square fitting method is adopted. Let $V\_i = \{<x_j, y_j, z_j> | j = 1, 2, \ldots m\}$ be the local coordinates of 2-ring neighboring vertices of vertex v_i. The objective function of the least square quadric surface fitting can be expressed as:

$$F = \sum_{j=1}^{m} \left\| \begin{pmatrix} x_j \\ y_j \\ ax_j^2 + bx_j y_j + cy_j^2 + ex_j + fy_j + g \end{pmatrix} - \begin{pmatrix} x_j \\ y_j \\ z_j \end{pmatrix} \right\|^2 \cdot w_j \qquad (3)$$

where w_j is the weighting function defined below. It is observed that the closer the neighboring vertex to v_i, the more it affects the surface shape. Therefore, w_j can be represented as

$$w_j = e^{-\|v_j - v_i\| / \max(\|v_1 - v_i\|, \ldots \|v_m - v_i\|)} \qquad (4)$$

Minimizing F gives rise to $\dfrac{\partial F}{\partial a} = 0$, $\dfrac{\partial F}{\partial b} = 0$, $\dfrac{\partial F}{\partial c} = 0$, $\dfrac{\partial F}{\partial e} = 0$, $\dfrac{\partial F}{\partial f} = 0$, and $\dfrac{\partial F}{\partial g} = 0$, that is,

$$\left(\sum_{j=1}^{m} w_j \begin{bmatrix} x_j^2 \\ x_j y_j \\ j_j^2 \\ x_j \\ y_j \\ 1 \end{bmatrix} \begin{bmatrix} x_j^2 \\ x_j y_j \\ j_j^2 \\ x_j \\ y_j \\ 1 \end{bmatrix}^T \right) \begin{bmatrix} a \\ b \\ c \\ e \\ f \\ g \end{bmatrix} = \sum_{j=1}^{m} w_j z_j \begin{bmatrix} x_j^2 \\ x_j y_j \\ j_j^2 \\ x_j \\ y_j \\ 1 \end{bmatrix} \qquad (5)$$

The linear equations can be solved with Gaussian elimination method [24], which yields the coefficients of the quadric surface. The vertex v_i is then projected onto the fitted quadric surface along its normal vector. Since its normal vector is exactly the z-axis in the local coordinate system, the coordinate of the projected point of v_i should be $(0, 0, g)$ in the local coordinate system. Its global coordinate is calculated by the coordinate transformation from local system to global system. Finally, the vertex v_i is updated with the global coordinate of its projection. The above steps are performed for every vertex, yielding what we call the *initial mesh smoothing*. Fig. 4 shows mesh smoothing of two simple models.

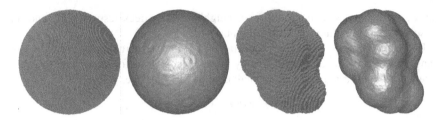

Fig. 4. Two examples of initial mesh smoothing. The first and third from left are the original models and the other two are meshes after smoothing.

2.2 Curvature Estimation and Labeling

Meshes generated in the initial smoothing step are much better than the original meshes but may not be good enough for subsequent applications if the original meshes are too bumpy. In our algorithm, an adaptive mesh smoothing algorithm is proposed to further smooth the meshes by taking advantage of the same surface fitting technique as seen in Section 2.1 but the size of the neighborhood considered adaptively changes according to the local geometric properties as explained below.

Many features in the real-world structures/models are usually defined on the basis of curvature variance of the surface meshes. When an input mesh, such as the cube-based meshes in Fig. 1, is too coarse, we cannot accurately calculate the curvature information. However, this becomes much less problematic after the initial mesh smoothing. To further smooth a mesh and preserve the features in the model, the same surface fitting and vertex projection schemes as in initial Section 2.1 are adopted, whereas the ring number of neighboring vertices for surface fitting is dynamically determined based on the curvature variance. In this subsection we introduce how the curvatures are analytically calculated and used to label the vertices.

In the initial mesh smoothing, each vertex is associated with a quadric surface using the least square fitting technique. Since each vertex has been adjusted to its projection in the normal direction, it is straightforward to calculate the curvature of the vertex on quadric surface by using the first and second fundamental forms of a surface. The analytical form of the quadric surface of a vertex v_i can be transformed to the parametric form as:

$$S = (u, v, h(u, v)) \tag{6}$$

The coefficients of the first and second fundamental forms of the quadric surface at v_i are calculated as:

$$\begin{cases} E = 1 + e^2 \\ F = ef \\ G = 1 + f^2 \end{cases}, \text{ and } \begin{cases} L = 2a \big/ \sqrt{e^2 + f^2 + 1} \\ M = b \big/ \sqrt{e^2 + f^2 + 1} \\ N = 2c \big/ \sqrt{e^2 + f^2 + 1} \end{cases} \tag{7}$$

The Gaussian and mean curvatures at v_i are then given respectively as follows:

$$K_{v_i} = \frac{LN - M^2}{2(EG - F^2)} = \frac{4ac - b^2}{(e^2 + f^2 + 1)^2} \tag{8}$$

$$H_{v_i} = \frac{EN - 2FM + GL}{2(EG - F^2)} = \frac{c + ce^2 + a + af^2 - bef}{(e^2 + f^2 + 1)^{3/2}} \tag{9}$$

The Gaussian and mean curvatures of a mesh face $face_i$ with vertices $v_1, v_2, \ldots, v_m$ are approximated as:

$$K_{face_i} = \frac{1}{m}\sum_{j=1}^{m} K_{v_j} \quad \text{and} \quad H_{face_i} = \frac{1}{m}\sum_{j=1}^{m} H_{v_j} \tag{10}$$

After the Gaussian (K) and mean (H) curvatures are obtained, all the mesh vertices and faces are labeled with the corresponding type. In particular, the signs of K and H define the surface type and the values of K and H define the surface sharpness. Besl and Jain [25] proposed eight fundamental surface types using the signs of K and H as shown in Table 1.

Table 1. Eight fundamental surface types

	$K > 0$	$K = 0$	$K < 0$
$H < 0$	Peak	Ridge	Saddle ridge
$H = 0$	N/A	Flat	Minimal surface
$H > 0$	Pit	Valley	Saddle valley

In our algorithm, four simplified types are considered: convex, flatten, minimal and concave. In Table 1, the convex type includes peak, ridge and saddle ridge surfaces; the flatten and minimal types only includes flat and minimal surfaces respectively; and the concave type contains pit, valley and saddle valley surfaces. These surface types are extended to mesh vertices. After the curvatures are calculated, the faces and vertices are labeled with the criterion in Table 2.

Table 2. Four surface types

Surface/Vertex type	Signs of H & K
Convex type	$H < 0$
Flatten type	$H = 0$
Minimal type	$H = 0$ && $K < 0$
Concave type	$H > 0$

2.3 Adaptive Mesh Smoothing

After curvature-based vertrex labeling, the whole mesh is segmented into four types of features, each of which consists of topologically adjacent vertices and faces with the same surface type. In order to further smooth the mesh model, the quadric surface fitting on each vertex is performed in a similar way as in Section 2.1, and the vertex is adjusted by projecting onto the fitted surface. Meanwhile, to preserve the features of the mesh model during the smoothing process, the neighboring vertices used in the surface fitting here should have the same type as that of the center vertex. As a result, these four types of features will be preserved after smoothing.

However, there are cases in which more than one feature may be seen in a segment having the same surface type. Consequently, if the vertex type was considered as the only criterion for definging the neighborhood of the vertex in the quadric surface fitting, many sub-features would be filtered. To address this issue, the standard deviation of curvatures within the neighborhood chosen is used as an additional constraint to determine the neighborhood size to preserve the sub-features in a mesh. The standard deviation of curvatures in a neighborhood of a vertex should be bounded by a pre-defined threshold.

Below we give the detailed algorithm of adaptive mesh smoothing. Let $V = \{v_1, v_2, ... v_n\}$ be the set of mesh vertices after initial mesh smoothing; $T = \{t_1, t_2, ... t_n\}$ the types of vertices; k the fixed ring number in the initial mesh smoothing and $tmpVerArray$ the temporary vertex array. k is set as 2 by default. For each vertex v_i,

Step 1. Search its k-ring neighboring vertices and calculate the standard deviation σ_k of curvatures of all such vertices;
Step 2. Set $tmpVerArray$ = NULL. For each vertex v_i in the set of the outermost (k-th) ring neighboring vertices,
    i) if the types t_i and t_j of v_i and v_j are same, add the v_j into $tmpVerArray$;
    ii) otherwise, go to step 4.
Step 3. Calculate the standard deviation σ_{k+1} of curvatures of k-ring neighboring vertices together with vertices in $tmpVerArray$.
    i) if σ_{k+1} is less than σ_k, add each vertex of $tmpVerArray$ into the neighboring vertices and update $k \leftarrow k+1$, $\sigma_k \leftarrow \sigma_{k+1}$; then go to step 2;
    ii) otherwise, go to step 4.
Step 4. Fit the k-ring neighboring vertices of v_i with a quadric surface and project v_i onto the surface to get the vertex v_i';
Step 5. Adjust v_i to v_i', i.e. $v_i \leftarrow v_i$'.

During the smoothing processes, we adjust mesh vertices by updating their coordinates, while the geometric and topological relationship of the mesh model remains unchanged, which reduces the space and time complexity of this smoothing algorithm.

2.4 Mesh Quality Improvement

When the input mesh has complicated features and the resolution of the mesh is not high enough to capture the features, the smoothing approaches described above may produce some unwanted errors or low-quality meshes, such as twisted polygons, very short edges, or very sharp angles. Below we briefly describe the strategies to handle these three cases.

- **Twisted meshes.** Normally, the four vertices in a quadrilateral should be arranged in order (e.g., counterclockwise), while in a twisted quadrilateral, the vertices are arranged in a "Z" shape. Fig. 5(a) shows the twisted meshes, $mesh_1$, $mesh_2$ and Fig. 5(b) gives the corresponding meshes after adjustment. In order to detect whether a quadrilateral is twisted or not, we use the normal vectors of the two triangles associated with the quadrilateral. In a regular quadrilateral the angle between the normal vectors n_{123}, n_{341} of the triangles $v_1v_2v_3$, $v_3v_4v_1$ is less than $90°$, as shown in Fig. 5(c). By contrast, the angle would be more than $90°$ in a twisted quadrilateral (see Fig. 5(d)). A twisted quadrilateral usually occurs in conjunction with its opposite quadrilateral sharing the same edge. Therefore, the common edge (v_3v_2 in Fig. 5(a)) between a pair of twisted quadrilaterals can be easily identified and its end points are swapped to cure this problem.

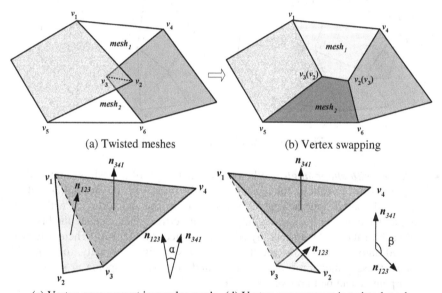

(a) Twisted meshes   (b) Vertex swapping

(c) Vertex arrangement in regular mesh   (d) Vertex arrangement in twisted mesh

Fig. 5. Illustration of handling twisted meshes

- **Meshes with short edges.** This occurs when the length of one edge of a quadrilateral is shorter than a pre-defined threshold (for example, see v_1v_2 in Fig. 6). The steps to correct this type of meshes are described as follows. For the end point v_1 of a short edge v_1v_2,,
  - Find the two associated edges v_1v_3, v_1v_4 in the meshes, $mesh_1$, $mesh_2$, containing v_1v_2,
  - Compute the bisector c_1 of $\angle v_3v_1v_4$ and a plane pln passing through c_1
    and the vector $\overrightarrow{v_1v_3} \times \overrightarrow{v_1v_4}$;

– Find the intersection point p_1 between *pln* and the edges of all the polygons incident to v_1;

For the other end point v_2, its intersection point p_2 can be computed in the same way. Then, the trisected points v'_1, v'_2 are determined on the polyline $\overline{p_1 v_1 v_2 p_2}$ such that the lengths of $p_1 v'_1$, $v'_1 v_1 v_2 v'_2$ and $v'_2 p_2$ are equal. Finally, v_1, v_2 are updated with v'_1, v'_2 such that the short edge is prolonged (Fig. 6).

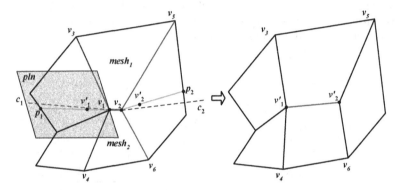

Fig. 6. Illustration of handling short edges

- **Meshes with shape angles.** When an internal angle of a mesh is smaller than a pre-defined threshold, the mesh needs to be improved. To this end, we adopt the angle-based approach as described in [26]. For the vertex v_1 with sharp angle, we consider the surrounding incident vertices and calculate the bisector of each of the angles formed by adjacent vertices on that ring. Then v_1 is projected onto each bisector (see the blue points in Fig. 7). Finally, the centroid of all these projection points is calculated and used to update v_1 so that the sharp angle can be improved (Fig. 7).

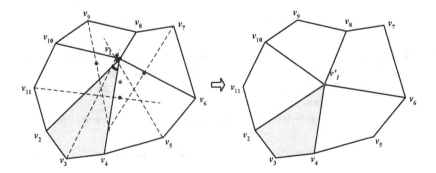

Fig. 7. Illustration of handling sharp angles

3 Implementation and Results

All algorithms described have been implemented in Visual C++ and OpenGL, running on a Pentium IV PC configuration with 3.0G Hz. Many 3D surface models have been tested and a couple of them are demonstrated here.

Fig. 8 shows the smoothing result of the 2CMP molecule, randomly chosen from the Protein Data Bank (http://www.rcsb.org/). The original mesh in Fig. 8(a) was generated using the approach described in [27]. The initial mesh smoothing gives the result as shown in Fig. 8(b). The curvature labeling is performed and shown in Fig. 8(c), where the patches in green and red are convex and concave surfaces respectively. A smoother mesh is achieved by the adaptive mesh smoothing, as seen in Fig. 8(d). However, there are still some defective polygons in the smoothed mesh, as shown with small dark dots in Fig. 8(e). The meshes after the quality improvement are shown in Fig. 8(f). To estimate the smoothness of a surface mesh, we calculate the curvature variation at each vertex as the maximal difference between the curvature of the vertex and the curvatures of the surrounding vertices. The average curvature variations over the entire mesh of 2CMP are 0.5032, 0.3529, and 0.0921 in the original mesh, the mesh after initial smoothing and the adaptively smoothed mesh respectively, suggesting that the proposed adaptive mesh smoothing method is effective in smoothing a surface mesh. Fig. 9 gives the result of another molecule called 2HAO, taken again from the Protein Data Bank. There are 72098 and 228132 vertices in the 2CMP and 2HAO models, and the computational time is 31.51s and 92.24s respectively.

As mentioned in the introduction, the marching cube method is able to produce smooth surface meshes, but many skinny triangular meshes are generated as well. For instance, the angle histograms of several molecular surface meshes generated by the marching cube method are shown in [28], where one can see a lot of very small (near 0^0) and large angles (near 180^0). To demonstrate the quality of the meshes , the angle histograms of the meshes generated with our method are plotted in Fig. 10, where the angles used are the internals angles of all the triangles obtained by dividing each quadrilateral into two triangles. It is interesting to observe that there are two obvious peaks in the histogram: one around $60°$ that corresponds to equilateral triangles and the other around $90°$ indicating that a large number of regular (equilateral) quadrilateral have been kept after the mesh smoothing.

Fig. 11 demonstrates the mesh smoothing results from 3D imaging data. Fig. 11(a) shows a cross section of the 3D cryo-electron microscopy reconstruction of the rice dwarf virus. The initial surface mesh, shown in Fig. 1(a), was generated using the 3D image segmentation algorithms [29]. Fig. 11(c) shows the surface mesh after applying the smoothing techniques as described in Section 2. Fig. 11(b) shows a cross section of the 3D electron tomographic reconstruction of the ventricle muscle cell. The initial surface mesh, illustrated in Fig. 1(b), was extracted using automatic image segmentation methods [30]. Fig. 11(d) shows the mesh after our smoothing algorithms. From both examples, we can see that our mesh smoothing approach can be used in conjunction with image segmentation for a variety of 3D biomedical imaging data.

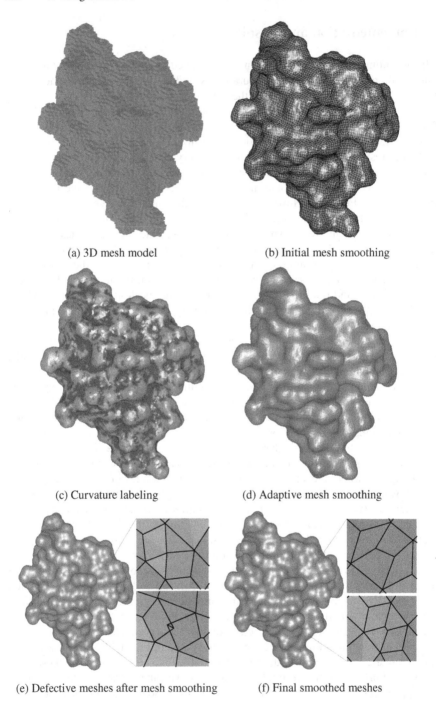

(a) 3D mesh model

(b) Initial mesh smoothing

(c) Curvature labeling

(d) Adaptive mesh smoothing

(e) Defective meshes after mesh smoothing

(f) Final smoothed meshes

Fig. 8. Example of mesh smoothing for the molecular model 2CMP

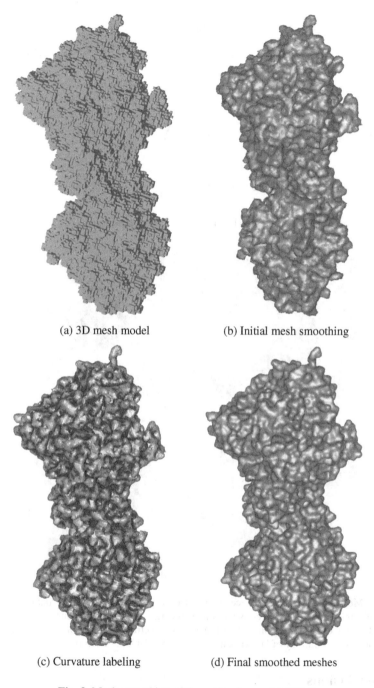

(a) 3D mesh model   (b) Initial mesh smoothing

(c) Curvature labeling   (d) Final smoothed meshes

Fig. 9. Mesh smoothing of the molecular model 2HAO

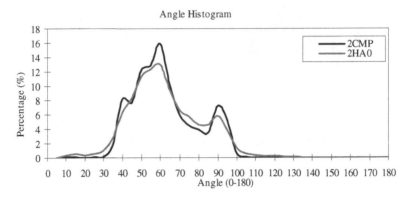

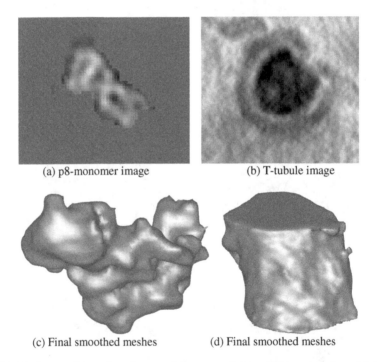

Fig. 10. Mesh angle histogram of models: 2CMP and 2HAO

(a) p8-monomer image   (b) T-tubule image

(c) Final smoothed meshes   (d) Final smoothed meshes

Fig. 11. Mesh smoothing of surface models extracted from 3D imaging data. (a) & (c): a cryo-electron microscopy reconstruction of the rice dwarf virus (courtesy of Dr. Wah Chiu, Baylor College of Medicine). (b) & (d): an electron tomographic reconstruction of ventricular cells (courtesy of Dr. Masahiko Hoshijima, UC-San Diego).

4   Conclusions

We have presented in this paper a novel mesh smoothing method based on quadric surface fitting for general mesh models. The fitting algorithm is combined with vertex projection, curvature estimation, and mesh labeling of four types. Based on the geometric characteristics of surfaces, the adaptive mesh smoothing is conducted to further smooth meshes while important features are well preserved. To improve the mesh quality, three adjustment methods are adopted to handle defective meshes.

Our mesh smoothing algorithm can handle a variety of meshes, including molecular models, imaging data and industrial surface meshes. As demonstrated in the results, our method can generate meshes with high quality, i.e., no sharp angle or short edge in the output meshes. This is particularly useful and sometimes necessary in such applications as surface reconstruction, visualization, and numerical simulation. While the examples demonstrated here are all quadrilateral meshes, our method can be easily modified to handle triangular or other types of surface meshes.

Due to the quadric model being used to approximate the local shape of a freeform surface, there may be some cases where our method does not work perfectly. Such cases include very sharp and thin features and meshes with too low resolutions to capture the features of interest. One of our future efforts is to detect these ill-posed regions and apply mesh subdivision techniques to increase the mesh resolution so that our method can work more effectively in these circumstances. There are still a few small angles persisting in the final meshes as can be seen in Fig. 10. We shall work on better methods to handle these cases during the vertex projection step. While sharp edges or corners are not commonly present in the biomedical examples we have tested, these features will be taken care of in our future work by making a good balance between sharp features and possible surface noises.

References

1. Field, D.A.: Laplacian smoothing and Delaunay triangulations. Communications in Applied Numerical Methods 4, 709–712 (1988)
2. Taubin, G.: A Signal Processing Approach to Fair Surface Design. In: Proceedings of SIGGRAPH 1995, pp. 351–358 (1995)
3. Desbrun, M., Meyer, M., Schröder, P., Barr, A.H.: Implicit Fairing of Irregular Meshes Using Diffusion and Curvature Flow. In: Proceedings of SIGGRAPH 1999, pp. 317–324 (1999)
4. Vollmer, J., Mencl, R., Müller, H.: Improved Laplacian smoothing of noisy surface meshes. In: Proceedings of Eurographics, pp. 131–138 (1999)
5. Ohtake, Y., Belyaev, A., Bogaeski, I.: Polyhedral Surface Smoothing with Simultaneous Mesh Regularization. In: Geometric Modeling and Processing, pp. 229–237 (2000)
6. Peng, J., Strela, V., Zorin, D.: A Simple Algorithm for Surface Denoising. In: Proceedings of IEEE Visualization 2001, pp. 107–112 (2001)
7. Kobbelt, L.: Discrete fairing. In: Proceedings of the 7th IMA Conference on the Mathematics of Surfaces, pp. 101–131. Springer, Cirencester (1996)
8. Kobbelt, L., Botsch, M., Schwanecke, U., Seidel, H.: Feature sensitive surface extraction from volume data. In: Proceedings of SIGGRAPH 2001 (2001)
9. Welch, W., Witkin, A.: Free-form shape design using triangulated surfaces. In: Proceedings of SIGGRAPH 1994, pp. 247–256. ACM Press, Orlando (1994)

10. Desbrun, M., Meyer, M., Schröder, P., Barr, A.H.: Anisotropic Feature-Preserving Denoising of Height Fields and Bivariate Data. In: Graphics Interface, pp. 145–152 (2000)
11. Taubin, G.: Linear anisotropic mesh filtering. IBM Research Technical Report. RC2213 (2001)
12. Liu, X., Bao, H., Heng, P., Wong, T., Peng, Q.: Constrained fairing for meshes. Computer Graphics Forum 20(2), 115–123 (2001)
13. Liu, X., Bao, H., Shum, H., Peng, Q.: A novel volume constrained smoothing method for meshes. Graphical Models 64, 169–182 (2002)
14. Tasdizen, T., Whitaker, R., Burchard, P., Osher, S.: Geometric surface smoothing via anisotropic diffusion of normals. In: Proceedings of IEEE Visualization, pp. 125–132 (2002)
15. Ohtake, Y., Belyaev, A., Seidel, H.-P.: Mesh Smoothing by Adaptive and Anisotropic Gaussian Filter. In: Vision, Modeling and Visualization, pp. 203–210 (2002)
16. Fleishman, S., Drori, I., Cohen-Or, D.: Bilateral Mesh Denoising. ACM Trans. Gr. (2003)
17. Zhang, H., Fiume, E.L.: Mesh Smoothing with Shape or Feature Preservation. In: Advances in Modeling, Animation, and Rendering, pp. 167–182 (2002)
18. Clarenz, U., Diewald, U., Rumpf, M.: Anisotropic geometric diffusion in surface processing. In: IEEE Visualization 2000, pp. 397–405 (2000)
19. Bajaj, C., Xu, G.: Anisotropic Diffusion on Surfaces and Functions on Surfaces. ACM Trans. Gr. 22(1), 4–32 (2003)
20. Jones, T., Durand, F., Desbrun, M.: Non-iterative, feature-preserving mesh smoothing. In: Proceedings of SIGGRAPH 2003, pp. 943–949. ACM Press, San Diego (2003)
21. Li, Z., Ma, L., Jin, X., Zheng, Z.: A new feature-preserving mesh-smoothing algorithm. Visual Comput. 25, 139–148 (2009)
22. Frey, P.J.: About surface remeshing. In: Proc. in 9th IMR, New-Orleans, pp. 123–136 (2000)
23. Milroy, M.J., Bradley, C., Vickers, G.W.: Segmentation of a wrap-around model using an active contour. Computer Aided Designed 29(4), 299–320 (1997)
24. Atkinson, K.A.: An Introduction to Numerical Analysis, 2nd edn. John Wiley & Sons, New York (1989)
25. Besl, P.J., Jain, R.: Segmentation through Variable-Order Surface Fitting. In: IEEE PAMI 1988, vol. 10(2), pp. 167–192 (1988)
26. Zhou, T., Shimada, K.: An angle-based approach to two-dimensional mesh smoothing. In: Proc. in 9th IMR, New-Orleans, pp. 373–384 (2000)
27. Yu, Z.: A list-based method for fast generation of molecular surfaces. In: The 31st International Conference of IEEE Engineering in Medicine and Biology Society (accepted, 2009)
28. Yu, Z., Holst, M., Cheng, Y., McCammon, J.A.: Feature-Preserving Adaptive Mesh Generation for Molecular Shape Modeling and Simulation. Journal of Molecular Graphics and Modeling 26(8), 1370–1380 (2008)
29. Yu, Z., Bajaj, C.L.: Computational approaches for automatic structural analysis of large bio-molecular complexes. IEEE/ACM Transactions on Computational Biology and Bioinformatics 5(4), 568–582 (2008)
30. Yu, Z., Holst, M., Hayashi, T., Bajaj, C.L., Ellisman, M.H., McCammon, J.A., Hoshijima, M.: Three-dimensional geometric modeling of membrane-bound organelles in ventricular myocytes: Bridging the gap between microscopic imaging and mathematical simulation. Journal of Structural Biology 164(3), 304–313 (2008)

Quality Improvement of Non-manifold Hexahedral Meshes for Critical Feature Determination of Microstructure Materials

Jin Qian[1], Yongjie Zhang[1], Wenyan Wang[1], Alexis C. Lewis[2], M.A. Siddiq Qidwai[2], and Andrew B. Geltmacher[2]

[1] Department of Mechanical Engineering, Carnegie Mellon University, 5000 Forbes Avenue, Pittsburgh, PA 15213, USA
[2] Multifunctional Materials Branch, Naval Research Laboratory, 4555 Overlook Avenue SW, Washington, DC 20375, USA

Abstract. This paper describes a novel approach to improve the quality of non-manifold hexahedral meshes with feature preservation for microstructure materials. In earlier works, we developed an octree-based isocontouring method to construct unstructured hexahedral meshes for domains with multiple materials by introducing the notion of *material change edge* to identify the interface between two or more materials. However, quality improvement of non-manifold hexahedral meshes is still a challenge. In the present algorithm, all the vertices are categorized into seven groups, and then a comprehensive method based on pillowing, geometric flow and optimization techniques is developed for mesh quality improvement. The shrink set in the modified pillowing technique is defined automatically as the boundary of each material region with the exception of local non-manifolds. In the relaxation-based smoothing process, non-manifold points are identified and fixed. Planar boundary curves and interior spatial curves are distinguished, and then regularized using B-spline interpolation and resampling. Grain boundary surface patches and interior vertices are improved as well. Finally, the local optimization method eliminates negative Jacobians of all the vertices. We have applied our algorithms to two beta titanium datasets, and the constructed meshes are validated via a statistics study. Finite element analysis of the 92-grain titanium is carried out based on the improved mesh, and compared with the direct voxel-to-element technique.

Keywords: Non-manifold hexahedral mesh, quality improvement, pillowing, B-spline curve, geometric flow, optimization, microstructure materials.

1 Introduction

Many researchers have incorporated simulated material microstructures into predictive finite element analysis tools: assigning properties such as grain shape, size, and crystallography based on measured distributions [3, 6, 10, 27]. Models based on simulated microstructures can provide important insights into macroscopic material behaviors. The identification of local behaviors and critical microstructural features requires the use of real, experimentally-measured microstructures as the simulation input. The collection and analysis of 3D microstructure data has been carried out by a number of groups [11, 18, 19, 34]. These 3D datasets, however, are typically quite large and complex. Incorporating these datasets into predictive models therefore requires advanced meshing techniques which can preserve the essential features with as few elements as possible.

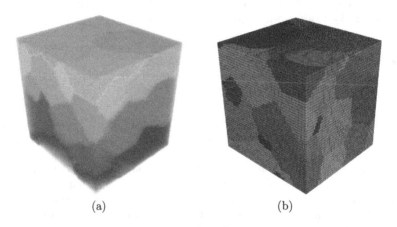

(a)   (b)

Fig. 1. Polycrystalline 92-grain beta titanium. (a) The volume rendering result; and (b) The mesh obtained through the voxel-to-brick transformation.

In its simplest form, the microstructural data is comprised of a 3D array of values which represent a property or identifier for each voxel as shown in Figure 1(a). People have used various ways to construct tetrahedral meshes for microstructure data; however, hexahedral meshes are preferred due to their superior performance in finite element analysis in terms of reduced error, smaller element number and improved reliability. A straightforward approach is to convert each voxel to a 3D brick element as shown in Figure 1(b). This direct method of meshing has the advantages of rapid mesh generation and automatic identification of grain boundaries [20]. However, the resulting mesh has a number of disadvantages for simulation of mechanical behavior. In particular, these meshes typically consist of a large number of elements, with a constant shape and volume. Additionally, direct conversion of voxel data to brick elements [26] results in a stair-casing effect at the boundaries.

In this paper, we choose an octree-based isocontouring algorithm to construct hexahedral meshes [40]. This algorithm is robust and works for arbitrarily complicated geometry and topology. The generated mesh may have a few elements with unsatisfied aspect ratios. Therefore, the mesh needs to be improved. In our quality improvement algorithm, all vertices are categorized into seven groups according to their relative locations in the individual grain and in the whole representative volume element (RVE) dataset. With this vertex classification, we apply different algorithms to different groups of vertices to improve the mesh quality. A comprehensive method based on pillowing, geometric flow and optimization techniques is developed. The shrink set in the modified pillowing technique is defined automatically as the boundary of each material region with the exception of local non-manifolds. In the relaxation-based smoothing process, non-manifold points are identified and fixed. Planar boundary curves and interior spatial curves are distinguished and regularized based on B-spline interpolation and resampling. Grain boundary surface patches and interior vertices are improved as well using geometric flow and the weighted averaging method. Finally, the local optimization method eliminates negative Jacobians of all the vertices, and improves the aspect ratio of the mesh.

We have applied our quality improvement algorithms to two beta titanium datasets, and the generated hexahedral meshes are validated via a statistics study. To compare our meshing technique to the voxel-to-brick technique, a 3D volume consisting of 92 grains is sampled from a reconstructed titanium microstructure [34]. A detailed comparison between these two meshes is made, and a series of finite element analysis are carried out. A number of significant advantages of our algorithm can be observed, particularly with respect to grain boundaries and triple junctions. These features are of particular interest in this model because they are the most likely sites for initiation of plastic flow in a polycrystalline metallic material.

The remainder of this paper is organized as follows: Section 2 summarizes related previous work. Section 3 reviews data acquisition and octree-based mesh generation. Section 4 talks about the classification of vertices and the corresponding criteria. Section 5 explains the detailed algorithm of hexahedral mesh quality improvement. Section 6 presents the results of the quality improvement and finite element analysis. Section 7 draws conclusions and outlines future work.

2 Previous Work

Image-Based Modelling for Microstructures: Image-based modelling has been employed to analyze the properties and to predict the performance of microstructure materials [36, 19, 21]. For example, serial sectioning and optical microscopy with periodic electron backscatter diffraction (EBSD) together with computer aided 3D reconstruction were used to reconstruct microstructures [28]. Some of these microstructures show significant difference in the 3D properties than traditional assumptions of the 2D information [16], and some other cases present that better prediction of material behavior can be obtained by simulating real 3D microstructure [5]. Large-scale image models have been created for linear elastic, isotropic continuum plastic and crystal plastic cases [20, 21]. In addition, the crystallography of each individual grain was proved to play an important role in the mechanical response of materials. Due to the limited computational resource, an abbreviated representative volume element (RVE) has been selected for modelling with certain criteria [3, 13].

Hexahedral Mesh Generation: Generally there are two meshing techniques for polycrystals: structured and unstructured. Usually block-structured grids are used because it is easy to implement. However, this technique produces non-conforming representation of the grain boundaries with large number of elements [3, 7, 23]. Another approach is the unstructured meshing which produces tetrahedral meshes conforming to the grain boundaries [41], for example, Delaunay triangulation and Voronoi tessellation. The Delaunay triangulation is based on a criterion called "empty sphere", which states that any node must not be contained in the circumsphere of any tetrahedra within the mesh [8]. Voronoi tessellation divides the domain into a set of polygonal regions, which are bounded by the perpendicular bisectors of the lines joining the points [17].

The direct method for hexahedral mesh generation includes five distinct groups: grid-based [29, 30, 31], medial surface [24, 25], plastering [2], whisker weaving [35] and isosurface extraction. The isosurface extraction method extracts the boundary surface and constructs the uniform and adaptive hexahedral meshes [38, 37]. Furthermore, this method has been extended to meshing domains with multiple materials [40]. These techniques create meshes with good aspect ratios in the interior volume; however, bad elements may be generated along the boundaries. Therefore, quality improvement is a very important step after mesh generation.

Quality Improvement: When two neighbouring hexes share two faces, a "doublet" is formed [22]. The pillowing technique was developed to remove doublets [32]. As the simplest and most straightforward method, Laplacian smoothing relocates the vertex to the average of the vertices connecting to it [8]. Based on the weighted-averaging method, several other smoothing techniques were developed. These methods are easy to implement and inexpensive, but they may invert or degrade the local elements [33]. Instead of relocating vertices based on a heuristic algorithm, an optimization technique is utilized to improve the mesh quality. This algorithm measures the quality of the surrounding elements of one node, and optimizes the mesh with respect to a certain objective function [14, 15]. Optimization-based smoothing yields better results, but it is more computationally expensive, and the optimization step length is sometimes hard to decide. Therefore, a combined Laplacian/optimization-based approach was recommended [9, 4].

Most of the previous improvement methods were designed for manifold meshes. In this paper, we will talk about a robust quality improvement approach for non-manifold hexahedral meshes with feature preservation.

3 Data Acquisition and All-Hex Mesh Generation

With the advent of advanced material characterization techniques, it becomes possible to quantify and analyze 3D metallic microstructure with the advanced material characterization techniques such as serial sectioning, X-ray tomography and X-ray diffraction. The composite material model discussed in this paper is the bcc single phase titanium alloy, beta-21s. A 3D microstructural reconstruction of the alloy is made using serial sectioning techniques and optical microscopy with periodic electron backscatter diffraction (EBSD), followed by computerized reconstruction. After segmentation and labeling by grain, the resulting 3D image consists of 2000 million data voxels, and it contains 4700 grains in which 2200 are interior. This methodology was outlined in detail in [28]. In order to reduce the computational cost, an abbreviated 92-grain representative volume element (RVE) is selected for modelling. This microstructural information of the RVE is further reduced by sampling every third voxel in the xy planes of the reconstruction, and every second voxel in the z direction. The simplified RVE represents 92 total grains while 16 are interior. Figure 1(a) shows the volume rendering result. A hexahedral mesh is created by transforming every voxel into one hexahedral element, as shown in Figure 1(b). This mesh is bumpy, and may affect the accuracy of the finite element analysis.

In this paper, we choose an octree-based method to construct quality hexahedral meshes conforming to all the grain boundaries [40]. One minimizer point is calculated for each cell to minimize the predefined Quadratic Error Function (QEF): $QEF(x) = \sum(n_i \cdot (x - p_i))^2$, where p_i and n_i are the position and normal vectors at the intersection points. We analyze each *material change edge* to construct all the material boundaries, and then analyze each interior grid point for all-hexahedral mesh generation. In this octree data structure, each grid point is shared by eight cells, and we use the calculated eight minimizers to construct a hexahedron. In this way, a conforming mesh for domains with multiple materials is constructed automatically. Our hexahedral meshing method provides a material index for each hexahedral element, which will be used in the following vertex classification.

4 Vertex Classification

Due to the non-manifold nature of the microstructure data, we first classify all the vertices into different groups, and then choose various methods to improve them. Here are some definitions in our vertex classification:

RVE boundary: *The RVE boundary is the outer boundary box of the dataset. It consists of 8 corners, 12 edges and 6 faces. During quality improvement, the RVE boundary needs to be preserved.*
Grain boundary surface patches: *The common surfaces of any two neighboring grains are referred as the grain boundary surface patches.*

Boundary curves: *The common curves of any two neighboring grain boundary surface patches are referred as the boundary curves. Depending on whether such curves lie on the six RVE planes or not, we distinguish them as the planar curves and the interior curves.*

Planar curves: *If a boundary curve is located on the RVE plane, then it is a planar curve. Planar curves start and end with planar non-manifold points.*

Interior curves: *If a boundary curve is inside the RVE volume, then it is an interior curve. These curves start and end with interior non-manifold points.*

In non-manifold microstructure materials, all planar and interior curves start with one non-manifold point, while end with another. Once we figure out all the non-manifold vertices in the dataset and then consider every curve they send out, we obtain all the information of the curves: where they start, where they pass by and where they end. The properties of these curves can be used to check the topology of the microstructure data. Based on the above definitions, we categorize all the vertices into seven groups as shown in Figure 2.

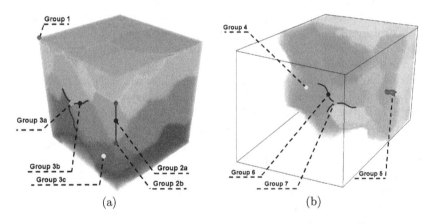

Fig. 2. Vertex classification. (a) Groups 1-3 on the RVE boundary; and (b) Groups 4-7 inside the RVE volume.

Group 1: The eight corners of the RVE box, which are fixed during the improvement in order to keep the RVE boundary.

Group 2: Vertices on the twelve edges of the RVE box, which only move along the edge. It has two sub-categories: Group 2a are the vertices inside one grain, which are smoothed only along the edge; and Group 2b are the vertices shared by two or more than two grains, which are fixed during the improvement.

Group 3: Vertices on the six faces of the RVE box, which are smoothed only on the plane. It has three sub-categories: Group 3a are the vertices shared by more than two grains, which are planar non-manifold points and they are fixed during the improvement; Group 3b are the planar curve vertices shared by two grains, which are smoothed along the tangent direction of the planar curve during the improvement; and Group 3c are the vertices of one grain, which can only move on the plane.

Group 4: Vertices inside one grain, which will be improved using the weighted-averaging method.

Group 5: Vertices located on the grain boundary surface patches shared by two grains, which are smoothed on the tangent plane of the grain boundary during the improvement.

Group 6: Vertices located on interior curves, which can only move along the tangent direction of the interior curve.

Group 7: Interior non-manifold vertices shared by more than two interior curves, which are fixed during the improvement.

For different groups, we improve the quality using different methods as discussed in the next section. Eight corners are fixed; vertices on the RVE planes/edges are moved only on the planes/edges; the planar/interior curves are faired and regularized so that they only move along the curves; and vertices on the grain boundary are smoothed only on the boundary tangent plane.

5 Quality Improvement of Non-manifold Hex Meshes

The hexahedral mesh generated from the 3D volumetric data may be noisy, have "doublets" in the boundary, or even have negative Jacobians. These will result in a poorly conditioned stiffness matrix, and affect the stability, convergence and accuracy of the finite element analysis. Therefore, quality improvement is an important step after mesh generation. Here we develop a comprehensive approach based on a modified pillowing technique, relaxation-based smoothing and optimization. There are four main steps in our algorithm: (1) A relaxation-based smoothing is implemented, which will smooth the grain boundaries while preserving surface features; (2) In order to eliminate doublets, a modified pillowing technique is developed to create a boundary layer; (3) Another smoothing is implemented to drag the shrink set inside and improve the mesh quality; (4) Finally, a local optimization is carried out to improve the worst Jacobian. During the relaxation-based smoothing, different schemes are used for various groups of vertices. A curve fairing method based on B-spline interpolation and resampling is utilized to regularize all the planar and interior curves. Each grain boundary surface patch is faired and regularized using tangent movements, and the interior vertices of each grain are smoothed using the weighted-averaging method.

5.1 Modified Pillowing for Non-manifold Boundaries

In the hexahedral mesh of microstructure materials, "doublets" exist along the grain boundaries. A doublet is formed by two neighboring hexahedra sharing two faces that have an angle greater than 180 degrees. In this situation, it is practically impossible to generate a reasonable Jacobian by relocating vertices. The pillowing technique was developed to remove doublets [22, 32], which was a sheet insertion operation. Figure 3(a) shows one doublet in 2D, and Figure 3(b) shows the pillowed layer along the boundary, then a smoothing operation is required to improve the mesh quality as shown in Figure 3(c). When doublets appear on both sides of the grain boundaries, two sheets need to be inserted. The main challenge in pillowing is how to automatically and efficiently define the shrink set.

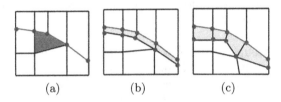

(a)   (b)   (c)

Fig. 3. One 2D doublet and the pillowing, smoothing results. (a) One doublet in the green element along the red boundary; (b) The inserted yellow pillowing layer; and (c) The improved mesh after smoothing.

In our algorithm, the pillowing process is applied to one grain by another. If the grain is interior to the RVE volume with closed boundary, we set the whole grain boundary as the shrink set, as shown in Figure 4(a). Note that the grain boundary may not be closed, i.e., it ends at the RVE boundaries, therefore, its shrink set is also open as shown in Figure 4(b). Sometimes, one grain boundary has local non-manifolds. For example, one boundary edge is shared by more than two faces on the boundary. In this situation, we first detect these non-manifold edges, then disable all the elements connecting to them. The shrink set for this grain is set as the grain boundary except the disabled elements, see Figure 4(c). After pillowing, the disabled elements are added back to the mesh, and the vertex classification is updated for the grain boundary.

5.2   Fairing and Regularization for Curves

There are many boundary curves in the microstructure data, which start and end at non-manifold points. Since noises exist on the curve, a curve fairing method based on the mean curvature flow is first carried out to smooth both the planar and interior curves. In order to regularize these curves while preserving geometric features at the same time, we fix the starting and ending non-manifold points (Group 3a, Group 7) of the curves, and move interior vertices (Group 3b, Group 6) only along the tangent direction using B-spline interpolation and resampling.

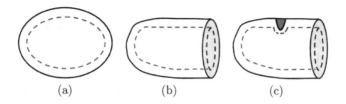

Fig. 4. One grain with its pillowed layer (red dashed lines). (a) An interior grain with closed boundary; (b) A grain with open boundary, the yellow region is on the RVE boundary; and (c) A grain with local non-manifolds (pink region), which are disabled and not pillowed.

Let $[x_0, x_1, x_2, \cdots x_n]$ be a set of points in 3D space, which represents a planar or interior curve. In order to fair this curve, we construct the mean curvature flow iteration:

$$\frac{dx_i}{dt} + \frac{x_i - x_{i-1}}{\|x_i - x_{i-1}\|} + \frac{x_i - x_{i+1}}{\|x_i - x_{i+1}\|} = 0, \tag{1}$$

where $i = 1, \cdots, n\text{-}1$. In this flow, the beginning and the ending points of the curve are fixed. This equation can be solved using the Euler scheme:

$$x_i^{(k+1)} = x_i^{(k)} + \tau * \left(\frac{x_i^{(k)} - x_{i-1}^{(k)}}{\|x_i^{(k)} - x_{i-1}^{(k)}\|} + \frac{x_i^{(k)} - x_{i+1}^{(k)}}{\|x_i^{(k)} - x_{i+1}^{(k)}\|} \right), \tag{2}$$

where $i = 1, \cdots n\text{-}1$. In this scheme, τ is the step size, the superscript k represents the iteration time, and $x_i^{(0)} = x_i$ is the original positions of the input points. The first and the last points are fixed, therefore $x_0^{(k)} = x_0^{(k+1)} = x_0$, and $x_n^{(k)} = x_n^{(k+1)} = x_n$. Then the points on the curves are relocated according to Equation 2. In order to preserve the essential features, the displacement of each vertex is projected to the tangent direction of the curve. Planar curves (Group 3b) are restricted to move only on the RVE planes.

Once the curve is faired, we use a regularization technique to make the points on the curve more evenly distributed via two steps: B-Spline interpolation and resampling. B-Spline interpolation is to construct a cubic non-uniform B-Spline curve, $C(t) = \sum_{j=0}^{m} N_{j,3}(t) P_j$, which passes through all the data points x_i ($i = 0, 1, \cdots, n$) on the original curve in the given order, satisfying $C(t_i) = x_i$ for all $i = 0, 1, \cdots, n$. P_j are the control points and $N_{j,3}$ are the B-Spline basic functions of degree 3. In this paper, the parameters are determined using the chord-length parameterization, since it is a good approximation to the arc-length parameterization. The total chord length is calculated using $L = \sum_{i=1}^{n} \|x_i - x_{i-1}\|_2$, and the chord length from x_0 to x_i

is $L_i = \sum_{j=1}^{i} \|x_j - x_{j-1}\|$. Suppose the domain of the B-Spline curve is $[0, 1]$, then the parameter t_k is calculated using $t_0 = 0$, $t_n = 1$ and $t_i = \frac{L_i}{L}$ for $i = 1, 2, \cdots, n - 1$. In this way, the knot vector of the non-uniform B-Spline curve is $\mathbf{t} = [0\ 0\ 0\ t_0\ t_1\ \cdots, t_{n-1}\ t_n\ 1\ 1\ 1]$. After we obtain the knot vector, the next step is to calculate the control points which satisfy the interpolation condition $x_i = C(t_i) = \sum_{j=0}^{m} N_{j,3}(t_i)P_j$, where $m = n + 2$. The number of control points is $n + 3$, in other words, we have $n + 3$ unknowns. According to the interpolation constraint, we have $n + 1$ equations, so we have to add another two conditions. In this paper, we use the natural end conditions, which means the curvature of the curve at the two ends are zero. In this case, we add two more equations: $C''(0) = 0$ and $C''(1) = 0$. Then the equations can be organized in the following format:

$$Md = e, \tag{3}$$

where

$$M = \begin{bmatrix} N_0''(t_0) & N_1''(t_0) & N_2''(t_0) & N_3''(t_0) & & \\ N_0(t_0) & N_1(t_0) & N_2(t_0) & N_3(t_0) & & \\ & N_1(t_1) & N_2(t_1) & N_3(t_1) & N_4(t_1) & \\ & & & \ddots & & \\ & N_{n-2}(t_{n-1}) & N_{n-1}(t_{n-1}) & N_n(t_{n-1}) & N_{n+1}(t_{n-1}) & \\ & & N_{n-1}(t_n) & N_n(t_n) & N_{n+1}(t_n) & N_{n+2}(t_n) \\ & & N_{n-1}''(t_n) & N_n''(t_n) & N_{n+1}''(t_n) & N_{n+2}''(t_n) \end{bmatrix}, \tag{4}$$

$$d = \begin{bmatrix} P_0\ P_1\ P_2\ \cdots P_n\ P_{n+1}\ P_{n+2} \end{bmatrix}^T, \tag{5}$$

$$e = \begin{bmatrix} 0\ x_0\ x_1\ \cdots x_{n-1}\ x_n\ 0 \end{bmatrix}^T. \tag{6}$$

After solving this linear system, we obtain all the control points of the B-spline curve, P_j for $j = 0, 1, \cdots, m$. Till now, the B-spline has been constructed. The second step is data point resampling. In this step, we aim to obtain $n + 1$ sample points on the curve we constructed, which are evenly distributed along the curve. After we get the new sampled data points, we move the curve points towards the new sampled data points. Here we choose the points whose parameters are equally spaced inside the domain, for example, we calculate $n + 1$ points on curve $C(t)$ at $t = 0$, $\frac{1}{n}$, $\frac{2}{n}$, $\cdots$, $\frac{n-1}{n}$, 1, as the sampling points. After this, we relocate the original data points to the obtained sample points iteratively. Figure 5(a) shows the complete set of the curves obtained from the 92-grain data, including the planar ones as well as the interior ones. Figure 5(b-c) show the curves on one RVE plane before and after smoothing and regularization.

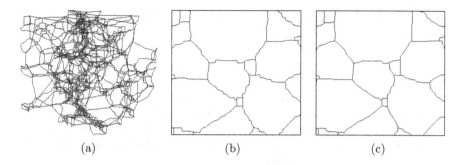

Fig. 5. Boundary curves fairing and regularization. (a) All the curves in the RVE volume; (b) The curves on one RVE plane; and (c) The improved curves.

5.3   Geometric Flow and Optimization

Vertices on grain boundary surface patches are improved using geometric flows [39]. The surface diffusion flow is defined as:

$$\frac{\partial \mathbf{x}}{\partial t} = \varDelta H(\mathbf{x})\mathbf{n}(\mathbf{x}) + v(\mathbf{x})\mathbf{T}(\mathbf{x}), \tag{7}$$

which denoises the surface and improves the aspect ratio while preserving geometric features. Here $\mathbf{x}$ is a surface point, $\varDelta$ is the Laplace-Beltrami (LB) operator, H is the mean curvature, $\mathbf{n}(x)$ is the unit normal vector at the node $\mathbf{x}$, $v(\mathbf{x})$ is the tangent velocity of the surface, and $\mathbf{T}(\mathbf{x})$ is the unit tangent vector at the node $\mathbf{x}$. The first term on the right represents the vertex movement in the normal direction of the surface, while the second term represents the movement on the tangent plane. Vertices are relocated by solving Equation 7 numerically. In addition, the volume of each grain is preserved due to the property of the surface diffusion flow. A discretized Laplace-Beltrami operator is computed numerically [39], and Gaussian integration points are used to calculate the quad area and the hex volume. Other interior vertices are relocated to the mass center of their neighboring elements using the volume-weighted averaging method.

After quality improvement using Equation 7, negative Jacobians may still exist in the mesh. To further improve the mesh quality, a local optimization method is applied to eliminate the negative Jacobians of all the vertices. Each hexahedron is mapped into a trilinear parametric space in terms of ξ, η and ζ. The Jacobian is the determinant of the Jacob matrix from the physical coordinate system to the parametric coordinate system,

$$J = \begin{pmatrix} \frac{\partial x}{\partial \xi} & \frac{\partial x}{\partial \eta} & \frac{\partial x}{\partial \zeta} \\ \frac{\partial y}{\partial \xi} & \frac{\partial y}{\partial \eta} & \frac{\partial y}{\partial \zeta} \\ \frac{\partial z}{\partial \xi} & \frac{\partial z}{\partial \eta} & \frac{\partial z}{\partial \zeta} \end{pmatrix}. \tag{8}$$

The optimization method starts with looping all the vertices to compute their Jacobians, and then the very vertex with the worst Jacobian is found and improved using a conjugate gradient method, in which the objective function is the Jacobian of that vertex. Then a new loop begins and the new worst region in the improved mesh is found and optimized in the same manner. We keep improving the worst Jacobian until it is greater than a pre-defined threshold. We choose the traditional definition of the Jacobian matrix using the finite element basis functions [40]. Generally if the eight corners of one hex all have positive Jacobians, then the Jacobian inside the hex is usually positive. In order to guarantee that the finite element analysis works properly, we can check the Jacobians at the Gaussian integration points inside each element and include them in our loop.

It should be noted that during these two steps, the vertex classification and the specific moving method towards different vertex groups need to be followed. For instance, when we relocate one vertex belonging to Group 5 (vertices on the common patch of two neighboring grain surfaces), the movement should be projected to the tangent direction of the common surface patch.

6   Finite Element Analysis and Results

Our techniques greatly improve the mesh quality, which were computed on a PC equipped with an Intel Q9600 CPU and 4GB DDR-II memories. Figure 6(a-b) show the mesh before and after quality improvement for the 92-grain dataset, and Figure 6(c-d) show the mesh for the 20-grain dataset. The improved meshes have good Jacobians and are smooth on boundaries. Table 1 shows the statistics before and after quality improvement of the mesh. Note that in the optimization, we use the Jacobian defined in the finite element method as the objective function; while in the statistics, we measure the mesh quality with the Jacobian defined by three edge vectors [14, 37, 40]. For the 92-grain data, the original mesh has 30 negative Jacobians. After the quality improvement, all the Jacobians are greater than 0.002. The condition number is also greatly improved, ranging from 1 to 465.3. For the 20-grain data, the worst Jacobian is 0.003 after quality improvement, while the worst condition number becomes 399.8. Figure 7 shows the histogram of the Jacobian and the condition number before and after the quality improvement.

Table 1. Statistics of meshes before and after quality improvement.

Mesh	Mesh Size (Vertex ♯, Elem ♯)	Jacobian (worst, best)	Condition ♯ (worst, best)	Vertex ♯ with Negative Jacobian
92-grain Original	(30600, 27720)	(-0.004, 1.0)	(12781.3, 1.0)	30
After Improvement	(54064, 49673)	(0.002, 1.0)	(465.4, 1.0)	0
20-grain Original	(32768, 29791)	(-0.02, 1.0)	(12549.2, 1.0)	8
After Improvement	(46368, 42505)	(0.003, 1.0)	(399.8, 1.0)	0

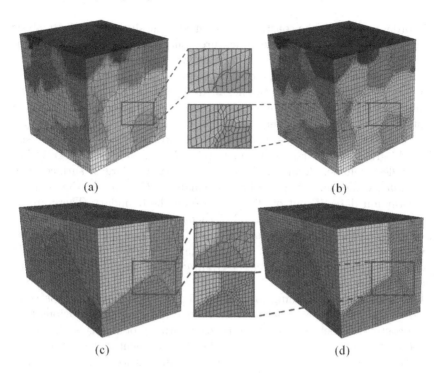

(a)   (b)

(c)   (d)

Fig. 6. The meshes before and after quality improvement. (a-b) The original (left) and improved (right) mesh for the 92-grain beta titanium data; and (c-d) The original (left) and improved (right) mesh for the 20-grain beta titanium data.

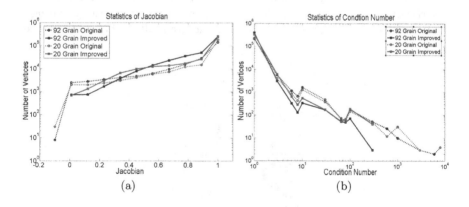

(a)   (b)

Fig. 7. Statistics of meshes before and after the quality improvement. (a) The histogram of the Jacobian; and (b) The histogram of the condition number.

In order to validate the constructed hexahedral meshes, a statistic analysis is carried out. Important geometry features, such as the number and neighbour of the non-manifold vertices, together with accuracy features, such as the surface area and the volume of each grain, are calculated and compared between the original data and the hexahedral mesh. In the 92-grain beta titanium data, 16 out of 92 grains are interior. There are 866 non-manifold points in total, and 602 of them are triple junctions shared by three grains, 250 are shared by four grains, 11 are shared by five, and 3 are shared by six. The 59th grain has the largest volume, $21518.7 \mu m^3$, and the 22th grain has the smallest volume, $12.5 \mu m^3$; the 59th grain has the largest surface area, $4830.6 \mu m^2$, while the 76th grain has the smallest, $37.1 \mu m^2$. In the 20-grain beta titanium data, 19 out of 20 grains are on the boundary. There are 85 non-manifold points in total, and 59 of them are triple junctions shared by three grains, 26 are shared by four. The 10th grain has the largest volume, $25843.7 \mu m^3$, and the 14th grain has the smallest volume, $0.6 \mu m^3$; the 15th grain has the largest surface area, $5041.2 \mu m^2$, while the 14th grain has the smallest surface area, $4.6 \mu m^2$.

The isocontouring meshing and quality improvement algorithm is compared to the direct voxel-to-element technique in detail. It is obvious that the isocontouring method leads to a more conforming boundary surface with less abrupt changes in geometry than the brick element mesh of the same microstructure. Figure 8 shows two meshes of three selected grains (Grains 49, 55, and 58) extracted from the 92-grain titanium data using these two techniques. In addition to the removal of the stair-casing at grain boundaries, the present meshing algorithm has the advantage of providing higher fidelity at the grain boundaries, where plasticity is most likely to initiate. As seen in Figure 8(b), the elements around the grain boundaries are more refined. This is particularly important in studies of the percolation of plasticity through a microstructure - a high-fidelity mesh at the grain allows for more precise

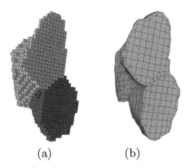

(a)   (b)

Fig. 8. Comparison of meshes of three selected grains (Grains 49, 55 and 58) generated using two different techniques. (a) Voxel-to-element transformation; and (b) Isocontouring mesh generation and quality improvement with one boundary layer.

quantification of the location of plastic activity, and its progression either into the center of the grain or across the boundary into the adjacent grain. Another significant advantage of the present meshing algorithm is the relatively small number of elements required to represent a 3D volume. The resulting volumetric meshes consist of 200,000 elements for the voxel-to-element technique [34], while the same volume meshed using the present algorithm consists of only 49,673 elements. The significant reduction in total number of elements while maintaining the accuracy of the microstructural representation allows for significantly larger volumes to be incorporated into simulations. This is particularly important in the study of structure-property correlations, where representative volumes must be considered in order to extract the global microstructural response from local phenomena.

To check the accuracy of our meshing and quality improvement algorithm as compared to the direct voxel-to-element technique, two finite element simulations of the mechanical response of the 92-grain volume are carried out, using meshes generated by the two techniques. The commercial software ABAQUS is used to simulate the mechanical response. Eight-noded brick elements (C3D8) are used for both the direct voxel-to-element mesh and the mesh generated and improved using our algorithm. A uniaxial tension in the z direction is applied to the z=0 plane, while the opposite plane is fixed. All the remaining free surfaces are constrained to move as planes only.

Elastic Finite Element Analysis: The elastic modulus changes with respect to the crystalline orientations and this causes higher stress and strain gradients near the grain boundaries, where local plasticity may initiate. Therefore, it is important to include the grain orientation information in the analysis. In the RVE data, the average orientation of each grain is measured and applied. Each grain is treated as an anisotropic grain with its own orientation. The parameters of the orthogonal titanium are shown in Table 2. Figure 9(a) shows the Mises Stress distribution in the RVE volume in response to an applied uniaxial strain of 0.6375% in the global y-direction, while Figure 9(b) shows the max principal logarithmic strain distribution. It is obvious that the grain boundaries have higher stress level. The analysis time for this elastic model is 64 seconds.

Table 2. Elastic material properties of the beta titanium for the orthogonal anisotropic model (Gpa).

C_{11}	C_{12}	C_{13}	C_{21}	C_{22}	C_{23}	C_{31}	C_{32}	C_{33}
97.7	82.7	97.7	82.7	82.7	97.7	37.5	37.5	37.5

Plastic Finite Element Analysis: In our plastic finite element analysis, the crystal behavior of the material is simulated using the framework of hypoelasticity and resolved shear stress developed in [1]. A User-Material (UMAT)

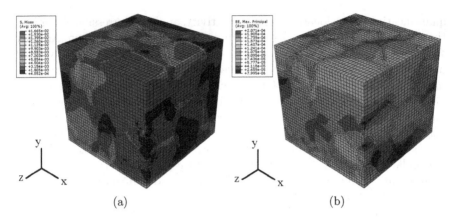

Fig. 9. Elastic finite element analysis result of the 92-grain titanium data. (a) The Mises stress distribution; and (b) The max elastic strain distribution.

Table 3. Plastic material properties of beta titanium for the single crystal constitutive model.

Material Parameter	Value
Family of slip systems (48 independent)	$s_1 = \langle 1\ 1\ 1 \rangle$, $n_1 = \{1\ 1\ 0\}$ $s_2 = \langle 1\ 1\ 1 \rangle$, $n_2 = \{1\ 1\ 1\}$ $s_3 = \langle 1\ 1\ 1 \rangle$, $n_3 = \{1\ 2\ 3\}$
Elastic moduli	$C_{11} = 97.7$Gpa, $C_{12} = 82.7$Gpa, $C_{44} = 37.5$Gpa
Shearing rate parameters	$m = 50$, $\gamma_o = 0.0023/s$
Hardening moduli parameters (Taylor hardening is assumed)	$h_{o1} = 1.5$Gpa, $h_{o2} = 1.98$Gpa, $h_{o3} = 1.64$Gpa $\tau_{o1} = \tau_{o2} = \tau_{o3} = 200$Mpa, $\tau_{s1} = \tau_{s2} = \tau_{s3} = 500$Mpa $q_1 = q_2 = q_3 = 1$

subroutine [12] is employed. The parameters of the user-subroutine are given in Table 3. Figure 10(a) shows a contour plot of the von Mises stress in response to an applied uniaxial strain of 0.6375% in the global y-direction. Figure 10(b) shows the global stress-strain response of the material under three different loading conditions: uniaxial tension in the x, y, and z directions using the two meshes. A very good agreement is obtained. The computation time for the crystal plasticity simulation using our mesh is reduced significantly, primarily due to the number of elements comprising the mesh. As noted previously, the voxel-to-element mesh consists of nearly four times the number of elements as the mesh generated using our algorithm. This results in a reduction of computation time by a factor of 10, from approximately 4 days to 8 hours, using the same computational resources (parallel computing on 156 processors). This significant reduction in mesh size and computation time,

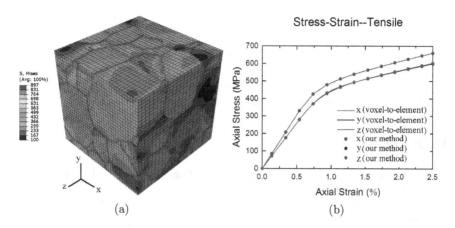

(a)   (b)

Fig. 10. Plastic finite element analysis result of the 92-grain titanium data. (a) The Mises stress distribution in the RVE; and (b) Stress-strain response for applied strain in three global directions using the two meshing algorithms.

while maintaining accuracy of the global response and increasing precision at the grain boundaries, makes the present meshing algorithm a superior method for mesh generation of microstructures.

7   Conclusion and Future Work

We have developed a comprehensive approach consisting of pillowing, geometric flow, regularization and optimization to improve the quality of non-manifold hexahedral meshes for microstructure materials. Vertices are distinguished and classified first, and then a pillowing technique considering the local non-manifold situation is carried out. Planar and spatial curves are identified and smoothed along the tangent plane of the curve with the non-manifold points fixed. Then, a regularization technique is developed to make the curve segments regular. In the following relaxation-based smoothing process, surface patches and interior vertices are improved. Finally, a local optimization method improves the worst Jacobian of the mesh. We have successfully applied our algorithm to two beta titanium data. Finite element analysis is carried out based on the improved mesh, and the result is compared with the voxel-to-element meshing technique.

For large datasets, the requirements of memory and CPU time increase significantly. We will optimize our data structure and develop parallel meshing algorithms to construct non-manifold hexahedral mesh efficiently for very large dataset. In the future, we will also apply the algorithm to more applications, and include the material properties into our mesh generation and quality improvement procedure.

Acknowledgement

This research was supported in part by ONR grant N00014-08-1-0653. The volume rendering results in this paper were generated using a software package (VolRover) developed by Prof. Chandrajit Bajaj's CVC group at The University of Texas at Austin, which is gratefully acknowledged.

References

1. Asaro, R.: Micromechanics of Crystals and Polycrystals. Division of Applied Sciences 23, 1–115 (1983)
2. Blacker, T., Myers, R.: Seams and Wedges in Plastering: a 3D Hexahedral Mesh Generation Algorithm. Engineering With Computers 2, 83–93 (1993)
3. Cailletaud, G., Forest, S., Jeulin, D., Feyel, F., Galliet, I., Mounoury, V., Quilici, S.: Some Elements of Microstructural Mechanics. Acta Materialia 27, 351–374 (2003)
4. Canann, S., Tristano, J., Staten, M.: An Approach to Combined Laplacian and Optimization-based Smoothing for Triangular, Quadrilateral and Quad-dominant Meshes. In: 7th International Meshing Roundtable, pp. 479–494 (1998)
5. Chawla, N., Ganesh, V., Wunsch, B.: Three-Dimensional (3D) Microstructure Visualization and Finite Element Modeling of the Mechanical Behavior of SiC Particle Reinforced Aluminum Composites. Scripta Materialia 51, 161–165 (2004)
6. Dawson, P., Mika, D., Barton, N.: Finite Element Modeling of Lattice Misorientations in Aluminum Polycrystals. Scripta Materialia 47, 713–717 (2002)
7. Diard, O., Leclercq, S., Rousselier, G., Cailletaud, G.: Evaluation of Finite Element Based Analysis of 3D Multicrystalline Aggregates Plasticity Application to Crystal Plasticity Model Identification and the Study of Stress and Strain Fields near Grain Boundaries. International Journal of Plasticity 21, 691–722 (2005)
8. Field, D.: Laplacian Smoothing and Delaunay Triangulations. Communications in Applied Numerical Methods 4, 709–712 (1988)
9. Freitag, L.: On Combining Laplacian and Optimization-based Mesh Smoothing Techniques. Trends in Unstructured Mesh Generation 220, 37–43 (1997)
10. Ghosh, S., Moorthy, S.: Three dimensional Voronoi Cell Finite Element Model for Microstructures with Ellipsoidal Heterogeneties. Computational Mechanics 34, 510–531 (2004)
11. Groeber, M., Ghosh, S., Uchic, M., Dimiduk, D.: A Framework For Automated Analysis and Simulation of 3D Polycrystalline Microstructures: Part 1: Statistical Characterization. Acta Materialia 56, 1257–1273 (2008)
12. Huang, Y.: A User-Material Subroutine Incorporating Single Crystal Plasticity in the ABAQUS Finite Element Program. Division of Applied Sciences 23, 3647–3679 (1991)
13. Kanit, T., Forest, S., Galliet, I., Mounoury, V., Jeulin, D.: Determination of The Size of the Representative Volume Element for Random Composites: Statistical and Numerical Approach. International Journal of Solids and Structures 40, 3647–3679 (2003)

14. Knupp, P.: Achieving Finite Element Mesh Quality via Optimization of the Jacobian Matrix Norm and Associated Quantities. Part II - a Framework for Volume Mesh Optimization and the Condition Number of the Jacobian Matrix. International Journal for Numerical Methods in Engineering 48, 1165–1185 (2000)
15. Knupp, P.: Hexahedral and Tetrahedral Mesh Untangling. Engineering with Computers 17, 261–268 (2001)
16. Kral, M., Spanos, G.: Three-Dimensional Analysis of Proeutectoid Cementite Precipitates. Acta Materialia 47, 711–724 (1999)
17. Kumar, S., Kurtz, S.: Simulation of Material Microstructure Using a 3D Voronoi Tesselation: Calculation of Effective Thermal Expansion Coefficient of Poly-crystalline Materials. Acta Metallurgica et Materialia 42, 3917–3927 (1994)
18. Lee, S., Rollett, A., Rohrer, G.: Three-Dimensional Microstructure Reconstruction Using FIB-OIM. Materials Science Forum, 915–920 (2007)
19. Lewis, A., Bingert, J., Rowenhorst, D., Gupta, A., Geltmacher, A., Spanos, G.: Two and Three Dimensional Microstructural Characterization of a Super-Austenitic Stainless Steel. Materials Science and Engineering 418, 11–18 (2005)
20. Lewis, A., Geltmacher, A.: Image-Based Modeling of the Response of Experimental 3D Microstructures to Mechanical Loading. Scripta Materialia 55(9), 81–85 (2006)
21. Lewis, A., Jordan, K., Geltmacher, A.: Determination of Critical Microstructural Features in an Austenitic Stainless Steel Using Image-Based Finite Element Modeling. Metallurgical and Materials Transactions A: Physical Metallurgy and Materials Science 39(5), 1109–1117 (2008)
22. Mitchell, S., Tautges, T.: Pillowing Doublets: Refining a Mesh to Ensure That Faces Share at Most One Edge. In: 4th International Meshing Roundtable, pp. 231–240 (1995)
23. Nygards, M.: Number of Grains Necessary to Homogenize Elastic Materials with Cubic Symmetry. Mechanics of Materials 35, 1049–1057 (2003)
24. Price, M., Armstrong, C.: Hexahedral Mesh Generation by Medial Surface Subdivision: Part i. International Journal for Numerical Methods in Engineering 38, 3335–3359 (1995)
25. Price, M., Armstrong, C.: Hexahedral Mesh Generation by Medial Surface Subdivision: Part ii. International Journal for Numerical Methods in Engineering 40, 111–136 (1997)
26. Qidwai, M., Lewis, A., Geltmacher, A.: Image-based Computational Modeling and Analysis of a Beta Titanium Alloy. Acta Materialia, in review (2009)
27. Rollett, A., Lee, S., Campman, R., Rohrer, G.: Three-Dimensional Characterization of Microstructure by Electron Back-Scatter Diffraction. Annual Review of Materials Research 37, 627–658 (2007)
28. Rowenhorst, D., Gupta, A., Feng, C., Spanos, G.: 3D Crystallographic and Morphological Analysis of Coarse Martensite: Combining EBSD and Serial Sectioning. Scripta Materialia 55, 11–16 (2006)
29. Schneiders, R.: A Grid-based Algorithm for the Generation of Hexahedral Element Meshes. Engineering with Computers 12, 168–177 (1996)
30. Schneiders, R., Weiler, F.: Octree-Based Generation of Hexahedral Element Meshes. In: 5th International Meshing Roundtable, pp. 205–216 (1996)
31. Shephard, M., Georges, M.: Three-Dimensional Mesh Generation by Finite Octree Technique. International Journal for Numerical Methods in Engineering 32, 709–749 (1991)

32. Shepherd, J.: Topologic and Geometric Constraint-Based Hexahedral Mesh Generation. University of Utah (2007)
33. Shontz, S., Vavasis, S.: A Mesh Warping Algorithm Based on Weighted Laplacian Smoothing. In: 12th International Meshing Roundtable, pp. 147–158 (2003)
34. Spanos, G., Rowenhorst, D., Lewis, A., Geltmacher, A.: Combining Serial Sectioning, EBSD Analysis, and Image-Based Finite Element Modeling. MRS Bulletin 33, 597–602 (2008)
35. Tautges, T., Blacker, T., Mitchell, S.: The Whisker-Weaving Algorithm: a Connectivity Based Method for Constructing All-hexahedral Finite Element Meshes. International Journal for Numerical Methods in Engineering 39, 3327–3349 (1996)
36. Youssef, S., Maire, E., Gaertner, R.: Finite Element Modeling of the Actual Structure of Cellular Materials Determined by X-ray Tomography. Acta Materialia 53, 719–730 (2005)
37. Zhang, Y., Bajaj, C.: Adaptive and Quality Quadrilateral/Hexahedral Meshing from Volumetric Data. Computer Methods in Applied Mechanics and Engineering 195, 942–960 (2006)
38. Zhang, Y., Bajaj, C., Sohn, B.: 3D Finite Element Meshing from Imaging Data. The special issue of Computer Methods in Applied Mechanics and Engineering on Unstructured Mesh Generation 194, 5083–5106 (2005)
39. Zhang, Y., Bajaj, C., Xu, G.: Surface Smoothing and Quality Improvement of Quadrilateral/Hexahedral Meshes with Geometric Flow. In: 14th International Meshing Roundtable, pp. 449–468 (2005)
40. Zhang, Y., Hughes, T., Bajaj, C.: An Automatic 3D Mesh Generation Method for Domains with Multiple Materials. Computer Methods in Applied Mechanics and Engineering (in press, 2009)
41. Zhao, Y., Tryon, R.: Automatic 3D Simulation and Micro-Stress Distribution of Polycrystalline Metallic Materials. Computer Methods in Applied Mechanics and Engineering 193, 3919–3934 (2004)

Automatic CAD Models Comparison and Re-meshing in the Context of Mechanical Design Optimization

Jean-Christophe Cuillière[1], Vincent François[1], Khaled Souaissa[1,2], Abdelmajid Benamara[2], and Hedi BelHadjSalah[2]

[1] ERICCA, Département de Génie Mécanique UQTR, Trois Rivières, Canada
cuillier@uqtr.ca, francois@uqtr.ca
[2] LGM, ENIM, Monastir, Tunisie
abdel.benamara@enim.rnu.tn

Abstract. A lot of research work has been focused on integrating FEA (finite elements analysis) with CAD (Computer Aided Design) over the last decade. In spite of improvements brought by this integration, research work remains to be done in order to better integrate all the operations led during the whole design process. The design process involves several modifications of an initial design solution and until now, in this context, the communication between CAD modules (dedicated to different tasks involved in the product design process) remains static. Consequently, there is a need for more flexible communication processes between CAD modules through the design cycle, if not through the product life cycle. Some approaches have been developed aiming at the reduction of the design process length when using FEA, and aiming at the automation of part's data transfer from one step of the process to the next one. Automatic re-meshing is one of these approaches. It consists in automatically updating the part's mesh around modifications zones, in the case of a minor change in the part's design, without the need to re-mesh the entire part. The purpose of this paper is to present a new tool, aiming at the improvement of automatic re-meshing procedures. This tool basically consists in automatically identifying and locating modifications between two versions of a CAD model (typically an initial design and a modified design) through the design process. The knowledge of these modifications is then used to fit portions of the initial design's mesh to the modified design (a process referred to as automatic re-meshing). A major benefit of the approach presented here is that it is completely independent of the description frame of both models, which is made possible with the use of vector-based geometric representations.

Keywords: Model comparisons/BREP/Vectorial space/Remeshing/ NURBS.

1 Introduction

The design process of mechanical parts, usually involves several modification of an initial design solution. This means that during the design and manufacturing cycle of a given part, the geometry can change several times. When the new version of a part's geometry has to be analyzed using FEA (Finite Element Analysis), the mesh usually needs to be completely built from scratch and no benefit is made of meshes corresponding with the analysis of previous versions of the design. This is obviously a great waste of time if we consider that a part's geometry changes very little from one design version to the next one. It is even a greater waste of time when analysis requires significant adaptive mesh refinement. In fact, repeating these tasks (mesh generation and adaptive refinement) for every new design alternative makes the design process very expensive with regard to processing time. In order to reduce the time of the design phase in product development, automatic updating of models and processes (such as mesh generation and FEA) could induce very important gains with regard to processing time. For example, instead of re-meshing entirely a modified model, it can be re-meshed only around modification zones, while partially preserving the former mesh. Also, when performing a FEA on a modified design, instead of solving the entire modified model, it would be very powerful to be able to solve the problem only in modifications zones and to retrieve results of previous analyses, in zones where the design has not been modified (Fig. 1).

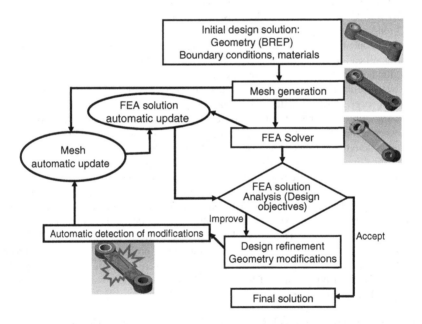

Fig. 1. The iterative FEA process in the context of design optimization

The implementation of these concepts in the context of the design process requires the development of a set of tools, which are (Fig. 1) :

1. A tool aimed at automating the identification and localization of modifications between different versions of a CAD model through the design progress.
2. A tool aimed at automatically retrieving elements from previous versions of a FEA model and only remeshing in modification zones.

3. A FEA solver able to retrieve results of previous analyses in zones where the design has not been modified and to restrict the calculation to modification zones.

This paper focuses on the presentation of our research work towards the development of the two first tools in the list just above.

2 Comparison between CAD Models

The automatic comparison between two versions of a given CAD model through the design process and the retrieval of identical shapes is not really a new subject of interest [1-11]. Nevertheless, it is sill a subject of interest and a subject of research investigations, especially because CAD models are in constant evolution and because this allows for the development of new functionalities and for the improvement of existing functionalities. Also, in the context of designing and storing huge numbers of digital product models, the interest for this type of functionalities is clearly growing. For example, in the aeronautics and automotive industries, where thousands of CAD models for manufactured parts are contained in product databases, being able to easily re-use stored design/manufacturing information would result in a much faster and more efficient design processit is.

Among functionalities mentioned just above, a key aspect is the ability to compare two versions of a CAD model, being fully independent of the frames in which these two versions are defined. In fact, many existing comparison approaches require that these versions are defined with respect to the same frame, which means that they are located and oriented the same way. In the context of our research and in the context of modern feature-based CAD systems, this comparison has to be fully independent from the definition frames of feature-based models.

The main concept underlying our approach is to base comparisons between models on a vector representation of 3D shapes. Basically, a solid geometry is modeled in any CAD system using a spatial type of representation, referred to as a BREP (Boundary REPresentation) [12]. This spatial representation is fundamentally based on the location of geometric entities (vertices, control points, curves, surfaces, etc.). As described in the next section, in our work, this BREP is added with vector representations of 3D shapes (the vectorial space, the metric tensor and the initia tensor) [11, 13-15]. Indeed, a very interesting property of these vector-based representations is that they can be derived into frame-independent quantities. Consequently, they provide us with meaningful data on which comparisons can be made independently from any definition frame.

2.1 Vectorial Space, Metric Tensor and Initia Tensor

The definition of the *vectorial space* of BREP entities (for example, curves and surfaces) is based on a point cloud, which is directly derived from the BREP entity control points associated with their NURBS (NonUniform Rational B-Splines) description [16]. The vectorial space is computed from this point cloud as a vector sheaf where each vector is defined by a pair of control points. The coordinates of control points (and incidentally, the components of vectors derived) are defined in a 4D space. The fourth coordinate of a given control point corresponds to the weight associated with it in its NURBS representation. If the native BREP entity is not a NURBS then equivalent NURBS parameters can be obtain from most CAD system. So the vectorial space is defined for every type of BREP entities.

Once these vectors are computed, a corresponding metric tensor is derived. Using the 4D coordinates mentioned earlier, any vector $\vec{V}$ can be written in a unique way as a linear combination of the four basis vectors $B = \left(\vec{e_1}, \vec{e_2}, \vec{e_3}, \vec{e_4} \right)$. The vectorial space of a BREP entity $\overline{\overline{\Gamma}}_\alpha = \left\{ \vec{V}_0, ..., \vec{V}_p \right\}$, which is defined in this basis just above, can be expressed as:

$$
\overline{\overline{\Gamma}}_\alpha = \begin{Bmatrix} \vec{V}_0 \\ \vec{V}_1 \\ . \\ . \\ . \\ \vec{V}_p \end{Bmatrix} = \begin{bmatrix} V_0^x & V_0^y & V_0^z & V_0^w \\ V_1^x & V_1^y & V_1^z & V_1^w \\ . & & & \\ . & . & . & . \\ . & & & \\ V_p^x & V_p^x & V_p^z & V_p^w \end{bmatrix} \begin{Bmatrix} \vec{e}_1 \\ \vec{e}_2 \\ \vec{e}_3 \\ \vec{e}_4 \end{Bmatrix}
\tag{1}
$$

Then, the metric tensor $\overline{\overline{G}}(\overline{\overline{\Gamma}}_\alpha)$ of this vector's set is defined as the tensor product:

$$
\overline{\overline{G}}(\overline{\overline{\Gamma}}_\alpha) = \overline{\overline{\Gamma}}_\alpha \otimes \overline{\overline{\Gamma}}_\alpha = \begin{Bmatrix} \vec{V}_0 \\ \vec{V}_1 \\ . \\ . \\ \vec{V}_p \end{Bmatrix} \otimes \begin{Bmatrix} \vec{V}_0 \\ \vec{V}_1 \\ . \\ . \\ \vec{V}_p \end{Bmatrix}
$$

$$
= \begin{bmatrix} V_0^x & V_0^y & V_0^z & V_0^w \\ V_1^x & V_1^y & V_1^z & V_1^w \\ . & . & . & . \\ . & . & . & . \\ V_p^x & V_p^x & V_p^z & V_p^w \end{bmatrix} \begin{bmatrix} V_0^x & V_0^y & V_0^z & V_0^w \\ V_1^x & V_1^y & V_1^z & V_1^w \\ . & . & . & . \\ . & . & . & . \\ V_p^x & V_p^x & V_p^z & V_p^w \end{bmatrix}^T
\tag{2}
$$

which can be written as:

$$
\overline{\overline{G}}(\overline{\overline{\Gamma}}_\alpha) = \overline{\overline{\Gamma}}_\alpha \otimes \overline{\overline{\Gamma}}_\alpha = \begin{bmatrix} \vec{V}_0 \cdot \vec{V}_0 & \vec{V}_0 \cdot \vec{V}_1 & . & . & \vec{V}_0 \cdot \vec{V}_p \\ \vec{V}_1 \cdot \vec{V}_0 & \vec{V}_1 \cdot \vec{V}_1 & . & . & \vec{V}_1 \cdot \vec{V}_p \\ . & . & . & . & . \\ . & . & . & . & . \\ \vec{V}_p \cdot \vec{V}_0 & \vec{V}_p \cdot \vec{V}_1 & . & . & \vec{V}_p \cdot \vec{V}_p \end{bmatrix}
\tag{3}
$$

The inertia tensor $\overline{\overline{I}}(\Gamma_\alpha)$ of a BREP entity α is defined as the inertia matrix of the entity's set of control points obtained when a unit point mass is attached to each control point:

$$\overline{\overline{I}}(\Gamma_\alpha) = \begin{bmatrix} \sum_{i=1}^{n} m_i.(y_i^2 + z_i^2 + w_i^2) & -\sum_{i=0}^{n} m_i.x_i.y_i & -\sum_{i=0}^{n} m_i.x_i.z_i & -\sum_{i=0}^{n} m_i.x_i.w_i \\ & \sum_{i=1}^{n} m_i.(x_i^2 + z_i^2 + w_i^2) & -\sum_{i=0}^{n} m_i.y_i.z_i & -\sum_{i=0}^{n} m_i.y_i.w_i \\ & & \sum_{i=1}^{n} m_i.(x_i^2 + y_i^2 + w_i^2) & -\sum_{i=0}^{n} m_i.z_i.w_i \\ & (sym) & & \sum_{i=1}^{n} m_i.(x_i^2 + y_i^2 + z_i^2) \end{bmatrix} \quad (4)$$

2.2 Comparison between CAD Models

The basic idea on which our approach is comparing to CAD models (and by the way identifying differences between them) by comparing metric and inertia tensors associated with entities of the two BREP models.

With regard to the following definitions of *similarity*, *identity* and *localized identity*, this comparison is performed through four consecutive steps (Fig. 2)

Definition 1: Two entities are similar if they have the same shape regardless of their size (Fig. 3a).

Definition 2: Two entities are identical if they are similar and if their dimensions are the same (Fig. 3b).

Definition 3: Two entities are identically localized if they are identical and if their location is the same relative to the same reference face in both models (Fig. 3c).

These four steps are the following:

Step 1: comparisons between metric tensors associated with topologic entities of the two BREPs being compared (typically edges and faces) result in a list of similar topologic entities between the two models.

Step 2: comparisons between the intertia tensors of similar topologic entities result in a list of identical topologic entities between the two models.

Step 3: A local frame is computed for each topologic entity tagged as identical. This local frame is derived from the principal direction vectors of the entity's inertia tensor. Then the coordinates of topologic entities' barycentre are calculated with regard to these local frames and comparisons based on these coordinates result in a list of topologic entities which are identified as localized identical.

Step 4: All the topologic entities of a given model which have not been identified as having a corresponding localized identical topological entity in the other model are tagged as modified.

Once modified topologic entities are identified between the two models, the process ends with the classification of these entities in the following three categories: new entities, erased entities and partially modified entities. This classification

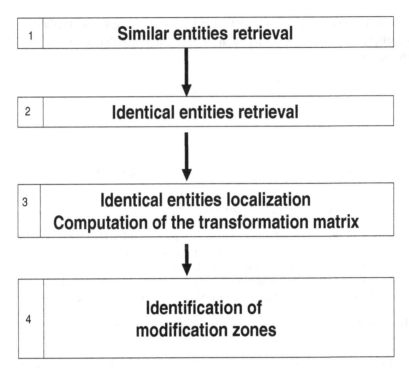

1	**Similar entities retrieval**
2	**Identical entities retrieval**
3	**Identical entities localization** **Computation of the transformation matrix**
4	**Identification of** **modification zones**

Fig. 2. The general framework of our comparison algorithm

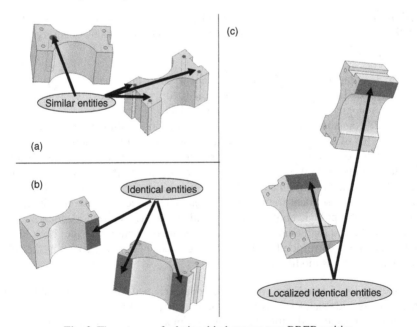

Fig. 3. Three types of relationship between two BREP entities

is basically performed using the same type of concepts as those described earlier (comparisons between metric and inertia tensors, considering local frames and co-ordinates of barycentres). It important to outline that for entities that are finally identified as partially modified, the modification zones are not explicitly identified at this stage of the process. We will see in the next paragraphs that these modification zones are in fact implicitly identified through the remeshing process itself.

Once these steps are completed and if necessary, the homogeneous transformation matrix between definition frames (expressing the relative position and orientation of the two models' definition frames) can be computed.

Fig. 4 illustrates results obtained applying this approach on two versions of a sample part through the design process.

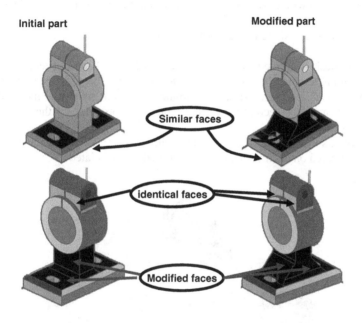

Fig. 4. A comparison result

3 Automatic Remeshing

3.1 Algorithm

The remeshing algorithm used in this study is an adaptation and improvement of a previous version developed by our research team [10, 17, 18]. This adaptation is closely related to the fact that the comparison process itself, has been improved, as described in the previous section.

In this new remeshing process, edges, faces and the volume are processed (namely remeshed) using the same generic scheme. This scheme is based on the following definition of *mesh blocks*.

Definition 4: A mesh block is a continuous subset of mesh elements (segments on edges (1D), triangles on faces (2D), and tetrahedrons inside the volume (3D)), all of which are related to the same topologic entity of the BREP model (respectively an edge, a face or a body).

As mentioned above, modification zones are not explicitly known. The basic principle of the method we are describing here is to define mesh blocks by cutting out the initial mesh. The process starts with one tetrahedron block (the initial mesh).

We have illustrated the various steps of the method using an elementary part (Fig. 5), i.e., a beam with a square section whose length has been decreased and to which grooves have been suppressed.

The steps in the remeshing process are as follows:

1. The initial mesh is applied to the modified model. Two layers of tetrahedrons (the minimum number of layers needed to guarantee the process will work properly and efficiently) are destroyed around entities of the initial model that have disappeared (the second list in the information provided by the comparison algorithms).
2. At this point, some vertices are associated with a mesh node and some are not. A node is created on each vertex to which no node is associated.

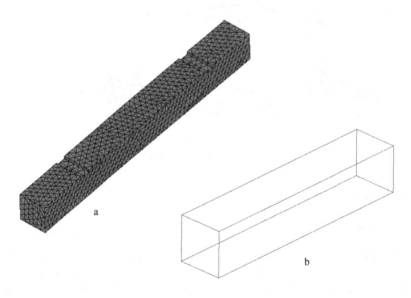

Fig. 5. Remeshing on a simple case a) The mesh of the initial part b) The modified part

3. Tetrahedrons of the initial mesh with a node B closer than a distance L to a node A created at step 2 are destroyed. L is calculated using the equation:

$$L = \max\left[2.E_A, 2.E_B, 4.\left|E_A - 2.E_B\right|\right] \tag{5}$$

Where

E_A is the nodal spacing function value at node A, and

E_B is the nodal spacing function value at node B.

4. 1D mesh blocks (segments) on edges are created (Fig. 6a).
5. As the 3D mesh of the modified part is partially known, each edge of the BREP structure is either without elements, completely meshed or partially meshed. Standard edge discretization procedures must be adapted for partially meshed edges. These edges are cut into sets of sub-edges, so that a sub-edge is either completely meshed or without elements. After this preliminary process, standard edge discretization procedures are applied to mesh sub-edges that remain without elements (Fig. 6b). Once these procedures are complete, all BREP edges are completely meshed (Fig. 6c). Segment blocks that cannot be inserted on edges of the modified model are eliminated.

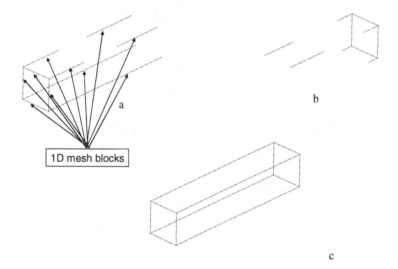

Fig. 6. a) 1D mesh blocks; b) edge mesh processed; c) edge mesh processed with edge mesh recovered

6. Tetrahedrons of the initial mesh with a node located inside a zone defined around segments created at step 5 are destroyed. The shape of this destruction zone is derived from a parabola with height L as defined at step 3 (Fig. 7a). The shape is very specific and has been designed so that a minimum number of tetrahedrons are destroyed and so that the convergence of advancing front automatic mesh generation is guaranteed.

7. 2D mesh blocks (triangles) on faces are created (Fig. 8a).
8. In the BREP structure, a face is either without elements, completely meshed or partially meshed. An adaptation of standard procedures is also necessary for partially meshed faces (Fig. 8b). The initialization front of our advancing front mesh generator is adapted. The front is initialized on the boundaries of triangles belonging to partially meshed faces. The advancing front mesh generator is then used to compute the entire face's mesh (Fig. 8c). Here again, triangle blocks that cannot be inserted on faces of the modified model are eliminated.

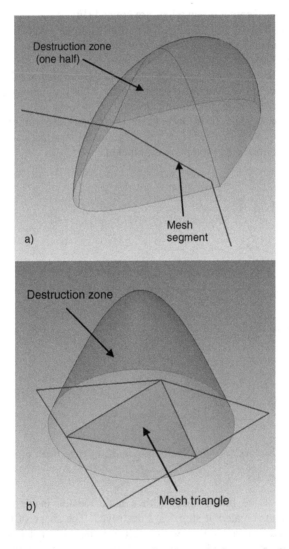

Fig. 7. Destruction zone a) for a mesh segment; b) for a mesh triangle.

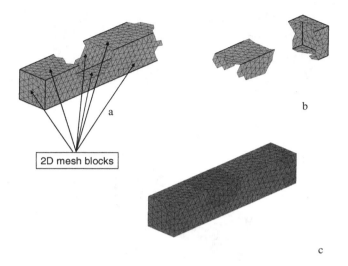

2D mesh blocks

Fig. 8. a) 2D mesh blocks; b) faces mesh processed; c) face mesh processed with face mesh recovered.

9.   Tetrahedrons of the initial mesh with a node located inside a zone (Fig. 7b) defined around triangles created at step 5 are destroyed. Here again, the shape of this destruction zone is derived from a parabola with height L as defined at step 3 (Fig. 7b). This shape is different from those used at step 6 (for segments) and has also been designed so that a minimum number of tetrahedrons are destroyed and the convergence of advancing front automatic mesh generation is guaranteed.

10. 3D mesh blocks (tetrahedrons) inside the volume are created (Fig. 9a).

11. In the BREP structure, a volume is either without elements, completely meshed or partially meshed. The same adaptation of standard procedures is also necessary for partially meshed volumes (Fig. 9b). The initialization front of our advancing front mesh generator is adapted. The front is initialized on the boundaries of tetrahedrons belonging to partially meshed volumes. The advancing front mesh generator is then used to obtain the entire volume's mesh (Fig. 9c). Mesh blocks that cannot be inserted inside the modified model are eliminated.

This last step has also been adapted from our previous work on automatic remeshing. The convergence of the remeshing process is improved because the destruction of tetrahedrons inherent to 3D advancing front mesh generation itself is coupled with the destruction of tetrahedrons performed at step 9 of the remeshing algorithm. Thus, convergence of the 3D advancing front process, which is a very sensitive problem, is achieved using a smaller value for parameter L. These

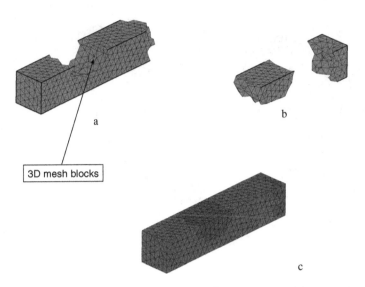

Fig. 9. a) 3D mesh blocks; b) volume mesh processed; c) volume mesh processed with volume mesh recovered.

results in improved efficiency of the remeshing since a larger number of initial tetrahedrons are retrieved.

The implementation of the remeshing process is made simpler by the fact that the majority of operations used here are the same as those used for meshing the initial solid. The efficiency (reduced CPU time) of the remeshing strategy is directly related to the importance and number of modification zones.

The finite elements fall into two categories:

- tetrahedrons retrieved from the mesh of the initial model.
- new tetrahedrons created by the remeshing procedure.

The procedure also generates a table of corresponding nodes in the initial and modified meshes.

3.2 Results

In Fig. 10 to Fig. 12, results of the comparison of BREP models and automatic remeshing are presented for three mechanical parts, to illustrate the method's potential. Matching results for initial and modified parts are once again identified using colour conventions. The colour of the triangular mesh indicates whether the face is identical, localized identical and/or partially modified. Black indicates that the face is erased or new. For each example, illustrations of elements that have been retrieved from the initial mesh, as well as elements that have been newly generated on the modified model, are provided.

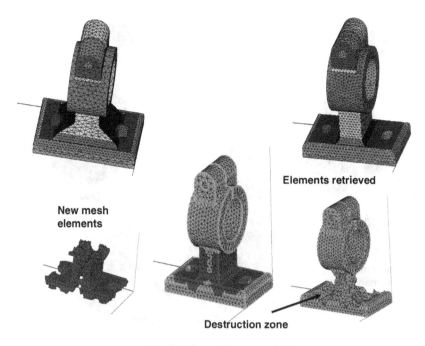

New mesh elements

Elements retrieved

Destruction zone

Fig. 10. Remeshing example 1

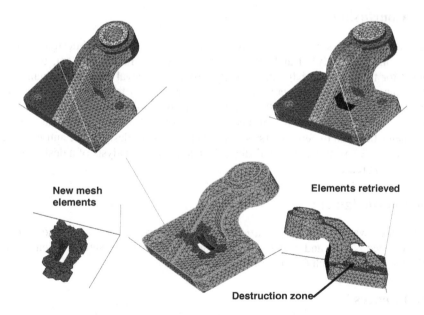

New mesh elements

Elements retrieved

Destruction zone

Fig. 11. Remeshing example 2

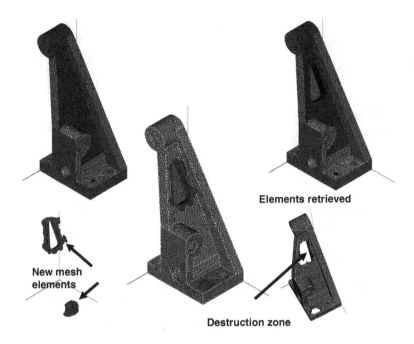

Fig. 12. Remeshing example 3

4 Conclusion

The design process typically involves many iterations of an initial design solution, which may be analyzed and refined numerous times. In this context, there is a need for tools allowing for the fast and efficient retrieval of results obtained in previous analyses at any stage of the design process. In this paper, we have presented a tool that automates the retrieval of modifications between different versions of a design, and the use of this information to automate the retrieval of finite elements between different versions of the analysis of a design. The automatic retrieval of FEA results between different versions of the analysis of a design is still an ongoing research.

Acknowledgments

This study was carried out as part of a project supported by research funding from the Quebec Nature and Technology Research Fund and the Natural Sciences and Engineering Research Council of Canada (NSERC).

References

[1] Iyer, N., Jayanti, S., Lou, K., Kalyanaraman, Y., Ramani, K.: Three-dimensional shape searching: state-of-the-art review and future trends. Computer-Aided Design 37(5), 509–530 (2005)

[2] Cardone, A., Gupta, S.K., Karnik, M.: A Survey of Shape Similarity Assessment Algorithms for Product Design and Manufacturing Applications. Journal of Computing and Information Science in Engineering 3(2), 109–118 (2003)

[3] Hong, T., Lee, K., Kim, S., Chu, C., Lee, H.: Similarity comparison of mechanical parts. Computer-Aided Design and Application 2(6), 759–768 (2005)

[4] Hilaga, M., Shinagawa, Y., Kohmura, T., Kunii, T.L.: Topology Matching for Fully Automatic Similarity Estimation of 3D Shapes. In: SIGGRAPH 2001, Los Angeles (2001)

[5] Osada, R., Funkhouser, T.A., Chazelle, B., Dobkin, D.P.: Matching 3D Models with Shape Distributions. In: Shape Modeling International. IEEE Computer Society Press, Los Alamitos (2001)

[6] Ohbuchi, R., Minamitani, T., Takei, T.: Shape-Similarity Search of 3D Models by using Enhanced Shape Functions. In: TPCG (2003)

[7] Saupe, D., Vranic, D.V.: 3D model retrieval with spherical harmonics and moments. In: Radig, B., Florczyk, S. (eds.) DAGM 2001. LNCS, vol. 2191, p. 392. Springer, Heidelberg (2001)

[8] Cicirello, V.A., Regli, W.C.: Machining Feature-Based Comparisons of Mechanical Parts. In: Shape Modeling International (2001)

[9] Ramesh, M., Yip-Hoi, D., Dutta, D.: Feature Based Shape Similarity Measurement for Retrieval of Mechanical Parts. Journal of Computing and Information Science in Engineering 1(3), 245–257 (2001)

[10] Francois, V., Cuilliere, J.-C.: 3D Automatic remeshing applied to model modification. Computer Aided Design 32(7), 433–444 (2000)

[11] Serre, P., Riviere, A., Clement, A.: Geometric Product Specification and Verification: Integration of Functionality. In: Analysis of fonctional geometrical specification, pp. 115–125. Kluwer, Dordrecht (2003)

[12] Mortenson, M.E.: Geometric Modeling (1985)

[13] Souaissa, K., Hadjsalah, H., Benamara, A., Cuillière, J.C., Francois, V.: Outil de recherche de formes CAO identiques. In: CSME 2008, Ottawa (2008)

[14] Serre, P., Riviere, A., Clement, A.: The clearence effect for assemblability of over-contrained mechanism. In: 8th CIRPP International Seminar on Computer Aided Tolerancing, Charlotte, USA (2003)

[15] Clement, A., Riviere, A., Serre, P., Valade, C.: The TTRS: 13 Constraints for Dimensionning and Tolerancing. In: 5th CIRPP International Seminar on Computer Aided Tolerancing, Toronto, Canada (1997)

[16] Piegl, L.A., Tiller, W.: The NURBS book (1997)

[17] Francois, V., Cuilliere, J.-C.: An a priori adaptive 3D advancing front mesh generator integrated to solid modeling. In: Recent Advances in Integrated Design and Manufacturing in Mechanical Engineering, pp. 337–346 (2003)

[18] Francois, V., Cuilliere, J.-C.: Automatic mesh pre-optimization based on the geometric discretization error. Advances in Engineering Software 31(10), 763–774 (2000)

Distance Solutions for Medial Axis Transform

Hao Xia and Paul G. Tucker

Whittle Laboratory, Department of Engineering, University of Cambridge,
1 JJ Thomson Ave, Camridge, CB3 0DY, England
{hx222,pgt23}@cam.ac.uk

Abstract. A method towards robust and efficient medial axis transform (MAT) of arbitrary domains using distance solutions is presented. The distance field, d, is calculated by solving the hyperbolic-natured Eikonal (or Level Set) equation. The solution is obtained on Cartesian grids. Both the fast-marching method and fast-sweeping method are used to calculate d. Medial axis point clouds are then extracted based on the distance solution via a simple criteria: the Laplacian or the Hessian determinant of d. These point clouds in 2D-pixel and 3D-voxel space are further thinned to curves and surfaces through binary image thinning algorithms. This results in an overall hybrid approach. As an alternative to other methods, the current $d-$MAT procedure bypasses difficulties that are usually encountered by pure geometric methods (e.g. the Voronoi approach), especially in 3D, and provides better accuracy than pure thinning methods. It is also shown that the $d-$MAT approach provides the potential to sculpt/control the MAT form for specialized solution purposes. Various examples are given to demonstrate the current approach.

Keywords: Eikonal equation, wall distance, medial axis transform, thinning.

1 Introduction

In physics, the nearest normal wall distance d is still a key parameter in many turbulence modeling and simulation approaches [10, 28] and also in peripheral applications incorporating additional solution physics [17, 29]. Such examples include explosive front, multiphase flow, and electrostatic particle force modeling. Also, in grid generation the near-wall isovalues of d can be used to form the boundary layer mesh [25, 30], while the far-field d contours can be used as a rapid means of evaluating computational interfaces on unstructured overset meshes with relative movements. Not only being useful in traditional physics, distance field has also been important in computer vision, modeling and computational physics.

In the general fields of shape analysis and solid modeling, including automated meshing, obtaining the medial axis transform (MAT) for a given geometry (or shape/domain) is regarded as an essential step [14, 19, 22, 21, 23]. Efficient and robust techniques are therefore required. Medial axis has been widely studied by researchers across the community. Several different types of methods exist, mainly including *thinning*, *Voronoi diagram*, *distance field*, and *hybrid methods*. A brief review on these methods can be found in [9]. Most widely discussed are the Voronoi diagram [6, 16, 3, 7] and distance field methods [1, 24, 27]. By nature, the Voronoi diagram is geometry-based and the distance field is often related to general differential equations. One of the advantages of differential equations is that their extension to 3D space is straightforward. On the other hand, although being quite intuitive, the Voronoi type methods often encounter increasing geometric and logical complexity and the approaches rigidity prevents MAT modification/customization.

In this paper, we propose a hybrid differential MAT approach based on the pixel/voxel distance field solution, namely $d-$MAT. Notice the advantage of such a hybrid method is to extract a well approximated medial point cloud on a properly calculated distance field, while pure thinning methods are fundamentally *discrete* and cannot ensure an accurate distance distribution. A similar approach has been suggested by Bouix [5]. However, the Laplacian or Hessian determinant criteria of the distance field proposed in this study seems simpler, more robust, and is independent on the thinning techniques. As shown later in this paper, the differential equation-based approach also provides a biased MAT, in other words the medial axis does not necessarily lie at the mid-point of the space, and it can be sculpted/controlled by the user.

To evaluate d, there are several methods. They can be broadly classified as: search procedures, integral approaches, and numerically solving differential equations. Crude search procedures often require $O(n_v n_s)$ operations where n_v and n_s correspond to the number of volume and surface node points [28]. This can be $O(n_v \sqrt{n_s})$ and $O(n_v \log n_s)$ operations, however, for complex geometries such specialized approaches are difficult to apply [28]. Differential equation-based methods have been discussed in detail in References [29, 28, 30]. Advantageously, they are naturally compatible with vector and parallel computer architectures. The focus here is on the solution of Eikonal differential equation within the framework of the integer Cartesian space using the fast-marching method (FMM) [25] or the fast-sweeping method (FSM) [31].

2 Solutions of Eikonal Equation

2.1 H-J/Eikonal Equation for d

To overcome the expense of calculating d, Sethian [25] considered viscosity solutions of the following general Hamilton-Jacobi equation, $\epsilon \to 0$,

$$\beta \frac{\partial \phi}{\partial t} + H(\nabla \phi, \mathbf{x}, \beta) = \epsilon \nabla^2 \phi \tag{1}$$

with $H(\nabla \phi, \mathbf{x}, \beta) = F|\nabla \phi| - (1 - \beta)$. A stationary Eikonal equation with hyperbolic nature is obtained when $\beta = 0$:

$$F|\nabla\phi| = 1 + \epsilon\nabla^2\phi \qquad (2)$$

where the dependent variable ϕ describes first arrival times of propagating wave fronts from boundaries, and $F(\mathbf{x})$ is the local speed function of these fronts. The wall distance is then simply $d = F\,\phi$, if $F \equiv const..$ With $\epsilon \to 0$, Eq. (2) can be solved by numerical schemes with just enough dissipation to gain an entropy (physical sensible) solutions [25, 28]. As shown by Tucker [28, 30], the right hand side Laplacian is useful and often employed to control the front propagation velocity. However, this potential is not explored here.

With $\epsilon = 0$, Eq. (2) is also the multidimensional form of the boundary value formulation of the front tracking problem, $F = \frac{dx}{d\phi}$, where $F > 0$. Details of the procedures to solve Eq. (2) with $\epsilon \equiv 0$ using FMM and FSM are given in Appendix.

2.2   Domain and Initialization

For simplicity and efficiency, uniform Cartesian grids[1] are used in this d calculation context, namely $\Delta x = \Delta y = \Delta z = h$. The numbers of grid points in x, y and z direction are $I + 1$, $J + 1$ and $K + 1$ respectively. The physical space is transformed and scaled into $x \in [0 : I]$, $y \in [0 : J]$ and $z \in [0 : K]$, so that no grid storage is needed and point locating is straightforward. To locate a given point (x_0, y_0, z_0) in the Cartesian grid, the floor function $\lfloor \cdot \rfloor$ is simply applied. For example, if $i = \lfloor x_0 \rfloor$, then $i \leq x_0 \leq i + 1$, and so are y_0, z_0.

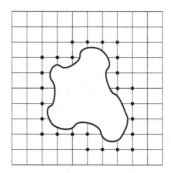

Fig. 1. A 2D schematic showing the wall boundary (or initial front) and the initialization of the solution ($F > 0$ for outward propagation).

Figure 1 shows the position of the wall boundary or initial front inside a Cartesian grid. This arbitrary curve can either be open or closed. To initialize, all immediate nearby grid points of the curve are marked as known (black

---

[1] Non-uniform Cartesian grids are particularly attractive when the geometric features have a broad range of scales.

(a)   (b)

Fig. 2. Distance field contours for (a) a cosine wave, and (b) a curved cross domain

dots in Figure 1) and the value of d on these points are evaluated exactly. The remaining points are marked as 'unknown' and the value of d on them set as ∞. Then, the whole domain is solved numerically with FMM or FSM.

2.3   Distance Solution Examples

Figure 2 shows the distance contours of example 1 and 2 solved on a uniform Cartesian grid with $I = J = 200$. In Example 1, frame (a), the boundary (initial front) is a piece of cosine curve $x \in [0, 4\pi]$, and in example 2, frame (b), the domain is the closed area of a curved cross profile. Both examples show the entropy conformed fronts propagating from the initial position in 2D space. The analogous compression from the concave feature and the expansion from the convex and sharp feature are also clear.

Example 3 is a slightly more complex multiply connected 2D domain. Various shapes (including a rectangle, a square, a triangle, an ellipse and a circle) are subtracted from the domain leaving internal voids with disparate d scales. A Cartesian grid with $I = J = 400$ is used. Both FMM and FSM demonstrate good efficiency to obtain accurate solutions with FMM using 4.7 seconds and a single 2^2-sweep FSM using 2.3 seconds on a single 2.33GHz Intel Xeon CPU core. Figure 3a shows the distance contour. Collisions of fronts are seen in the central area. A center line cut (shown as the dash-dot line) is made through the domain. Figure 3b compares numerical solutions together with the exact d distribution on the center line. The FMM and FSM solutions are close to each other and agree with the analytical solution.

Example 4 is a 3D cubic box, where a variety of shape subtractions are made. FSM is applied to get the solution on a $I = J = K = 200$ uniform 3D Cartesian grid. Figure 4 plots the d contours at three cut planes. The predicted maximum d is within 0.05% of the exact solution.

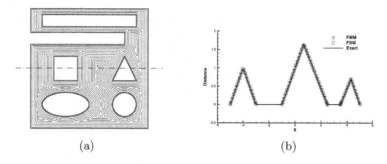

(a)   (b)

Fig. 3. Distance field contours for example 3

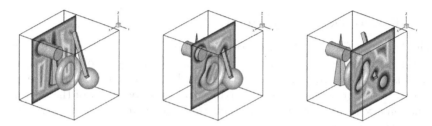

Fig. 4. Distance field obtained for the cubic box domain at various planes. The contour range evenly spreads from 0 (blue) to d_{max} (red).

3   Medial Axis Transform via Distance Field: $d-$MAT

The medial axis of a shape provides a compact representation of its features and their connectivity. As a result, researchers have discovered and are still exploring its use in many fields, such as topology recognition for grid generation [26, 23]. The medial axis is defined when the shape is embedded in an Euclidean space and is endowed with a distance function. Therefore, an expedient route is to efficiently obtain d (as discussed in previous sections) followed by the medial axis construction. (Notice that this approach is different from the pure geometric Voronoi-diagram based approach, in which the equal distance nature of the Voronoi-diagram is a key.) In 3D, a sphere is called *medial* if it meets S, the domain boundary, only tangentially in at least two points. The medial axis M is defined as the closure of the set of centers of all medial spheres. Figure 5 is an illustration in 2D. Only the medial axis inside the domain is considered here. Informally, the medial axis of a surface in 3D is the set of all points that have more than one closest point on the surface. They are often called the medial axis transform (MAT) for that 3D bounded domain.

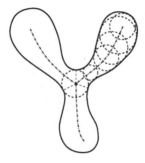

Fig. 5. Schematic of the medial axis for a 2D domain defined by a closed curve

3.1   Feature Detection Criteria

The key step for the MAT construction here is to detect the medial axis feature in a given d field. Notice that the unique property of a medial axis point is that it has equal distance to multiple boundaries. Hence, in space, the medial axis represents the 'local maxima' or non-smoothness of the distance function $d(\mathbf{x})$. For example, in one dimension the medial axis is a single point as shown in Figure 6.

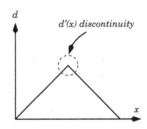

Fig. 6. An example of 1D distance field and its medial axis (vertex)

In multi-dimensions, one of the general formulations to represent this discontinuity in gradient or 'slope jump' is the Laplacian, $\nabla^2 d$. In smooth regions, $\nabla^2 d \equiv 0$. In the vicinity of the local maxima of d, $\nabla^2 d \rightarrow -\infty$. Hence, with the numerical approximation of d, the medial axis area can be identified by specifying a criteria such as $\nabla^2 d < -\epsilon$, where positive small number ϵ is a user specified threshold.

Another way to detect the medial axis feature is through the Hessian matrix determinant. The Hessian matrix, $H_{ij}(d) = \frac{\partial^2 d}{\partial x_i \partial x_j}$, is a square matrix of second-order partial derivatives of a function. It is the true 'second order' derivative in multi-dimensional space. The determinant of the Hessian matrix has been widely used in computer vision [15] (e.g. the 'blob detection') and shock wave detection in computational fluid dynamics [2, 20].

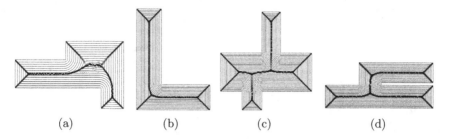

(a)   (b)   (c)   (d)

Fig. 7. Distance field and marked medial axes for simple domains: (a, b) Laplacian based medial axes and (c, d) Hessian based.

Figure 7 gives geometries where d distance contours and medial axes (in thicker line) are shown. For the medial axis identification, Figures 7(a) and (b) use the Laplacian criteria while 7(c) and 7(d) apply the Hessian matrix criteria. The results are very similar. We note that generally the Laplacian is simpler and cheaper to calculate. Figure 8 gives Laplacian based medial

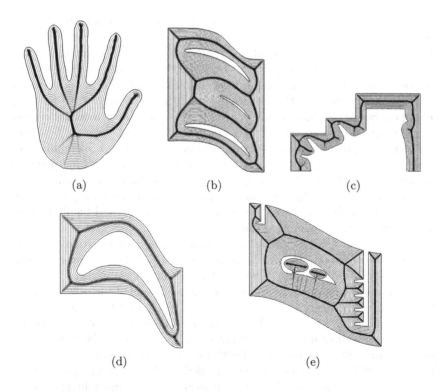

(a)   (b)   (c)

(d)   (e)

Fig. 8. Distance field and marked medial axes for complex domains

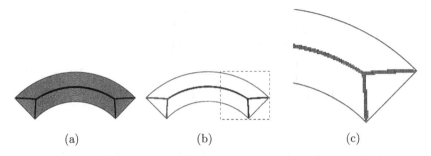

Fig. 9. Point cloud represented medial axis for a bended duct: (a) d contours and marked MA, (b) Underlying point cloud of the MA, (c) Zoom-in.

axes for more complex geometries. These include a hand profile, compressor passage with two blades and one splitter, a turbine blade shroud, a turbine blade in a passage, and a test section with an airfoil and other devices.

3.2   Thinning and Representation

It is very often still not enough to just have the medial axis as a finite thickness curve or surface. The ultimate aim of MAT is to reduce the marked area to thin curves/surfaces and possibly to represent them in a parametric form. This requires a 'thinning' operation. As shown, for example, in Figure 9 once the medial axis area is marked, by say application of $\nabla^2 d < -\epsilon$ one only obtains a point cloud of the underlying mesh nodes. However, to build up the topology, data reduction and connectivity of the point cloud are needed. Because the complex medial axis area does not represent a simple bounding line or surface, surface reconstruction methods or point cloud simplification methods cannot be applied directly, particularly due to branching points of the medial axis. Hence, here, shape recognition methods are considered. A typical procedure is to first convert the point cloud into a binary image and then thin the 'width' of the band (or shell) in the image, and finally transfer back to medial axis skeleton of the domain.

Notice that the term 'thinning' is a morphological operation that is used to remove selected foreground pixels/voxels from 2D/3D binary images, somewhat like erosion or opening. The result of thinning can be regarded as the minimum set of the topology preserving representation of the original shape. It can be used for many applications, and is often useful for skeletonization. Thinning is normally only applied to binary images, and produces another binary image as output. In the current case, because the uniform Cartesian grids are used, they are conformed with binary (black & white) images. The marked medial axis points can be regarded as 'black' pixels and the rest are 'white'. Therefore, the thinning methodologies for binary images can be applied here to thin medial axis point clouds.

Thinning in $\mathbb{Z}^2$

Thinning methodologies for 2D binary images have been extensively studied since 1980s (see [13] for a review). In this study, the algorithm described in [11] is adopted. Consider a pixel at p and its eight neighbors, see Figure 10. The following notations are used. The pixels $x_1, x_2, \ldots, x_8$ are the 8-neighbors of p and are collectively denoted by $N(p)$. They are said to be 8-adjacent to p. The pixels, x_1, x_3, x_5, x_7 are the 4-neighbors of and 4-adjacent to p. The value of each pixel can either be 0 or 1 meaning 'white' and 'black'.

x_4	x_3	x_2
x_5	p	x_1
x_6	x_7	x_8

Fig. 10. Pixel adjacencies of $N_8(p)$ in $\mathbb{Z}^2$, where the value at each pixel can either be 0 or 1

The thinning algorithm is generally described as: p is deleted (i.e. value changed from 1 to 0), if only if all the following conditions are satisfied:

C1: $X_H(p) = 1$,
C2: $2 \leq \min[n_1(p),\, n_2(p)] \leq 3$,
C3: $(x_2 \vee x_3 \vee x_8) \wedge x_1 = 0$ for odd iterations; $(x_6 \vee x_7 \vee x_4) \wedge x_5 = 0$ for even iterations.

where $\vee$, $\wedge$ are logical 'OR', 'AND' operators, and

$$X_H(p) = \sum_{i=1}^{4} b_i \quad \text{and} \quad b_i = \begin{cases} 1, & \text{if } (x_{2i-1} = 0) \text{ and } (x_{2i-1} \text{ or } x_{2i+1} = 1) \\ 0, & \text{otherwise} \end{cases}$$

$$(3)$$

$$n_1(p) = \sum_{k=1}^{4} x_{2k-1} \vee x_{2k} \quad \text{and} \quad n_2(p) = \sum_{k=1}^{4} x_{2k} \vee x_{2k+1} \qquad (4)$$

A note to this algorithm is that it preserves the connectivity of the pixels, or to say: no two remaining pixels after thinning are 'disconnected' if they were neighbors before the thinning. Therefore, if p and x_1 are the remaining pixels in Figure 10, the local connectivity is straightforward, because x_1 is either directly or diagonally adjacent to p. A simple $N_8(p)$ neighborhood check restores the full connectivity. p is identified as a medial axis branching point if $\sum N_8(p) > 2$; for p, where $\sum N_8(p) = 1$, p is an end point of a medial axis; and finally for any p with $\sum N_8(p) = 2$, p is a middle point in a medial axis curve.

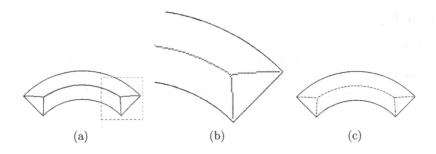

Fig. 11. The bended duct after applying thinning: (a) Single-pixel-wide point cloud, (b) Zoom-in, (c) Rebuilt connectivity and splined MA curves.

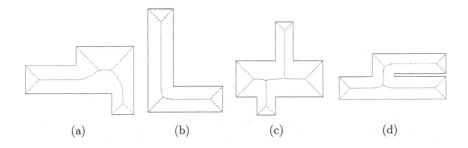

Fig. 12. Fina medial axes for simple domains previously shown in Figure 7

The typical procedure of applying this algorithm can be described with the aid of Figures 9 & 11. First, the binary image is obtained from the marked medial axis point cloud by one of the feature detecting criteria on the d solution, as in Figure 9b & c. Secondly, the binary image is thinned using the above pixel thinning algorithm, shown in Figure 11a & b. Thirdly, the pixels are transferred from $\mathbb{Z}^2$ back to $\mathbb{R}^2$. Finally, and optionally, the MAT points can be splined in smooth curves, as in frame (c). Figures 12 & 13 show the results of pixel thinning applied to the cases previously shown in Figures 7 & 8.

Thinning in $\mathbb{Z}^3$

It is much more complicated to thin a binary image with $p \in \mathbb{Z}^3$. The recent study of Palágyi [18] shows that a robust (both topology-preserving and maximum thinning) surface-thinning algorithm is possible. It deals with the full 26-adjacent voxels to p, see Figure 14. This is an analogous extension from the 2D thinning algorithms but different in the underlying details. We adopt this method in this study to thin the equivalent $\mathbb{Z}^3$ binary image representing the marked medial axis point cloud in 3D.

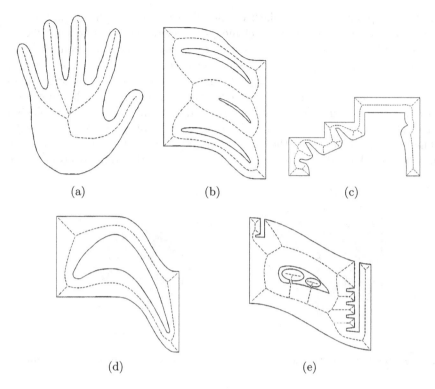

(a)   (b)   (c)

(d)   (e)

Fig. 13. Final medial axes for more complex domains previously shown in Figure 8

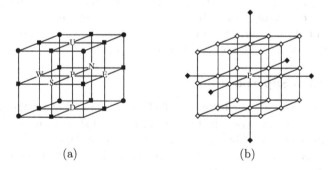

(a)   (b)

Fig. 14. Voxel adjacencies of $N_{26}(p)$ in $\mathbb{Z}^3$

Consider a voxel p. Its directly connected 6-adjacent neighbors are denoted as **U, N, E, S, W** and **D**, shown in Figure 14a, also denoted as $N_6(p)$. The set $N_{18}(p)$ contains the set $N_6(p)$ and the 12 points marked "black square". The set $N_{26}(p)$ contains the set $N_{18}(p)$ and the 8 points marked "black circle". Whether p is deletable depends on $N_{26}(p)$ marked "diamond" and six additional points marked "black diamond", see Figure 14b. An important step is to construct structuring elements to delete the extra voxels. Due to the extended neighbor dependency, this procedure, although becoming more complex than its 2D counterpart, still seems manageable. A complete set of 114 structuring elements are suggested in [18]. Details of this algorithm can be found in the original paper. A note to the application of this thinning algorithm is that in our study the extracted medial point cloud by the Laplacian criteria is already quite thin, usually a few voxels thick, say $N_{th} \sim 4$ voxels. The iterations involved in the above thinning is at most $N_{th}/2 \sim 2$, and this number is independent on the size of the image, i.e. the total voxels in space.

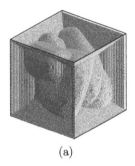 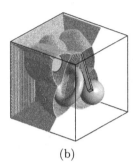

(a)   (b)

Fig. 15. Full and partial view of the 3D medial axis surface through voxel thinning

For example, the 3D medial surfaces, shown in Figure 15, are obtained using this approach. The FSM d solution of the $200 \times 200 \times 200$ Cartesian grid in $\mathbb{Z}^3$ is calculated first. The voxels with local maxima of d are marked using the Laplacian criteria, and finally thinned into the shown complex medial surface inside the domain, between objects.

3.3   Solution Superposition

One useful feature of the current $d-$MAT approach is that it allows solution superposition. For instance, in Figure 16, frames (a), (b) and (c) are three independent solutions. The first is simply a square domain. The other two are solid points at two different locations placed in an infinite domain. After applying the minimum superposition:

$$d = \min(d_a, d_b, d_c) \tag{5}$$

where d_a, d_b and d_c are the independent solutions, the solution shown in frame (d) is obtained. The solution is the same as if it was solved with the three boundary conditions imposed simultaneously. This decomposition is particularly useful because different speed functions can be applied to different solutions before the superposition. For example, Figure 17 represents a different pattern of the distance field after the superposition of

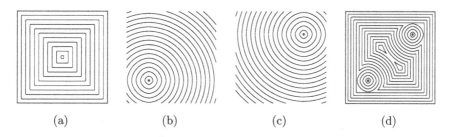

(a)   (b)   (c)   (d)

Fig. 16. Eikonal d solution superposition for a square domain and two points in infinite domains: (a) Square domain solution; (b, c) Point source solutions; and (d) Superposition of solutions

Fig. 17. Solution superposition for a square box and two points distance field solution with $F_a : F_b : F_c = 1 : 4 : 2$

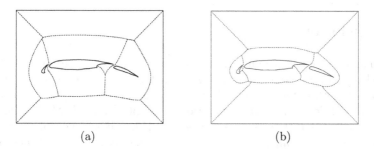

(a)   (b)

Fig. 18. Standard and biased medial axes for a multi-element airfoil domain

three solutions with different speeds, in which case the speed functions are $F_a : F_b : F_c = 1 : 4 : 2$.

With this useful feature, one can build biased medial axes by setting different speed functions to superimposed solutions. Figure 18(a) shows the standard medial axes of the flow domain of a multi-element airfoil, and frame (b) demonstrates the biased medial axes where front speed from the far field

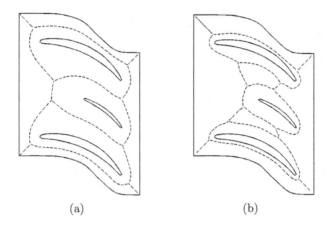

Fig. 19. Standard and biased medial axes for a compressor passage with two blades and one splitter

boundaries is three times faster than the one from the airfoil. Sometimes to create more sensible domain decomposition, it is desirable to have such a biased medial axis. For instance, in Figure 19, the biased medial axes may provide more sensible information for domain division. We also note that to further sculpt the medial axis (see Reference [30]) other d dependent functions can be added to the differential Eikonal type equation. This sculpting can provide a d−MAT for high quality hexehedral meshing. This is left as future work.

4 Conclusion

Fast marching and sweeping methods of the Eikonal equation have been applied to obtain distances, d, in arbitrary domains. A continuous distance field based method for creating the medial axis transform d−MAT has been proposed and demonstrated with various examples. This d−MAT approach combines fast Eikonal equation solutions with the pixel/voxel thinning. The link between the distance solution and the medial axes are established upon the use of simple criteria of $\nabla^2 d$ and $|H_{ij}(d)|$. It has been found to be a robust alternative to the classical pure geometric or pure thinning methods. Solution superposition and boundary dependent front speeds have been found useful to generate the biased medial axis, providing more general information than the standard medial axis. Because the differential equation does not pose any particular challenge when extended from 2D to 3D, it is a promising approach for medial axes evaluation in geometrically complex domains. The differential equation basis allows for flexible d−MAT customization through the addition of source terms that are functions of d itself. Future work will be focused on the representation of the 3D thinned medial surfaces which will involve building the connectivity through local voxel neighborhood, re-sampling the voxels and splining the surfaces.

Acknowledgments. This work is funded by UK Engineering and Physical Sciences Research Council (EPSRC) under grant number EP/E057233/1. We would like to thank Professor Kálmán Palágyi from University of Szeged for sharing the implementation of the 3D thinning algorithm. Helpful discussions with Dr Neil Dodgson from the Computer Laboratory at Cambridge University are also appreciated.

A   Finite Difference Solution Procedures

A.1   Fast-Marching Method

The gradient $\nabla\phi$ in Eq. (2) is discretized by Godunov's upwind difference scheme. This correctly chooses the physically vanishing weak viscosity solution. The following form is suggested by Osher [17] and Sethian [25],

$$
\left(
\begin{array}{l}
\max(D_{ijk}^{-x}\phi - D_{ijk}^{+x}\phi,\ 0)^2\ + \\
\max(D_{ijk}^{-y}\phi - D_{ijk}^{+y}\phi,\ 0)^2\ + \\
\max(D_{ijk}^{-z}\phi - D_{ijk}^{+z}\phi,\ 0)^2
\end{array}
\right) = \frac{1}{F_{ijk}^2}
\tag{6}
$$

where D_{ijk}^- and D_{ijk}^+ are the first-order backward and forward finite difference operators given by:

$$
\begin{array}{ll}
D_{ijk}^{-x}\phi = \dfrac{\phi_{i,j,k} - \phi_{i-1,j,k}}{\Delta x}, & D_{ijk}^{+x}\phi = \dfrac{\phi_{i+1,j,k} - \phi_{i,j,k}}{\Delta x} \\[2ex]
D_{ijk}^{-y}\phi = \dfrac{\phi_{i,j,k} - \phi_{i,j-1,k}}{\Delta y}, & D_{ijk}^{+y}\phi = \dfrac{\phi_{i,j+1,k} - \phi_{i,j,k}}{\Delta y} \\[2ex]
D_{ijk}^{-z}\phi = \dfrac{\phi_{i,j,k} - \phi_{i,j,k-1}}{\Delta z}, & D_{ijk}^{+z}\phi = \dfrac{\phi_{i,j,k+1} - \phi_{i,j,k}}{\Delta z}
\end{array}
\tag{7}
$$

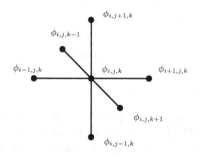

Fig. 20. Finite difference stencil at (i, j, k)

The stencil of grid points involved is shown in Figure 20. Although schemes with higher order accuracy are available, Refs. [25, 28] suggest the first-order scheme is sufficient for d calculation. Through substitution of the above equation Eq. (6) can be rewritten as:

$$\max\left(\frac{\phi - \phi_1}{\Delta x}, 0\right)^2 + \max\left(\frac{\phi - \phi_2}{\Delta y}, 0\right)^2 + \max\left(\frac{\phi - \phi_3}{\Delta z}, 0\right)^2 = \frac{1}{F_{ijk}^2} \quad (8)$$

where

$$
\begin{aligned}
\phi &= \phi_{i,j,k} \\
\phi_1 &= \min(\phi_{i-1,j,k}, \ \phi_{i+1,j,k}) \\
\phi_2 &= \min(\phi_{i,j-1,k}, \ \phi_{i,j+1,k}) \\
\phi_3 &= \min(\phi_{i,j,k-1}, \ \phi_{i,j,k+1})
\end{aligned} \quad (9)
$$

In a way similar to the Dijkstra algorithm [8], the idea of the Fast-Marching Method (FMM) is to introduce an order to the selection of grid points. This order is based on a causality criteria, where the arrival time ϕ at any point depends only on the neighbors that have smaller values. Therefore, the FMM relies on propagating the information in one direction from smaller values of ϕ to larger ones. This is also convenient for Eq. (8), since it can be further reduced to a standard quadratic equation,

$$\left(\frac{\phi - \phi_1}{\Delta x}\right)^2 + \left(\frac{\phi - \phi_2}{\Delta y}\right)^2 + \left(\frac{\phi - \phi_3}{\Delta z}\right)^2 = \frac{1}{F_{ijk}^2} \quad (10)$$

The general algorithm can be described as follows. The front constructs a narrow band of trial points, distinct from the accepted (known) points and the far (unknown) points. Among the current trial points, the point, e.g. denoted as A, with smallest ϕ is moved from 'trial' status to 'known'. Each neighbor of A is added to trial points if it was 'unknown' before, and the quadratic equation (10) is solved for each neighboring point of A. Therefore, a new narrow band of trial points is formed. A recursive procedure is performed until there are no more 'unknown' points. The key to this algorithm is to find the smallest ϕ among the trial points. To store the trial points A special min-heap data structure [4] is suggested in [25], in which the worst case of finding the smallest ϕ has the complexity of $O(\log N)$. This is significantly improved compared with a crude search, which is of $O(N)$.

A.2   Fast-Sweeping Method

To avoid implementing any complex data structure essentially needed in the marching methods, the Fast-Sweeping Method (FSM) [31] is also considered. For discretization, FSM shares the same finite difference scheme with FMM, i.e. the first-order Godunov's scheme Eq. (6). However, the FSM does not have the FMM's convenience of reducing Eq. (8) to a standard quadratic equation. A solution procedure of Eq. (8) is suggested in [31] by seeking the solution of the following general form,

$$\max(x - a_1,\, 0)^2 + \max(x - a_2,\, 0)^2 + \cdots + \max(x - a_n,\, 0)^2 = b^2 \tag{11}$$

where n is the number of dimensions. The above equation is re-organized so that $a_1 \leq a_2 \leq \cdots \leq a_n$. When $n = 2$, the unique solution of the above equation is:

$$\bar{x} = \begin{cases} a_1 + b, & \text{if } a_1 + b \leq a_2, \\ \dfrac{a_1 + a_2 + \sqrt{2b^2 - (a_1 - a_2)^2}}{2}, & \text{otherwise.} \end{cases} \tag{12}$$

and for $n \geq 3$ a recursive procedure, detailed in [31], is required to find $\bar{x}$, in which an integer p, $1 \leq p \leq n$, is sought such that $a_p < \bar{x} \leq a_{p+1}$ is the unique solution of

$$(x - a_1)^2 + (x - a_2)^2 + \cdots + (x - a_p)^2 = b^2 \tag{13}$$

FSM involves Gauss-Seidel iterations with alternating sweep orderings. At each grid point $x_{i,j,k}$ whose value is not initialized as known (or fixed), the solution of Eq. (8), denoted by $\bar{\phi}_{i,j,k}$, is computed from the most recent values of its neighbors $\phi_{i\pm1,j,k}$, $\phi_{i,j\pm1,k}$ and $\phi_{i,j,k\pm1}$, and then the value at $x_{i,j,k}$ is updated by:

$$\phi_{i,j,k}^{new} = \min(\phi_{i,j,k}^{old},\, \bar{\phi}_{i,j,k}) \tag{14}$$

The whole domain is repeatedly swept with 2^n alternating orderings. For example when $n = 3$, these orderings are

$$\{i = 0 : I \text{ or } I : 0\} \times \{j = 0 : J \text{ or } J : 0\} \times \{k = 0 : K \text{ or } K : 0\}$$

Notice that often just one 2^n sweep cannot guarantee a converged solution for an arbitrary speed function $F(\mathbf{x})$, especially those with complex characteristics [31, 12]. However, in practice, for constant F or those with simple characteristics one 2^n sweep is sufficient for an accurate solution.

References

1. Ahuja, N., Chuang, J.H.: Shape Representation Using a Generalized Potential Field Model. IEEE Trans. Pattern Anal. Mach. Intell. 19(2), 169–176 (1997)
2. Ait-Ali-Yahia, D., Habashi, W.G., Tam, A., Vallet, M., Fortin, M.: A directionally adaptive methodoloty using an edge-based error estimate on quadrilateral grids. International Journal for Numerical Methods in Fluids 23(7), 673–690 (1998)
3. Ang, P.Y., Armstrong, C.G.: Adaptive Curvature-Sensitive Meshing of the Medial Axis. In: Proceedings of the 10th International Meshing Roundtable, pp. 155–165 (2001)

4. Atkinson, M.D., Sack, J.R., Santoro, B., Strothotte, T.: Min-max heaps and generalized priority queues. Commun. ACM 29(10), 996–1000 (1986)
5. Bouix, S., Siddiqi, K.: Divergence-Based Medial Surfaces. In: Proceedings of the Sixth European Conference on Computer Vision, pp. 603–618 (2000)
6. Chou, J.J.: Voronoi diagrams for planar shapes. IEEE Computer Graphics and Applications 15(20), 52–59 (1995)
7. Dey, T.K., Zhao, W.: Approximate medial axis as a voronoi subcomplex. In: Proceedings of the seventh ACM symposium on Solid modeling and applications, pp. 356–366 (2002)
8. Dijkstra, E.: A Note on Two Problems in Connection with Graphs. Numerische Mathematik 1, 269–271 (1959)
9. Du, H., Qin, H.: Medial axis extraction and shape manipulation of solid objects using parabolic PDEs. In: Proceedings of the ninth ACM symposium on Solid modeling and applications, pp. 25–35 (2004)
10. Fares, E., Schröder, W.: A differential equation for approximate wall distance. International Journal for Numerical Methods in Fluids 39, 743–762 (2002)
11. Guo, Z., Hall, R.W.: Parallel thinning with two-subiteration algorithms. Commun. ACM 32(3), 359–373 (1989)
12. Jeong, W.-K., Whitaker, R.T.: A Fast Iterative Method for Eikonal Equations. SIAM J. Sci. Comput. 30(5), 2512–2534 (2008)
13. Lam, L., Lee, S.-W., de Suen, C.Y.: Thinning Methodologies-A Comprehensive Survey. IEEE Trans. Pattern Anal. Mach. Intell. 14(9), 869–885 (1992)
14. Li, T.S., Mckeag, R.M., Armstrong, C.G.: Hexahedral meshing using midpoint subdivision and integer programming. Comput. Methods Appl. Mech. Engrg. 124, 171–193 (1995)
15. Mikolajczyk, K., Schmid, C.: Scale & Affine Invariant Interest Point Detectors. Int. J. Comput. Vision 60(1), 63–86 (2004)
16. Näf, M., Székely, G., Kikinis, R., Shenton, M.E., Kübler, O.: 3D Voronoi skeletons and their usage for the characterization and recognition of 3D organ shape. Comput. Vis. Image Underst. 66(2), 147–161 (1997)
17. Osher, S., Sethian, J.: Fronts propagating with curvature dependent speed: algorithms based on Hamilton-Jacobi f ormulations. Journal of Computational Physics 79, 12–49 (1988)
18. Palágyi, K.: A 3D fully parallel surface-thinning algorithm. Theor. Comput. Sci. 406(1-2), 119–135 (2008)
19. Price, M.A., Armstrong, C.G.: Hexahedral mesh generation by medial surface subdivision: part II. solids with flat and concave edges. International Journal for Numerical Methods in Engineering 40, 111–136 (1997)
20. Qin, N., Liu, X.: Flow feature aligned grid adaptation. International Journal for Numerical Methods in Engineering 67(6), 62–72 (2006)
21. Quadros, W.R., Owen, S.J., Brewer, M., Shimada, K.: Finite Element Mesh Sizing for Surfaces Using Skeleton. In: Proceedings of the 13th International Meshing Roundtable, pp. 389–400 (2004)
22. Quadros, W.R., Ramaswami, K., Prinz, F.B., Gurumoorthy, B.: LayTracks: A New Approach To Automated Quadrilateral Mesh Generation using MAT. In: Proceedings of the 9th International Meshing Roundtable, pp. 239–250 (2000)
23. Rigby, D.L.: TopMaker: A Technique for Automatic Multi-Block Topology Generation Using the Medial Axis, NASA Report: NASA/CR-2004-213044 (2004)
24. Rumpf, M., Telea, A.: A continuous skeletonization method based on level sets. In: Proceedings of the symposium on Data Visualisation, p. 151 (2002)

25. Sethian, J.: Level Set Methods and Fast Marching Methods: Evolving Interfaces. In: Computational Geometry, Fluid Mechanics, Computer Vision, and Materials Science. Cambridge University Press, Cambridge (1999)
26. Sheehy, D.J., Armstrong, C.G., Robinson, D.J.: Shape Description By Medial Surface Construction. IEEE Transactions on Visualization and Computer Graphics 2(1), 62–72 (1996)
27. Sud, A., Otaduy, M.A., Manocha, D.: DiFi: Fast 3D Distance Field Computation Using Graphics Hardware. In: Proceedings of Eurographics 2004, pp. 557–566 (2004)
28. Tucker, P.G.: Differential equation-based wall distance computation for DES and RANS. J. Comput. Phys. 190(1), 229–248 (2003)
29. Tucker, P.G., Rumsey, C.L., Bartels, R.E., Biedron, R.T.: Transport Equation Based Wall Distance Computations Aimed at Flows with Time-Dependent Geometry, NASA Report, NASA-TM-2003-212680 (2003)
30. Tucker, P.G., Rumsey, C.L., Spalart, P.R., Bartels, R.E., Biedron, R.T.: Computation of Wall Distances Based on Differential Equations. AIAA Journal 43(3) (2005)
31. Zhao, H.: A Fast Sweeping Method for Eikonal Equations. Mathematics of Computation 74(250), 603–627 (2004)

Automatic Non-manifold Topology Recovery and Geometry Noise Removal

Aurélien Alleaume

Distene SAS, Pôle Teratec-Bard 1, Domaine du Grand rué 91680
Bruyères-le-Chatel, France
`aurelien.alleaume@distene.com`

Abstract. This paper presents an effective method for automatic topology recovery of non-manifold geometries. Mixing topology recovery and geometry noise detection/removal allowed us to achieve effectively the automation and robustness required by such methods. We developped our method on CAD boundary representation (BREP) geometries in the context of surface mesh generation, but it could also be directly applied to discrete STL or mesh geometries. Its reliability and efficiency has been validated on a variety of complex manifold and non-manifold CAD geometries.

1 Introduction

Even today, legacy or imported CAD models often lack a large amount of topological information (patch connectivity) or contain some geometrical noise entities (like micro-curves) which are often due to translations involved when exchanging data between CAD systems. These can prevent a conforming boundary representation of the surface assembly from being built straightaway.

This especially becomes a problem for modern parametric meshing methods like [5] when it comes to generate a conforming surface mesh.

Even though great improvements were allowed by modern file formats such as STEP, these issues still did not totally disappear especially on large geometries made of complex assemblies.

Let us recall that boundary representation models are composed of two parts:

1. Geometry: surfaces, curves and points.
2. Topology: faces, coedges, edges and vertices. A face being a connex component of a surface bounded by coedges; a coedge a bounded interval of a curve. Edges are bounded by vertices and associate coedges accross patches to define how faces are connected. Vertices associate points to define how edges are connected. Higher level entities such as lumps and shells are often defined but are beyond the scope of the method we propose.

The problem of topology recovery is to compute coherent topological edges and vertices data from :

1. the given geometry,
2. the topology of faces and coedges.

In other words, points and curves must be associated to one another to define how faces are linked together. This might also require some curves to be split as it will be illustrated later on figure 1.

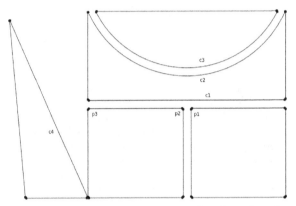

Let us consider that this whole figure is smaller than the tolerance ε : p_1 and p_2 will be identified. p_1 will also be projected on c_1. p_1 will not be projected on c_2 because $face(c_2) \in F(p_1)$. Last, neither p_1 will be projected on c_3 nor p_3 on c_4 because the reverse projection condition contained in $C_{point}()$ is not satisfied.

Fig. 1. Vertex projection and association

Methods for topology recovery have been proposed since the early 90's. Tolerance driven methods which associate all geometrical entities within a given threshold quickly show their limits as the CAD geometries are often multiscale, making the global tolerance adjustment complex, case dependent and even sometimes impossible.

In [2], a method to recover the topology of a mesh based CAD is proposed. It uses heuristics to determine wrong associations of edges, thus removing the burden of tolerance tuning. The method proposed by [3] to repair a discrete polyhedral CAD geometry is to first compute a shared-vertex polygonal representation and then merge each polygon edge with the most appropriate edge. While providing more automation, these methods only apply to discrete mesh based CAD geometries and do not remove the need for a human intervention on complex models.

The approach recently described in [1] based on scoring function to pair curves together extends these methods to continuous CAD models, providing more automation and robustness. Unfortunately, it applies only to manifold geometries and it is still not adapted to models containing geometry noise such as micro-curves.

Our experience with CAD models and topology reconstruction is that, without effective geometry noise detection and removal, a reconstruction method cannot be effective on complex or dirty models. This led us to another approach whose basic keys are:

1. to **mix topology recovery with geometry noise detection and removal**. Since the geometrical noise detection is a multiscale problem, it is mainly made through topological considerations without requiring another extraneous tolerance threshold. This allows us to detect automatically noise, without removing small features.
2. to **associate geometrical entities (points then curves) as long as it makes sense from a geometrical/topological non-manifold point of view** and the optionally specified tolerance is not reached.

One of the major strengths of our method is its generality. Unlike [2] it is not based on several specific heuristics but only on a couple of criteria: one for vertex association and one for edge association. Moreover it provides greater automaticity without restricting the field of application to manifold geometries like [1]. Last but not least, non-manifold geometry features do not require any special treatment but are treated as the general case.

2 Algorithm Description

2.1 Pre-requisites

We suppose that the intra-face topology is known or has already been reconstructed. In other words, vertex association is already achieved inside each face of the input geometry. We concentrate here on the reconstruction of the inter-face topology which is the most complex part of the problem.

To simplify notations, we will use the term *curve* instead of *coedge* from now on.

Definition 1.
For a curve c, we will note $\hat{c}$ the edge corresponding to c, i.e. the set of curves associated to c.

For a point p, we will note $\hat{p}$ the vertex corresponding to p, i.e. the set of points associated to p.

We will also define order(p) as the number of points in $\hat{p}$ and order(c) as the number of curves in $\hat{c}$.

In the illustrations of this section, points associated together as a vertex are surrounded by a dashed line ellipsis.

2.2 The Optional ε Tolerance Parameter

The only parameter the user can optionally tune in the algorithm is the tolerance ε. It defines the size of the smallest feature or gap the user wants to preserve. Even if some modelling artifacts or features are smaller than this tolerance, it will not be a problem for our reconstruction algorithm.

Moreover, in our method, the effects of a given ε are overriden by the following rules:

- a geometric detail smaller than ε will be kept unless it prevents conformity achievement;
- no vertex/edge association above ε will be made
- a vertex/edge association below ε will not be made if it is in contradiction with the local geometrical/topological configuration.

In practice, this ε parameter is only optional and practically never needs to be tuned because our method is mostly driven by the local geometry and topology.

2.3 Pre-processing

Since a great part of the algorithm is based on the notion of proximity between geometrical entities, curves must be discretized and accelerating structures need to be built in order to achieve a good implementation of this method. The algorithm will mainly need:

- to localize and enumerate close curves in the space,
- to enumerate the edges connected to each vertex.

2.4 Vertex Association and Projection

This step is the heart of the method: it detects which vertex associations or projections make sense or not, given the local geometry/topology. It also removes a part of the geometry noise using a collapsing technique.

Definition 2.
For a vertex v, let us define :

- $C(v)$ *the set of curves c such that v is an extremity of c*
- $\mathcal{F}(v)$ *the set of faces f such that $v \in f$*

Definition 3.
For a point p and a curve c we define:

- $face(c)$ *as being the face c belongs to.*
- *the distance between p and c as $d(p, c) = min_{a \in c} d(p, a)$ where $a \in c$ means that a is a point of c.*
- $P_c(p) \in c$ *the projection of p on c, i.e. a point such that $d(p, c) = d(p, P_c(p))$*

Let us remark that in this definition:

- $P_c(p)$ is not necessarily an orthogonal projection;
- $P_c(p)$ always exists (we consider a finite space) but is not necessarily unique. In this latter case, any choice of minimum can be made.

Definition 4. *For a curve c, we will say that 2 points $p_1 \in c$ and $p_2 \in c$ are compatible if $d_c(p_1, p_2)$, the distance between p_1 and p_2 along c, verifies: $d_c(p_1, p_2) \leq min\{\varepsilon, \frac{1}{10} \times length(c)\}$.*

Let us remark that in this definition the choice of the constant $\frac{1}{10}$ might seem arbitrary, but this choice does not actually have a strong impact on the algorithm behavior.

Definition 5. *For a set of faces F, we note $P_F(p)$ the projection (in the sense of definition 3) of the point p on all the curves of the faces in F.*

We can now introduce the criterion upon which the vertex association is decided:

Definition 6. *Let p be a point and c a curve. We define the proposition $C_{point}(p, c)$ as: $d(p, c) \leq \varepsilon$ and $P_{\mathcal{F}(\hat{p})}(P_c(p))$ is compatible with p.*

The last condition of this definition can be seen as a form of reflexivity condition and can be interpreted as "the reverse projection of the projection of p is topologically equivalent to p".

We will use the condition C_{point} to decide whether merging two points or projecting a point on a curve makes sense or not from a geometrical/topological point of view. The distance condition will prevent us from merging entities beyond the tolerance while the reverse projection condition will prevent us from creating faces overlaps or partially collapsing faces.

The algorithm 1 to project and associate vertices is illustrated on some common cases on figure 1. The reader should note that geometry noise removal happens when we remove (if it exists) the smallest edge between the two points we are about to identify.

2.5   Edge Association

The edge association step heavily depends on the results of the previous vertex association step: for all curves sharing the same vertex extremities, it determines whether they should be associated as a single edge or not.

Definition 7.
For an edge e, let $\mathcal{F}(e)$ be the set of faces f such that $e \in f$.

For 2 curves c_1 and c_2 we define the distance between c_1 and c_2 as:

$$d(c_1, c_2) = max\{max_{a \in c_1} d(a, c_2), max_{b \in c_2} d(b, c_1)\}$$

```
foreach point p in the model do
    list = ∅;
    foreach curve c such that d(p, c) ≤ ε do
    |   list += c;
    end
    sort list in the ascending order relatively to the function d(p, .);
    foreach curve c in list (in order) do
        if face(c) ∉ F(p̂) then
            if C_point(p, c) then
                p₁ = the first extremity of c ;
                p₂ = the second extremity of c ;
                if p and p₁ are compatible then
                |   remove (if exists) the smallest edge between p̂ and p̂₁;
                |   identify p and p₁ ;
                else if p and p₂ are compatible then
                |   remove (if exists) the smallest edge between p̂ and p̂₂;
                |   identify p and p₂ ;
                else
                |   split the curve c at P_c(p);
                |   identify P_c(p) and p;
                end
            else
                // merging further does not make topological sense
                end of loop on curve c;
            end
        end
    end
end
```

Algorithm 1. The vertex projection and association algorithm

Definition 8. *For 2 curves c_1 and c_2 we define the proposition $C_{curve}(c_1, c_2)$ as: $d(c_1, c_2) \leq \varepsilon$ and $\mathcal{F}(\hat{c_1}) \cap \mathcal{F}(\hat{c_2}) = \emptyset$.*

The condition C_{curve} is the criterion upon which the edge association is decided. The distance condition will prevent us from merging curves beyond the tolerance while the empty intersection condition will prevent us from creating faces overlaps and from partially collapsing faces.

This edge association step is described in algorithm 2 and illustrated on a simple case on figure 2. For clarity concerns, we did not detail the simple extra treatment which can be added to deal correctly with some face overlaps.

2.6   Topology Post-processing

While constructing vertices association, we may have removed already some geometry noise, but there might still be some micro-curves preventingthe

```
foreach curve c₁ in the model do
    list = ∅;
    foreach curve c₂ such that c₁ and c₂ share the same extremities do
        | list += c₂;
    end
    sort list in the ascending order relatively to the function d(., c₁);
    foreach curve c₂ in list (in order) do
        if C_curve(c₁, c₂) then
            | identify curves c₁ and c₂;
        else
            | // merging further for c₁ would not make topological sense
            | end of loop on c₂;
        end
    end
end
```

Algorithm 2. The edge association algorithm

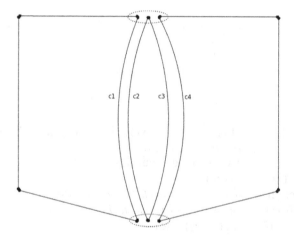

Let us consider that this whole pattern is smaller than the
tolerance ε: curves c_1 and c_2 will be merged, as well as c_3 and c_4
but c_1 will not be identified to c_3 because $\mathcal{F}(\hat{c}_1) \cap \mathcal{F}(\hat{c}_3) \neq \emptyset$ (c_1
and c_2 have been identified before).

Fig. 2. Edge association

global conformity to be achieved, as illustrated on figure 3. The goal of the
post-processing which is described in algorithm 3 is to identify and collapse
them.

The reader can remark that at the end of the post-processing, any face
whose curves have all been removed is removed.

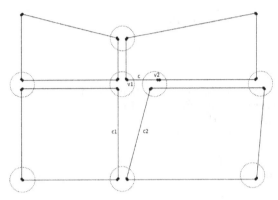

Let us consider that c is a small curve preventing conformity. It will be removed, vertices v_1 and v_2 will be merged, as well as curves c_1 and c_2

Fig. 3. Noise post-processing

---

list $= \emptyset$;
foreach *curve c in the model* **do**
    if $order(c) == 1$ *and* $length(c) \leq \varepsilon$ **then**
        | list += e;
    end
end
sort list in the ascending order relatively to the length of the curves;
foreach *curve c in list (in order)* **do**
    let v_1 and v_2 be the vertex extremities of c;
    if $order(c) == 1$ **then**
        foreach *curve $c_1 \in \mathcal{C}(v_1)$ such that $c_1 \neq c$* **do**
            foreach *curve $c_2 \in \mathcal{C}(v_2)$ such that $c_2 \neq c$* **do**
                if $C_{curve}(c_1, c_2)$ **then**
                    identify vertices v_1 and v_2;
                    remove curve c;
                    if *c_1 and c_2 share the same extremities* **then**
                        | identify curves c_1 and c_2;
                    end
                    continue with the next iteration on c;
                end
            end
        end
    end
end
remove any face whose curves have all been removed;

Algorithm 3. The topology post-processing algorithm

3   Applications

We implemented our algorithm in PreCAD, a module aimed at preparing CAD data for surface meshing. The CAD is first imported from a CAD engine, such as for example Opencascade [9] or ACIS [10]; the topology of each face is then reconstructed separately. Our algorithm is then applied to recover the global topology.

After this processing, the CAD is meshed with BLSURF [5]. To test our method deeper, all the input topology information for vertices and edges was discarded from the input CAD. Last but not least, the capacity to change the optional tolerance parameter was not used for any of these cases. Our goal was indeed to assess a fully automatic approach.

Figures 4, 5, 6, 7 and 8 show typical examples of geometries we are able to reconstruct automatically with our method.

We tested our method on a large test base of 800 more or less complex IGES, STEP and SAT cases with successful results. Table 1 gives some statistics on a representative set of industrial geometries. It presents:

- Input faces/curves: the number of faces and curves the input CAD geometries is made of.
- Removed faces/curves: the number of faces and curves that our algorithm decided to collapse because they were considered as noise, preventing conformity achievement.
- Free edges: the number of edges that remain unassociated after the process. One can expect this count to be null for a closed manifold surface.
- Non-manifold edges: the number of topological edges that have a class with at least 3 geometrical curves.

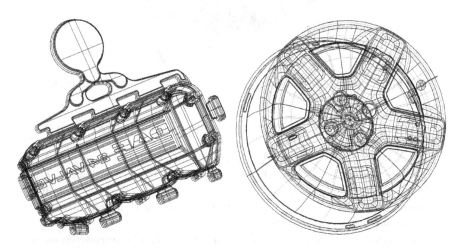

Fig. 4. Fad and Wheel geometries made of respectively 1,328 and 1,447 faces (courtesy of Fathi El-Yafi)

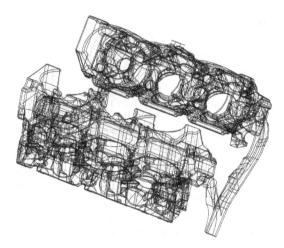

Fig. 5. V6 engine part made of 7,268 faces (courtesy of Fathi El-Yafi)

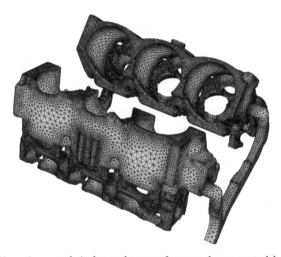

Fig. 6. V6 engine patch-independent surface mesh generated by BLSURF

- Time: CPU time required by the computation of the topology on an AMD Opteron 2.4 GHz.
- Valid volume: "yes" means that a volume mesh can be generated from the derived surface mesh with the surface constrained Delaunay mesher Tetmesh-GHS3D (see [6]). In this case, the surface mesh defines a volume and does not self-intersect. We can thus consider with a high probability the reconstruction as a success.
- Reconstruction: tells whether the reconstruction can be considered as successful or not. This is mainly determined by examining free and non-manifold (higher order) edges and visually checking in the generated surface

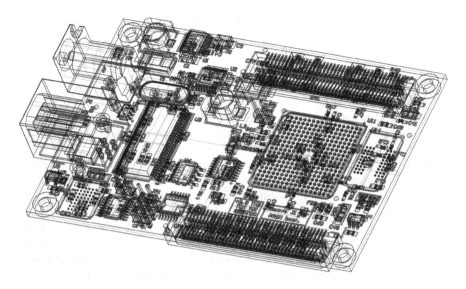

Fig. 7. Circuit board non-manifold multi-scale geometry made of 31,705 faces

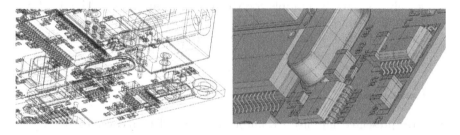

Fig. 8. Circuit board details

Table 1. Results

CAD Geometry	Mold	EBox	Fad	Wheel	V6	VX	Circuit
Input faces	381	2,325	1,328	1,447	7,268	18,700	31,705
Input curves	2,201	11,134	15,985	10,874	40,015	100,068	158,460
Removed faces	0	0	0	0	16	4	0
Removed curves	9	0	16	126	1,717	89	14
Free egdes	4	0	0	0	0	0	0
Non-manifold edges	0	162	0	0	0	3,620	524
Time	0.5s	1.25s	2.01 s	1.74s	7.72s	35.52s	28.01s
Valid volume	no	no	yes	yes	no	no	no
Reconstruction	failure	success	success	success	success	success	success

mesh that the computed topology is correct and conforms to expectations in these area.

These results lead to several remarks:

1. Even on large, complex and noisy geometries, the method is fast and really reliable, without any parameter tuning.
2. As expected geometry noise is removed in varying quantities, (from 0 to 4.2 percent) depending on the quality of the input case (see removed curves/faces rows).
3. Non-manifold (higher order) edges are correctly reconstructed in the non-manifold geometries (EBox, VX and Circuit). As expected, non-manifold edges are not created in manifold geometries.
4. The "Mold" case is considered as a failure because two curve associations were beyond the default tolerance ε and thus were rejected, leaving four free edges. In other words, this case could be successfully processed when changing the ε parameter. This shows that even though tolerance tuning issues have been drastically reduced, the human intervention cannot be totally eliminated from the reconstruction process when the geometry is too far from the intended topology.
5. Even when the reconstruction is successful, the derived surface mesh is not always directly suitable for volume mesh generation. This is often due to a wrong CAD design or to missing CAD boolean operations. To correct this kind of default, a face imprinting step like described in [8] should be carried out after topology recovery.

4   Conclusion

We have presented an effective automatic method to recover topology information while removing geometry noise from non-manifold geometries. Its robustness and efficiency has been shown on a wide range of complex CAD geometries.

Even though we are now able to reconstruct successfully the topology from complex geometries, we still cannot guarantee that the derived surface mesh defines a volume, especially on large geometries made of complex assemblies. That is why our next concern will be to apply our approach to automatic face imprinting techniques.

Acknowledgement. This work was partially funded by the Pôle System@tic's EHPOC project.

References

1. Renaut, E., Borouchaki, H., Laug, P.: Extraction of surface topological skeletons. In: Proceedings of the Numerical geometry, grid generation and high performance computing Conference(NUMGRID 2008), pp. 256–263 (2008)

2. Weihe, K., Willhalm, T.: Reconstructing the Topology of a CAD Model: A Discrete Approach. In: Burkard, R.E., Woeginger, G.J. (eds.) ESA 1997. LNCS, vol. 1284, pp. 500–513. Springer, Heidelberg (1997)
3. Barequet, G., Kumar, S.: Repairing CAD models. In: Proceedings of the 8th conference on Visualization (VIS 1997), pp. 363–370 (1997)
4. Mezentsev, A., Woehler, T.: Methods and algorithms of automated CAD repair for incremental surface meshing. In: Proceedings of the 8th International Meshing Roundtable, pp. 299–309 (1999)
5. Laug, P., Borouchaki, H.: BLSURF Mesh Generator for Composite Parametric Surfaces. INRIA Technical Report RT-0235 (1999)
6. George, P.L., Hecht, F., Saltel, E.: Automatic mesh generator with specified boundary. In: Computer Methods in Applied Mechanics and Engineering, pp. 269–288 (1991)
7. Merazzi, S., Gerteisen, E.A., Mezentsev, A.: A generic CAD-mesh interface. In: Proceedings of the 9th International Meshing Roundtables (2000)
8. Clark, B.W., Hanks, B.W., Ernst, C.D.: Conformal Assembly Meshing with Tolerant Imprinting. In: Proceedings of the 17th International Meshing Roundtable, pp. 267–280 (2008)
9. Open CASCADE S.A.S., Open CASCADE Technology, 3D modeling & numerical simulation, http://www.opencascade.org
10. Spatial Corporation, 3D ACIS Modeling, http://www.spatial.com

A New Procedure to Compute Imprints in Multi-sweeping Algorithms*

Eloi Ruiz-Gironés, Xevi Roca, and Josep Sarrate

Laboratori de Càlcul Numèric (LaCàN),
Departament de Matemàtica Aplicada III,
Universitat Politècnica de Catalunya,
Jordi Girona 1-3, E–08034 Barcelona, Spain
Tel.: 34-93-401 69 11; Fax: 34-93-401 18 25
{eloi.ruiz,xevi.roca,jose.sarrate}@upc.edu

Abstract. One of the most widely used algorithms to generate hexahedral meshes in extrusion volumes with several source and target surfaces is the multi-sweeping method. However, the multi-sweeping method is highly dependent on the final location of the nodes created during the decomposition process. Moreover, inaccurate location of inner nodes may generate erroneous imprints of the geometry surfaces such that a final mesh could not be generated. In this work, we present a new procedure to decompose the geometry in many-to-one sweepable volumes. The decomposition is based on a least-squares approximation of affine mappings defined between the loops of nodes that bound the sweep levels. In addition, we introduce the concept of computational domain, in which every sweep level is planar. We use this planar representation for two purposes. On the one hand, we use it to perform all the imprints between surfaces. Since the computational domain is planar, the robustness of the imprinting process is increased. On the other hand, the computational domain is also used to compute the projection onto source surfaces. Finally, the location of the inner nodes created during the decomposition process is computed by averaging the locations computed projecting from target and source surfaces.

Keywords: Finite element method, mesh generation, hexahedral mesh, multi-sweeping, computational domain.

1 Introduction

Extrusion geometries often appear in numerical simulation processes. These volumes are usually created using CAD packages that allow to extrude a surface along a sweep path. These one-to-one geometries are delimited by a *source surface*, a *target surface* and a series of *linking sides*, see Fig. 1.

---

* This work was partially sponsored by the Spanish *Ministerio de Ciencia e Innovación* under grants DPI2007-62395, BIA2007-66965 and CGL2008-06003-C03-02/CLI.

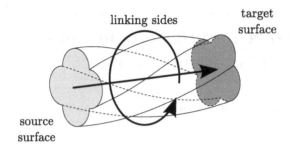

Fig. 1. Classification of surfaces defining a one-to-one sweep volume

In the last years, several methods have appeared to mesh *many-to-many* sweep volumes [1, 2, 3, 4]. That is, extrusion volumes that contain many source and target surfaces. These methods rely on the decomposition of the volume into sub-volumes that are meshable by one-to-one [5, 6, 7, 8, 9] or many-to-one [10] sweep techniques. Then, each sub-volume is meshed separately. In general, the decomposition of a many-to-many sweeping is performed by projecting target surfaces onto the corresponding source surfaces. Then, the projected target surfaces and source surfaces are imprinted in order to determine the decomposition of the volume. Finally, each sub-volume is meshed separately using a many-to-one sweep scheme. In fact, each sub-volume is further decomposed into barrels, and each barrel is meshed using a one-to-one sweep scheme.

However, the algorithm is highly dependent on the location of the inner nodes created during the decomposition process. The quality of the imprints is also affected by the location of inner nodes. Inaccurate locations may lead to erroneous imprints. Thus, a low quality mesh with inverted elements may be generated. Usually, this effect appears when there are non-planar sweep levels or highly curved surfaces in the geometry. In this work, we present a new algorithm to decompose a many-to-many geometry that overcomes these drawbacks.

The method is based on a least-squares approximation of an affine mapping defined between the loops of nodes that bound the sweep levels according to [8]. The decomposition is performed using a two-step procedure. First, the target surfaces are projected in the sweep direction to the source surfaces. In the second step, the nodes on the source surfaces are projected back to the target surfaces. Then, the final location of inner nodes is the weighted average of the position of nodes projecting them from target and from source surfaces. During the first step of the decomposition process it is necessary to compute the imprints between the source and target surfaces. The main contribution of this work is to define the computational domain of a loop of nodes as a planar representation of them. Since the computational domain is planar, the imprinting process becomes more robust. In addition, we also use the computational domain to project inner nodes onto source surfaces.

2   The Multi-sweeping Method

This section presents an outline of the proposed multi-sweeping method. The algorithm is performed in five steps:

(i) Surfaces classification.
(ii) Linking sides meshing.
(iii) Loop face projection and imprinting.
(iv) Loop edge meshing and volume decomposition into many-to-one sub-volumes.
(v) Meshing all many-to-one sub-volumes.

Although this work is focused on the decomposition steps (iii) and (iv), we also present an outline of the whole multi-sweeping method. The first step of the algorithm is the classification of the surfaces as source, target and linking sides. The classification is accomplished using the procedure presented in [11]. This procedure is performed as follows. First, it finds a non-submappable surface and classifies it as source surface. Then, the algorithm proceeds to classify the adjacent surfaces in an advancing front manner depending on the angle between the adjacent surfaces. When the surfaces of the geometry are classified, the linking sides are meshed using the submapping algorithm [12, 13]. Figure 2(a) shows a simple multi-sweep geometry with its linking sides meshed.

Once the linking sides are meshed the decomposition process begins. The standard decomposition process starts at the target surfaces and ends at the source surfaces. In contrast, the final meshing process starts at the source surfaces of the many-to-one sub-volumes and ends at the target surfaces.

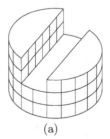

(a)   (b)

Fig. 2. (a) Mesh generated on the linking sides using the submapping method. (b) A simple multi-sweep geometry with a row of sweep nodes.

3 Basic Definitions

In this section we provide four basic definitions that will be used through this work. We briefly review the concepts of sweep node, loop geometry engine and control loop previously introduced in [1]. In addition, we introduce the new concept of computational domain for multi-sweeping.

3.1 Sweep Node

Sweep nodes are a computational structure that stores the vertical connectivity of the linking sides mesh. Each sweep node contains a pointer to the next sweep node in the sweep direction and another pointer to the previous sweep node in the counter-sweep direction (some of these pointers may be null). Figure 2(b) presents a multi-sweep geometry with a row of sweep nodes. Note that there are sweep nodes that do not point to a sweep node directly above or below.

3.2 Loop Geometry Engine

The loop geometry engine is a data structure that represents source, target and mid-level faces as loops of sweep nodes. In addition, a graph is used in order to define the topology of such loop faces. For instance, Fig. 3(a) shows a surface represented by: *i)* a loop face; *ii)* a loop wire that describes the boundary of the loop face; *iii)* an ordered list of four loop edges that defines the loop wire, and *iv)* four loop vertices that define the initial and final points of each loop edge. Figure 3(b) details this relationship using the topology graph. For instance, the loop wire uses loop edge 1 and, conversely, loop edge 1 is included in the loop wire.

Loop vertices are represented by a sweep node that provides information about its location. Loop edges are defined by an ordered list of sweep nodes.

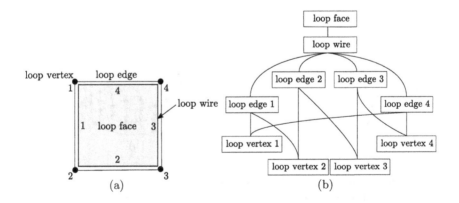

Fig. 3. (a) A loop face. (b) Associated topology graph.

The list of nodes provides a discretization of the loop edge. Loop wires are defined by a closed loop of loop edges. Loop faces are represented by a loop wire that defines the outer boundary and several loop wires that define inner boundaries.

The structure of the loop geometry engine is similar to a geometry engine except that not all the elements in the loop geometry have an underlying geometrical representation. This engine provides both geometrical and topological information about the loop geometry. In addition, this structure is responsible for creating loop geometry and maintaining the topology graph during the whole algorithm, see [1] for more details.

3.3 Control Loop

Control loops are the loops of sweep nodes that bound each of the sweep layers of the volume. While one-to-one sweep volumes have a single control loop for each layer, the number of control loops in multi-sweeping volumes may differ from layer to layer, see Fig. 4(a). The main objective of the control loops is to define the projection between two consecutive sweep layers. To this end, control loops are composed by a loop of sweep nodes that define the projection (black dots in Fig. 4(a)) and, for implementation purposes, a list of sweep nodes to project (white dots in Fig. 4(a)).

In a given level of the decomposition process we may have both source and target loop faces. In these cases, given the sets of target and source loop faces, $(\tau_i)_{i=1,...,n}$ and $(\sigma_i)_{i=1,...,m}$, respectively, the control loops are computed as:

(i) Compute provisional target control loops, L_t, as follows:
- The loops of nodes of control loops are defined as the Boolean union, $L_t = \bigcup_{i=1,...,n} \tau_i$, see [14].
- The nodes to project are those that belong to more than one loop face.

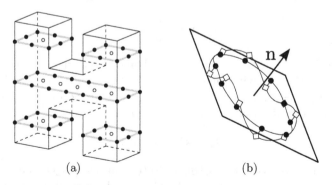

(a)   (b)

Fig. 4. (a) Five control loops in a multi-sweep geometry. (b) Physical sweep nodes (white squares) and computational sweep nodes (black circles).

(ii) Compute provisional source control loops, L_s, using the procedure detailed in step (i).

(iii) We collect the nodes of control loops in L_s (both the nodes that describe the loops and the nodes to project) that are not included in L_t. Then we insert these nodes into the corresponding control loops in L_t.

3.4   Computational Domain

It is important to point out that control loops in real geometries are usually non-planar. Therefore, the imprinting operations needed in the decomposition process may lead to inaccurate representations due to tolerance definitions. To overcome this drawback we introduce the concept of computational domain.

The *computational domain* of a given control loop is a planar representation of it. Figure 4(b) shows a curved control loop in the physical domain (white squares) and its representation in the computational space (black circles). The computational domain is used to project the sweep nodes through the volume and to perform the imprinting between the surfaces of the geometry. Since the control loops in the computational domain are planar, the proposed imprinting process is more robust, see Sect. 4. The main reason is that skew lines (neither parallel nor concurrent) do not exist in a bi-dimensional space. Therefore, computing the intersection of segments in the computational domain is less affected by tolerance errors than in the physical domain. In addition, the *winding number* algorithm [15] can be used to test if a node is inside a loop of nodes. This algorithm loses its accuracy and robustness for non-planar loops of nodes.

In reference [8] the *pseudo-area* vector, $\mathbf{a}$, of a loop of points $\{\mathbf{x}_i\}_{i=1,\ldots,n}$ is defined as

$$\mathbf{a} = \sum_{i=1}^{n} \mathbf{x}_i \times \mathbf{x}_{i+1},$$

where $\mathbf{x}_{n+1} = \mathbf{x}_1$. The *pseudo-normal* vector is defined as

$$\mathbf{n} = \frac{\mathbf{a}}{||\mathbf{a}||}. \tag{1}$$

The construction of the computational domain of a given control loop is performed in the following manner. First, we compute the pseudo-normal of the loop nodes that defines the control loop, $\mathbf{n}$. Second, we define the computational position of $\mathbf{x}_i$, for $i = 1, \ldots, n$ as

$$\overline{\mathbf{x}}_i = \mathbf{x}_i - \langle \mathbf{x}_i, \mathbf{n} \rangle \mathbf{n}, \tag{2}$$

where $\langle \cdot, \cdot \rangle$ denotes the dot product. Note that we project points $\mathbf{x}_i$, for $i = 1, \ldots, n$, on the plane defined by the pseudo-normal vector. From the

3D representation of the computational domain given by (2), it is straight-forward to compute a 2D representation, in (ξ, η) coordinates, of the computational domain.

Two remarks on the proposed method have to be made. First, each control loop has its own computational domain, even when control loops are located in the same sweep level. Second, in order to construct the computational domain, we need to compute the pseudo-normal vector. This vector can be defined if a control loop is set. Therefore, our method can be applied to all extrusion geometries such that all levels are bounded by control loops. Note that periodic surfaces such as cylinders or spheres that expand from inside to outside, or vice versa, do not meet this condition. Thus, the proposed method can not be applied to these cases because the control loop is not set. In these situations, we can always split the geometry in two sub-volumes such that the control loops are properly defined.

4   Loop Face Projection and Imprinting

4.1   Loop Face Projection

Starting at target surfaces, we compute the associated loop faces as detailed in Sect. 3.2. Then, we compute the corresponding control loops according to Sect. 3.3. When the control loops are constructed, the algorithm proceeds to build their associated computational domain, see Sect. 3.4.

To project the sweep nodes between layers we use the sweeping scheme presented in [7, 8]. It is important to point out that the projection of sweep nodes is performed both in the physical domain and in the 3D representation of the computational domain. The computational and physical locations are stored for each sweep node. The projection in the physical space is performed in order to capture the shape of the target and/or source surfaces. The projection in the computational space is performed in order to obtain accurate and robust imprints. When a sweep node is projected to the next level, the projected sweep node becomes the next sweep node of the original one. Conversely, the original sweep node is the previous sweep node of the projected one.

In each level we check if new target or source loop faces have to be added to the existing control loops. If this is the case we update the control loops according to Sect. 3.3. For each one of these control loops an imprinting process is performed, as detailed in Sect. 4.2.

Once all the source surfaces are reached the geometry is completely decomposed. However, the location of the inner sweep nodes created during the decomposition process has to be improved. To this end, the source surfaces are projected back to the target surfaces and the final location of inner sweep nodes is computed as a weighted average of the computed locations projecting from target and source surfaces, see Sect. 4.5.

Fig. 5. Representation of the imprinting process in the physical domain. (a) Two target loop faces (white) and a source loop face (grey). (b) Target loop edges intersection. (c) Projection of target sweep nodes on source faces. (d) Collapse of target sweep nodes and source sweep nodes.

4.2   Loop Face Imprinting Pre-process

Before starting the actual imprinting process, target and source loop faces of the same control loop have to be pre-processed. It is important to point out that each one of the following steps is performed in the 3D representation of the computational domain. That is, using $\bar{\mathbf{x}}$ coordinates. Figure 5 shows a representation of the imprinting pre-process in the physical domain. Figure 5(a) shows two target loop faces (white) and one source loop face (grey) ready to be pre-processed. First, the loop edges of target loop faces are intersected with each other in order to obtain a conformal model (Fig. 5(b)). Second, the nodes that define the target loop faces are projected to the corresponding source loop faces (Fig. 5(c)). Section 4.3 further details the procedure to project those nodes. Finally, we search a sweep node on a target loop face and a sweep node on a source loop face that are closer than a given tolerance. Then the target node is collapsed with the source node (Fig. 5(d)). This tolerance is defined as $h/5$, where h is the prescribed element size. On the one hand, if the tolerance parameter is too big, we may collapse nodes that are too far and the imprinting process may fail. On the other hand, if the tolerance is too small, we may miss some collapsing nodes.

4.3   Mapping of Sweep Nodes from the Computational Domain to a Source Surface

This section is focused on a new procedure to project sweep nodes of target loop faces to source loop faces of the same control loop. For a given source loop face, σ, we need to detect which nodes of target loop faces lie inside σ. Since loop faces are planar in the computational domain we use the winding number algorithm, see Fig. 6 for a graphical representation of this algorithm. Given a test point, p, the algorithm adds the angles between p and two consecutive points in the loop. If the result equals 0, the point is outside. Else, if the result equals 2π, the point is inside. That is, the algorithm counts the number of turns around the test point. This algorithm is robust when dealing with planar loops of nodes. Note that in 3D, the number of turns is not well defined and, for this reason, the winding number algorithm may fail.

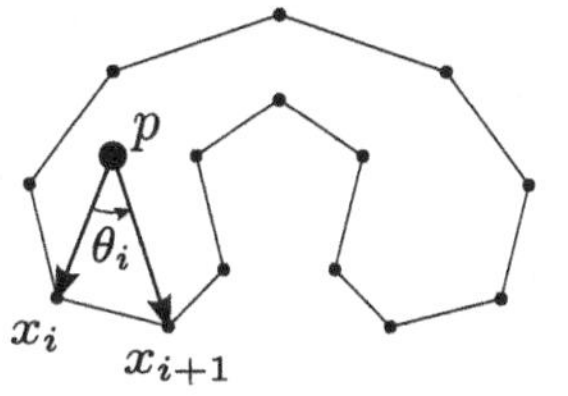

$$\sum_{i=1}^{N} \theta_i = \begin{cases} 0 & \text{point is outside} \\ \\ 2\pi & \text{point is inside} \end{cases}$$

Fig. 6. Graphical representation of the winding number algorithm

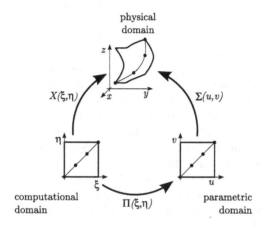

Fig. 7. Mapping of sweep nodes from the computational domain to a source surface

Given a list of sweep nodes inside a source loop face, we have to project them on the geometrical surface. Let (ξ, η) and (x, y, z) be two coordinate systems of the 2D representation of the computational domain and the physical domain, respectively (see Fig. 7). Therefore, we need to compute a mapping $X(\xi, \eta)$ to project the sweep nodes of the target loop faces onto the physical domain. Note that all the source loop faces represent a geometric surface. Hence, instead of projecting the sweep nodes directly from the computational space, (ξ, η), to the physical space, (x, y, z), we propose to compute a mapping $\Pi(\xi, \eta)$ to project the nodes to the parametric space of the surface. Then, using the parameterization of the surface, $\Sigma(u, v)$, we finally project the nodes to the physical domain. That is

$$X = \Sigma \circ \Pi.$$

Note that the mapping $\Pi(\xi, \eta)$ is unknown. In order to compute the mapping $\Pi(\xi, \eta)$, we approximate it by an affine mapping $\widetilde{\Pi}(\xi, \eta)$. This affine mapping is determined by means of a least-squares approximation of a linear mapping between the 2D computational domain representation of the source loop face and its representation in the parametric space of the source surface, see [6] for details. That is we approximate $X(\xi, \eta)$ as

$$X \approx \Sigma \circ \tilde{\tilde{\Pi}}.$$

4.4   Loop Face Imprinting

The result of the imprinting process between target loop faces, $(\tau_i)_{i=1...n}$, and source loop faces, $(\sigma_i)_{i=1...m}$, is stored in three lists of loop faces: Θ_o, Θ_t and Θ_s. List Θ_o contains the loop faces that come from intersections of a target and a source loop face. The loop faces included in Θ_o are called *overlap* loop faces. List Θ_t contains the sections of target loop faces that do not intersect a source loop face. These faces are the new target loop faces that replace the old target loop faces, $(\tau_i)_{i=1...n}$. Finally, list Θ_s contains the section of source loop faces that do not intersect a target loop face. This list contains the new source loop faces that replace the old source loop faces, $(\sigma_i)_{i=1...m}$. The loop face imprinting operation is based on segment intersections, see [1, 14] for more details. The main difference of the presented method is that all of the calculations to obtain the imprints are performed in the 3D representation of the computational domain. Hence, the imprinting process becomes more accurate and robust as stated in Sect. 4.3.

In addition, we create a new graph that relates the new loop faces created in the imprinting process and the original loop faces from which they come from. We call this graph the *loop face partitioning graph*. This graph allows us to recover the set of loop faces in which a loop face is decomposed. Moreover, it also permits us to recover the loop faces in which a given loop face is included. Figure 8(a) shows the imprinting process of a target loop face, T, and a source loop face, S. In this case, $\Theta_o = \{O\} = \{T \cap S\}$, $\Theta_t = \{T'\} = \{T - S\}$ and $\Theta_s = \{S'\} = \{S - T\}$. Figure 8(b) shows the corresponding loop face partitioning graph. Note that loop face O is a section of S and T. Conversely, T is partitioned in loop faces T' and O.

In order to illustrate the differences between the imprinting process in the physical and computational domain, Fig. 10(a) shows a curved loop face, T, and a planar loop face, S in the physical domain. The intersection between

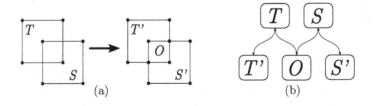

(a)   (b)

Fig. 8. (a) Imprinting process of a target loop face, T, and a source loop face, S. (b) Associated loop face partitioning graph.

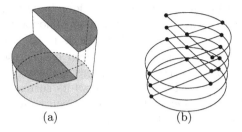

Fig. 9. A multi-sweep geometry and its correspondent loop faces. (a) Wire frame model with two target faces (light grey) and two source loop faces (dark grey). (b) Loop faces.

the segments that define these loop faces in the physical domain are marked using a white circle. Figure 10(b) shows the Boolean difference between S and T computed in the physical domain. Note that we do not obtain the desired result. Figure 10(c) shows the representation of the previous loop faces in the computational domain. When the intersections are calculated in the computational domain, two additional nodes are obtained. Therefore, we obtain the correct representation of $S - T$, see Fig. 10(d).

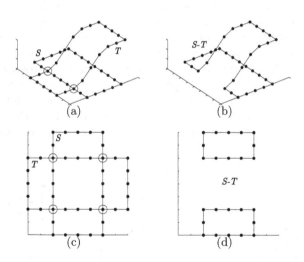

Fig. 10. Imprinting process in the physical and computational domains. (a) Intersection vertices of loop faces S and T performed in the physical domain. (b) Boolean difference between S and T performed in the physical domain. (c) Intersection vertices of loop faces S and T performed in the computational domain. (d) Boolean difference between S and T performed in the computational domain.

When the imprinting ends, the new target loop faces, Θ_t, are collected in order to create the new control loops. These control loops are used to project the new target loop faces to the next level, see Sect. 4.1. The projection and imprinting process is iterated until the last sweep level is reached. Figure 9(a) presents a multi-sweep geometry with two target surfaces (light grey) and two source surfaces (dark grey). Figure 9(b) shows its corresponding loop faces when the imprinting process ends. Note that the source surfaces are split due to the diameter that divide the lower surface.

4.5 Final Location of Inner Sweep Nodes

Once all the source surfaces are reached and the geometry is decomposed, we have to improve the final location of inner sweep nodes. To this end, we project back inner sweep nodes from source surfaces to target surfaces. The final location of inner sweep nodes is computed as a weighted average of the computed locations projecting from target and source surfaces.

It is important to point out that once all the source surfaces are reached, they are always bounded by overlap loop faces. Therefore, given an overlap loop face, ω, obtained in the imprinting process, the sweep nodes that have to be mapped to the previous level in the physical domain are the ones in ω such that their previous sweep node lies inside the volume (not on the boundary).

For instance, Fig. 11 presents a cylindrical geometry with a planar bottom surface split in two parts, and a non-planar top surface. Sweep nodes on the source surface that have to be projected to the previous level are marked with white squares. Figure 11(a) shows the computed location of inner sweep nodes projecting from target surface. These nodes do not reproduce the shape of the top surface. Figure 11(b) presents the computed location of inner sweep nodes projecting from source surface. These nodes do not reproduce the shape of the planar bottom surface. However, Fig. 11(c) shows the final location of inner sweep nodes computed as a weighted average of the location

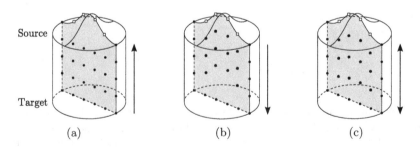

Fig. 11. Projection of inner sweep nodes. (a) Projecting from the target surface. (b) Projecting from the source surface. (c) Averaging the location computed projecting from target and source surfaces.

Algorithm 1. Inner nodes creation

1: **function** CreateInnerNodes(LoopFace ω, ListOfSweepNodes *nodes*)
2:     **while** There are nodes to project **do**
3:         Int *projectionLevel* $\leftarrow 0$
4:         ControlLoop $CL_t, CL_s \leftarrow$ ObtainTargetAndSourceControlLoops(ω)
5:         Projector *projector* $\leftarrow$ createProjector(CL_s, CL_t)
6:         **for all** Node *node* $\in$ *nodes* **do**
7:             Point *qSource* $\leftarrow$ projectNode(*projector, node*)
8:             SweepNode *previousNode* $\leftarrow$ getPreviousNode(*node*)
9:             Point *qTarget* $\leftarrow$ getPositionFromTarget(*previousNode*)
10:           Int *depthLevel* $\leftarrow$ getDepth(*previousNode*)
11:           Int $N \leftarrow$ *depthLevel* + *projectionLevel*
12:           Point $q \leftarrow$ (*projectionLevel*/N) *qTarget* + (*depthLevel*/N) *qSource*
13:           setPosition(*previousNode, q*)
14:           **if** hasToBeProjected(*nextNode*) **then**
15:             *node* $\leftarrow$ *previousNode*
16:           **else**
17:             remove(*node, nodes*)
18:           **end if**
19:         **end for**
20:         Int *projectionLevel* $\leftarrow$ *projectionLevel* + 1
21:     **end while**
22: **end function**

Algorithm 2. Selection of target and source control loop

Return : ControlLoop CL_T, CL_S
1: **function** ObtainTargetAndSourceControlLoops(LoopFace ω)
2:     LoopFace $\sigma \leftarrow$ obtainContainingLoopFace(ω)
3:     **if** isNull(σ) **then**
4:         LoopFace $\tau \leftarrow$ previousLoopFace(ω)
5:     **else**
6:         LoopFace $\tau \leftarrow$ previousLoopFace(S)
7:     **end if**
8:     $CL_t \leftarrow$ getControlLoop(τ)
9:     $CL_s \leftarrow$ nextControlLoop(CL_t)
10:     $\omega \leftarrow \tau$
11: **end function**

---

computed projecting from source and target surfaces. Note that the final location reproduces the shape of both cap surfaces.

For each overlap loop face, ω, and a list of sweep nodes in ω that have to be projected, *nodes*, Algorithm 1 presents a procedure that computes their final location. First, at Line 4, the algorithm finds a source control loop, CL_s, in the current sweep level and a target control loop, CL_t, in the previous sweep level that define the projection between the two levels (see Algorithm 2).

Note that Algorithm 2 updates the reference overlap loop face, ω. Then, at Line 5 of Alg. 1, the node projector is constructed as detailed in [8]. Next, for each node in *nodes*, the algorithm computes the projection from the source surface, Line 7. Then, at Lines 8 and 9, the previous node and its position computed projecting from target faces is recovered. Next, at Line 12 the final position of the previous node is obtained by interpolating the positions obtained from the target and source projections. Note that the *depth level* of a sweep node is defined as the number of times this node has been projected from a target loop face. Finally, the algorithm checks if the previous node has to be projected, Line 14. If so, the node is updated. Else, the node is removed from the list of nodes to project. The procedure is iterated until all nodes are re-located in the physical domain.

5   Loop Edge Meshing and Volume Decomposition

When the imprinting process ends, it is necessary to re-mesh the loop edges in order to ensure that each loop face contains an even number of intervals. In this work, we solve an integer linear problem to assign an even number of intervals using the lp_solve library [16]. However, applying this strategy to all loop edges leads to a large integer linear problem, especially for small element sizes in the sweep direction. In order to reduce the computational cost of the integer linear problem, we propose to impose an even number of intervals to the source loop faces obtained during the imprinting process. Then, the information is propagated in the sweep direction. Hence, the new integer linear problem is

$$
\begin{aligned}
&\min \sum_{e \in \mathcal{E}} n_e \\
&\text{subject to:} \\
&\sum_{e \in \sigma} n_e = 2n_\sigma \text{ for all source loop face } \sigma, \\
&n_e \geq N_e \qquad \text{for all source loop edges } e \in \mathcal{E},
\end{aligned}
\tag{3}
$$

where n_e is the number of intervals of loop edge e, $\mathcal{E}$ is the set of loop edges contained in the source loop faces, N_e is a lower bound for n_e and $2n_\sigma$ is the number of intervals of loop face σ.

Once all loop edges are re-meshed, we decompose the geometry into sub-volumes that can be meshed using a many-to-one sweep scheme. The main idea consists on collecting every loop face *stacked* on a target face. Figure 12 presents a multi-sweep geometry decomposed into two sub-volumes that can be meshed using a many-to-one sweep method. In this work, all the resulting sub-volumes are meshed by the same least-squares projection procedure that we have already used to project inner nodes during the decomposition process, see [7, 8].

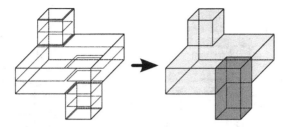

Fig. 12. Multi-sweep volume decomposed in many-to-one sub-volume

6 Examples

This section presents three examples of meshes that have been generated using the presented algorithm. The user assigns an element size and the algorithm automatically decomposes the geometry. Then, each sub-volume is meshed using a many-to-one sweep scheme. In the figures that illustrate these examples we mark the sweep direction with an arrow.

The first example illustrates the advantages of the proposed method to compute the location on inner sweep nodes created during the decomposition process. It presents a twisted and curved cylinder in which the cap faces are not planar, see Fig. 13. Note that the bottom surface is divided in two parts. The lower surfaces are classified as target and the upper surface is classified as source. Figure 13(a) presents the nodes created during the decomposition process using only the information of target surfaces. Note that there are inner nodes near the source surface that are located outside the geometry. This will lead to inverted elements during the meshing process. Figure 13(b)

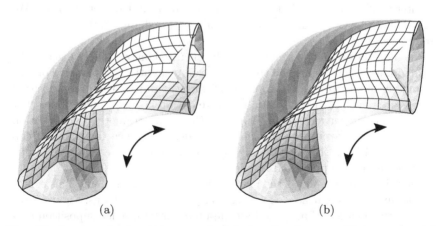

(a)   (b)

Fig. 13. Inner nodes location computed during the decomposition process. (a) Projecting nodes from target surfaces. (b) Interpolating nodes projected from source and target surfaces.

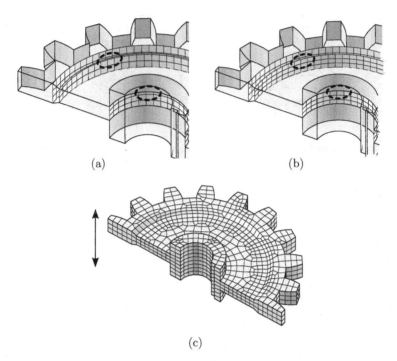

(a)   (b)

(c)

Fig. 14. Inner nodes location computed during the decomposition process (a) Projecting nodes from target surfaces. (b) Interpolating nodes projected from source and target surfaces. (c) Mesh generated using the multi-sweeping method on the half of a gear.

presents the inner nodes location computed using the proposed procedure by interpolating the nodes projected from source and target surfaces. We obtain inner layers of nodes that reproduce the shape of the source and target surfaces.

The second example presents the mesh generated on one half of a gear. The loops of nodes that define the control loops are slightly non-planar. For this reason, the nodes projected from target surfaces during the decomposition process are deviated (Fig. 14(a)). Thus, we obtain layers of inner nodes that almost intersect between them. Applying the proposed method the location of inner nodes is accurate and the inner nodes reproduce the shape of cap surfaces, see Fig. 14(b). The final mesh generated using the proposed multi-sweeping method is presented in Fig. 14(c).

The third example shows the mesh generated on a crankshaft. This example contains curved surfaces as well as non-planar control loops. Figure 15(a) presents the location of inner nodes computed during the decomposition process. Although the control loops are highly non-planar, the position of inner nodes is not deviated. Figure 15(b) presents the final mesh generated using the multi-sweeping algorithm.

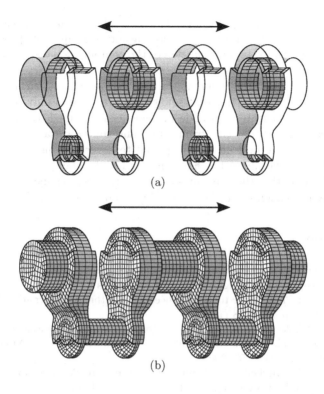

Fig. 15. (a) Inner nodes location computed during the decomposition process using the presented method. (b) Final mesh generated using the multi-sweeping method on a crankshaft.

7 Conclusions

It is well known that the element quality of the meshes generated using the multi-sweeping method is heavily affected by the position of the inner nodes created during the decomposition process. Inaccurate location of inner nodes can produce low quality elements or even inverted elements during the meshing process. In practice, this effect is caused by non-planar sweep levels or curved surfaces in the geometry. To overcome this drawback, we have presented a new algorithm that automatically decomposes the geometry. The geometry decomposition is achieved by computing a least-squares approximation of an affine mapping between the control loops of consecutive layers.

First, we compute the geometry decomposition advancing from target to source surfaces. This decomposition is performed computing a least-squares approximation in the physical space and the 3D representation of the computational space. The projection in the physical space is used to compute a first approximation to the location of inner nodes. The projections in the 3D representation of the computational space are used to carry out the imprinting procedure. Since the computational domain is planar, the robustness of the imprinting process is increased and more accurate results are produced. In addition, we use the 2D representation of the computational space to project the inner nodes onto the source surfaces.

Second, we project the inner nodes mapped on sources surfaces back to the target surfaces using the least-squares approximation only in the physical domain. An accurate final node placement is obtained by averaging the locations computed projecting from target and source surfaces.

The proposed method has been applied to several industrial geometries and in all cases the algorithm has generated high quality meshes. Finally, it is worth to notice that the presented multi-sweeping method has been successfully implemented in the ez4u meshing environment [17].

References

1. White, D.R., Saigal, S., Owen, S.J.: CCSweep: automatic decomposition of multi-sweep volumes. Engineering with computers (20), 222–236 (2004)
2. Blacker, T.: The Cooper tool. In: Proceedings of the 5th International meshing roundtable, pp. 217–228 (1996)
3. Lai, M., Benzley, S.E., White, D.R.: Automated hexahedral mesh generation by generalized multiple source to multiple target sweeping. Int. J. Numer. Methods. Eng. (49), 261–275 (2000)
4. Shepherd, J., Mitchell, S.A., Knupp, P., White, D.R.: Methods for Multisweep Automation. In: Proceedings of 9th International Meshing Roundtable, pp. 77–87 (2000)
5. Knupp, P.: Next-generation sweep tool: a method for generating all-hex meshes on two-and-one-half dimensional geometries. In: Proceedings of the 7th International meshing roundtable, pp. 505–513 (1998)
6. Roca, X., Sarrate, J., Huerta, A.: Mesh projection between parametric surfaces. Communications in Numerical Methods in Engineering (22), 591–603 (2006)
7. Roca, X., Sarrate, J.: A new least-squares approximation of affine mappings for sweep algorithms. Engineering with Computers (to appear)
8. Roca, X., Sarrate, J.: An Automatic and General Least-Squares Projection Procedure for Sweep Meshing. In: Proceedings of the 15th International meshing roundtable, pp. 487–506 (2006)
9. Staten, M.L., Canann, S.A., Owen, S.J.: BMSweep: Locating Interior Nodes During Sweeping. Engineering with Computers (15), 212–218 (1999)
10. Scott, M.A., Benzley, S.E., Owen, S.J.: Improved many-to-one sweeping. Int. J. Numer. Methods. Eng. (63), 332–348 (2006)
11. White, D.R., Tautges, T.: Automatic scheme selection for toolkit hex meshing. Int. J. Numer. Methods. Eng. (49), 127–144 (2000)
12. White, D.R.: Automatic Quadrilateral and hexahedral meshing of pseudo-cartesian geometries using virtual subdivision, Master thesis, Brigham Young University (1996)
13. Ruiz-Gironés, E., Sarrate, J.: Generation of structured meshes in multiply connected surfaces using submapping. In: Advances in engineering software (to appear)

14. White, D.R., Saigal, S.: Improved imprint and merge for conformal meshing. In: Proceedings of the 11th International meshing roundtable, pp. 285–296 (2002)
15. O'Rourke, J.: Computational geometry in C. Cambridge University Press, Cambridge (2000)
16. Lpsolve (2008), `http://sourceforge.net/projects/lpsolve`
17. Roca, X., Sarrate, J., Ruiz-Gironés, E.: A graphical modeling and mesh generation environment for simulations based on boundary representation data, Congresso de Métodos Numéricos em Engenharia (2007)

Defeaturing CAD Models Using a Geometry-Based Size Field and Facet-Based Reduction Operators

William Roshan Quadros and Steven J. Owen

*Sandia National Laboratories, Albuquerque, New Mexico, U.S.A
{wrquadr,sjowen}@sandia.gov

Abstract. We propose a method to automatically defeature a CAD model by detecting irrelevant features using a geometry-based size field and a method to remove the irrelevant features via facet-based operations on a discrete representation. A discrete B-Rep model is first created by obtaining a faceted representation of the CAD entities. The candidate facet entities are then marked for reduction by using a geometry-based size field. This is accomplished by estimating local mesh sizes based on geometric criteria. If the field value at a facet entity goes below a user specified threshold value then it is identified as an irrelevant feature and is marked for reduction. The reduction of marked facet entities is primarily performed using an edge collapse operator. Care is taken to retain a valid geometry and topology of the discrete model throughout the procedure. The original model is not altered as the defeaturing is performed on a separate discrete model. Associativity between the entities of the discrete model and that of original CAD model is maintained in order to decode the attributes and boundary conditions applied on the original CAD entities onto the mesh via the entities of the discrete model. Example models are presented to illustrate the effectiveness of the proposed approach.

Keywords: defeaturing, feature suppression, CAD simplification, facet reduction.

1 Introduction

CAD technology has evolved significantly in recent decades, facilitating complex and detailed modeling in the early design stages of computational simulation. This has brought new challenges in the pre analysis stages including the defeaturing of irrelevant or unwanted features prior to mesh generation. This process frequently requires extensive user interaction to eliminate unwanted features and misalignments between parts. Therefore, automatic defeaturing procedures that reduce user time and increase the success ratio of mesh generation are needed. Defeaturing also reduces the degrees of freedom, thereby decreasing the analysis time and memory usage.

---

* Sandia is a multiprogram laboratory operated by Sandia Corporation, a Lockheed Martin Company for the United States Department of Energy's National Nuclear Security Administration under contract DE-AC04-94AL85000.

Defeaturing, sometimes known as feature suppression or simplification, is intended to address a wide range of characteristics, typically included for design purposes, that may not be relevant or desired for finite element (FE) simulation. Some of these geometric features often present in single body solid models may include small features such as holes, pegs, slots, fillets, chamfers, slender/sliver surfaces and thin-wall regions. Retaining the features described here may result in smaller mesh sizes at these localized regions which can lead to mesh size transition and mesh quality issues. This might lead to ill-conditioning of the FE model and using excessive computing power may not help. In addition, many practical hex meshing algorithms such as sweeping and mapping, which are sensitive to topological features can often fail while accounting for these anomalies.

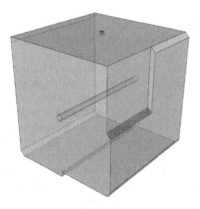

Fig. 1a. CAD model with chamfer, hole, steps, pegs, fillet, and imprint

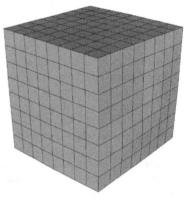

Fig. 1b. Hex mesh of the defeatured model

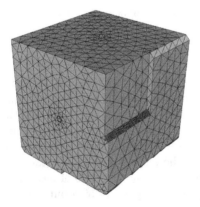

Fig. 1c. 27,181 tets with average element quality 0.8411 on original model

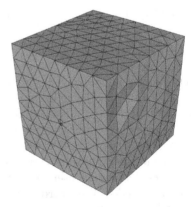

Fig. 1d. 3,879 tets with average element quality 0.8615 on defeatured model

Figure 1a shows a simple model with small features including multiple steps, chamfers, imprinted curves, pegs, and holes, which would make the hex meshing of this model difficult. Figure 1b shows a defeatured representation that has been reduced to a simple cube. In this case a hexahedral mesh can be generated on the model. Figure 1c shows the unreduced model with a tet mesh applied and the reduced model in Figure 1d also with a tet mesh. The number of tet elements without and with defeaturing are 27,181 and 3,879 respectively, significantly reducing the degrees of freedom in the mesh. The average element quality is also higher in the defeatured model as the small features, that would have required smaller element sizes, are no longer present - consequently avoiding large mesh size transitions.

It should be noted that the proposed approach does not directly address the CAD clean-up and repair issues. Instead we assume that the input CAD data is geometrically and topologically valid. While the focus of this work is on the defeaturing of correctly resolved solid models, the authors believe that the proposed framework can be extended to also address common CAD repair issues.

2 Background

Third party geometry kernels in CUBIT [4] such as ACIS, Granite, or Catia provide a rich set of geometry query and modification capabilities; however, they typically do not provide tools to automatically control defeaturing of CAD models for mesh generation. As a consequence, this work is intended to address the defeaturing of CAD models, particularly those that are generated in commercial 3D solid modelers (such as SolidWorks, Pro/Engineer, or UniGraphics) and are then imported into an independent mesh generation tool such as CUBIT. Built into many 3D solid modelers is a convenient feature or parameter-based representation of the model that provides capability to simply turn off unwanted features or modify parameters to simplify the model. It is clear that analysts should take advantage of parameter-based defeaturing whenever available prior to exporting to a third party mesh generation tool. In practice, however, the full design model containing features irrelevant to the simulation, is used as the basis for mesh generation. For the cases in which this work addresses, the model is provided without feature information and with only the boundary representation (B-Rep) topology and NURBS-based geometry definition. As a consequence, the users usually spend significant amount of time in manually identifying unwanted features and applying appropriate modification operations to simplify the model prior to mesh generation.

Techniques for feature suppression and defeaturing have been widely studied with many approaches presented in the literature. A recent study by Thakur et. al [14] surveys CAD model simplification techniques and classifies existing methods into four categories: techniques based on surface entity based operators, volume entity based operators, explicit feature based operators, and dimension reduction operators. Based on Thakur's classification, surface-based operations are most applicable to use within the context of the required application. Techniques for surface-based defeaturing operations can be categorized based upon whether operations are applied directly to the continuous NURBS-based geometry description or whether they utilize discrete representation of the domain.

NURBS-based defeaturing, such as those described by Clark [3] have the advantage of accurately redefining the topology and geometry. While geometry kernels can assist by providing the necessary geometric operations, identifying the numerous topologic configurations where defeaturing should occur in a consistent and robust manner is an open-ended problem. Inoue et al. [10] reported a face clustering technique for FE mesh generation similar to virtual topology [1]. The approach iteratively merged the model faces to obtain face cluster regions having sufficiently large area compared to the mesh element size. Authors defined metrics for mesh area, boundary smoothness, and surface flatness and used them for ranking the faces for merging. This approach is efficient for 2-manifold surface models and does not defeature topological features like holes. Focault et al. [8] proposed a hypergraph-based Mesh Constraint Topology (MCT) for simplification. Composite topological entities of MCT are defined as the union of Riemannian surfaces/curves constituting the reference model. Graph-based operators then perform delete, insert, collapse, and merge of MCT entities. Indeed, Thakur et al. [14] provides a comprehensive list of literature where authors have attempted to enumerate various cases of defeaturing directly on NURBS-based solid models.

Alternatively, several methods have been proposed that perform defeaturing in a discrete domain. For example, Gao et al. [9] use a feature recognition strategy to identify features for suppression in the continuous model. Following removal of the features a Delaunay triangulation approach is used to fill gaps or holes that may be left. Unfortunately, feature-based methods can often lack completeness as defeaturing only a subset of features, such as holes or fillets may miss other regions that are not identified as features such as narrow gaps or misalignments in an assembly. Fine et al. [6] used operators based on vertex removal and spherical error zone concept to transform the input polyhedral geometry while preserving it within an envelope. This envelope is obtained from a mechanical criterion which can be based either on an a posteriori error estimator or on a priori estimation.

Other discrete methods perform defeaturing as part of the meshing procedure or as a post-process to meshing. Mobley et al. [11] propose a method that does not directly modify the geometry, but rather defines operations within the surface meshing algorithm that ignores or combines features. Borouchaki and Laug [2] perform operations on the triangles of the mesh following the initial meshing operation of individual surfaces. Simplification is performed by identifying surfaces to be combined and then optimizing local mesh quality while ignoring the boundary between identified regions. Foucault et al. [7] extended the advancing front method to composite geometry. Dey et al.[5]

defined a priori error metrics based on element quality, and edge-collapse operations were iteratively performed on poor quality elements as a post-meshing operation.

Performing the reduction operations on the mesh itself can however adversely influence the mesh quality. If, for example sliver features are present in the model, mesh sizes in the initial mesh will be significantly finer than is needed in the final mesh. Aggressive collapse operations are then needed in order to achieve the desired size and quality once the slivers have been identified. In contrast, the proposed method preforms reduction operations on a facet-based discrete model that represents geometry. This defeatured discrete model will later be used as the underlying geometry for a meshing algorithm. The quality of the triangles in the facet representation is not critical provided the underlying geometry is reasonably represented. As a result, the reduction operations do not have to concentrate on maintaining a quality triangulation, as would be required with other methods that perform defeaturing during or after mesh generation.

Performing defeaturing on the mesh is however attractive because it leaves the original model untouched for subsequent meshing operations for alternate element resolutions or for application of boundary conditions. However, one of the drawbacks is that the identification of features to be suppressed as well as specific defeaturing operations must be integrated directly within the meshing algorithm. For CAD-based tools such as CUBIT, which includes numerous options for surface meshing, it would be advantageous to perform defeaturing operations independent of any specific meshing algorithm, allowing for any surface meshing procedure to utilize defeaturing without special modifications. Thus, our method provides the advantages of mesh-based defeaturing without the associated drawbacks.

Following a review of the various approaches for defeaturing, the following list summarizes the criteria for which our automatic defeaturing framework has been developed:

1. A strategy for identifying features for suppression should utilize only the topology and geometry of the B-Rep NURBS model and should not consider any additional data such as feature data.
2. The procedure should be applicable for any surface mesh generation algorithm, including paving, triangle advancing front and Delaunay methods. Specialized changes to individual meshing algorithms to accommodate defeaturing should be avoided.
3. The final mesh should maintain associativity with the original model. The attributes and boundary conditions applied on the original B-Rep model should be mapped to the final mesh.
4. The algorithm should not be based on any specific proprietary geometry kernel as multiple geometry kernels exist within the CUBIT environment.
5. Finally, the extent of automatic feature suppression must be user controllable. Individually identifying features for preservation or for suppression should also be provided.

The authors assert that the proposed defeaturing algorithm meets the criteria described above.

3   Defeaturing Algorithm

Here we briefly discuss the steps involved in defeaturing irrelevant features of a NURBS based B-Rep CAD model via a discrete model. A robust method for automatically detecting the irrelevant features is very critical to the success of the proposed approach. A geometry-based mesh sizing function is used as the field to automatically identify the irrelevant features that need to be defeatured. With the proposed method, the original CAD model is not altered as the actual defeaturing is performed on a discrete model of the original CAD model. The following illustrates the basic steps used in the defeaturing algorithm.

Figure 2a shows a typical industrial CAD model containing common features such as holes, fillets, blends, and steps. Note that the zoom view shows a small step which is an unintended feature. Such a small step would result in poor element quality if retained. This model also contains multiple tiny holes. Assuming these features are not significant to the simulation, they can significantly increase the element and node count in the final mesh, resulting in increased analysis time and memory usage.

Figure 2b shows a discrete representation of the input CAD model. First, the facets of each surface are obtained and stitched to form a water tight model. The B-Rep topology is then embedded in the facet-based discrete model, establishing the associativity between facet entities and the original B-Rep topological entities.

Figures 2c and 2d show the defeatured model represented via the facet-based B-Rep model. Note that the holes and steps are no longer seen in the defeatured model. To accomplish defeaturing, two main steps are used: (1) Identification of irrelevant features on the discrete model; and (2) Performing facet-based reduction operations to remove irrelevant features.

Reduction operations are first performed on lower order entities followed by higher order. For example, the facet entities associated with curves are first collapsed before collapsing the facet entities associated with surfaces. Following each reduction operation, care is taken to ensure a valid B-Rep is maintained and that the mapping between the topological entities of the original CAD model and that of the reduced discrete model is updated. This is essential for decoding meshing attributes such as meshing schemes, size specifications and boundary conditions on the discrete model. Thus, the defeatured model can be updated and meshed using the attributes defined on the original solid model.

Figures 2e and 2f show the mesh on the defeatured and the original models, respectively. Mesh generators in CUBIT are capable of using the discrete B-Rep model as the basis for their geometry. As a result it is not currently necessary to convert the defeatured model back into a NURBS B-Rep

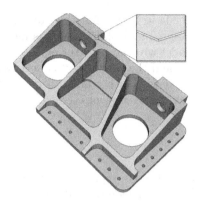

Fig. 2a. Original B-Rep CAD Model

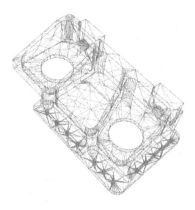

Fig. 2b. Discrete B-Rep Model

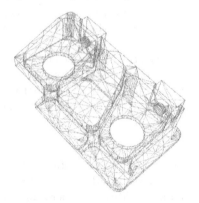

Fig. 2c. Defeatured Discrete Model

Fig. 2d. Defeatured Discrete Model

Fig. 2e. Tet mesh on Defeatured Model

Fig. 2f. Tet mesh on Original Model

format if only the mesh is desired. Note that in Figure 2e that the fine mesh around the holes is no longer seen in the defeatured model. Also as the steps are defeatured the element quality improves significantly. The minimum tet quality in Figure 2e is 0.038 compared to 0.004 in Figure 2f. Also note that the number of tet elements in the defeatured model is 4,975 which is much less than the 11,328 tet elements in the model with no defeaturing. As the mapping between the entities of the original and the defeatured models has been maintained, the mesh can be associated with either or both discrete and continuous models. We now discuss each of the steps of the defeaturing procedure in more detail.

3.1 Obtaining a Discrete Model

The first step of the procedure is to obtain a discrete representation of the CAD model. Requirements of the discrete model include the following:

1. The deviation of the facets from the original CAD model should be proportional to the user defined mesh size. For example, a small mesh size would require a proportionately small facet size so that curvature can adequately be captured in the final mesh.
2. The discrete representation should be watertight. This implies that each adjacent surface shares common facet edges with its neighbors.
3. Relationships should be maintained between the original B-rep entities and the discrete entities. This is to ensure that the final mesh will be correctly associated with the original CAD model once the procedure is complete.

In most cases it is convenient to obtain the discrete representation from a third party CAD kernel such as ACIS or Granite. Building a watertight faceted representation of the model may require additional operations to stitch adjacent surfaces to ensure surfaces share common edges. Additionally, to ensure correct associativity between the original B-Rep entities and the mesh entities that will be generated, the data to link back to the original B-Rep must be generated at this point and maintained throughout the procedure.

For the current work, the mesh-based geometry (MBG) [12] definition as proposed by Owen et al. [12] is used to build and represent the discrete representation once it is extracted from the CAD model. The MBG definition can also represent non-CAD-based models; for example models initially defined by triangle facets, such as STL formats, or by extracting geometry from a legacy FE mesh.

3.2 Detection of Features for Suppression

The identification of features for suppression, is a key component of the proposed method and is critical to the success of defeaturing. Feature identification is achieved using mesh sizing function work of the author [13]. This work describes a systematic approach to reveals the geometric factors that completely capture the geometric complexity for controlling element sizes. Irrelevant features are defined as those entities where the mesh size determined via geometric factors falls below the user specified threshold value ε_*.

Below are the three main steps involved in detecting irrelevant features and are subsequently discussed:

1. Identify geometric factors of input solid model.
2. Measure geometric factors using a set of tools.
3. Mark irrelevant regions on the discrete model using the tools.

Identification of Geometric Factors

To begin the procedure, the input CAD model is first decomposed into disjoint subsets, defined as dimensionally-based groupings of geometric entities. Let an input CAD solid model S contain N surfaces, F_i, where $i = 1, 2, ...N$, which are curvature-continuous and do not intersect at the interior. We can then define S in terms of the sum of its disjoint subsets as:

$$S = in(S) + \sum_{n=1}^{N} in(F_n) + \sum_{m=1}^{M} in(C_m) + \sum_{l=1}^{L} V_l \tag{1}$$

where the disjoint subsets of solid S are, the interior of the solid, $in(S)$, the interior of each surface, $in(F_n)$, the interior of each curve, $in(C_m)$, and vertices, V_l. As the subsets are disjoint, the geometric complexity of each subset is independent of the others.

Table 1 tabulates the full list of geometric factors that are used to measure the complexity of each disjoint subset. Geometric factors identified on the disjoint subsets include 3D proximity, 2D proximity, surface curvature, curve curvature, 1D proximity, and curve twist. The user can optionally provide threshold values for the size field computed using these geometric factors. ε_{3D}, ε_{2D}, ε_{sc}, ε_{1D}, ε_{cc} and ε_t represent user defined threshold values for 3D proximity, 2D proximity, surface curvature, 1D proximity, curve curvature, and twist respectively. If not explicitly identified by the user, default values for ε_* are assigned.

Measuring Geometric Factors

The tools needed to measure the geometric factors of each disjoint subset are also shown in table 1 and figure 3. As shown in Figure 3, a 3D-skeleton and 2D-skeleton are used to measure 3D-proximity and 2D-proximity, respectively. 3D Skeletons, as described in [13], are computed using a grassfire approach that progressively marches over a PR-octree decomposition of the input solid model. A distance function which approximates the closest distance between an octree node and the boundary surfaces of the solid at each octree node, is updated as the grassfire is propagated from the octree nodes closest to the boundary towards the interior. The skeleton of the solid model

Table 1. Geometric factors, tools, and checks for disjoint subsets

Subset of CAD Model	Geometric Factors	Tools for Measuring Geometric Factor	Geometric Check
$in(S)$	3D proximity	3D skeleton distance (d_{3D})	$2d_{3D} < \varepsilon_{3D}$
$in(F_n)$	2D proximity	2D skeleton distance (d_{2D})	$2d_{2D} < \varepsilon_{2D}$
	surface curvature	min. principal radius of curvature (r_{min})	$r_{min} < \varepsilon_{sc}$
$in(C_m)$	1D proximity	curve length (l)	$l < \varepsilon_{1D}$
	curve curvature	radius of curvature (r_c)	$r_c < \varepsilon_{cc}$
	curve twist	torsion (t)	$f(t, r_c) < \varepsilon_t$

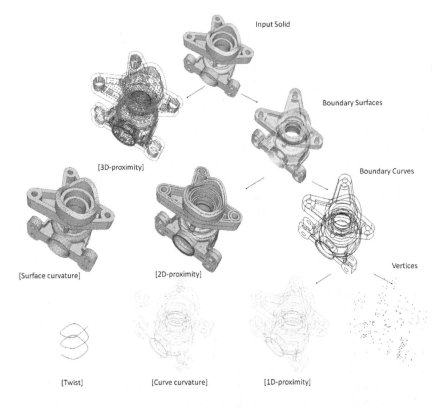

Fig. 3. Tools proposed to measure geometric factors of disjoint subsets

is approximated from the points that are generated where the opposing fronts collide. The distance function at the skeleton points can approximate one-half the local thickness or 3D-proximity between different combinations of geometric entities in a computationally efficient manner.

While an important geometric factor, the current implementation neglects 3D-proximity as only surface defeaturing is performed in the current work. As a result, only the geometric factors that capture the complexity of $in(F_n)$, $in(C_m)$, and V_l are considered. A more complete implementation should consider 3D-proximity to avoid inadvertently collapsing thin volumes that would otherwise not be detected with any other geometric factors. Future work may require a tetrahedral volumetric discrete representation to adequately perform volume defeaturing via tetrahedral collapse operations by considering 3D proximity. Currently, only surface defeaturing is performed using all the other geometric factors.

The 2D skeleton, also described in [13] measures 2D-proximity between boundary curves and vertices of a given surface. For a general surface the 2D-skeleton can be extracted in a similar manner to the 3D-skeleton by utilizing the same grassfire approach by progressively marching from the boundary curves towards the interior on the discrete model. To improve computational efficiency, on planar regions a chordal axis transform is used to extract the 2D-skeleton. Local edge swap operators are performed to remove illegal edges to obtain a more accurate 2D-skeleton. Figure 4b shows an example of the 2D-skeleton points extracted on multiple surfaces of the input solid. The red dots indicate where the smallest local thickness has been detected.

Other tools for measuring geometric factors, as illustrated in table 1, include minimum principal radius of curvature on a surface, r_{min}, curve length, l, radius of curvature of a curve, r_c and torsion, t. These values can be computed directly from the CAD solid model by querying the underlying geometry kernel, or else approximated from the discrete model. Since direct queries are computationally more expensive, approximating these values from the discrete representation is generally sufficient.

Identifying Features to be Suppressed

The geometric checks, as defined by the skeleton and other tools outlined in table 1 are used to identify irrelevant features for suppression. These tools calculate the maximum mesh size at a given point on a facet entity of the discrete model. A complete size field function $s = f(x, y, z)$ obtained by interpolating the distance at the skeleton points and other tools is shown in figure 4c with a color code. If the field value (maximum mesh size) is less than the minimum threshold value ε_* then that facet entity is marked for suppression. Note that the entities specifically identified by the user for preservation are ignored and never marked for suppression.

For example, unwanted features such as narrow regions and slender surfaces can be identified for reduction using the 2D-skeleton tool. In this case, the local thickness, or twice the skeleton distance $(2d_{2D})$ is used in calculating

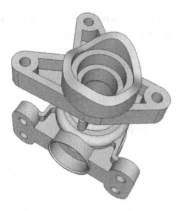

Fig. 4a. CAD model with narrow regions

Fig. 4b. 2D-skeleton of each surface with color coded local thickness

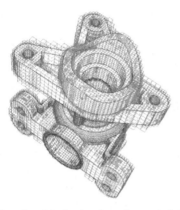

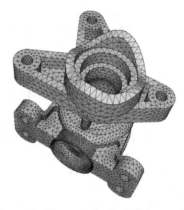

Fig. 4c. Mesh size based on skeleton and other tools

Fig. 4d. Fine mesh at skeleton points with smaller local thickness

maximum mesh size due to 2D-proximity. If the computed mesh size due to the 2D-skeleton is less than ε_{2D} then the facet entities associated with that skeleton point are marked for reduction. This ensures that by defeaturing these facets, mesh elements with sizes below the size ε_{2D} will not exist in the final mesh.

Similarly, other tools reveal the local maximum mesh size which in turn influences the detection of irrelevant features for a given ε. As the geometric checks are performed, the facet entities corresponding to the irrelevant features are detected and marked for suppression. For example, the 3D-skeleton tool, when used, can detect thin-walled regions via the 3D-proximity check; surface and curve curvature based checks can detect facet entities near high curvature regions such as fillets, blend patches, and holes; and 1D-proximity can detect facet edges associated with tiny curves and small features such as pegs and indentations via curve length checks.

3.3 Suppressing Features

Once all irrelevant features have been identified, operations to suppress the features can be performed. This is accomplished by using a standard edge collapse operator on all edges marked for suppression as ilustrated in figure 5. To accomplish this, the topological entities of the discrete B-Rep model are visited in a dimension-based order. For example, facet edges associated with the curves are collapsed prior to collapsing those associated with the interior of the surface. As the collapse operations are performed the associativity between the geometric entities of the discrete model and the facet entities are updated. This may require that following any given edge collapse that the resulting edges may need to maintain associativity to multiple geometric entities. This generalized one-to-many, child-parent associativity between facet entity and original CAD B-Rep entity is maintained and updated throughout the procedure.

Following any operation, local checks and updates are made to the reduced facet-based B-Rep to ensure a valid topological configuration is always

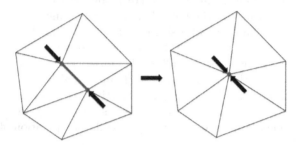

Fig. 5. The standard edge collapse operation

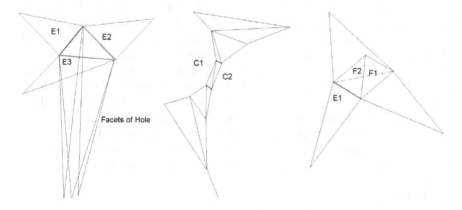

Fig. 6a. Single curve overlap   **Fig. 6b.** Two curves overlap   **Fig. 6c.** Single surface overlap

maintained. Figures 6a to 6c illustrate three specific cases that may make the discrete model invalid after an edge collapse operation.

Figure 6a shows an edge collapse at a cylindrical hole. In this case, the end circle of the cylindrical hole has been reduced to the three edges E1, E2, and E3, which are associated with one circular curve. After collapsing edge E1, the edges E2 and E3, although separate edges will share the same end vertices. To resolve this case, edges E2 and E3 are merged and represented as a single edge in the facet model. This ensures that there will be only one facet edge incident on any given pair of facet vertices.

Figure 6b shows that collapsing two small facet edges (shown in red) whose end vertices are incident on two curves C1 and C2 will result in a facet-edge associated with both the curves C1 and C2. This results in an invalid topology in the discrete B-Rep model even though the geometry continues to be watertight. To make the topology valid, both the curves C1 and C2 are split at their common overlapping edges to form a third curve C3. The facet edges at the overlapping region are then associated with a new curve C3 instead of associating with both C1 and C2. Although C3 is created as a new distinct curve entity in the discrete model, associativity back to its original geometry owners in the CAD B-Rep is maintained.

Figure 6c shows the facets of a single surface with edge E1 as one of the base edges of a triangular pyramid. Performing a collapse operation on edge E1, which is not an edge of F1 or F2 would cause the two facets F1 and F2 to overlap. Similar to the case described in figure 6a, although the facets are unique entities, they would share common vertices, creating a non-manifold or dangling facet. To avoid this case, the facets F1 and F2 are first merged and then subsequently deleted as part of the collapse operation of edge E1.

3.4   Meshing the Defeatured Model

Once all features identified for suppression have been reduced, the discrete model can be used as the basis for any of the surface and volumetric meshing tools in the CUBIT tool suite. Because associativity has been maintained throughout the reduction procedures, mapping between the original and the defeatured model is used to translate the user defined meshing schemes and mesh sizes prior to mesh generation. Since small features have now been removed from the discrete model, the resulting mesh in most cases will be higher quality with fewer degrees of freedom than if the model had not been defeatured. Using the same mapping between the original CAD model and discrete model, any boundary conditions specified in terms of the original CAD entities can be decoded onto the final FE mesh entities.

4   Results

The proposed approach has been implemented in C++ in CUBIT and has been tested on a limited number of industrial CAD models. We illustrate the effectiveness of the proposed method on two selected single part examples.

While this work focusses on single part models, work is underway to extend the framework to support volume defeaturing on assembly models.

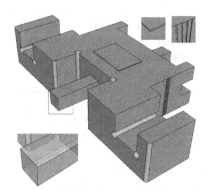

Fig. 7a. Original B-Rep CAD Model

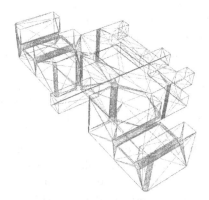

Fig. 7b. Discrete Model

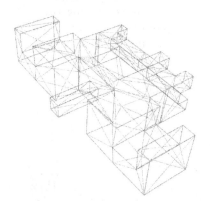

Fig. 7c. Defeatured Discrete Model

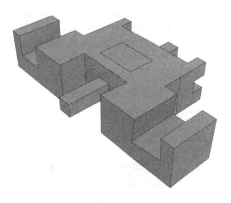

Fig. 7d. Defeatured B-Rep Model

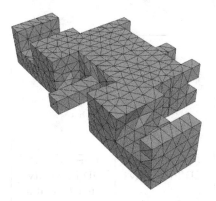

Fig. 7e. Tet mesh on Defeatured B-Rep Model

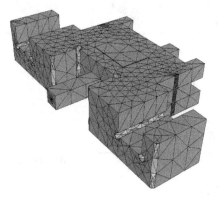

Fig. 7f. Tet mesh on Original B-Rep Model

Figures 7a to 7d show the different stages of defeaturing on a simple industrial part. Note that the model shown in Figure 7a and Figure 7b contains unintended features such as a long cylindrical hole, fillets at both convex and concave edges, multiple slots, and a protruded slab. Figures 7c and 7d show the defeatured model. Features have been removed based upon geometric factors that are controlled by a user specified threshold value of 1.0. Figure 7e shows a tetrahedral mesh on the defeatured model that has 3,490 tets with minimum element quality of 0.29 and average element quality of 0.825. In contrast, Figure 7f is a tetrahedral mesh on the model without defeaturing. The number of tet elements increases significantly to 51,109 and the minimum and average quality falls to 0.008 and 0.75, respectively. Both figures 7e and 7d use the same automatic triangle and tetrahedral meshing algorithms with the same default mesh settings.

Table 2 shows the effect of geometric factors on defeaturing. In CUBIT geometric factors can be controlled by issuing commands to enable a geometric factor. As more geometric factors are added, more facets get qualified for defeaturing and are marked for collapse. Note that using the geometric factors users can remove a specific type of feature. For example, adding the curve length check will remove the small peg at the top of the model. Subsequently adding the 2D-proximity criteria resulted in removing the chamfer. This is

Table 2. Controlling geometric factors to influence defeaturing

	No Features Removed	Small Curves Removed	Chamfer Removed	Hole & Fillet Removed
Geometric Model				
Mesh				
Geom. Factors	None	Curve Length	Curve Length 2D Proximity	Curve Length 2D Proximity Curve Curvature Surface Curvature

because the distance between the opposite long edges of the chamfer fell below the user specified threshold of 0.6. Similarly, adding curve and surface curvature-based checks removed the facets associated with the central hole and a fillet. In general, the geometric factors determine the mesh size at the facet entities first and then mark the facet entities for deletion only if the mesh size falls below the user specified threshold value.

5 Conclusion

This paper introduces a new facet-based approach to defeaturing CAD models by automatic identification of irrelevant features via a geometry-based size field. Robust detection of irrelevant features is achieved by using geometric factors that control the size field. The geometric factors are identified by a systematic analysis of the geometric complexity of the input CAD model. These geometric factors are also used as user controls to target the selection of irrelevant features, where the field value or mesh size, goes below a user specified threshold value. The actual defeaturing is performed on a facet-based discrete B-Rep model as it is less complex and more robust than performing it on the original NURBS model. It also has the advantage of leaving the original CAD model untouched so that multiple alternative defeatured representations can be derived based upon the needs of the simulation. As the meshing algorithms themselves are independent of the actual defeaturing procedures, any mesh generation scheme can be applied to the defeatured model without modification. The proposed method has been demonstrated on a limited set of industrial models and has proven effective in reducing user time, degrees of freedom, and mesh quality issues.

We have limited the initial work described here to feature suppression on single part CAD models. Extension of these procedures to assembly models will be a necessary next step. The proposed framework can also be extended to perform imprinting [15] and CAD repair via the discrete B-Rep model. Another potentially valuable area for study would be the introduction of physics-based sizing properties into the field function that drives the criteria for identification of features to be suppressed. This would provide a mechanism to automatically retain features in the model where important physics is occurring, but defeature where it is not.

References

1. Blacker, T.D., Sheffer, A., Clements, J., Bercovier, M.: Using Virtual Topology to Simplify the Mesh Generation Process. Trends in Unstructured Mesh Generation, vol. AMD-220, pp. 45–50 (1997)
2. Borouchaki, H., Laug, P.: Simplification of composite parametric surface meshes. Engineering with Computers 20, 176–183 (2004)
3. Clark, B.W.: Removing Small Features with Real CAD Operations. In: Proceedings of the 16th International Meshing Roundtable, pp. 183–198 (2007)

4. CUBIT Geometry and Mesh Generation Toolkit, Sandia National Laboratories (2009), http://cubit.sandia.gov
5. Dey, S., Shephard, M.S., Georges, M.K.: Elimination of the Adverse Effects of Small Model Features by the Local Modification of Automatically Generated Meshes. Engineering with Computers 13, 134–152 (1997)
6. Fine, L., Remondini, L., Leon, J.C.: Automated Generation of FEA Models Through Idealization Operators. International Journal for Numerical Methods in Engineering 49(1-2), 83–108 (2000)
7. Foucault, G., Cuilliere, J.C., Francois, V., Leon, J.C., Maranzana, R.: An Extension of the Advancing Front Method to Composite Geometry. In: Proceedings of the 16th International Meshing Roundtable, seattle (2007)
8. Foucault, G., Cuilliere, J.C., Francois, V., Leon, J.C., Maranzana, R.: Adaptation of CAD Model Topology for Finite Element Analysis. Computer Aided Design 40(2), 176–196 (2008)
9. Gao, S., Zhao, W., Yang, F., Chen, X.: Feature Suppression Based CAD Mesh Model Simplification. In: Proceedings IEEE International Conference on Shape Modeling and Applications, NY, June 4-6 (2008)
10. Inoue, K., Itoh, T., Yamada, A., Furuhata, T., Shimada, K.: Face clustering of a large-scale CAD model for surface mesh generation. Computer Aided Design 33(3), 251–261 (2001)
11. Mobley, A.V., Carroll, M.P., Canann, S.A.: An Object Oriented Approach to Geometry Defeaturing for Finite Element Meshing. In: Procceddings 7th International Meshing Roundtable, pp. 547–563 (1998)
12. Owen, S.J., White, D.R.: Mesh-Based Geometry. International Journal for Numerical Methods in Engineering 58(2), 375–395 (2003)
13. Quadros, W.R.: A Computational Framework for Generating 3D Finite Element Mesh Sizing Function via Skeletons, Ph.D. Thesis, Carnegie Mellon University, Pittsburgh, PA, U.S.A (2005)
14. Thakur, A., Banerjee, A.G., Gupta, S.K.: A Survey of CAD Model Simplification Techniques for Physics-based Simulation Applications. Computer Aided Design 41(2), 65–80 (2009)
15. White, D.R., Saigal, S., Owen, S.J.: An Imprint and Merge Algorithm Incorporating Geometric Tolerances for Conformal Meshing of Misaligned Assemblies. International Journal for Numerical Methods in Engineering 59, 1839–1860 (2004)

Towards Exascale Parallel Delaunay Mesh Generation*

Nikos Chrisochoides, Andrey Chernikov, Andriy Fedorov, Andriy Kot,
Leonidas Linardakis, and Panagiotis Foteinos

Center for Real-Time Computing
The College of William and Mary
Williamsburg, VA 23185
{nikos,ancher,fedorov,kot,leonl01,pfot}@cs.wm.edu

Abstract. Mesh generation is a critical component for many (bio-)engineering applications. However, *parallel* mesh generation codes, which are essential for these applications to take the fullest advantage of the high-end computing platforms, belong to the broader class of adaptive and irregular problems, and are among the most complex, challenging, and labor intensive to develop and maintain. As a result, parallel mesh generation is one of the last applications to be installed on new parallel architectures. In this paper we present a way to remedy this problem for new highly-scalable architectures. We present a multi-layered tetrahedral/triangular mesh generation approach capable of delivering and sustaining close to 10^{18} of concurrent work units. We achieve this by leveraging concurrency at different granularity levels using a hybrid algorithm, and by carefully matching these levels to the hierarchy of the hardware architecture. This paper makes two contributions: (1) a new evolutionary path for developing multi-layered parallel mesh generation codes capable of increasing the concurrency of the state-of-the-art parallel mesh generation methods by at least 10 orders of magnitude and (2) a new abstraction for multi-layered runtime systems that target parallel mesh generation codes, to efficiently orchestrate intra- and inter-layer data movement and load balancing for current and emerging multi-layered architectures with deep memory and network hierarchies.

1 Introduction

The complexity of programming adaptive and irregular applications on architectures with hierarchical communication networks of processors is an order of magnitude higher than on sequential machines, even for parallel mesh generation algorithms/codes which can be mapped directly on multi-layered architectures. Automatically exploiting concurrency for irregular and adaptive computation like Delaunay mesh generation is more complex than exploiting concurrency for regular (or array-based) and non-adaptive computations.

---

* This material is based upon work supported by the National Science Foundation under Grants No. CCF-0833081, CSR-0719929, and CCS-0750901 and by the John Simon Guggenheim Foundation.

Static analysis can not be used for adaptive and irregular applications like parallel mesh generation [27]. In [1, 33] we introduced a speculative (or optimistic) method for parallel Delaunay mesh generation which was recently adopted by the parallel compilers community [28, 35] to study abstractions for parallelization of adaptive and irregular applications. This technique has two major problems for high-end computing: (1) although it works reasonably well for the shared memory model, it is communication intensive for distributed memory machines; and (2) its concurrency can be limited by the problem size at the faster (and thus smaller) shared memory layer of the hierarchy.

In this paper we address both problems using a hybrid multi-layer approach which is based on a decoupled approach [29] at the larger (and slower) layers, an extension of an out-of-core weakly coupled method [25, 26] at the intermediate layers, and a speculative or optimistic but tightly-coupled method [1] at the faster (shared memory) layers (i.e., multi-core). The out-of-core layer utilizes additional disk storage and makes it possible to free the main memory for the storage of data used only in the current computation. In addition, we extend our runtime system [3] to efficiently manage both intra- and inter-layer communication in the context of data migration due to load balancing and migration of data/tasks between layers and between nodes across the same layer.

We expect that this paper can have an impact in two different areas: (1) *Mesh Generation:* we present the first highly scalable parallel mesh generation method capable to provide and sustain concurrency on the order of 10^{18}. (2) *Engineering Applications:* for the first time we provide unprecedented scalability for large-scale field solvers for applications like the direct numerical simulations of turbulence in cylinder flows with very large Reynolds numbers [18] and coastal ocean modeling for predicting storm surge and beach erosion in real-time [43]. In these applications three-dimensional simulations are conducted using two-dimensional meshes in the xy-plane which are replicated in the z-direction in the case of cylinder flows or using bathe-metric contours in the case of coastal ocean modeling. In addition, this method can be extended for Advancing Front Techniques. The approach we develop is independent of the geometric dimension (2D or 3D) of the mesh. Although the mesh-generation-specific domain decomposition has been developed only for 2D, a similar argument applies to 3D with the use of alternative decompositions, e.g., graph partitioning implemented in the Zoltan package [16].

This paper is organized as follows. In Section 2 we review the related prior work. In Section 3 we describe the organization of our Multi-Layered Runtime System. In Section 4 we present the proposed Multi-Layered Parallel Mesh Generation algorithm. In Section 5 we put the runtime system and the parallel mesh generation algorithm together. Section 5.1 contains our preliminary performance data, and Section 6 concludes the paper.

2   Background

In this section we present an overview of parallel mesh generation approaches related to the method we present in this paper. In addition we review parallel runtime systems related to our runtime system PREMA (Parallel Runtime Environment for Multicomputer Applications) which we extend to handle multi-layered applications.

2.1   Related Work in Parallel Mesh Generation

There are three conceptually different approaches to mesh generation. *Delaunay meshing* methods (see [19] and the references therein) use the Delaunay criterion for point insertion during refinement. *Advancing front meshing* techniques (see e.g. [38]) build the mesh in layers starting from the boundary of the geometry. Some of the advancing front methods use the Delaunay property for point placement, but no theoretical guarantees are usually available. *Adaptive space-tree meshing* (see e.g. [32]) is based on adaptive space subdivision (e.g., adaptive octree, or body-centric cubic lattice), and can be flexible in the definition of the meshed object geometry (e.g., implicit geometry representation). Certain theoretical guarantees on the quality of the mesh created in such a way are provided by some of the methods in this group.

A comprehensive review of parallel mesh generation methods can be found in [14]. In this section we review only those methods related to parallel Delaunay mesh generation. The problem of parallel Delaunay *triangulation* of a specified point set has been solved by Blelloch et al. [4]. A related problem of streaming triangulation of a specified point set was solved by Isenburg et al [20]. In contrast, Delaunay *refinement* algorithms work by inserting additional (so-called Steiner) points into an existing mesh to improve the quality of the elements. In Delaunay mesh refinement, the computation depends on the input geometry and changes as the algorithm progresses. The basic operation is the insertion of a single point which leads to the removal of a poor quality tetrahedron and of several adjacent tetrahedra from the mesh and to the insertion of several new tetrahedra. The new tetrahedra may or may not be of poor quality and, hence, may or may not require further point insertions. We and others have shown that the algorithm eventually terminates after having eliminated all poor quality tetrahedra, and in addition the termination does not depend on the order of processing of poor quality tetrahedra, even though the structure of the final meshes may vary [11, 12, 29]. Therefore, the algorithm guarantees the quality of the elements in the resulting meshes.

The parallelization of Delaunay mesh refinement codes can be achieved by inserting multiple points simultaneously. If the points are far enough from each other, as defined in [11], then the sets of tetrahedra influenced by their insertion are sufficiently separated, and the points can be inserted independently. However, if the points are close, then their insertion needs to be serialized because of possible violations of the validity of the mesh or of the

Delaunay property. One way to address this problem is to introduce runtime checks [28, 33] which lead to the overheads due to locking [1] and to rollbacks [33]. Another approach is to decompose the initial geometry [30] and apply decoupled methods [19, 29]. The third approach presented in [8, 9, 11] is to use a judicious way to choose the points for insertion, so that we can guarantee their independence and thus avoid runtime data dependencies and overheads. In [9] we presented a scalable parallel Delaunay refinement algorithm which constructs uniform meshes, i.e., meshes with elements of approximately the same size and in [11] we developed an algorithm for the construction of graded meshes. The work by Kadow and Walkington [22, 23] extended [4, 5] for parallel mesh generation and further eliminated the sequential step for constructing an initial mesh, however, all potential conflicts among concurrently inserted points are resolved sequentially by a dedicated processor [22].

In summary, in parallel Delaunay mesh generation methods we can explore concurrency at three levels of granularity: (i) *coarse-grain* at the subdomain level, (ii) *medium-grain* at the cavity level (this is a common abstraction for many different mesh generation methods), and (iii) *fine-grain* at the element level. The fine-grain can only increase the concurrency by a factor of three or four in two or in three dimensions, respectively. However, a detailed profiling of our codes revealed that up to 24.5% of the cycles is spent on synchronization operations, for both the protection of work-queues and for tagging each triangle upon checking it for inclusion in a cavity. Synchronization is always limited among the two or three threads co-located on the same core, and memory references due to synchronization operations always hit in the cache. However, the massive number of processed triangles results in a high percentage of cumulative synchronization overhead. We will revisit the fine-grain level when there is better hardware support for synchronization.

2.2   Related Work in Parallel Runtime Systems

Because of the irregular and adaptive nature of parallel mesh generation we wish to optimize, we restrict our discussion in this section to software systems which *dynamically* balance application workload and we use the following six important criteria: (1) *Support for data migration.* Migrating processes or threads adds to the complexity of the runtime system, and is often not portable. Migrating data, and thereby implicitly migrating computation is a more portable and simple solution. (2) *Support for explicit message passing.* Message passing is a programming paradigm that developers are familiar with, and the Active Messages [42] communication paradigm we use is a logical extension to that. Explicit message passing is also attractive because it does not hide parallelism from the developer. (3) *Support for a global namespace.* A global name-space is a prerequisite for automatic data migration; applications need the ability to reference data regardless of where it is in the parallel system. (4) *Single-threaded application model for inter-layer interactions*. Presenting the developer with a single-threaded communication model between layers greatly reduces application code complexity and development effort. (5) *Automatic load balancing*. The runtime system should migrate data or computation transparently and without intervention from the application. (6) *Customizable data/load movement/balancing*. It cannot be said that there is a "one size fits all" load balancing algorithm; different algorithms perform well in different circumstances. Therefore, developers need the ability to easily develop and experiment with different application- and machine-specific strategies without the need to modify their application code.

Systems such as the C Region Library (CRL) [21] implement a shared memory model of parallel computing. Parallelism is achieved through accesses to shared regions of virtual memory. The message passing paradigm we employ explicitly presents parallelism to the application. In addition, PREMA does not make use of copies of data objects, removing much of the complexity involved with data consistency and read/write locks. In [17, 41] the authors propose the development of component-based software strategies and data structure neutral interfaces for large-scale scientific applications that involve mesh manipulation tools.

Zoltan [15] and CHARM++ [24] are two systems with similar characteristics to PREMA. Zoltan provides graph-based partitioning algorithms and several geometric load balancing algorithms. Because of the synchronization required during load balancing, Zoltan behaves in much the same way as other stop-and-repartition libraries, whose results are presented in [2]. CHARM++ is built on an underlying language which is a dialect of C++, and provides extensive dynamic load balancing strategies. However, the *pick-and-process* message loop guarantees that entry-point methods execute "sequentially and without interruption" [24]. This may lead to a situation in which coarse-grained work units may delay the reception of load balancing messages, negating their usefulness, as was seen with the single-threaded PREMA results presented in [2]. The Adaptive Large-scale Parallel Simulations (ALPS) library [7] is based on a parallel octree mesh redistribution and targets hexahedral finite elements, while we focus on tetrahedral and triangular elements.

3 Multi-layered Runtime System

The application we target (parallel mesh generation) naturally lends itself to a hierarchical partitioning of work (specifically: domain, subdomain, independent subdomain region, and cavity). At the first two levels of this hierarchy, we use the concept of *mobile object*, or Mobile Work Unit (MWU), as an abstraction for work partitioning. MWU is a container, which is not attached to a specific processing element, but, as its name suggests, can migrate between address spaces of different nodes. Work processing is facilitated by means

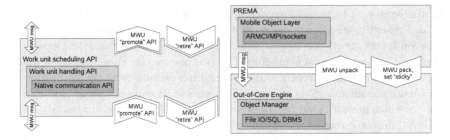

Fig. 1. Left: an abstraction for the hierarchical design of one runtime system layer. The layers are arranged vertically, such that the arrows represent the transfer of data between the adjacent layers. Right: a 2-layer instantiation of the proposed design which we tested using traditional out-of-core parallel mesh generation methods [25, 26].

of sending *mobile messages*, which are directed to MWUs. As we showed in [3], this abstraction is extremely convenient for the development of mesh generation codes, and is indispensable for one of the most challenging problems in parallel mesh generation: dynamic data/load movement/balancing.

Deep memory and network architecture hierarchies are intrinsic to the state-of-the-art High Performance Computing (HPC) systems. Based on our experience, MWU abstraction is effective in handling data movement, work distribution and load-balancing across a single layer of the HPC architecture hierarchy (among the nodes and disk storage units), while large-to-small work subdivision vertically aligns with the hierarchy of the architecture: mesh subdomains, for meshes with over 10^{18} elements, can be too large to fit in memory, while cavities can be processed concurrently at the level of a CPU core at a lower communication/synchronization cost. The objective of the multi-layered runtime system design is to provide communication and flow control support to leverage the hierarchical structure of both the application work partitioning and HPC architecture.

In our previous work on runtime systems we explored various possibilities for the design and the implementation of load-balancing on a Cluster of Workstations (CoW) [3]. In this paper, our design approach is based upon three levels of abstraction, as shown in Fig. 1(left). At the lowest level, there is *native communication* infrastructure, which is the foundation for implementing the concept and basic *MWU handling* routines (migration and MWU-directed communication). Given the ability to create and migrate MWUs, the *scheduling framework* implements high-level logic by monitoring the status of the system and the available objects, and rearranges them accordingly across the processing elements horizontally, or moving them up and down the vertical hierarchy. An important feature of the design is the MWU-directed communication. The life cycle of an MWU is determined by the messages (mostly, work requests) it receives from other MWUs and processing elements, and the status of the system. Depending on its status, availability of work, as well as the degree and nature of concurrency which can be achieved, an MWU can be "retired" to a lower level (characterized by lower degree of concurrency, when no work is pending for MWU, or when there are no resources to keep it at the current layer), or "promoted" to an upper layer (e.g., due to availability of resources or request for fast synchronization due to unresolved dependencies).

As a specific example of how multi-layered design can be realized, we implemented a two-layered framework based on the abstract design presented above (see Fig. 1, right). The top layer is an expanded version of the PREMA system [3]. The native communication can be either one among ARMCI [34], MPI or TCP sockets. The abstraction of mobile work units is realized by MOL [13], and high-level MWU scheduling is determined by the dynamic load-balancing policies implemented within the Implicit Load-balancing Library [3]. Overall, this layer is responsible for the maintenance of a balanced work distribution across a single layer of nodes.

4   Multi-layered Parallel Mesh Generation

Figure 2 presents the pseudo-code for the multi-layered (hybrid) parallel mesh generation algorithm. It starts with the initial Planar Straight Line Graph (PSLG) $\mathcal{X}$ which defines the domain Ω and the user-defined bounds on circumradius-to-shortest edge length ratio and on the size of the elements. First, we apply a Domain Decomposition procedure [30] to decompose Ω into N non-overlapping subdomains: $\Omega = \bigcup_{i=1}^{N} \Omega_i$ with the corresponding PSLGs $\mathcal{X}_i$, where N is the number of computational clusters. Then the boundary of each Ω_i is discretized using the Parallel Domain Delaunay Decoupling (PD^3) procedure [29] such that subsequent refinement is guaranteed not to introduce any additional points on subdomain boundaries. Next each subdomain represented by $\mathcal{X}_i$ is loaded onto a selected node from cluster i. Then $\{\mathcal{X}_i\}$ are further decomposed using the same method [30] into even smaller subdomains. However, in this case the boundaries of the subdomains are not discretized since PD^3 uses the worst case theoretical bound on the smallest edge length, which generally leads to over-refined meshes in practice. Instead, we use Parallel Constrained Delaunay Meshing (PCDM) algorithm/software [10] which at the cost of some communication introduces points on the boundaries as needed. Specifically, we use its out-of-core implementation (OPCDM) [26]. In addition we take advantage of the shared memory offered by multi-core systems and use the multi-threaded algorithm/implementation we presented in [1]. The meshes produced by the Multithreaded PCDM (MPCDM) algorithm are not constrained by the artificial subdomain boundaries and therefore generally have an even smaller number of elements than the meshes produced by the PD^3 algorithm.

SCALABLEPARALLELDELAUNAYMESHGENERATION($\mathcal{X}$, $\bar{\rho}$, $\bar{A}$)
Input: $\mathcal{X}$ is the PSLG which defines the domain Ω
$\qquad$ $\bar{\rho}$ is the upper bound on circumradius-to-shortest edge length ratio
$\qquad$ $\bar{A}$ is the upper bound on element size
Output: A distributed Delaunay mesh $\mathcal{M}$ which respects the bounds $\bar{\rho}$ and $\bar{A}$
1 $\quad$ Use MADD($\mathcal{X}$, N) to decompose the domain into subdomains
$\qquad$ represented by $\{\mathcal{X}_i\}$, $i = 1, \ldots, N$, where N is the number of clusters
2 $\quad$ Use PD^3($\{\mathcal{X}_i\}$, $\bar{\rho}$, $\bar{A}$), to refine the boundaries of $\mathcal{X}_i$
3 $\quad$ Load each of the $\mathcal{X}_i$, $i = 1, \ldots, N$, to a node n_i in cluster i
4 $\quad$ **do** on every node n_i simultaneously
5 $\qquad$ Use MADD($\mathcal{X}_i$, M_i) to decompose each subdomain
$\qquad\qquad$ into even smaller subdomains $\mathcal{X}_{ij}$, $j = 1, \ldots, M_i$
6 $\qquad$ Distribute the subdomains $\mathcal{X}_{ij}$, $j = 1, \ldots, M_i$, among P_i nodes in cluster i
7 $\qquad$ **do** on every node in cluster i simultaneously
8 $\qquad\qquad$ Use OPCDM($\{\mathcal{X}_{ij}\}$, $\bar{\rho}$, $\bar{A}$) to refine the subdomains
9 $\qquad$ **enddo**
10 $\quad$ **enddo**

OPCDM($\{\mathcal{X}_k\}$, $\bar{\rho}$, $\bar{A}$)
11 $\quad$ Let Q be the set of subdomains that require refinement
12 $\quad$ $Q \leftarrow \{\mathcal{X}_k\}$, $Q_o \leftarrow \emptyset$
13 $\quad$ **while** $Q \cup Q_o \neq \emptyset$
14 $\qquad$ $\mathcal{X} \leftarrow$ SCHEDULE(Q, Q_o)
15 $\qquad$ MPCDM($\mathcal{X}$, $\bar{\rho}$, $\bar{A}$)
16 $\qquad$ Update Q (the operation of finding any new subdomains that need
$\qquad\qquad$ refinement, e.g., after receiving messages, and inserting them into Q)
17 $\quad$ **endwhile**

MPCDM($\mathcal{X}$, $\bar{\rho}$, $\bar{A}$)
18 $\quad$ Construct $\mathcal{M} = (V, T)$ an initial Delaunay triangulation of $\mathcal{X}$
19 $\quad$ Let *PoorTriangles* be the set of poor quality triangles in T
$\qquad$ with respect to $\bar{\rho}$ and $\bar{A}$
20 $\quad$ **while** *PoorTriangles* $\neq \emptyset$
21 $\qquad$ Pick $\{t_i\} \subseteq$ *PoorTriangles*
22 $\qquad$ **do** using multiple threads simultaneously
23 $\qquad\qquad$ Compute the set of Steiner points $P = \{p_i\}$ corresponding to $\{t_i\}$
24 $\qquad\qquad$ Compute the set of Steiner points $P' \subseteq P$ which encroach upon constrained edges
25 $\qquad\qquad$ $P \leftarrow P \setminus P'$
26 $\qquad\qquad$ Replace the points in P' with the corresponding segment midpoints
27 $\qquad\qquad$ Compute the set of cavities $C = \{\mathcal{C}(p) \mid p \in P \cup P'\}$,
$\qquad\qquad\qquad$ where $\mathcal{C}(p)$ is the set of triangles whose circumscribed circles include p
28 $\qquad\qquad$ **if** C create conflicts
29 $\qquad\qquad\qquad$ Discard a subset of C and the corresponding points from $P \cup P'$
$\qquad\qquad\qquad$ such that there are no conflicts
30 $\qquad\qquad$ **endif**
31 $\qquad\qquad$ BOWYERWATSON(V, T, p), $\forall p \in P \cup P'$
32 $\qquad\qquad$ REMOTESPLITMESSAGE(p), $\forall p \in P'$
33 $\qquad$ **enddo**
34 $\qquad$ Update *PoorTriangles*
35 $\quad$ **endwhile**

SCHEDULE(Q, Q_o)
36 $\quad$ **while** $Q \neq \emptyset$
37 $\qquad$ $\mathcal{X} \leftarrow$ **pop**(Q)
38 $\qquad$ **if** $\mathcal{X}$ is *in-core* **return** $\mathcal{X}$ $\quad$ **else** SCHEDULETOLOAD($\mathcal{X}$), **push**(Q_o, $\mathcal{X}$) $\quad$ **endif**
39 $\quad$ **endwhile**
40 $\quad$ $\mathcal{X} \leftarrow$ **pop**(Q_o)
41 $\quad$ **if** $\mathcal{X}$ is *in lower-layer or out-of-core* LOAD($\mathcal{X}$) **endif**
42 $\quad$ **return** $\mathcal{X}$

BOWYERWATSON(V, T, p)
43 $\quad$ $V \leftarrow V \cup \{p\}$
44 $\quad$ $T \leftarrow T \setminus \mathcal{C}(p) \cup \{(p\xi) \mid \xi \in \partial\mathcal{C}(p)\}$,
$\qquad$ where $(p\xi)$ is the triangle obtained by connecting point p to edge ξ

Fig. 2. The multi-layered parallel mesh generation algorithm.

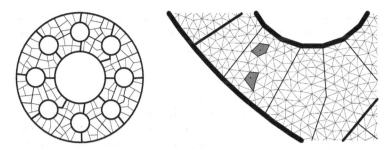

Fig. 3. (Left) Thick lines show the decoupled decomposition of the geometry into 8 high level subdomains which are assigned to different clusters. Medium lines show the boundaries between the subdomains assigned to separate nodes within a cluster. Thin lines show the boundaries between individual subdomains assigned to the same node. **(Right)** Parallel expansion of multiple cavities within a single subdomain using the MPCDM algorithm.

4.1 Domain Decomposition Step

We use the Medial Axis Domain Decomposition (MADD) algorithm/software we presented in [30]. MADD can produce domain decompositions which satisfy the following three basic criteria: (1) The boundary of the subdomains create good angles, i.e., angles no smaller than a given tolerance Φ_o, where the value of Φ_o is determined by the application which uses the domain decomposition. (2) The size of the separator should be relatively small compared to the area of the subdomains. (3) The subdomains should have approximately equal size, area-wise. This approach is well suited for both uniform and graded domain decomposition. Before the subdomains become available for further processing by the PCDM method they are discretized using the pre-processing step from PD^3 [29, 31] which guarantees that any Delaunay algorithm can generate a mesh on each of the subdomains in a way that does not introduce any new points on the boundary of the subdomains (i.e., the algorithm terminates and can guarantee conformity and Delaunay properties without the need to communicate with any of the neighbor subdomains).

4.2 Parallel Delaunay Mesh Generation Step

We use two different approaches, for different layers of the multi-layered architecture: (1) combine a coarse- and medium-grain (speculative-based) approach which is designed to run on a multi-core processor and (2) combine coarse- and coarser-grain which is designed after the traditional out-of-core PCDM method, for a multi-processor node as well as a cluster of nodes. First we describe the in-core PCDM method [10]. The PSLGs for all subdomains are triangulated in parallel using well understood sequential algorithms, e.g., described in [36, 39]. Each triangulated subdomain contains the collections of the constrained edges, the triangles, and the points. For the point insertion, we use the Bowyer-Watson (B-W) algorithm [6, 44]. The constrained (boundary) segments are protected by diametral lenses [37], and each time a segment is encroached, it is split in the middle; as a result, a *split* message is sent to the neighboring subdomain [10]. PCDM is designed to run on multi-processor nodes and clusters of nodes, i.e., it uses the message passing paradigm. Each process lies in its own address space and uses its own copy of a custom memory allocator. Second, the time corresponding to low aggregation decreases as we increase the number of processors; this can be explained by the growth of the utilized network and, consequently, the aggregate bandwidth. Similar studies for new HPC architectures need to be repeated and this parameter will be adjusted accordingly i.e., this parameter is machine specific.

Next we describe the two variations of PCDM we use for the multi-layered algorithm of Figure 2. First, we use the *Out-of-Core (OPCDM) approach (line 8 of the hybrid algorithm)* [26] which utilizes the bottom layer of the HPC architectures, i.e., the processing units with the large storage devices. Before processing a subdomain (using MPCDM) in the main loop we check whether the next subdomain in queue is in-core and mark it as sticky if it is or post a non-blocking load request for that subdomain. Second, after all bad triangles for a subdomain are processed we check whether the next subdomain in queue is in-core. If it is not we push it back in queue and examine the next. If we cannot find an in-core subdomain we load the next subdomain in queue with a blocking call. It should be noted that the Run-Time System (RTS) marks subdomains with multiple incoming messages as sticky and may attempt to prefetch them. Additionally, when processing incoming messages (when the application is polling), the RTS first executes messages addressed to in-core subdomains regardless of the order in which messages were received (the order of the messages sent to the same subdomain is preserved). The execution order of the subdomains does not affect neither correctness/quality nor termination for our algorithm.

Second, the *Multithreaded (MPCDM) approach (line 15 of the multi-layered algorithm)* [1] which targets the top layer of the HPC architecture, i.e., utilizes the fastest processing unit (hardware supported threads of cores). The threads create and refine individual cavities concurrently, using the B-W algorithm. MPCDM is synchronization-intensive mainly because threads need to tag each triangle while working on a cavity, to detect conflicts during concurrent cavity triangulation. Each subdomain is divided up into distinct areas (in order to minimize conflicts and overheads due to rollbacks), and the refinement of each area is assigned to a single thread. The decomposition is performed by equipartitioning — using straight lines as separators (strip-partitioning) that form a rectangular parallelogram enclosing the subdomain. Despite being straightforward and computationally inexpensive, this type of decomposition can introduce load imbalance between threads for irregular subdomains. The load imbalance can be alleviated by dynamically adjusting the position of the separators at runtime. The size of the queues (private and shared — of triangles that intersect the thread-separator) of bad quality triangles is proportional to the work performed by each thread. Large differences in the populations of queues of different threads at any time during the refinement of a single subdomain are a safe indication of load imbalance. Such events are, thus, used to trigger the load balancing mechanism. Whenever the population of the queues of a thread becomes larger than (100 / Number of Threads)% compared with the population of the queues of a thread processing a neighboring area, the separator between the areas is moved towards the area of the heavily loaded thread.

5 Putting It All Together

In this Section we present the highlights of the implementation for the multi-layered algorithm. The following implementation details are pertinent to the description of the runtime system, which we discussed previously: (1) hierarchical decomposition of work into MWUs, (2) interaction of the algorithm implementation with those units (via run-time system API), and (3) the management of MWUs by the run-time system.

The construction and the registration of the MWUs with the runtime system take place immediately after the decomposition of the input domain in line 5 of the algorithm, see Figure 2. A subdomain has dependencies on the neighboring subdomains, which share a common boundary, and may require coordination in order to process points inserted at that boundary. After the subdomains are defined, their movement, work processing, and communication (i.e., delivery of the Split messages) are handled transparently by the runtime system. The work processing is implemented in two mobile message handlers: subdomain refinement and split point processing subroutines.

We approach the issue of load-balancing across the nodes by using the dynamic load-balancing framework of PREMA [3]. *Intra-layer object migration* is triggered by the imbalance of work assigned to different subdomains due to different levels of refinement, different domain geometry, and, consequently, different rates of split messages arriving at each subdomain. *Inter-layer migration* of the MWUs is required for the efficient memory utilization, and the ability of the given layer to handle larger problem sizes. Scheduling of the MWUs between the PREMA and the OoCS follows the scheme described in the previous Section. The complex issue we will have to resolve, for truly (i.e., greater than two layers of processors) multi-layered architectures like the HTMT Petaflops design [40], is how to handle guaranteed delivery of the mobile messages in the causal order. With current two-layered architectures this is not a problem.

5.1 Preliminary Data

In this Section we report some of the preliminary results for the implementations of the three individual levels of the proposed hybrid algorithm: Domain Decomposition, Coarse+medium granularity (PCDM) and Coarse+coarser granularity (OPCDM). We evaluated the performance of the Domain Decomposition procedure on the fastest platform we had in our availability (dual Intel Pentium 3.6GHz). For the evaluation of the performance of the upper two levels of the algorithm (coarse+medium and coarse+coarser, i.e., traditional out-of-core) we used a cluster consisting of four IBM OpenPower 720 nodes. The nodes are interconnected via a Gigabit Ethernet network. Each node consists of two 1.6 GHz Power5 processors, which share eight GB of main memory. Each physical processor is a chip multiprocessor (CMP) integrating two cores. Each core, in turn, supports simultaneous multithreading (SMT) and offers two execution contexts. As a result, eight threads can be executed concurrently on each node. The two threads inside each core share a 32 KB, four-way associative L1 data cache and a 64 KB, two-way associative L1 instruction cache. All four threads on a chip share a 1.92 MB, 10-way associative unified L2 cache and a 36 MB 12-way associative off-chip unified L3 cache. The results for each of the three levels are as follows:

Domain Decomposition: Given the Chesapeake Bay model, we can sequentially decompose it using MADD into two subdomains in less than 0.5 seconds. This model is defined by 13,524 points and has 26 islands (i.e., quite complex geometry and resolution), see Figure 4. These two subdomains can be distributed to two cores and decomposed in parallel into four subdomains in less than 0.5 seconds. If we continue this way by building a logical binary tree over 10^{12} cores, the model can be decomposed into 10^{12} (or approximately 2^{40}) coarse grain subdomains in less than 40 seconds, assuming that half of this time is spent on communication. All subdomains satisfy the properties required by the Parallel Constrained Delaunay Mesh (PCDM) generation algorithm which we apply on each of these subdomains.

Coarse+medium granularity: On the medium grain level, the PCDM method can expose up to 8×10^5 potential concurrent cavity expansions per subdomain [1]. This level of the algorithm was evaluated (see Table 1) on the pipe model, see Figure 3. In each configuration we generate as many triangles as possible, given the available physical memory and the number of MPI processes and threads running on each node. The times reported for parallel PCDM executions include pre-processing time, domain decomposition, MPI bootstrap time, data loading and distribution, and the actual computation (mesh generation) time. We compare the execution time of parallel PCDM with that of the sequential execution of PCDM and with the execution time of Triangle [36], the best known sequential implementation for Delaunay mesh generation which has been heavily optimized and manually fine-tuned. For sequential executions of both PCDM and Triangle the reported time includes data loading and mesh generation time. On a single processor, we can significantly improve the performance attained by using a single core, compared with the coarse-grain only implementation. In the fixed problem size, it proves 29.4% faster than coarse-grain when one MPI process

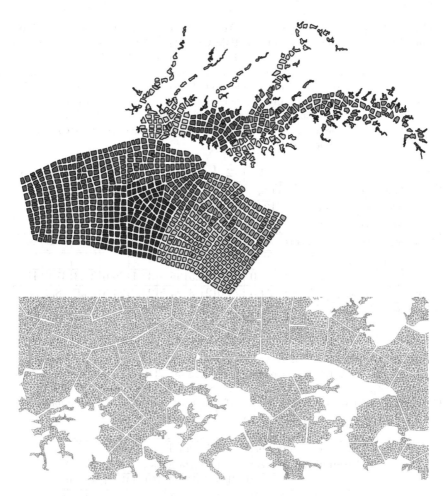

Fig. 4. (Top) The Chesapeake Bay model decomposed into 1024 subdomains that are mapped onto eight clusters of a multi-layered architecture. The assignment of subdomains to clusters is shown with different colors. The use of PD^3 eliminates communication between clusters, however, the use of the multi-layered PCDM in each of the original subdomains requires inter-layer communication and some synchronization at the top level. (Bottom) Part of the Chesapeake Bay model meshed in a way that satisfies conformity and Delaunay properties; thus, correctness and termination can be mathematically guaranteed.

is executed by a single core and 10.2% faster when two MPI processes correspond to each core (one per SMT context). In the scaled problem size the corresponding performance improvements are in the order of 31% and 12.7% respectively. Moreover, coarse+medium grain PCDM outperforms on a single core the optimized, sequential Triangle by 15.1% and 13.7% for the fixed and

Table 1. Execution times (in sec.) of the coarse grain and the coarse+medium grain PCDM in 2D on a cluster of four IBM OpenPower 720 nodes. As a sequential reference we use either the single-thread execution time of PCDM or the execution time of the best known sequential mesher (Triangle). Triangle quality in all tests is fixed to $20°$ degrees minimum angle bound. We present coarse-grain PCDM results using either one MPI process per core (`Coarse`) or one MPI process per SMT execution context (`Coarse (2/core)`). 60M triangles are created in the fixed problem size experiments. 15M triangles correspond to each processor core in the scaled problem size experiments.

Cores	1	2	4	6	8	10	12	14	16
Triangle Fixed	114.7								
Coarse Fixed	124.1	63.8	32.5	23.3	18.0	14.6	12.8	10.8	10.7
Coarse Fixed (2/Core)	97.4	49.0	21.2	16.3	12.2	10.1	9.1	7.9	8.3
Coarse+Medium Fixed	87.5	44.7	22.8	16.7	12.9	10.6	9.4	9.1	8.0
Triangle Scaled	28.4								
Coarse Scaled	31.0	32.2	32.5	35.6	37.1	36.6	38.3	37.6	41.8
Coarse Scaled (2/Core)	24.5	25.0	21.3	24.5	24.2	24.3	25.5	28.3	28.1
Coarse+Medium Scaled	21.4	22.5	22.8	25.5	26.7	27.1	27.8	29.9	30.4

Table 2. Normalized speed (on a cluster of 4 IBM OpenPower 720 nodes) of the PCDM in 2D with virtual memory and the OPCDM for problems that have memory footprint twice as large as the available physical memory. OPCDM(d) and OPCDM(b) refer to the experiments performed with the disk object manager and the database object manager respectively.

Mesh size, $\times 10^6$ triangles	number of nodes	Normalized speed, $\times 10^3$ triangles per second		
		PCDM	OPCDM(d)	OPCDM(b)
158.25	8(1)	242.45	156.22	160.11
316.50	16(2)	240.54	160.20	165.06
633.07	32(4)	239.82	157.67	161.08

scaled problem sizes respectively. On the fine grain level, the element-level concurrency allows us to process three or four elements concurrently (in 2D and 3D respectively), bringing the total potential concurrency to over 10^{18}.

Coarse+coarser granularity: Our evaluation (see Table 2) demonstrated that OPCDM is an effective solution for solving very large problems on computational resources with limited physical memory. We are able to generate meshes that otherwise would require 10 times the number of nodes using in-core implementation. The performance of the implementation was evaluated in 2D in terms of mesh generation speed[1]. We define per-processor mesh generation (normalized) speed as the average number of elements generated by a single processor over a unit time period, and it is given by $V = \frac{N}{T \times P}$, N is the number of elements generated, P is the number of processors in the configuration and T is execution time. We observe that the overhead introduced by the out-of-core functionality is not large: the per-processor mesh generation speed is only 33% slower for the meshes that fit completely in-core. At the same time, for the cases when we do use out-of-core functionality, up to 82% of disk I/O is overlapped with the computation.

6 Conclusions

We presented a multi-layered mesh generation algorithm capable to quickly generate and sustain in the order of 10^{18} of concurrent work units with granularity large enough to amortize overhead for hardware threads on current multi-threaded architectures. In addition we presented a multi-layered communication abstraction and its implementation on current 2-layered multi-core architectures. We used the resulting runtime system to implement a multi-layered parallel mesh generation code on IBM OpenPower 720 nodes (two-layered HPC architecture). The parallel mesh generation method/software mathematically guarantees termination, correctness, and quality of the elements. The mathematical guarantees are crucial for the size of problems we target, because even a single failure to solve a small subproblem my require the recomputation of the whole problem. Our implementation indicates that: (1) we pay very small overhead to generate very large number of concurrent work units, (2) intra-layer communication overhead is very small [10], (4) very large percentage (more than 80%) of inter-layer communication can be tolerated, (5) synchronization required only at the highest level where there is very fast hardware support, (5) work load balancing can be handled transparently with small overhead [3] at the coarse-grain layer (6) load balancing at the medium-grain layer can be handled easily and with low overhead within the application and (7) our out-of-core subsystem allows us to significantly decrease the processing times due to the reduction of wait-in-queue delays. However, the more complex multi-core and multi-CPU multi-layered designs will demand new hierarchical location management directories and policies, which will be a major future research effort (out of the scope of this paper) related to the system design.

Acknowledgments. We thank the anonymous reviewers for detailed comments which helped us improve the manuscript. We thank Professor Harry Wang and Dr. Mac Sisson from Virginia Institute of Marine Science for providing the data for the shoreline of the Chesapeake Bay.

---

[1] To date, there is no agreed upon standard to evaluate the performance of out-of-core parallel mesh generation codes. The existing metrics for in-core parallel algorithms are not sufficient for this task.

References

1. Antonopoulos, C.D., Ding, X., Chernikov, A.N., Blagojevic, F., Nikolopoulos, D.S., Chrisochoides, N.P.: Multigrain parallel Delaunay mesh generation: Challenges and opportunities for multithreaded architectures. In: Proceedings of the 19th Annual International Conference on Supercomputing, pp. 367–376. ACM Press, New York (2005)
2. Barker, K., Chrisochoides, N.: An evalaution of a framework for the dynamic load balancing of highly adaptive and irregular applications. In: Supercomputing Conference. ACM, New York (2003)
3. Barker, K., Chernikov, A., Chrisochoides, N., Pingali, K.: A load balancing framework for adaptive and asynchronous applications. IEEE Transactions on Parallel and Distributed Systems 15(2), 183–192 (2004)
4. Blelloch, G.E., Hardwick, J.C., Miller, G.L., Talmor, D.: Design and implementation of a practical parallel Delaunay algorithm. Algorithmica 24, 243–269 (1999)
5. Blelloch, G.E., Miller, G.L., Talmor, D.: Developing a practical projection-based parallel Delaunay algorithm. In: Proceedings of the 12th Annual ACM Symposium on Computational Geometry, Philadelphia, PA, May 1996, pp. 186–195 (1996)
6. Bowyer, A.: Computing Dirichlet tesselations. Computer Journal 24, 162–166 (1981)
7. Burstedde, C., Ghattas, O., Stadler, G., Tu, T., Wilcox, L.C.: Towards adaptive mesh PDE simulations on petascale computers. In: Proceedings of Teragrid (2008)
8. Chernikov, A.N., Chrisochoides, N.P.: Practical and efficient point insertion scheduling method for parallel guaranteed quality Delaunay refinement. In: Proceedings of the 18th Annual International Conference on Supercomputing, Malo, France, pp. 48–57. ACM Press, New York (2004)
9. Chernikov, A.N., Chrisochoides, N.P.: Parallel guaranteed quality Delaunay uniform mesh refinement. SIAM Journal on Scientific Computing 28, 1907–1926 (2006)
10. Chernikov, A.N., Chrisochoides, N.P.: Algorithm 872: Parallel 2D constrained Delaunay mesh generation. ACM Transactions on Mathematical Software 34(1), 1–20 (2008)
11. Chernikov, A.N., Chrisochoides, N.P.: Three-dimensional Delaunay refinement for multi-core processors. In: Proceedings of the 22nd Annual International Conference on Supercomputing, Island of Kos, Greece, pp. 214–224. ACM Press, New York (2008)
12. Paul Chew, L.: Guaranteed-quality triangular meshes. Technical Report TR89983, Cornell University, Computer Science Department (1989)
13. Chrisochoides, N., Barker, K., Nave, D., Hawblitzel, C.: Mobile object layer: a runtime substrate for parallel adaptive and irregular computations. Adv. Eng. Softw. 31(8-9), 621–637 (2000)
14. Chrisochoides, N.P.: A survey of parallel mesh generation methods. Technical Report BrownSC-2005-09, Brown University (2005); Also appears as a chapter in Bruaset, A.M., Tveito, A.: Numerical Solution of Partial Differential Equations on Parallel Computers. Springer, Heidelberg (2006)

15. Devine, K., Hendrickson, B., Boman, E., John, M.S., Vaughan, C.: Design of dynamic load-balancing tools for parallel applications. In: Proc. of the Int. Conf. on Supercomputing, Santa Fe (May 2000)
16. Devine, K.D., Boman, E.G., Riesen, L.A., Catalyurek, U.V., Chevalier, C.: Getting started with zoltan: A short tutorial. In: Proc. of 2009 Dagstuhl Seminar on Combinatorial Scientific Computing, Also available as Sandia National Labs Tech. Report SAND2009-0578C
17. Diachin, L., Bauer, A., Fix, B., Kraftcheck, J., Jansen, K., Luo, X., Miller, M., Ollivier-Gooch, C., Shephard, M.S., Tautges, T., Trease, H.: Interoperable mesh and geometry tools for advanced petascale simulations. Journal of Physics: Conference Series 78(1), 12015 (2007)
18. Dong, S., Lucor, D., Karniadakis, G.E.: Flow past a stationary and moving cylinder: DNS at Re=10,000. In: Proceedings of the 2004 Users Group Conference (DOD_UGC 2004), Williamsburg, VA, pp. 88–95 (2004)
19. George, P.-L., Borouchaki, H.: Delaunay Triangulation and Meshing. Application to Finite Elements. HERMES (1998)
20. Isenburg, M., Liu, Y., Shewchuk, J., Snoeyink, J.: Streaming computation of Delaunay triangulations. ACM Transactions on Graphics 25(3), 1049–1056 (2006)
21. Johnson, K., Kaashoek, M., Wallach, D.: CRL: High-performance all-software distributed shared memory. In: 15th Symp. on OS Prin (COSP15), December 1995, pp. 213–228 (1995)
22. Kadow, C.: Parallel Delaunay Refinement Mesh Generation. PhD thesis, Carnegie Mellon University (2004)
23. Kadow, C., Walkington, N.: Design of a projection-based parallel Delaunay mesh generation and refinement algorithm. In: 4th Symposium on Trends in Unstructured Mesh Generation, Albuquerque, NM (July 2003), http://www.andrew.cmu.edu/user/sowen/usnccm03/agenda.html
24. Kalé, L., Krishnan, S.: CHARM++: A portable concurrent object oriented system based on C++. In: Proceedings of OOPSLA 1993, pp. 91–108 (1993)
25. Kot, A., Chernikov, A., Chrisochoides, N.: Effective out-of-core parallel Delaunay mesh refinement using off-the-shelf software. In: Proceedings of the 20th IEEE International Parallel and Distributed Processing Symposium, Rhodes Island, Greece (April 2006). http://ieeexplore.ieee.org/search/wrapper.jsp?arnumber=1639361
26. Kot, A., Chernikov, A.N., Chrisochoides, N.P.: Out-of-core parallel Delaunay mesh generation. In: 17th IMACS World Congress Scientific Computation, Applied Mathematics and Simulation, Paris, France, Paper T1-R-00-0710 (2005)
27. Kulkarni, M., Pingali, K., Ramanarayanan, G., Walter, B., Bala, K., Chew, L.P.: Optimistic parallelism benefits from data partitioning. In: Architectural Support for Programming Languages and Operating Systems (2008)
28. Kulkarni, M., Pingali, K., Walter, B., Ramanarayanan, G., Bala, K., Chew, L.P.: Optimistic parallelism requires abstractions. SIGPLAN Not. 42(6), 211–222 (2007)
29. Linardakis, L., Chrisochoides, N.: Delaunay decoupling method for parallel guaranteed quality planar mesh refinement. SIAM Journal on Scientific Computing 27(4), 1394–1423 (2006)

30. Linardakis, L., Chrisochoides, N.: Algorithm 870: A static geometric medial axis domain decomposition in 2D Euclidean space. ACM Transactions on Mathematical Software 34(1), 1–28 (2008)
31. Linardakis, L., Chrisochoides, N.: Graded Delaunay decoupling method for parallel guaranteed quality planar mesh generation. SIAM Journal on Scientific Computing 30(4), 1875–1891 (2008)
32. Mitchell, S.A., Vavasis, S.A.: Quality mesh generation in higher dimensions. SIAM Journal for Computing 29(4), 1334–1370 (2000)
33. Nave, D., Chrisochoides, N., Chew, L.P.: Guaranteed–quality parallel Delaunay refinement for restricted polyhedral domains. In: Proceedings of the 18th ACM Symposium on Computational Geometry, Barcelona, Spain, pp. 135–144 (2002)
34. Nieplocha, J., Carpenter, B.: Armci: A portable remote memory copy library for distributed array libraries and compiler runtime systems. In: Proceedings RTSPP IPPS/SDP 1999 (1999) ID: bib:Nieplocha
35. Scott, M., Spear, M., Dalessandro, L., Marathe, V.: Delaunay triangulation with transactions and barriers. In: Proceedings of 2007 IEEE International Symposium on Workload Characterization (2007)
36. Shewchuk, J.R.: Triangle: Engineering a 2D Quality Mesh Generator and Delaunay Triangulator. In: Lin, M.C., Manocha, D. (eds.) FCRC-WS 1996 and WACG 1996. LNCS, vol. 1148, pp. 203–222. Springer, Heidelberg (1996)
37. Shewchuk, J.R.: Delaunay refinement algorithms for triangular mesh generation. Computational Geometry: Theory and Applications 22(1–3), 21–74 (2002)
38. Shöberl, J.: NETGEN: An advancing front 2d/3d-mesh generator based on abstract rules. Computing and Visualization in Science 1, 41–52 (1997)
39. Si, H., Gaertner, K.: Meshing piecewise linear complexes by constrained Delaunay tetrahedralizations. In: Proceedings of the 14th International Meshing Roundtable, San Diego, CA, pp. 147–163. Springer, Heidelberg (2005)
40. Sterling, T.: A hybrid technology multithreaded computer architecture for petaflops computing 1997. TY: STD; CAPSL Technical Memo 01, Jet Propulsion Library, California Institute of Technology, California (January 1997)
41. To, A.C., Liu, W.K., Olson, G.B., Belytschko, T., Chen, W., Shephard, M.S., Chung, Y.W., Ghanem, R., Voorhees, P.W., Seidman, D.N., Wolverton, C., Chen, J.S., Moran, B., Freeman, A.J., Tian, R., Luo, X., Lautenschlager, E., Challoner, A.D.: Materials integrity in microsystems: a framework for a petascale predictive-science-based multiscale modeling and simulation system. Computational Mechanics 42, 485–510 (2008)
42. von Eicken, T., Culler, D., Goldstein, S., Schauser, K.: Active messages: A mechanism for integrated communication and computation. In: Proceedings of the 19th Int. Symp. on Comp. Arch., pp. 256–266. ACM Press, New York (1992)
43. Walters, R.A.: Coastal ocean models: Two useful finite element methods. Recent Developments in Physical Oceanographic Modeling: Part II 25, 775–793 (2005)
44. Watson, D.F.: Computing the n-dimensional Delaunay tesselation with application to Voronoi polytopes. Computer Journal 24, 167–172 (1981)

On the Use of Space Filling Curves for Parallel Anisotropic Mesh Adaptation

Frédéric Alauzet[1] and Adrien Loseille[2]

[1] INRIA Paris-Rocquencourt, Projet Gamma, Domaine de Voluceau, BP 105, 78153 Le Chesnay cedex, France
`Frederic.Alauzet@inria.fr`
[2] CFD Center, Dept. of Computational and Data Sciences, College of Science, MS 6A2, George Mason University, Fairfax, VA 22030-4444, USA
`aloseill@gmu.edu`

Abstract. Efficiently parallelizing a whole set of meshing tools, as required by an automated mesh adaptation loop, relies strongly on data localization to avoid memory access contention. In this regard, renumbering mesh items through a space filling curve (SFC), like Hilbert or Peano, is of great help and proved to be quite versatile. This paper briefly introduces the Hilbert SFC renumbering technique and illustrates its use with two different approaches to parallelization: an out-of-core method and a shared-memory multi-threaded algorithm.

Keywords: Space filling curves, shared-memory multi-threaded parallelization, out-of-core parallelization, anisotropic mesh adaptation, mesh partition.

1 Introduction

The efficient use of computer hardware is crucial to achieve high performance computing. No matter how clever an algorithm might be, it has to run efficiently on available computer hardwares. Each type of computer, from common PCs to fastest massively parallel machines, has its own shortcomings that must be accounted for when developing both algorithms and simulation codes. The wish to develop efficient parallel codes is thus driven by several requirements and practical considerations: the problem at hand that need to be solved, the required level of accuracy and the available computational power. The main motivation of this paper is to take advantage of today's ubiquitous multi-core computers in mesh adaptive computations.

Indeed, mesh adaptation is a method developed to reduce the complexity of numerical simulations by exploiting specifically the natural anisotropy of physical phenomena. Generally, it enables large complex problem to be solved in serial. However, the present hardware evolution suggests the parallelization of mesh adaptation platforms. Since 2004, first Moore's law corollary has plummeted from the 40% yearly increase in processor frequency, that it has enjoyed for the last 30 years, to a meager 10%. As for now, speed improvement can only be achieved through the multiplication of processors, now called cores, sharing the same memory within a single chip.

Space filling curves (SFCs) are mathematical objects that enjoy nice proximity in space properties. These properties made then very useful in computer science and scientific computing. For instance, they have been used for data reordering [26, 31], dynamic partitioning [29], 2D parallel mesh generation [8] or all of these in the context of Cartesian adapted meshes [1].

In this paper, we present a straightforward parallelization of all softwares of a mesh adaptation platform where the pivot of the strategy is the **Hilbert space filling curve**. This strategy must be efficient in the context of highly anisotropic adapted meshes for complex real-life geometries. This platform is highly heterogeneous as it contains several software components that have different internal databases and that consider different numerical algorithms. It generally involves a flow solver, an adaptive mesh generator or an adaptive local remesher, an error estimate software and a solution interpolation (transfer) software. Two classes of parallelization are given.

The first one is an intrusive parallelization of the code using the **pthreads** paradigm for **shared-memory** cache-based parallel computers. One of the main assets of this strategy resides in a slight impact on the source code implementation and on the numerical algorithms. This strategy is applied to the flow solver and to the error estimate software. Parallelization is at the loop level and requires few modifications of the serial code. However, to be efficient this approach requires a subtle management of cache misses and cache-line overwrite to enable correct scaling factor for loop with indirect addressing. The key point is to utilize a Hilbert space filling curve based renumbering strategy to minimize them.

The second one is an **out-of-core** parallelization that considers the software as a black box. This approach is applied to a local remesher and the solution transfer software. It relies on the used of the Hilbert SFC to design a fast and efficient mesh partitioning. This partitioning method involves a correction phase to achieve connected partitions which is mandatory for anisotropic mesh adaptation. The mesh partitioner is coupled with an adequate management of the software on each partition. In this case, the code can be run in parallel on the same computer or in a distributed manner on an heterogeneous architecture.

As regards the meshing part, global mesh generation methods, such as Delaunay or Frontal approaches, are still hard to parallelize even if some solutions have already been proposed [7, 17, 20, 23]. Therefore, a local remeshing approach which is easier to parallelize thanks to its locality properties has been selected over a global mesher. The key point is how to adapt the partition borders [4, 10, 11, 18, 25, 27].

We illustrate with numerical examples that this methodology coupling anisotropic mesh adaptation, cache miss reduction and, out-of-core and pthreads parallelization can reduce the complexity of the problem by several orders of magnitude providing a kind of "high performance computing" on nowadays multi-core personal computers.

This paper is outlined as follow. Section 2 recalls our mesh adaptation platform and Section 3 describes the test cases. Then, in Section 4, we present the Hilbert space filling curve based mesh renumbering. The shared-memory and the out-of-core parallelizations with their application to each stage of the mesh adaptation loop are introduced in Sections 5 and 6, respectively.

2 A Brief Overview of the Mesh Adaptation Platform

In the context of numerical simulation, the accuracy level of the solution depends on the current mesh used for its computation. And, for mesh adaptation, the size prescription, *i.e.,* the metric field, is provided by the current solution. This points out the non-linearity of the anisotropic mesh adaptation problem. Therefore, an iterative process needs to be set up in order to converge both the mesh and the solution, or equivalently the metric field and the solution. For stationary simulations, an adaptive computation is carried out *via* a mesh adaptation loop inside which an algorithmic convergence of the pair mesh-solution is sought. At each iteration, all components of the mesh adaptation loop are involved successively: the flow solver, the error estimate, the adaptive mesh generator and the solution interpolation stage. This procedure is repeated until the convergence of the mesh-solution pair is reached.

Our implementation of the mesh adaptation platform considers an independent dedicated software for each stage of the adaptation loop. As compared to the strategy where only one software contains all the stages of the mesh adaptation, we can highlight the following disadvantages and advantages. The main drawback is that between two stages, one software writes the data (e.g. the mesh and the fields) out-of-core and the next software reads them back and builds its internal database. This results in a larger part devoted to I/O as compared to the all-in-one approach. But, the CPU time for the I/O is generally negligible with respect to the global CPU time. The advantage of the proposed strategy is its flexibility. Each software can be developed independently with its own programming language and its own optimal internal database. For instance, the flow solver can keep a static database, the mesh generator can use specific topological data structures such as the elements neighbors, etc. Consequently, we may expect a higher efficiency in memory and in CPU time for each software. Moreover, each software is interchangeable with another one, only the I/O between the different softwares need to be compatible.

The mesh adaptation platform described in this paper involves the flow solver Wolf [5], Metrix for the error estimate [21], the local remesher Mmg3d [12] and Interpol for the solution transfer [6].

3   The Considered Test Cases

The efficiency of all the presented algorithms will be analyzed independently on the same list of test cases in their own dedicated sections. The efficiency is demonstrated thanks to CPU times and speed-ups, the speedup being the ratio between the CPU time in parallel and the CPU time in serial. The list of test cases is composed of uniform, adapted isotropic and anisotropic meshes for a wide range of number of tetrahedra varying from 40 000 to 50 000 000:

- uniform mesh: a transonic flow around the **M6 wing** [13] and **Rayleigh-Taylor instabilities** (IRT) [3]
- adapted isotropic mesh: a blast in a **city** [3]
- adapted highly anisotropic mesh: supersonic flows around Dassault-Aviation **supersonic business jet** (SSBJ) [21] and a NASA **spike** geometry, and a transonic flow around Dassault-Aviation **Falcon** business jet.

Meshes associated with these test cases are shown in Figure 1 and their characteristics are summarized in Table 1. Note that the SSBJ and the spike test cases involve very high size scale factor and highly anisotropic adapted meshes. For instance, for the SSBJ, the minimal mesh size on the aircraft is 2mm and has to be compared with a domain size of 2.5km.

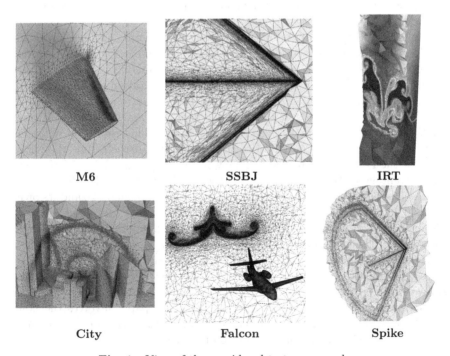

| M6 | SSBJ | IRT |
| City | Falcon | Spike |

Fig. 1. View of the considered test cases meshes

Table 1. Characteristics of all test cases

Case	Mesh kind	# of vertices	# of tetrahedra	# of triangles
M6	uniform	7 815	37 922	5 848
IRT	uniform	74 507	400 033	32 286
City	adapted isotropic	677 278	3 974 570	67 408
Falcon	adapted anisotropic	2 025 231	11 860 697	164 872
SSBJ	adapted anisotropic	4 249 176	25 076 962	334 348
Spike	adapted anisotropic	8 069 621	48 045 800	182 286

All the runs have been done on a 2.8 GHz dual-chip Intel Core 2 Quad (eight-processor) Mac Xserve with 32 GB of RAM.

4   The Hilbert Space Filling Curve

The notion of space filling curves (SFCs) has emerged with the development of the concept of the Cantor set [9]. Explicit descriptions of such curves were proposed by Peano [28] and Hilbert [15]. SFCs are, in fact, fractal objects [22]. A complete overview is given in [30]. A SFC is a continuous function that, roughly speaking, maps a higher dimensional space, e.g. $\mathbb{R}^2$ or $\mathbb{R}^3$, into a one-dimensional space:

$$h : \{1, \ldots, n\}^d \mapsto \{1, \ldots, n^d\} .$$

These curves enjoy strong local properties making them suitable for many applications in computer sciences and scientific computing. The **Hilbert SFC** is a continuous curve that fills an entire square or cube. For the Hilbert SFC, we have [26]:

$$|h(i) - h(j)| < \sqrt{6} \, |i - j|^{\frac{1}{2}} \quad \text{for} \quad i, j \in \mathbb{N} .$$

The Hilbert SFC for the square or the cube is generated by recursion as depicted in Figure 2. Its discrete representation depends on the level of recursion. In our use of the Hilbert SFC, the curve is not explicitly constructed but its definition is used to calculate an index associated with each mesh entity by means of the recursive algorithm. In other words, in three dimensions, the SFC provides an ordered numeration of a virtual Cartesian grid of size 2^{3p} where p is the depth of recursion in which our computational domain is embedded. The index of an entity is then obtained by finding in which cube the entity stands. In the following, we will present how these indices are used to renumber a mesh or to partition it.

4.1   Mesh Entities Renumbering

The Hilbert SFC can be used to map mesh geometric entities, such as vertices, edges, triangles and tetrahedra, into a one dimensional interval. In numerical

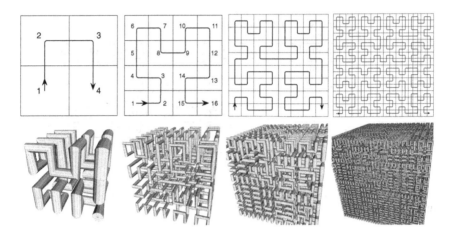

Fig. 2. Representation of the Hilbert curve in 2D and 3D after several recursions. Top, the 2D Hilbert SFC of the square after recursions 1, 2, 3 and 4. Bottom, the 3D Hilbert SFC of the cube after recursions 2, 3, 4 and 5.

Table 2. CPU time in seconds for sorting the vertices of the test cases meshes.

Case	M6	IRT	City	Falcon	SSBJ	Spike
SFC construction CPU in sec	0.004	0.038	0.338	1.002	1.894	3.632
Quicksort CPU in sec	0.001	0.010	0.104	0.974	0.668	1.518
Global CPU in sec	0.005	0.048	0.442	2.076	2.562	5.150

applications, it can be viewed as a mapping from the computational domain Ω onto the memory of a computer. The local property of the Hilbert SFC implies that entities which are neighbors on the memory 1D interval are also neighbors in the domain Ω. But, the reverse may not be true. Neighbors in the volume may be separated through the mapping. This approach has been applied to reorder efficiently Cartesian grids [1, 29] and its used for unstructured tetrahedral meshes has been indicated in [31]. Note that a large varieties of renumbering strategies exist which are commonly used in scientific computing [19] or in meshing [32]. Here, we apply and analyze such renumbering to unstructured, isotropic or anisotropic, adapted meshes.

First, the index is computed for each entity, this operation has a linear complexity. Then, the mesh entities have to be reordered to obtain the renumbering. This sort is done with standard C-library sorting routine such as quicksort, hence the $O(N \log(N))$ complexity of our method. Table 2 sums-up the CPU time for sorting all the test cases meshes on the Mac Xserve. Figure 3 illustrates an unstructured mesh of a scramjet (left) and how the vertices have been reordered in memory by following the line (right), this is the Hilbert SFC.

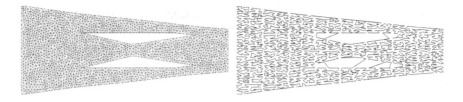

Fig. 3. Left, unstructured mesh of a scramjet. Right, the Hilbert SFC (red line) associated with the vertices. It represents how the vertices are ordered in memory.

Table 3. Number of CPU cycles required for typical operations on Intel Core 2.

Operations	mult/add	div	sqrt	cache miss L1	cache miss L2	mutex	condwait
Cycles	1	10	10	13	276	6 240	12 480

Renumbering strategies have a significant impact on the efficiency of a code. This is even more crucial for numerical methods on unstructured meshes. We can cite the following impact:

- reducing the number of cache misses during indirect addressing loops
- improving matrix preconditioning for linear system resolutions
- reducing cache-line overwrites during indirect addressing loops which is fundamental for shared memory parallelism, see Section 5.1
- may provide implicit hashing of the data.

Let us focus on the first item : cache misses that are due to indirect addressing. They occur when data are required for a computation and those data do not lie within the same cache line. For instance, such situation is frequent while performing a loop through the tetrahedra list and requesting for vertices data. It is worth mentioning that the cost of a cache miss is far more important than typical operations used in numerical applications, see Table 3.

To effectively reduce cache misses, all the mesh entities must be reordered and not, for instance, only the vertices. In our approach, the Hilbert SFC based renumbering is used to sort the vertices as their proximity depends on their position in space even for anisotropic meshes. As regards the topological entities, e.g. the edges, the triangles and the tetrahedra, the Hilbert SFC based renumbering can be applied by using their center of gravity. However, as they are topological entities, we prefer to consider a topological renumbering strategy based on vertex balls which also provides an implicit hashing of the entities. Note that similar wave renumbering strategies are described in [19]. This strategy reduces by 90% the number of cache misses[1] of the flow solver (Section 5.2) as compared to an unsorted mesh generated with a Delaunay-based algorithm.

---

[1] This statistic has been obtained with the Apple Shark code profiler.

For the test cases of Sections 5 and 6, we obtain in serial a speed-up up to 2.68, *i.e.,* up to almost three time faster, when all the entities are reordered as compared to the unsorted mesh. More precisely, speed-ups of 1.06, 1.62, 2.54 and 2.68 are obtained for the M6, IRT, City and Falcon test cases, respectively.

5   Exploiting Space Filling Curves for Efficient Shared-Memory Multi-threaded Parallelization

5.1   A Shared Memory Multi-threaded Parallelization

Our approach is based on posix standard threads (known as *pthreads*) thus taking advantage of *multi-core* chips and *shared memory* architectures supported by most platforms. Such an approach presents two main advantages: data do not need to be explicitly partitioned as with MPI based parallelism and its implementation requires only slight modifications to a previously developed serial code.

Symmetric parallelization of loops. In this case, a loop performing the same operation on each entry of a table is split into many sub-loops. Each sub-loop will perform the same operation (hence the name symmetric parallelism) on equally-sized portions of the main table and will be concurrently executed. It is the scheduler job to make sure that two threads do not write simultaneously on the same memory location. To allow for a fine load balancing, we split the main table into a number of blocks equal to 16 times the number of available processors.

Indirect memory access loops. Using meshes and matrices in scientific computing leads inevitably to complex algorithms where indirect memory accesses are needed. For example, accessing a vertex structure through a tetrahedron leads to the following instruction:

```
TetTab[i]->VerTab[j];
```

Such a memory access is very common in C, the compiler will first look for tetrahedron i, then vertex j, thus accessing the data indirectly. In this case, after splitting the main tetrahedra table into sub-blocks, tetrahedra from different blocks may point to the same vertices. If two such blocks were to run concurrently, memory write conflict would arise. To deal with this issue, an asynchronous parallelization is considered instead of a classic gather/scatter technique usually developed on distributed memory architectures, *i.e.,* each thread writes in its own working array to avoid memory conflict and then the data are merged. Indeed, this asynchronous parallelization has the following benefits: there is no memory overhead and for all the test cases on 8 cores this method was 20 to 30 % faster than the gather/scatter method. The main difficulty is then to minimize the synchronization costs that are expensive in CPU cycles, cf. Table 3. To this end, the scheduler will carefully choose the set of concurrently running blocks so that they share no common tetrahedra and no common vertices. In case no compatible blocks are to be found, some threads may be left idling, thus reducing the degree of parallelization. This method can even lead to a serial execution in the case of a total incompatibility between blocks.

The collision probability of any two blocks sets the mesh inherent parallelization factor. Meshes generated by advancing front or Delaunay techniques feature a very low factor (2 or 3 at best), while octree methods far much better. Renumbering the mesh elements and vertices as described in Section 4.1 dramatically enhances the inherent parallelism degree (by orders of magnitude). Applying such renumbering is *de facto* mandatory when dealing with indirect memory access loops. The block collision statistics for the test cases of Section 3 on 8 processors without and with the renumbering strategy of Section 4.1 are reported in Table 4.

As regards the scheduler cost, the operation of locking/unlocking a thread needs one mutex and one condwait, see Table 3. As these two operations are needed when launching and when stopping a thread, the resulting cost is approximatively 37 000 CPU cycles which is very expensive as compared to standard floating point operations or cache misses timings.

This approach has been implemented in the *LP2* library [24]. The purpose of this library is to provide programmers of solvers or automated meshers in the field of scientific computing with an easy, fast and transparent way to parallelize their codes. Thus, we can implement directly in parallel.

A sketch of the modifications of a serial code parallelized with the LP2 is given in Figure 4. Left, the dependencies of the tetrahedra array with respect to the vertices array are set. Right, the modification of the routine Solve which processes a loop on tetrahedra. This routine is called in parallel with two additional parameters iBeg and iEnd that are managed by the LP2. This illustrates the slight modifications that occur for the serial code. For instance,

Table 4. Collision percentage between blocks of entities when the list is not sorted or sorted.

Cases		Edges List		Tetrahedra List		Triangles List	
		Avg	Max	Avg	Max	Avg	Max
M6	no sort	2.96%	6.10%	7.25%	9.83%	0.35%	0.61%
	sort	0.94%	1.57%	0.95%	1.61%	0.37%	0.61%
IRT	no sort	24.06%	35.94%	45.21%	55.36%	1.56%	2.71%
	sort	1.00%	2.03%	1.00%	1.79%	0.42%	0.92%
City	no sort	73.90%	98.64%	93.29%	99.51%	1.21%	2.78%
	sort	1.09%	3.95%	1.10%	3.60%	0.25%	0.97%
Falcon	no sort	99.55%	100.00%	99.65%	100.00%	0.15%	1.02%
	sort	1.81%	4.63%	1.81%	4.24%	0.19%	0.59%

```
BeginDependency(Tetrahedra,Vertices);
for (iTet=1; iTet<=NbrTet; ++iTet) {
  for (j=0; j<4; ++j) {
    AddDependency( iTet, Tet[iTet].Ver[j] );
  }
}
EndDependency(Tetrahedra,Vertices);
```

```
Solve(Tetrahedra,iBeg,iEnd) {
  for (iTet=iBeg; iTet<=iEnd; ++iTet) {
    // .... same as serial
  }
}
```

Fig. 4. Modification of the serial code for shared-memory parallelization.

for our flow solver (see Section 5.2) most of (98.5 %) the explicit resolution part has been parallelized in less than one day.

5.2   Parallelizing the Flow Solver

The parallelized flow solver is a vertex-centered finite volume scheme on unstructured tetrahedral meshes solving the compressible Euler equations. To give a brief overview, the HLLC approximate Riemann solver is used to compute the numerical flux. The high-order scheme is derived according to a MUSCL type method using downstream and upstream tetrahedra. A high-order scheme is deduced by using upwind and downwind gradients leading to a numerical dissipation of 4th order. To guarantee the TVD property of the scheme, a generalization of the Superbee limiter with three entries is considered. The time integration is an explicit algorithm using a 5-stage, 2-order strong-stability-preserving Runge-Kutta scheme. We refer to [5] for a complete description.

A shared-memory parallelization of the finite volume code has been implemented with the pthreads paradigm described in Section 5.1. It uses the entities renumbering strategy proposed in Section 4.1. It took only 1 day to parallelize most of the resolution part with less than 2% of the resolution remaining in serial. More precisely, 6 main loops of the resolution have been parallelized:

- the time step evaluation which is a loop on the vertices without dependencies
- the boundary gradient[2] evaluation which is a loop on the tetrahedra connected to the boundary
- the boundary conditions which is a loop on the boundary triangles
- the flux computation which is a loop on the edges
- the source term computation which is a loop on the vertices without dependencies
- the update (advance) in time of the solution which is a loop on the vertices without dependencies.

The speed-ups, as compared to the serial version, obtained for each test case from 2 to 8 processors are summarized in Table 5. These speed-ups

---

[2] For this numerical scheme, the element gradient used for the upwinding can be computed on the fly during the flux evaluation.

Table 5. Speed-ups of the flow solver as compared to the serial version for all the test cases from 2 to 8 processors.

Cases		M6	IRT	City	Falcon	SSBJ	Spike
Speed-up	1 Proc	1.000	1.000	1.000	1.000	1.000	1.000
	2 Proc	1.814	1.959	1.956	1.961	1.969	1.975
	4 Proc	3.265	3.748	3.866	3.840	3.750	3.880
	8 Proc	5.059	6.765	7.231	6.861	7.031	7.223

contains the time of I/Os which is negligible for the flow solver. These results are very satisfactory for the largest cases for which an enjoyable speed-up around 7 is attained on 8 processors. The slight degradation observed between 4 and 8 processors is in part due to a limitation of the current hardware of the Intel Core 2 Quad chip. However, the speed-ups are lower for the smallest case, the M6 wing with only 7 815 vertices. This small test case with a light amount of work points out the over-cost of the scheduler for pthreads handling which is a weakness of the proposed approach. More precisely, we recall that launching and stopping a thread cost approximatively 37 000 CPU cycles. As the parallelization is at the loop level, obtaining correct speed-ups requires that the cost of handling each thread must remain negligible compared to the amount of work they are processing. But, this is not the case for the M6 test case. Indeed, if we analyze the time step loop, on 8 processors, this loop deals with 977 vertices each requiring 100 CPU cycles. Consequently, the management of the thread costs the equivalent of 38% of the total cost of the operations.

The speed-up of a parallel code is one criteria, but it is also of utmost importance to specify the real speed of a code. As regards flow solvers, it is difficult to compare their speed as the total time of resolution depends on several parameters such as the RK scheme (for explicit solver), the mesh quality that conditions the time step, the chosen CFL, etc. Therefore, for the same computation, high variations could appear. To specify the relative speed of the solver, we choose to provide the CPU time per vertex per iteration:

$$speed = \frac{CPU\ time}{\#\ of\ vertices \times \#\ of\ iterations}.$$

For the test cases on the Mac Xserve, the serial speed of the solver varies between 3.4 and 3.87 microseconds (μs) while a speed between 0.47 μs (for the spike) and 0.76 μs (for the M6) is obtained on 8 processors. To give an idea, a speed of 0.5 μs is equivalent to performing one iteration in half a second for a one million vertices mesh. In conclusion, the faster a routine, the harder it is to parallelize it with a satisfactory speed-up.

The flow solver has also been run on a 128 processors SGI Altix computer at the Barcelona Supercomputing Center to make a preliminary analysis on using a large number of processors. Such a machine uses a ccNUMA architecture suffering from high memory latency. In this context, minimizing main

Table 6. Speed-ups of the flow solver on a 128 processors SGI Altix computer.

# of proc	1	2	4	8	16	32	64	100
Speed-up	1.000	1.954	3.235	6.078	9.815	17.258	26.649	36.539

Table 7. Speed-ups of the error estimate code as compared to the serial version for all the test cases from 2 to 8 processors. Upper part, speed-up with respect to the whole CPU time. Lower part, speed-up of the parallelized part.

Cases		M6	IRT	City	Falcon	SSBJ	Spike
Total CPU in sec.	1 Proc	0.226	3.365	29.315	101.14	252.76	433.91
	2 Proc	1.44	1.58	1.47	1.42	1.53	1.33
Speed-up	4 Proc	1.77	2.19	1.76	1.71	2.10	2.01
	8 Proc	1.85	2.69	2.19	2.05	2.66	2.61
Gradation CPU in sec.	1 Proc	0.131	2.417	18.414	59.79	186.47	299.45
	2 Proc	1.87	1.89	1.88	1.90	1.92	1.89
Speed-up	4 Proc	3.19	3.55	3.44	3.53	3.62	3.47
	8 Proc	3.85	6.00	5.84	6.12	6.59	5.90

memory accesses through Hilbert SFC based renumbering, is all the more important. The considered test case is a large anisotropic adapted mesh containing almost 400 million tetrahedra. The speed-ups from 1 to 100 processors are given in Table 6. These first results are very encouraging and we are thus confident for the obtention of good speed-ups up to 128 processors in a near future.

5.3 Parallelizing the Error Estimate

The error estimate software will use the same parallelization methodology as the flow solver. This stage of the mesh adaptation platform is very inexpensive. In our experience, its cost is between 1 and 2 % of the whole CPU time. For instance, computing the metric with the estimate of [21] coupled with the mesh gradation algorithm of [2] for the spike test cases cost 7 minutes in serial (I/O included).

However, if the CPU time is carefully analyzed, we observed that the error estimate represents 3% of the CPU, the mesh gradation 69%, and the I/O plus building the database 27%. Consequently, we have only parallelized the mesh gradation algorithm as the error estimate computation is extremely inexpensive[3]. The speed-ups obtained for the whole CPU time (upper part) and for the mesh gradation phase (lower part) are given in Table 7. Concerning, the mesh gradation phase nice speed-ups are obtained for all the test cases. The impact on the whole CPU time is a speed-up between 2 and 2.6 on 8 processors.

---

[3] Its parallelization is not expected before running on 100 processors.

6 Exploiting Space Filling Curves for Efficient Out-of-Core Parallelization

6.1 A Fast Mesh Partitioning Algorithm

Mesh partitioning is one of the crucial task for distributed-memory parallelism. A critical issue is the minimization of the inter-processors communications, *i.e.,* the interfaces between partitions, while keeping well-balanced partitions. Indeed, these communications represent the main over-head of the parallel code. This problem is generally solved using graph connectivity or geometric properties to represent the topology of the mesh. These methods have now attained a good level of maturity, see ParMETIS [16].

However, in our context of anisotropic parallel mesh adaptation, the goal is different. The mesh partitioning is considered for purely distributed tasks without any communication, but for meshing or local remeshing purposes each partition must be connected. This requirement is not taken into account by classical partitioner. Therefore, we aim at designing the fastest possible algorithm that provides well-balanced connected partitions. We choose a Hilbert SFC based mesh-partitioning strategy applied to unstructured meshes, such as the ones proposed in [1, 29] which have been applied to Cartesian grids. This strategy is then improved to handle highly anisotropic unstructured meshes.

A Hilbert SFC based mesh partitioning. As the Hilbert SFC provides a one-dimensional ordering of the considered three-dimensional mesh, partitioning the mesh is simply equivalent to partitioning a segment. Given a 3D input mesh, the algorithm to create k subdomains is threefold:

1. get the index on the Hilbert SFC of each vertex/tetrahedron gravity center as stated in Section 4.1. An implicit Hilbert SFC is built
2. sort the vertices/tetrahedra list to get a new ordered list of elements
3. subdivide uniformly the sorted list of elements, *i.e.,* the Hilbert SFC, into k sublists. The sublists are the subdomains.

This algorithm is extremely fast and consumes little memory. The partitioning is equivalent to a renumbering plus a sort.

If this algorithm works perfectly for structured grids or uniform meshes, it needs, as standard partitioner, to be corrected when dealing with anisotropic meshes to obtain connected partitions. Indeed, two consecutive elements on the Hilbert SFC are close in the domain but they may be linked by a single vertex or edge, see Figure 5, and not by a face. In this case, the resulting subdomains are non-connected. A correction phase is then mandatory to ensure that each partition is connected. This is done in two steps. The first step consists in detecting each connected component of each subdomain thanks to a coloring algorithm [14]. A first non-colored tetrahedron initializes a list. Then, the first tetrahedron of the list is colored with the connected component index and removed from the list, and its non-colored neighbors (tetrahedra adjacent by a face) are added to the list. The process is repeated

Fig. 5. Case where two consecutive elements in Hilbert SFC based numbering are linked by a vertex leading to the creation of a non-connected partition.

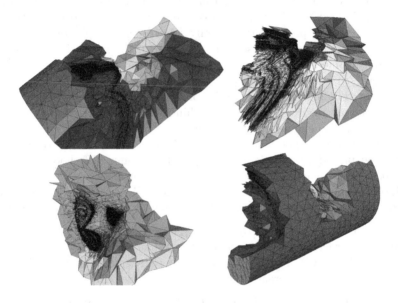

Fig. 6. Some partitions of a complex anisotropic adapted mesh.

until the list is empty. This algorithm requires as supplementary data structure the neighbors tetrahedra table. If each subdomain is composed of only one connected component then no correction is needed. Otherwise, in the second step, elements are reattributed to neighboring partitions in order to ensure that all partitions are connected. Figures 6 shows some partitions of a complex anisotropic adapted mesh.

Domain gathering. Domain gathering is done on the fly by reading each partition one after another and by updating a global table of vertices lying on interfaces. Consequently, only all the interface meshes are loaded in memory and no partition elements need to be kept in memory. The partition meshes are read and written element by element on the fly. Therefore, the whole mesh is never allocated. It results that this operation can gather a large number of partitions while requiring little memory. The key point is the algorithm to recover the one-to-one mapping between two interfaces. It consists in using a wave front approach based on the topology of the interface meshes to map

Table 8. Number of tetrahedra per partition for the spike test case.

1	2	3	4	5	6	7	8
6 005 604	6 006 014	6 006 041	6 005 838	6 005 737	6 005 026	6 005 757	6 005 783

Table 9. Statistics for partitioning into 8 blocks. CPU times is in seconds and memory is in MB.

Case	I/O CPU	Partition CPU	Total CPU	Max memory	Size variation
M6	0.10	0.09	0.191	5	7.68%
IRT	0.86	1.25	2.114	47	3.17%
City	10.89	12.05	22.94	469	0.04%
Falcon	50.47	41.68	92.15	1396	0.22%
SSBJ	98.43	89.35	187.78	2947	0.04%
Spike	120.22	168.35	288.57	5608	0.02%

one after another each vertex [19]. This algorithm is purely topologic. It is thus exact and not subjected to any floating point precision.

Numerical experiments. Let us analyze the spike test case for 8 partitions. The time to create the 8 connected partitions and to write the corresponding meshes is about 288s. The maximal memory allocated in this case is 5.6GB. The partitions are well-balanced, indeed the size in balance between partitions is no more than 0.02% as summarized by Table 8. We now give the detailed CPU time for each phase of the domain decomposition algorithm:

- Reading input data: 69s
- Create an initial Hilbert SFC based partition: 91s
- Create neighboring structure: 24s
- Correct partitions: 52.2s
- Writing output data: 50s.

As regards the partitions gathering, the algorithm consumes very low amount of memory as only the interfaces of the meshes are stored. For this example, the complete gathering step requires 42s and the maximal allocated memory is 156MB.

Results obtained for all the test cases are summarized in Table 9. All the CPU times are in seconds and the maximum allocated memory is in MB. An excellent load balancing is obtained for all the cases, except for the two smallest ones.

6.2   Parallelizing the Local Adaptive Remesher

The strategy to parallelize the meshing part is a out-of-core parallelization that uses the adaptive mesh generator as a black box. This method is completely distributed without any communications, thus it can be applied to shared-memory or heterogeneous architectures. The advantage of this method is its simplicity and its flexibility, as any adaptive mesh generators can be used. Here, the local remesher of [12] is utilized. The drawback is an I/O and build database over-cost.

In the context of parallel anisotropic mesh adaptation, the objectives are different from the solver ones. Apart from the traditional scaling of the parallel algorithm, the interest is in the possibility of improving the serial remeshing algorithm by:

- reducing the cache misses for efficiency
- reducing the scale factors for robustness purposes
- improving the local quadratic search algorithms that could occur in Delaunay-based mesh generators.

Previous points are necessary to foresee the generation of highly anisotropic meshes with dozens of million of elements.

The main difference between different parallelizations of local remeshing algorithms resides in how the partitions interfaces are handled. In some parallel adaptation implementations connectivity changes are performed in the interior of the partition and migration is used to make border regions interior [4, 10, 11, 18, 25]. In [27], tetrahedron connectivity changes are performed on purely-local and border tetrahedra in separate operations without migration. This difficulty generally comes from the use of a all-in-one method. Here, as each software is independent and a local remeshing strategy is employed, the necessity of remeshing partitions interfaces is not strictly necessary. Indeed, the parallel remeshing algorithm can be thought as an iterative procedure. The only constraint is then to ensure that from one step to another the boundaries of interfaces change to adapt them. Consequently, reducing the size of the interfaces is no more the most critical issue.

On the contrary, we prefer to generate well-balanced partitions for anisotropic remeshing. Note that in the context of remeshing, well-balanced partitions does not mean having the same number of vertices or elements. Indeed, the estimate time CPU of a mesh generator depends more of the current operation: insertions, collapses or optimization. The CPU time of these operations is not always linear with the number of vertices of the input mesh. We did not propose yet any improvements to deal with these non linearities.

The proposed method is a divide and conquer strategy using the mesh partitioner of Section 6.1 which is given by the following iterative procedure:

1. Split the initial mesh: each partition is renumbered using Hilbert SFC based strategy
2. Adapt each partition in parallel with only vertex insertion, collapses, swaps
3. Merge new adapted partitions and split the new adapted mesh with random interfaces: each partition is renumbered using Hilbert SFC based strategy
4. Optimize each partition in parallel with swaps and vertices movement

5. Merge new adapted partitions
6. return to 1.

Generally, two iterations are performed. Using this technique makes the anisotropic remeshing time satisfactory as compared to the flow solver CPU time in the adaptive loop. However, it is very difficult to quantify the CPU time of the meshing part as it depends on a large number of parameters, for instance, do we coarse the mesh, optimize it or insert a lot of vertices, etc.

Our experience on a large number of simulations with dozens million of elements shows that managing efficiently the cache misses leads to acceleration between 2 and 10 in serial. As regards the out-of-core parallelization, after adequate renumbering, satisfactory speed-ups are obtained. The speed-ups for the City and SSBJ test case are given in Table 10. These speed-ups are coherent as the partitions are balanced with respect to their size and do not take into account the future work of the mesh generator. Sometimes, the remeshing of one of the partitions is twice more costly than the remeshing of any of the other ones. This degrades considerably the speed-up. It can be improved by increasing the number of partitions for a fixed number of processors. For instance, for the SSBJ test case on 8 processors and 32 partitions the speed-up increases to 6.

Overall, this strategy combining cache miss management and out-of-core parallelization can provide speed-ups up to 40 on 8 processors (the speed-up may even be greater than the number of processors) as compared to the original code alone. But large fluctuations in the obtained speed-ups are observed and are highly dependent on the considered case.

Table 10. Speed-ups of the local remesher as compared to the serial version.

Cases	1 Proc	2 Proc	4 Proc	8 Proc
City	1.00	1.56	2.43	2.61
SSBJ	1.00	1.36	2.37	4.50

6.3 Parallelizing the Solution Interpolation

After the generation of the new adapted mesh, the solution interpolation stage consists in transferring the previous solution fields obtained on the background mesh onto the new mesh. This stage is also very fast if cache misses are carefully managed thanks to the Hilbert SFC based renumbering. For instance, the solution fields of the spike test case are interpolated in 107 seconds. Detailed CPU times are 47s for the I/Os and sort, 15s for building the database and 45s for the interpolation method. We notice that the I/Os and building the database are taking more than 50% of the CPU time. Thus, the expected speed-ups for a parallel version are limited.

The algorithm to efficiently parallelize the interpolation method of [6] with the pthreads paradigm is equivalent to partition the domain. But, partitioning is slower than the interpolation. This way has thus not been chosen. Nevertheless, this stage can be parallelized in the context of mesh adaptation with a distributed out-of-core strategy. The clue point is that the new mesh has already been partitioned and renumbered for mesh adaptation. Therefore, before merging all partitions, the interpolation can be applied in parallel to each new adapted partition separately. The over-cost of partitioning and gathering the mesh is already included in the mesh adaptation loop. Otherwise, it will be faster to run in serial. However, the expected speed-ups are limited by the I/Os and building the database associated with the background mesh which is not partitioned.

This method has been applied to all the test cases. Each pair mesh-solution of Section 3 are interpolated on a new (different) mesh of almost the same size, *i.e.*, a size variation of less than 10%. The CPU time in seconds for each case in serial is given Table 11. In parallel, no gain is observed for the smallest cases: the M6 and the IRT. For larger test cases, speed-ups between 1.3 and 2 are obtained on 2 processors and they are moderately higher with 4 processors. Unfortunately, CPU time degrades for 8 processors. This is mainly due to the fact that I/Os degrade because eight process run concurrently on the same computer while requesting access to the disk at the same time. Fortunately, this effect diminishes (or cancels) during an adaptive computations as the interpolation on each partition immediately follows the mesh adaptation. Indeed, the mesh adaptation of each partition finishes at different time.

Table 11. CPU times in seconds to interpolate the solution fields in serial.

Cases	M6	IRT	City	Falcon	SSBJ	Spike
CPU time in sec.	0.081	0.88	14.69	48.45	56.51	107.48

7   Conclusion

In this paper, we have presented a first step in the parallelization of the mesh adaptation platform. It has been demonstrated that the use of the Hilbert SFC authorizes a cheap and easy parallelization of each stage of the mesh adaptation platform. The parallelization can be shared-memory multithreaded or out-of-core. The Hilbert SFC is the core of the renumbering strategy and the mesh partitioner. It also importantly reduces the code cache misses leading to important gain in CPU time. As already mentioned in [31], many code options that are essential for realistic simulations are not easy to parallelize on distributed memory architecture, notably local remeshing, repeated h-refinement, some preconditioners, etc. We think that this strategy can provide an answer even if it is not the optimal one.

The weaknesses of this approach are I/Os and build database over-cost, especially on the fastest stages as the error estimate or the interpolation. The I/Os time is incompressible, it depends on the hardware. Indeed, solutions exist, like fast RAIDs. Improving the building database part require to parallelize complex algorithm such as hash table. The other point of paramount importance for the proposed shared-memory multi-threaded parallelization is the cost of locking/unlocking thread which can be prohibitive for a loop with a little amount of work.

In spite of that the proposed parallel adaptive methodology provides a kind of "high performance computing" on nowadays multi-core personal computers by reducing the complexity of the problem by several orders of magnitude.

Several improvements of the proposed approach are still in progress.

Regarding the shared-memory parallelization, the scheduler has to be parallelized to keep its cost constant whatever the number of processors and, at the loop level, some algorithm can be enhanced to improve their scalability. The out-of-core parallelization can be enhanced by parallelizing the mesh partitioner and by deriving a fine load-balancing that takes into account the future work on each partition of the local remesher thanks to the metric specification. And finally, for fast codes, the error estimate and the interpolation, we will have to tackle the problem of parallelization of database construction which involves hash tables.

References

1. Aftosmis, M., Berger, M., Murman, S.: Applications of space-filling curves to cartesian methods for CFD. AIAA Paper 2004-1232 (2004)
2. Alauzet, F.: Size gradation control of anisotropic meshes. Finite Elem. Anal. Des. (2009) doi:10.1016/j.finel.2009.06.028
3. Alauzet, F., Frey, P., George, P.-L., Mohammadi, B.: 3D transient fixed point mesh adaptation for time-dependent problems: Application to CFD simulations. J. Comp. Phys. 222, 592–623 (2007)
4. Alauzet, F., Li, X., Seol, E.S., Shephard, M.: Parallel anisotropic 3D mesh adaptation by mesh modification. Eng. w. Comp. 21(3), 247–258 (2006)
5. Alauzet, F., Loseille, A.: High order sonic boom modeling by adaptive methods. RR-6845, INRIA (February 2009)
6. Alauzet, F., Mehrenberger, M.: P1-conservative solution interpolation on unstructured triangular meshes. RR-6804, INRIA (January 2009)
7. Alleaume, A., Francez, L., Loriot, M., Maman, N.: Large out-of-core tetrahedral meshing. In: Proceedings of the 16th International Meshing Roundtable, pp. 461–476 (2007)
8. Behrens, J., Zimmermann, J.: Parallelizing an unstructured grid generator with a space-filling curve approach. In: Bode, A., Ludwig, T., Karl, W.C., Wismüller, R. (eds.) Euro-Par 2000. LNCS, vol. 1900, pp. 815–823. Springer, Heidelberg (2000)
9. Cantor, G.: über unendliche, lineare punktmannigfaltigkeiten 5. Mathematische Annalen 21, 545–586 (1883)

10. Cavallo, P., Sinha, N., Feldman, G.: Parallel unstructured mesh adaptation method for moving body applications. AIAA Journal 43(9), 1937–1945 (2005)
11. DeCougny, H.L., Shephard, M.: Parallel refinement and coarsening of tetrahedral meshes. Journal for Numerical Methods in Engineering 46(7), 1101–1125 (1999)
12. Dobrzynski, C., Frey, P.J.: Anisotropic Delaunay mesh adaptation for unsteady simulations. In: Proceedings of the 17th International Meshing Roundtable, pp. 177–194. Springer, Heidelberg (2008)
13. Frey, P.J., Alauzet, F.: Anisotropic mesh adaptation for CFD computations. Comput. Methods Appl. Mech. Engrg. 194(48-49), 5068–5082 (2005)
14. Frey, P., George, P.-L.: Mesh generation. Application to finite elements, 2nd edn. ISTE Ltd and John Wiley & Sons, Chichester (2008)
15. Hilbert, D.: über die stetige abbildung einer linie auf ein flächenstück. Mathematische Annalen 38, 459–460 (1891)
16. Karypis, G., Kumar, V.: A fast and high quality multilevel scheme for partitioning irregular graphs. SIAM Journal on Scientific Computing 20(1), 359–392 (1998)
17. Larwood, B.G., Weatherill, N.P., Hassan, O., Morgan, K.: Domain decomposition approach for parallel unstructured mesh generation. Int. J. Numer. Meth. Engng. 58(2), 177–188 (2003)
18. Lepage, C., Habashi, W.: Parallel unstructured mesh adaptation on distributed-memory systems. AIAA Paper 2004-2532 (2004)
19. Löhner, R.: Applied CFD techniques. An introduction based on finite element methods. John Wiley & Sons, Ltd., New York (2001)
20. Löhner, R.: A parallel advancing front grid generation scheme. Int. J. Numer. Meth. Engng 51, 663–678 (2001)
21. Loseille, A., Dervieux, A., Frey, P., Alauzet, F.: Achievement of global second-order mesh convergence for discontinuous flows with adapted unstructured meshes. AIAA paper 2007-4186 (2007)
22. Mandelbrot, B.B.: The Fractal Geometry of Nature. W.H. Freedman and Co., New York (1982)
23. Marcum, D.: Iterative partitioning for parallel mesh generation. In: Tetrahedron Workshop, vol. 2 (2007)
24. Marechal, L.: The LP2 library. A parallelization framework for numerical simulation. Technical Note, INRIA (2009)
25. Mesri, Y., Zerguine, W., Digonnet, H., Silva, L., Coupez, T.: Dynamic parallel adaption for three dimensional unstructured meshes: Application to interface tracking. In: Proceedings of the 17th International Meshing Roundtable, pp. 195–212. Springer, Heidelberg (2008)
26. Niedermeier, R., Reinhardt, K., Sanders, P.: Towards optimal locality in mesh-indexings. Discrete Applied Mathematics 7, 211–237 (2002)
27. Park, M., Darmofal, D.: Parallel anisotropic tetrahedral adaptation. AIAA Paper 2008-0917 (2008)
28. Peano, G.: Sur une courbe, qui remplit toute une aire plane. Mathematische Annalen 36, 157–160 (1890)
29. Pilkington, J., Baden, S.: Dynamic partitioning of non-uniform structured workloads with spacefilling curves. IEEE Transactions on Parallel and Distributed Systems 117(3), 288–300 (1996)
30. Sagan, H.: Space-Filling Curves. Springer, New York (1994)

31. Sharov, D., Luo, H., Baum, J., Löhner, R.: Implementation of unstructured grid GMRES+LU-SGS method on shared-memory, cache-based parallel computers. AIAA Paper 2000-0927 (2000)
32. Shontz, S., Knupp, P.: The effect of vertex reordering on 2D local mesh optimization efficiency. In: Proceedings of the 17th International Meshing Roundtable, pp. 107–124. Springer, Heidelberg (2008)

Mesh Insertion of Hybrid Meshes

Mohamed S. Ebeida[1], Eric Mestreau[2], Yongjie Zhang[1], and Saikat Dey[2]

[1] Department of Mechanical Engineering, Carnegie Mellon University,
Pittsburgh, PA, USA
msebeida@andrew.cmu.edu, jessicaz@andrew.cmu.edu

[2] Code 7130, Physical Acoustics Branch, Naval Research Lab,
Washington, DC, USA
eric.mestreau.ctr@nrl.navy.mil, saikat.dey.ctr.in@nrl.navy.mil

Abstract. A mesh insertion method is presented to merge a tool mesh into a target mesh. All the entities of the tool mesh are preserved in the output mesh while some of the entities of the target mesh are modified or eliminated in order to obtain a topologically conforming mesh. The algorithm can handle non-manifold surfaces formed of quadrilaterals and/or triangles as well as volumetric meshes based on hexahedra, prisms, pyramids and/or tetrahedra. Lower order elements such as beams can also be taken into consideration. A robust 2-steps advancing front algorithm is introduced to fill the narrow gap between the two mesh objects to obtain a complete crack-free connection. An efficient mesh data structure is developed to optimize the search operations and the intersection tests needed by the algorithm. Several application examples are provided to show the strength of the presented algorithm.

Keywords: Hybrid meshes, mesh data structure, advancing front methods, mesh insertion.

1 Introduction

Many engineering applications require two or more materials interacting with each other, for example, multiphase flows, fluid-structure interaction, and structural analysis of complex objects. Mesh generation of sophisticated models is a time consuming process. For large models, meshes are sometimes generated in independent pieces. The user may also have to deal with legacy models for which only discretized parts exist. In order to obtain a connected mesh appropriate for numerical simulations, it is sometimes needed to imprint the tool mesh into the target mesh, hence the need for such algorithm described in this paper. One can consider the case of a piece of equipment (tool mesh) that needs to be inserted into a large ship model. The equipment model is provided by its manufacturer as a discretized model. In order to numerically analyze the equipment connected to the infrastructure, a mesh insertion procedure is required.

After removing the undesired entities of the target mesh, connecting the tool mesh to the remaining part of the target mesh can be achieved using two possible approaches: advancing front methods [1, 2, 3, 4] and Constrained Delaunay Tetrahedralizations (CDT) [5, 6, 7]. The advancing front approach starts with a given surface "initial front". Elements are then created on the front toward the interior, preserving the domain boundaries. However, the reliability of many commercial advancing-front mesh generators is still under investigation. In some cases the algorithm fails and asks the user to modify the surface mesh without providing adequate reasons. This issue is more prevalent when the void to be filled has sharp features or contains narrow regions. Unfortunately the void obtained during the mesh insertion algorithm is narrow and has many sharp features, even if the involved mesh models have smooth boundaries. The advancing front method has some additional limitations: slow computational speed due to geometric search during the process and the relatively low quality of resulting meshes.

Delaunay tetrahedralizations methods [8, 9, 10, 11, 12] utilize the idea of an empty sphere for each created tetrahedron and hence generate elements with optimal quality for a given set of vertices. However, Delaunay tetrahedralization always generates a convex mesh independent of the modeled domain. In order to solve this problem, CDT are utilized to generate a mesh that respects the boundaries of the modeled domain. CDT algorithms, also have drawbacks: more sensitivity to numerical error than most geometric algorithms, and the connectivity of the input surface cannot be easily preserved. Hence, CDT algorithms cannot be implemented efficiently for 3D complex domains, especially if this domain contains narrow regions and sharp features.

In an attempt to avoid the limitations of both methods, a combination of the Delaunay and the advancing front approaches is commonly used [13, 14]. This algorithm starts with a Delaunay triangulation of a set of boundary nodes, which is used as a background mesh. New nodes are then added using the advancing front approach. This combined approach can increase the efficiency of the algorithm and produce high quality meshes. However, the surface recovery in 3D is often the weakest point.

In addition to the limitations mentioned above for both approaches, each of them is designed to handle tetrahedral meshes only. Owen and Saigal [15] proposed an advancing front algorithm that generates all-hex meshes but it starts with an initial tetrahedral mesh that respects the boundaries of the domain. Staten et al. [16] developed another advancing front algorithm to handle all-quad initial front, however this method is limited to simple domains. Unfortunately the void between the tool and the target meshes during a mesh insertion algorithm is narrow and might be surrounded by hybrid surfaces with sharp features. Recently, Ito et al. [17] presented an interesting method to accommodate small devices into a baseline mesh. However, this method preserves only the geometry of these devices and hence it is not suitable for some applications such as fluid-structure interaction.

In this paper, a new algorithm is presented, which converts two overlapping hybrid meshes into a conforming connected crack-free mesh. This methodology locally modifies the entities of one of the input meshes (Target Mesh) so that the two meshes can be merged into a conforming mesh across the interface. The algorithm utilizes an efficient, hybrid advancing front method to fill the void between the two meshes in a countable number of operations. In order to increase the efficiency of this algorithm, an optimal data structure is introduced and utilized in the query operations required during the advancing front procedure.

The remaining of this paper is organized as follows: In section 2, an optimal non-manifold hybrid mesh data structure (NHMD) is presented. The mesh insertion algorithm is then described in Section 3. In Section 4, several examples of mesh insertion are presented. Finally, Section 5 provides some concluding remarks along with current and future efforts.

2   An Optimal Non-manifold Hybrid Mesh Data Structure (NHMD)

In this section we present an optimal data structure for handling hybrid unstructured mesh models that might contain non-manifold entities. The requirements in choosing a data structure for the implementation of the mesh insertion algorithm are:

1. It should be able to handle hybrid mesh models with or without non-manifold surfaces;
2. It must occupy the least possible amount of storage; and
3. The query operations corresponding to any mesh entity (node, edge, face, element) should be executed in a constant time independent of the mesh size.

Mesh models contain a finite number of element types: lines, triangles, quadrilaterals, tetrahedra, pyramids, prisms and hexahedra. Each element is specified using its list of nodes. Figure 1 shows the local indices of the nodes associated with various types of elements in the mesh data structure. This list can be used to determine faces and edges of that element. For example the tetrahedron in Figure 1(d) is defined using the node sequence $\{1, 2, 3, 4\}$ and contains four faces given by $\{1, 2, 3\}$, $\{1, 3, 4\}$, $\{2, 1, 4\}$, and $\{3, 2, 4\}$. Each of these faces can be uniquely identified based on the index of their parent element and an additional local index for each one of them. The associated edges can be similarly identified. The developed data structure utilizes this information to optimize the required storage.

In a hybrid unstructured mesh model, the minimum amount of information to be stored is the connectivity matrix C. This matrix is sparse and a non-zero entry c_{ij} is one if the element i contains the node j. Sparse matrices are usually stored using three arrays. However, the connectivity matrix needs only two arrays (or STL vectors) since the value of all the non-zero entries

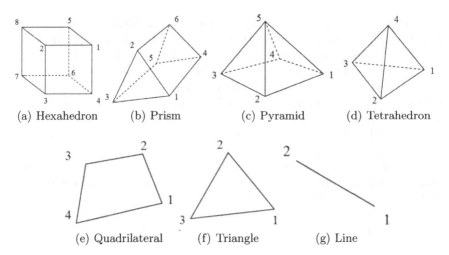

(a) Hexahedron   (b) Prism   (c) Pyramid   (d) Tetrahedron

(e) Quadrilateral   (f) Triangle   (g) Line

Fig. 1. The connectivity information associated with the main element types defined in the current implementation of the mesh data structure.

are always one. The number of non-zero entries is denoted nnz. Moreover, if the rows of this matrix are sorted based on the element type, and the number of elements in each type is stored, the entries of this matrix can be stored using one array with nnz entries. This array, C_I, together with numbers of elements in each type, represent the minimum storage size that can be used to hold the associated connectivity information of a hybrid mesh.

In order to have an efficient traverse algorithm we need to store the connectivity matrix C as well as its transpose C^T. For this purpose we store two more arrays, C_J and C_K. The size of C_J is the same as C_I while the size of C_K is the same as the number of nodes, N, in the associated mesh. Hence the total storage size for the connectivity information is $2\,nnz + N + 7$. In order to have a complete representation of a 3-dimensional mesh we store the coordinates of the associated nodes using three more arrays, each one has N entries. Note that in the mesh data structure C_I lists the nodes of the different elements within the mesh, while C_J lists the indices of the entries of C_I sorted based on the associated node index. In other words

If

$$k_i = C_J\,[i] \quad \text{and} \quad \mathrm{k}_{i+n} = C_J\,[i+n]\,,$$

then

$$C_I\,[k_i] \leq C_I\,[k_{i+n}] \quad \text{for all} \quad 0 \leq i, n \leq i + n < nnz.$$

Following this procedure, one can easily retrieve the neighboring elements around any given node. In order to optimize this process, C_K stores the minimum index of the entries in C_J associated with a given node. For example if a node i has two neighboring elements, then i would appear twice in C_I, the first location is at $C_J[C_K[i]]$ and the second location is at $C_J[C_K[i] + 1]$.

In this data structure we have seven types of elements and we chose to list them in the same order presented in Figure 1. For each element type we need to store the number of nodes, edges, and faces associated with that type, as well as the local indices of each entity associated with this element type. This information is used in the traverse algorithm to get the connectivity from an element to its bounding entities.

This sorting eliminates the need to store the element indices of the connectivity matrix C and hence we can save extra nnz entries. The element index of any entry k can be found using the following algorithm:

Algorithm 1. Retrieve the element index i_e of a given entry k in C_I

Set num_e = 0 , itype = 1 and num = Num_Elements[itype] * Num_Nodes[itype];
while $k >$ num **do**
   num_e = num_e + Num_Elements[itype];
   itype = itype + 1;
   num = num + Num_Elements[itype] * Num_Nodes[itype];
end while
num = num - Num_Elements[itype] * Num_Nodes[itype];
i_e = num_e + size_t ((k - num) / Num_Nodes[itype]);

Algorithm 1 is very efficient as it loops over a limited number of element types, currently $1 \leq itype \leq 7$. Num_Elements and Num_Nodes stores the number of elements and the number of nodes associated with each element type. Using this algorithm allows to retrieve the neighboring elements of any node in the mesh independent of the mesh topology.

In this mesh data structure, nodes play a vital role in the traverse algorithm. For example, to retrieve the neighboring faces of a given edge $\mathcal{E}$, one starts by retrieving all the neighboring elements of the two corner nodes of that edge, using C_J and C_K. Then we get the faces of these elements and collect those faces that contain $\mathcal{E}$. Note that each face (as well as each edge) is identified by a unique global index, which can be mapped easily to the index of its parent element. The complexity of this algorithm is a function of the mesh quality and is independent of the mesh size.

The mesh data structure is demonstrated using a simple mesh in Figure 2. This mesh contains seven nodes, three elements and eight edges. One of the nodes, V_4, is not associated with any element. this node is an *isolated* node. The transpose of of the associated connectivity matrix is presented in Figure 2(c). The stored arrays of the data structure corresponding to that mesh are given by:

$$\text{Num\_Elements} = [0\ 0\ 0\ 0\ 1\ 1\ 1]^T, C_I = [6\ 7\ 3\ 5\ 7\ 2\ 3\ 2\ 1]^T,$$

$$C_J = [9\ 6\ 8\ 3\ 7\ 4\ 1\ 2\ 5]^T, C_K = [1\ 2\ 4\ 6\ 6\ 7\ 8]^T$$

Note that $C_K[4] = C_K[5]$, this means that V_4 does not exist in C_I, in other words it is an isolated node. Also $C_K[4] - C_K[3] = 2$ indicates that V_3 exists twice in C_I, hence this node has two neighboring elements. The indices of these two locations are given by $C_J[C_K[3]] = 3$ and $C_J[C_K[3] + 1] = 7$. Algorithm 1 can be utilized to identify the element indices corresponding to these two locations. In order to retrieve an edge we start with its global index and retrieve its parent element. For example the edge e_6 is in the second edge of the first triangle listed in C_I. This triangle is defined using the node list (7 2 3), hence the required edge connects the nodes V_2 and V_3. Note that duplicated edges (such as e_2 and e_7) does not represent any storage problem, since we identify these edges implicitly through their parent elements. The same thing applies for duplicated faces in volumetric meshes.

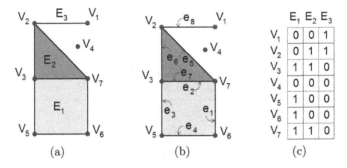

(a)   (b)   (c)

Fig. 2. A simple mesh utilized in demonstration of the mesh data structure. The associated entities are shown in (a) and (b). The transpose of the associated connectivity matrix is presented in (c).

This data structure was tested using a 2.0 GHz processor with a 2.0 GB RAM and it was capable of handling a mesh size up to 10 million nodes without using the swap space and up to 20 million nodes using the swap space. The mesh used for this test is an all-hexahedral model. The number of nodes is approximately equal to the number of elements. The time required to generate the mesh and the associated data structure is 45 seconds for the first case (10 million nodes) and 100 seconds for the second case (20 million nodes).

3 Mesh Insertion Algorithm

The mesh insertion algorithm starts by the detection, and the removal of the undesired entities of the tool mesh. These entities might be overlapping with some of the tool mesh elements or they can be located inside a closed surface associated with the tool mesh. The remaining elements of the target mesh are then connected to the tool mesh using a robust advancing front

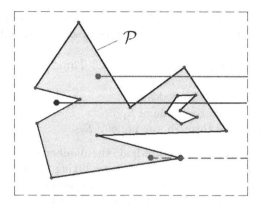

Fig. 3. Detection of nodes located inside a closed region using its bounding polynomial, $\mathcal{P}$. A (green) line connecting an internal node to the scaled bounding box has an odd number of intersections with $\mathcal{P}$, while a similar (blue) line for an external node has an even number of intersections. A case that might cause miscounting is illustrated using the (dotted green) line.

algorithm. During this step the advancing front algorithm may vary based on the type of elements of the target mesh. Two different approaches are used, the first one is used with the one-dimensional and two-dimensional elements (lines, triangles, and quadrilaterals) and the second one is useful for the three-dimensional elements (tetrahedra, pyramids, prisms and hexahedra). In the latter case, a novel advancing front algorithm is introduced to handle the narrow region between the tool and the target meshes.

3.1   Detection of the Undesired Entities of the Target Mesh

In order to mark the undesired entities of the target mesh, a method is implemented to detect whether a given node is located inside a closed triangular surface or not. The implemented algorithm starts by constructing a bounding box of $\mathcal{S}$. This bounding box is then scaled around its center with some factor $f > 1.0$. A line segment, $\mathcal{L}$, is then constructed by projecting the input node to one of the sides of that box. In our implementation we work with a projection in the direction of positive x-axis. Finally we count the number of intersections that $\mathcal{L}$ have with $\mathcal{S}$. If the number is odd, then this node is inside $\mathcal{S}$, otherwise it is not. The two-dimensional version of this idea is demonstrated in Figure 3. To count this number, we loop over all the elements of $\mathcal{S}$ and test for the intersection with $\mathcal{L}$. Using a line aligned with the x-axis simplifies the operations required for this test to a large extent. However, three special cases has to be handled correctly to have a robust algorithm:

1. $\mathcal{L}$ intersects an element of $\mathcal{S}$ at one of its three edges. The intersection point will be counted twice, since each edge in $\mathcal{S}$ is shared by two elements.

2. $\mathcal{L}$ intersects an element of $\mathcal{S}$ at one of its three nodes. The intersection point will be counted n times, where n represents the number of neighbor elements to that node.
3. $\mathcal{L}$ is tangent to a given element, $\mathcal{E}$, in $\mathcal{S}$. Three cases can occur in this situation:
   a) $\mathcal{L}$ passes through two nodes of $\mathcal{E}$.
   b) $\mathcal{L}$ passes through one node of $\mathcal{E}$.
   c) $\mathcal{L}$ does not pass through any node of $\mathcal{E}$.

In the implementation of this method, the number of intersections is adjusted to handle all these situations. After splitting the nodes of the target mesh into internal and external nodes using this method, one can easily eliminate the undesired elements from the target mesh. Figure 4 shows the output of this process using a submarine (triangular tool mesh) inserted in an ocean (hexahedral target mesh). Another example is presented for a mesh insertion of a cylinder (hexahedral tool mesh) in a non-manifold surface (triangular mesh. This example is demonstrated in Figure 5.

This detection method deals only with a triangular closed surfaces. For a volumetric mesh, the outer boundaries, $\mathcal{B}$, are extracted and all the quadrilaterals on $\mathcal{B}$ are split into triangles. In some cases, we might want to eliminate

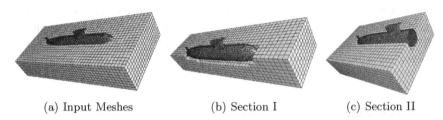

(a) Input Meshes   (b) Section I   (c) Section II

Fig. 4. Eliminating the undesired entities from the target mesh during the mesh insertion of a submarine (triangular tool mesh) in an ocean (hexahedral target mesh). Two cross-sections are utilized to show the interior of the target mesh.

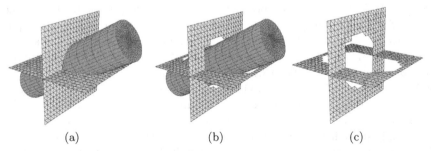

(a)   (b)   (c)

Fig. 5. Eliminating the undesired entities from the target mesh during the mesh insertion of a cylinder (hexahedral tool mesh) in a non-manifold surface (triangular target mesh). Two views are utilized to show the remaining part of the target mesh with and without the cylinder.

an extra layer of elements to remove the issues linked to narrow gaps when dealing with the advancing front method. Upon completion of this step, all the new boundary entities of the target mesh are marked. These entities will be connected to the surface of the tool mesh.

3.2   Connecting One-Dimensional and Two-Dimensional Entities to the Tool Mesh

A simple yet effective advancing front algorithm is now presented to connect the one-dimensional and two-dimensional marked entities of the target mesh to the surface of the tool mesh. For one-dimensional elements (Lines), the associated marked nodes are simply connected to the closest nodes on the surface of the tool mesh. Figure 6 shows the results of this step using a mesh insertion of a cylinder(hexahedral tool mesh) in a network of orthogonal lines (target mesh).

For two-dimensional entities, a watertight surface is introduced to connect the remaining part of the target mesh, $\mathcal{M}_t$, to the tool mesh $\mathcal{M}_T$. The proposed algorithm should handle non-manifold cases. A *terminal* mesh entity is an entity lying on the interface between the eliminated and the remaining parts of $\mathcal{M}_t$. A node, $\mathcal{N}^*$, that exists in more than two terminal edges is denoted as a *non-manifold node*. A *non-manifold edge* is an edge with one non-manifold node. A watertight triangular mesh is created to connect each terminal edges to the surface of the tool mesh. This triangular mesh is constructed in two steps. First, triangular elements are created using each terminal edge and its closest node $\mathcal{N}_j$ on $\mathcal{S}_T$. If the terminal edge is a non-manifold one, $\mathcal{N}_j$ is the closest node on $\mathcal{S}_T$ to the non-manifold node of that edge. Otherwise, $\mathcal{N}_j$ is the closest node on $\mathcal{S}_T$ to the center of that edge. Then for each pair of neighbor triangles that meet at a node $\mathcal{N}_i$ in $\mathcal{M}_t$ and have two points $\mathcal{N}_j$, $\mathcal{N}_k$ on $\mathcal{S}_T$, the shortest path, connecting $\mathcal{N}_j$ to $\mathcal{N}_k$ along the edges of $\mathcal{S}_T$, is extracted. A triangle is then constructed for each edge in that

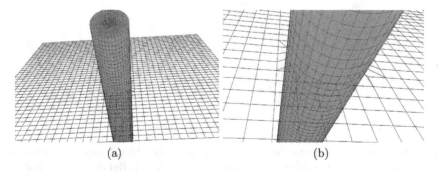

(a)   (b)

Fig. 6. Mesh insertion of a cylinder (hexahedral tool mesh) in a network of line (beam) elements (target mesh). Two views are utilized to show of the final conforming mesh.

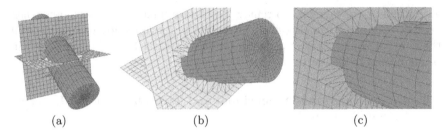

Fig. 7. Three differnet views for the output of the mesh insertion algorithm using a cylinder (tool mesh) and two perpendicular quadrilateral planes (target mesh with a non-manifold surface).

path using that edge and $\mathcal{N}_i$. Figure 7 shows the output of this algorithm for a mesh insertion of a hexahedral cylinder (tool mesh) and two perpendicular quadrilateral planes (target mesh with a non-manifold surface).

3.3   Connecting Three-Dimensional Entities to the Tool Mesh

Connection of three dimensional entities of the target mesh to the tool mesh turns out to be a challenging problem for many reasons: First, the three-dimensional void, $\mathcal{V}$, entrapped between both meshes is narrow and contains many sharp features even if the two input meshes have smooth surfaces. These two properties represent a real challenge for any advancing front algorithm. Figure 8 demonstrates the surfaces surrounding this void during a mesh insertion of a cylinder (Tool Mesh) in a box (Target Mesh). Eliminating the undesired entities of the target mesh modifies its boundary. The introduced part of this boundary contains many sharp features. Moreover, the surface surrounding the void between the two meshes might be hybrid.

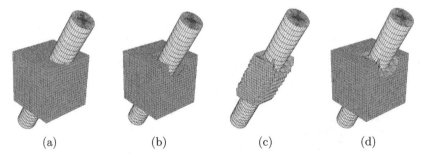

(a)   (b)   (c)   (d)

Fig. 8. Mesh insertion of a hexahedral cylinder (Tool Mesh) with a hexahedral box (Target Mesh). The input meshes are presented in (a). Removal of the desired entities is shown in (b). A rough surface surrounds the narrow void between the two meshes is shown in (c) and (d).

To eliminate the sharp features from the surface surrounding $\mathcal{V}$, an offset copy, $\mathcal{S}'$, of the boundary surface, $\mathcal{S}$, of the tool mesh is generated and then extracted using the elements of the target mesh. Hence the void between the tool mesh and the target mesh is trapped between the extracted surface, $\mathcal{S}''$ and $\mathcal{S}$. Note that these two surfaces are almost parallel, so the advancing front algorithm should be easier to construct with the required guarantee for the execution time and it should produce elements with much better quality.

The algorithm used to create $\mathcal{S}'$ is to loop over the nodes of $\mathcal{S}$, calculate the average normal, $\mathbf{n}_{av}$, of its neighboring faces and the minimum length, d_m, of its neighboring edges. First $\mathcal{S}'$ is created as an identical copy of $\mathcal{S}$ then each node is duplicated in the direction of its normal vector with a distance of 0.5 d_m. In the case of self intersection, this distance is recursively reduced by 50% till this problem is resolved. In some cases associated with sharp corners, a node is displaced to the interior of $\mathcal{S}$. This issue is solved by projecting the displaced point back to the surface of $\mathcal{S}$. Then that projection is extended until it intersects again with $\mathcal{S}$ and place the new node in the middle of these two intersections. If a second intersection does not exist, the projection line is extended by 20% and the new node is placed at its free end. Figure 9 illustrates the output of this process using a triangular mesh of a submarine model.

The final step before applying the advancing front algorithm is to extract the the offset surface, $\mathcal{S}'$, using the elements of the target mesh. This is

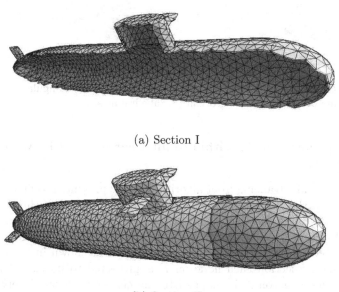

(a) Section I

(b) Section II

Fig. 9. Generating an offset copy (yellow) of the surface of the tool mesh (blue). Two sections of the ouput mesh are utilized.

achieved by projecting the terminal nodes to the surface onto S'. The projection direction is based on the average normal direction of the neighbor terminal faces of the projected node. Figure 10 illustrates the output of this process using a triangular mesh of a submarine and a hexahedral mesh of the surrounding water.

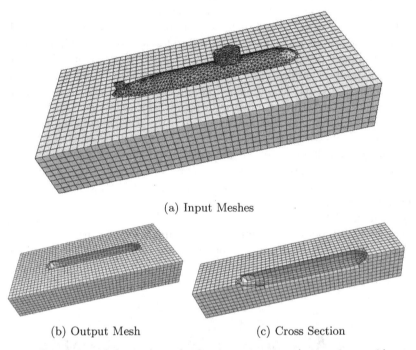

(a) Input Meshes

(b) Output Mesh   (c) Cross Section

Fig. 10. Extraction of the surface of a floating submarine (triangular mesh) using the elements of the surrounding ocean (hexahedral mesh).

An advancing front algorithm to fill a narrow void between two parallel surfaces

A robust hybrid advancing front algorithm is now presented to fill the narrow void between the two parallel surfaces S and S''. This algorithm is demonstrated using mesh insertion of a sphere (tetrahedral tool mesh) in a box (hexahedral target mesh) shown in Figure 11. First the boundaries, Γ, of S'' are projected to the surface, S, of the tool mesh as shown in Figure 12(a). During this projection each node on Γ is projected to its closest node on S. Then each edge on Γ is mapped to the closest path on S between the projection of its two end nodes. Hence, each edge on the boundaries of S'' is mapped to either a node, an edge or a sequence of edges on S. The mapped polyline, $\mathcal{P}$ is closed and formed by the edges of S. All the edges in $\mathcal{P}$ have the same direction. This can be used to extract the desired part of S as shown in

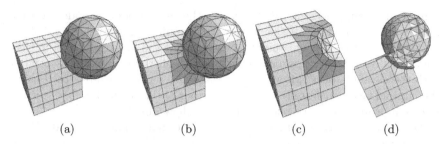

Fig. 11. Mesh insertion of a tetrahedral sphere (Tool Mesh) with a hexahedral box (Target Mesh). The input meshes are presented in (a). The output crack-free mesh is shown in (b) and (c) with and without the tool mesh. A section in (d) shows the interior elements of the final mesh.

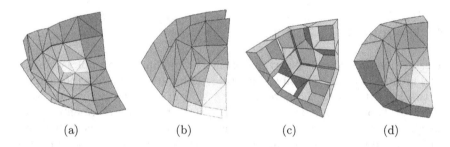

Fig. 12. Connecting the terminal entities of the hexahedral box to the surface of the tetrahedral sphere. In (a) the boundaries of S'' are projected to some edges forming a closed (blue) polyline in S. The input surfaces of the advancing front algorithm are presented in (b). Figure (c) shows the internal faces connecting S'' and S'''. The output of the advancing front algorithm is illustrated in (d).

Figure 12(b). This part is denoted S'''. At this point we have two hybrid surfaces, S'' and S''' that are almost parallel and separated by a narrow region. These two surfaces are the input of the following advancing front algorithm.

The starting point of the advancing front algorithm is to map each point on S'' to the closest nodes on S'''. Then the edges of S'' are connected to the edges and nodes on S''' using quadrilaterals and triangles. This is achieved by looping over the edges in S'' and extracting the shortest path that connects the mapping of its end nodes on S'''. If that path contains a single node, a triangle is constructed using that node and the edge under consideration. If that path is an edge on S''', a quadrilateral is constructed using the two edges. Otherwise, a series of triangles are constructed using the two edge nodes on S'' and the nodes on that path. The internal faces, connecting S'' to S''' are illustrated in Figure 12(c). After this step each face on S'' is completely surrounded by those internal faces. We loop over the faces on S'', and collect the surrounding faces as well as the faces from S'' that closes the surface. The collected faces are then checked if they can form one of our three-dimensional elements. If that is not the case, a point is added to that local void and used to fill the void with pyramids and tetrahedra. This is achieved by connecting this point to each one of the faces surrounding that void. Finally, a similar approach is followed to fill the local voids associated with the elements of S''', if any. Note that these local voids are closed using the internal faces generated earlier. The output of this algorithm for the mesh insertion problem of a tetrahedral sphere is shown in Figure 11.

Note that this advancing front algorithm does not require any kind of iterative loops. It is therefore executed after a countable number of operations. Also if the elements of S'' and S''' were closed in size, the algorithm tends to fill the voids with one of the primitive elements defined in the mesh data structure. This results in fewer elements to fill the void and hence better quality.

4   Examples

This section presents the behaviour of the method on two examples of industrial interest. The first example, illustrated in Figure 13 presents the merging of two stern tubes to the bare hull of a ship in order to analyse the stiffness and displacements of each part. The stern tubes are represented using hexahedral elements while the hull is given by a quadrilateral mesh. The second example illustrates merging a submarine shell (triangular tool mesh) in a hexahedral mesh representing the surrounding ocean. The output mesh is

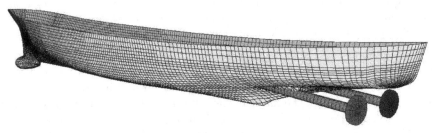

(a) A ship model after the insertion of two stern tubes

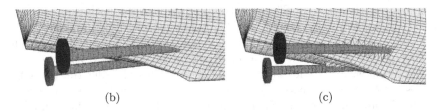

(b)   (c)

Fig. 13. Mesh insertion of two stern tubes (tool mesh with hexahedral elements) into the hull of a ship (target mesh with quadrilateral elements), and zoom-in views for both meshes before (b) and after (c) applying the mesh insertion algorithm.

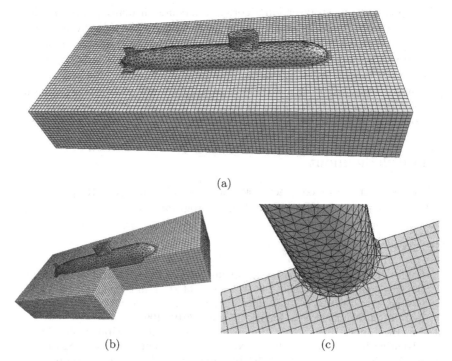

Fig. 14. Mesh insertion of a submarine shell (triangular tool mesh) into the surrounding ocean (target mesh with hexahedral elements). The interior elements are illustrated in figures (b) and (c) using two cross-sections.

illustrated in Figure 14 and can be used to study the fluid structure interaction between the hydrodynamic forces acting on the floating submarine and the tension forces generated in its shell.

5   Conclusion and Future Work

A new computational method for merging two hybrid meshes into a conforming mesh that preserves all the entities of one of them has been presented. An optimal data structure for hybrid meshes that might contain non-manifold surfaces has been developed. The optimality of this data structure is based on the storage requirements and the computational efficiency of the various query operations. We also presented a novel robust advancing front algorithm to fill the narrow void between the two meshes.

Many meshing algorithms require entire assemblies to be meshed at once in order to have conforming meshes between components [16, 18]. This requirement is relaxed by creating conforming meshes between assembly components after each component is meshed individually using the mesh insertion algorithm.

The presented mesh insertion algorithm depends in many parts on calculating the shortest path between two nodes along the edges of a given surface. The implemented method to perform this step might fail if the two nodes occur on two different sides of a narrow region. One way to fix this problem is to refine the target elements associated with this problem. However, we are currently investigating other solutions as well. We are currently working on the performance charts required to show the strength of the proposed data structure.

Acknowledgement

This work is funded by contract N0017308-C-6011 from Naval Research Laboratory as part of the HPCMP CREATE program.

References

1. Lohner, R., Parikh, P.: Generation of three-dimensional unstructured grids by the advancing front method. Int. J. Numer. Meth. Fluids 8, 1135–1149 (1998)
2. George, P.L., Seveno, E.: The advancing front mesh generation method revisited. Int. J. Numer. Meth. Engng. 37, 3605–3619 (1994)
3. Jin, H., Tanner, R.I.: Generation of unstructured tetrahedral meshes by the advancing front technique. Int. J. Numer. Meth. Engng. 36, 1805–1823 (1993)
4. Ito, Y., Shih, A., Soni, B.: Reliable isotropic tetrahedral mesh generation based on an advancing front method. In: 13th International Meshing Roundtable, pp. 95–105 (2004)
5. Chew, L.P.: Constrained Delaunay triangulations. Algorithmica 4, 97–108 (1989)
6. Shewchuk, J.R.: Constrained Delaunay tetrahedralizations and provably good boundary recovery. In: 11th International Meshing Roundtable, pp. 193–204 (2002)
7. Cohen-Steiner, D., Colin, E., Yvinec, M.: Conforming Delaunay triangulations in 3D. In: 18th Annual Symposium on Computational Geometry, pp. 199–208 (2002)
8. Dey, T.K., Bajaj, C.L., Sugihara, K.: On good triangulations in three dimensions. Int. J. Comput. Geom. & App. 2, 75–95 (1992)
9. Miller, G.L., Talmor, D., Teng, S.-H., Walkington, N.: A Delaunay based numerical method for three dimensions: generation, formulation, and partition. In: 27th Annual ACM Symposium on the Theory of Computing, Las Vegas, Nevada, pp. 683–692 (1995)
10. Shewchuk, J.R.: Tetrahedral mesh generation by Delaunay refinement. In: 14th Annual Symposium on Computational Geometry, Minneapolis, Minnesota, pp. 86–95 (1998)
11. Shewchuk, J.R.: Mesh generation for domains with small angles. In: 16th Annual Symposium on Computational Geometry, Hong Kong, pp. 1–10 (2000)
12. Edelsbrunner, H., Li, X.-Y., Miller, G., Stathopoulos, A., Talmor, D., Teng, S.-H., Ungor, A., Walkington, N.: Smoothing and cleaning up slivers. In: 32nd Annual Symposium on the Theory of Computing, Portland, Oregon, pp. 273–278 (2000)

13. Mavriplis, D.J.: An advancing front Delaunay triangulation algorithm designed for robustness. J. of Comput. Phys. 117, 90–101 (1995)
14. Marcum, D.L., Weatherill, N.P.: Unstructured grid generation using iterative point insertion and local reconnection. AIAA J. 33, 1619–1625 (1995)
15. Owen, S.J., Saigal, S.: H-Morph: an indirect approach to advancing front hex meshing. Int. J. Numer. Meth. Engng. 49, 289–312 (2000)
16. Staten, M.L., Owen, S.J., Blacker, T.D.: Unconstrained paving & plastering: a new idea for all hexahedral mesh generation. In: 14th International Meshing Roundtable, pp. 399–416 (2005)
17. Ito, Y., Murayama, M., Yamamoto, K., Shih, A.M., Soni, B.K.: Efficient computational fluid dynamics evaluation of small device locations with automatic local remeshing. AIAA Journal 47, 1270–1276 (2009)
18. Zhang, Y., Bajaj, C., Sohn, B.-S.: 3D finite element meshing from imaging data. Comput. Meth. in Appl. Mech. Engng. 194, 5083–5106 (2005)

Tensor-Guided Hex-Dominant Mesh Generation with Targeted All-Hex Regions

Ved Vyas and Kenji Shimada

Department of Mechanical Engineering
Carnegie Mellon University, Pittsburgh, PA, USA
ved@cmu.edu, shimada@cmu.edu

Abstract. In this paper, we present a method for generating hex-dominant meshes with targeted all-hex regions over closed volumes. The method begins by generating a piecewise-continuous metric tensor field over the volume. This field specifies desired anisotropy and directionality during the subsequent meshing stages. Meshing begins with field-guided tiling of individual structured hexahedral fronts wherever suitable and in regions of interest (ROI). Then, the hexahedral fronts are incorporated into an existing hex-dominant meshing procedure, resulting in a good quality hex-dominant mesh. Presently, many successful hex meshing methods require significant preprocessing and have limited control over mesh directionality and anisotropy. In light of this, hex-dominant meshes have gained traction for industry analyses. In turn, this presents the challenge of increasing the hex-to-tet ratio in hex-dominant meshes, especially in ROI specified by analysts. Here, a novel three-part strategy addresses this goal: generation of a guiding tensor field, application of topological insertion operators to tile elements and grow fronts towards the boundary, and incorporation of the fronts into a hex-dominant meshing procedure. The field directionality is generated from boundary information, which is then adjusted to specified uniform anisotropy. Carefully placed streamsurfaces of the metric field are intersected to shape new elements, and the insertion operators maintain mesh integrity while tiling new elements. Finally, the effectiveness of the proposed method is demonstrated with a non-linear, large deformation, finite element analysis.

1 Introduction

In this paper, we present a method for generating hex-dominant meshes with targeted all-hex regions over closed volumes. A three-part strategy is employed to achieve this: generation of a guiding metric field, application of topological insertion operators to insert elements and grow the hex fronts outward from starting interior points, and indirect incorporation of the fronts into a final hex-dominant mesh. The field governs element directionality and anisotropy over the volume for the tiling and final meshing stages. During the tiling stage, insertion operators maintain mesh integrity and attempt to prevent the mesh from marching into itself as they place new hexes. New elements are formed from the intersections of strategically placed stream-surfaces of the metric field. The nodes of the fronts are then incorporated into a rectangular bubble-packing process [24] as fixed bubbles, from which a tetrahedral and then final hex-dominant mesh are obtained.

The method proposed in this work possesses many salient features. First, a new method is described for constructing geometry-based tensor fields with boundary sensitivity and user-input anisotropy. These fields can be applied to any meshing algorithm which is suitably equipped. Second, both the tiling and bubble-packing steps produce valid meshes that conform to these fields. The tiling process is also flexible enough to start on the interior, and with the help of the field and boundary conformity operations, capture the boundary. Finally, the incorporation of the tiled hex fronts into the hex-dominant process allows targeting of all-hex regions in ROIs.

Hexahedral meshes are commonly preferred in many types of 3D Finite Element Analyses. However, robust all-hexahedral mesh generation has not yet been achieved with the guarantees, quality, and control of arbitrary anisotropy/directionality provided tetrahedral and hex-dominant meshing schemes.

Current practical methods for direct hexahedral mesh generation are at most semi-automatic, requiring model decomposition and application of topology specific methods such as (sub)mapping and sweeping among others [1]. Other methods, such as the grid based methods of Schneiders et al. [2] or Zhang et al. [3], are only subject to topology constraints if boundary features (for instance curves and vertices) need to be captured. Although they do provide some degree of anisotropy and size gradation using templates, there is no mechanism for controlling arbitrary directionality. Furthermore, 3D advancing front methods such as Plastering [4] start at the boundary while placing elements. Unfortunately, this leads to unmeshable voids and stitching problems between fronts. Unconstrained plastering is a promising extension that aims to avoid these difficulties, but is still under development [5].

In light of this, hexahedral-dominant meshing schemes have gained traction for semi- and fully automatic mesh generation. Rectangular bubble-packing, by Yamakawa and Shimada [24], is an indirect hex-dominant method that is capable of generating high quality hex-dominant meshes. The resulting meshes conform to a specified tensor field and have good hexahedral volume ratios. The method begins by packing rectangular solid cells and then generating a tetrahedral mesh using the cells' centers. This is followed by merging groups of tetrahedra into good hexahedral elements. Optionally, tets can be merged into prisms and possibly pyramids to produce a fully conformal mesh. That is, hanging edges which are present in a mixed hex-tet mesh will be eliminated through the use or prisms and pyramids. Alternatively, special subdivision templates have been developed to produce conformal meshes without pyramids [29].

From the inherent drawbacks and strengths of the methods mentioned here, the authors have reached the observation that a successful hex-dominant meshing strategy should include:

1. Using fields to control directionality and anisotropy.
2. Tailoring such fields to be boundary aligned and to incorporate additional directionality and anisotropic size specifications.
3. Meshing from the inside out by starting at a suitable location(s).
4. Deciding marching directions for meshing ahead of time (via the field).
5. Meshing element by element in such a way that the mesh remains topologically valid and does not march into itself.
6. Complementing difficult regions for all-hex with good quality hex-dominant regions.

Some of these points, such as the need for a boundary sensitive overlay for grid-based methods, have been noted by Blacker for all-hex meshing [6]. Furthermore, field generation has been performed for application to other methods such as BubbleMesh [7, 8, 9].

The method presented here attempts to utilize these desired characteristics through field generation and application of topological insertion operators to insert elements and grow the mesh outward from a starting interior point. The metric field incorporates boundary normal information and user-input uniform anisotropy, and is capable to providing information on element shaping and marching directions for meshing. This allows for a relatively simple tiling algorithm based on topological insertion operators. The insertion operators maintain mesh integrity while attempting to prevent the mesh from marching into itself. They also govern the shaping and via specific intersections of streamsurfaces of the eigenvector fields that are obtained from the metric field.

Due to the difficulties posed by applying this process to entire volumes, the proposed method attempts to incorporate the traced hex elements into a hex-dominant mesh. The resulting mesh is boundary conformal, approximately metric conformal, good quality, and has increased hex-to-tet ratios in the tiled regions. Therefore, predominantly hex meshes can be created in ROIs, for instance where boundary conditions are applied or important stresses/quantities are present.

Presently, the proposed method operates on closed volumes with a facet-based boundary representation equipped with sufficient guiding feature curves. It can be applied to analytic-surface based CAD models using the graphics facets and feature curve extraction. In addition, in this paper only fields of uniform anisotropy are considered. The basic approach, however, can be extended to non-uniform anisotropic fields.

The paper is organized as follows: Section 2 discusses related work that leads up to and supports the proposed method. Section 3 presents an overview of the methods used, and also covers preliminaries. Generation of metric tensor fields is described in Section 4, and meshing via hex tiling and packing is covered in Section 5. Section 6 contains results and our discussion, and the paper is concluded with a brief look at further work in Section 7.

2   Related Work

Riemannian metric tensors have become a popular means of anisotropic mesh control [7, 8, 10]. In the context of 3D mesh generation, such second-order tensors can be represented as 3×3, symmetric, positive-definite matrices in a local coordinate system. Given a metric tensor M, the dot (inner) product of two vectors $\mathbf{x}$ and $\mathbf{y}$ can be written as $\mathbf{x}^{\mathrm{T}}\mathsf{M}\mathbf{y}$ [11]. This can be used to obtain the norm of a vector under the metric via $\left(\mathbf{x}^{\mathrm{T}}\mathsf{M}\mathbf{x}\right)^{1/2}$, as well as the angle between two vectors. Additionally, the length l of a parametric curve $\mathbf{r}(t)$ where $t \in [t_0, t_1]$, can be expressed as:

$$l = \int_{t_0}^{t_1} \left[M_{ij} \frac{dr^i}{dt} \frac{dr^j}{dt} \right]^{1/2} dt = \int_{t_0}^{t_1} \left[\dot{\mathbf{r}}(t)^{\mathrm{T}} \mathsf{M}\left(\mathbf{r}(t)\right) \dot{\mathbf{r}}(t) \right]^{1/2} dt \,, \qquad (1)$$

where M can vary spatially and $i, j = 1, 2, 3$.

A metric tensor represented by a matrix M has a decomposition given by the spectral theorem [12]:

$$\mathsf{M} = \mathsf{Q}\mathsf{\Lambda}\mathsf{Q}^{\mathrm{T}} \,. \qquad (2)$$

Here, Q is an orthogonal eigenvector basis and $\mathsf{\Lambda}$ is a diagonal eigenvalue matrix. Using the ellipsoid interpretation of the metric, the eigenvalues are the inverse-squares of the ellipsoid's axis lengths.

These properties have made metrics an appealing method for representing mesh directionality (eigenvector directions) and anisotropy (corresponding eigenvalues). Researchers have applied metrics to generate and adapt meshes [7, 13].

There is a body of work directed toward generating directional fields on surfaces for the purpose of texture synthesis, painterly rendering, and surface meshing; refer to [14, 15].

Other algorithms relevant to this work include advancing front methods and grid-based methods, as previously mentioned. In addition to this, our underlying element-shaping strategy is inspired by the previous work of Alliez et al. [16] and Tchon et al. [17] on remeshing and pseudo-mesh generation using tensor fields and streamlines, respectively.

The proposed work is unique in that it assigns the tasks of element shaping and choosing marching directions to a field that is generated prior to meshing. Additionally, insertion operators have been developed that allow systematic insertion of elements in a valid manner, while meshing from the inside-out. Finally, by placing tiled fronts throughout the volume, this method is able to combine with existing methods to produce good quality hex-dominant meshes with augmented hex-to-tet ratios where desired.

3   Technical Approach and Preliminaries

The overall strategy is described in this section, which can be decomposed into three main steps:

1. Metric Field Generation: An iterative technique is used to solve for metric tensors on the nodes of a background mesh. The technique produces a boundary-aligned field, which is then adjusted to uniform anisotropy. With appropriate interpolation, this provides a piecewise-linear field over the volume.

2. Field-Guided Hex Tiling: The element shaping strategy attempts to form unit hexes that are aligned with the metric eigenvector streams and approximately adhere to the lengths encoded in the metric eigenvalues. This is achieved by creating new hex nodes at the intersections of approximately unit length streamsurface triplets. The meshing process begins by inserting a seed hex on the interior of the volume. During tiling, insertion operators determine how new hexes are extended from the current mesh. This continues until the boundaries are reached, at which point new elements are locally snapped to boundary surfaces, feature curves, and feature vertices to enable boundary conformity.

3. Hex-Dominant Mesh Generation: Packing of rectangular solid cells is augmented to incorporate the tiled hex fronts. The hex nodes are packed as fixed cells, and with some specific pre- and post-processing, the process is continued to obtain a hex-dominant mesh.

Figure 1 below depicts some of the main steps of the whole process, in the order described above.

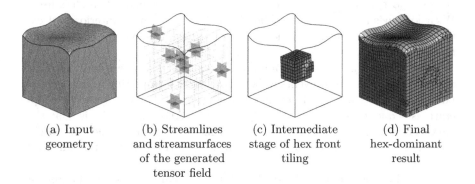

(a) Input   (b) Streamlines   (c) Intermediate   (d) Final
geometry   and streamsurfaces   stage of hex front   hex-dominant
             of the generated   tiling   result
             tensor field

Fig. 1. Overview of the whole process

Before elaborating on each of these steps, several techniques and constructs that support field generation and meshing need to be discussed.

3.1   Metric Field Representation, Support, and Operations

The metrics used in this work are stored as symmetric 3×3 matrices with directionality and anisotropy encoded through the eigenvectors and eigenvalues. This is convenient for tasks such as length measurement and interpolation. However, streamline generation requires the eigen-decomposition in order to use the eigenvector fields. Therefore, both representations are employed at different times. The eigen-decomposition is obtained via the Jacobi rotation method [18], and 2 is used to go back to the matrix form.

During meshing it is often necessary to query eigenvector directions, find the best eigenvector corresponding to an input direction, and to relate the eigenvectors of two closely aligned metrics. In this work, eigenvectors are specified by an index ($e = 1, 2, 3$) and sign ($\sigma = \pm 1$) with respect to a given metric. The eigenvector (index-sign pair with respect to a specific metric) that best corresponds to an input direction is taken to be the one that has the largest dot product with it. For example, if the input direction is the normal of a patch of mesh faces, this facilitates finding the best eigenvector to trace streamlines in the normal direction.

When relating two metrics, one of two methods is used. The first consistently orders eigenvectors based on eigenvalue magnitudes, so associativity between the eigenvectors of both metrics is determined by eigenvalue magnitude. This is used to define the three eigenvector fields which are traced during tiling. The second is employed during field generation to snap the eigenvectors of one tensor to the eigenvectors of another. It compares directions as mentioned before to determine the associativity.

The generated metric fields are piecewise-linear in component space. That is, the six independent components of the tensor field vary linearly over space. As a support, a convex, constrained Delaunay background tet mesh is used. With known surface normals, a flood fill is used to mark interior and exterior tets. Additionally, the metric tensors are stored on the nodes of the background mesh.

Point location in the background mesh is performed by walking [19]. The convexity requirement prevents walks from falsely terminating at non-convex boundaries. It is a suitable choice, considering that most of the field operations sequentially access face-adjacent tets. Another benefit is that the barycentric coordinates can be reused during interpolation. Once the target point is located, point inside/outside can be determined from the status of its enclosing tetrahedron.

Linear interpolation of the enclosing tet's nodal tensors (in component space) yields the local metric components. This is a relatively inexpensive interpolation kernel, but note that it can lead to undesirable properties such as swelling [20] and locally isotropic tensors. The latter case leads to umbilics in the tensor field [21]. The eigenvector directions change very rapidly in the proximity of umbilics, which is a problem when forming elements there. Therefore, tiling avoids such regions.

Streamlines are generated by specifying a starting point, an eigenvector index-sign pair (σ, e), and a desired length under the metric field. Fourth-order, fixed step-size Runge-Kutta [22] is used to generate successive points along a streamline while accumulating the metric-lengths of the new segments [12]. Due to the sign ambiguity of eigenvectors, the direction that most closely matches the established marching direction is chosen; similar considerations are made in [17]. The process is stopped once the measured length meets the desired length, or if the integrator is stalled. Depending on the use, either the entire polyline or just the endpoint is stored.

The next step is to generate streamsurfaces along two eigenvector directions, (σ_1, e_1) and (σ_2, e_2), with nominal lengths l_1 and l_2, respectively (see Figure 2). Primary streamlines are traced according to these initial directions and lengths from the initial point, $\mathbf{p}_0$. The primary streamlines are then re-sampled to have $n + 1$ points including the original first and last points.

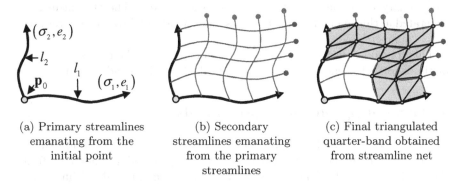

(a) Primary streamlines emanating from the initial point

(b) Secondary streamlines emanating from the primary streamlines

(c) Final triangulated quarter-band obtained from streamline net

Fig. 2. Formation of streamsurfaces $(n = 4)$

From each resampled point on the (σ_1, e_1) streamline, a sufficiently long streamline is generated in the local direction corresponding to (σ_2, e_2), and vice-versa. These secondary streamlines should be sufficiently long to account for rapidly converging/diverging eigenvector directions.

Due to the nature of these fields, it is not necessary for the secondary streamlines to intersect. Therefore, an optimization process is used to find the closest points on each pair of streamlines. Let $\mathbf{poly}_1$ and $\mathbf{poly}_2$ be arrays of m_1 and m_2 points, respectively, that contain the points from each streamline. Then the following parameterization provides a piecewise-continuous representation of each $(i = 1, 2)$:

$$\mathbf{p}_i(u) = \mathbf{poly}_i[u_0](1 - u') + \mathbf{poly}_i[u_0 + 1](u'), \tag{3}$$

where $u_0 = floor(u)$, $u' = u - u_0$, and $u \in [0, m_i - 1)$. One unit interval in the parameter space spans one line segment. Defining separate parameters for each curve, s and t, we can formulate an objective function from the squared-distance between the two curves:

$$f = \|\mathbf{p}_1(s) - \mathbf{p}_2(t)\|^2 . \tag{4}$$

A steepest descent scheme with adaptive step-sizing is used to find the parameters $t^\star$ and $s^\star$ that minimize this function. The gradient of f can be written as:

$$\nabla f = 2\left([\mathbf{p}_1(s) - \mathbf{p}_2(t)] \cdot \dot{\mathbf{p}}_1(s), [\mathbf{p}_1(s) - \mathbf{p}_2(t)] \cdot \dot{\mathbf{p}}_2(t)\right)^{\mathrm{T}} . \tag{5}$$

Central differencing is used to evaluate the tangent vectors of the two curves, $\dot{\mathbf{p}}_1(s)$ and $\dot{\mathbf{p}}_2(t)$. The average of the locations evaluated from the optimal parameters is returned as the closest point.

The net of closest points forms a grid, which is split into triangles to build the streamsurface. In practice, the portion of the net within some length along the primary directions is not generated to lower cost. The streamsurfaces generated in this manner also occupy a single "quadrant," so they are referred to as "quarter-bands."

This concludes discussion of the preliminaries, and in the next sections the methods for field generation and hexahedral tiling will be addressed.

4   Generating Metric Tensor Fields

The key to the proposed method is to separate the tasks of controlling element shape, selecting marching directions, and obtaining boundary alignment from the meshing algorithm. This allows for a simpler meshing algorithm, while incurring more effort during this stage. The following subsections describe a "form-fitting" approach to first generate individual surfaces fields, and then the final boundary-aligned volume field. It is designed to match the intuitive notion of how directionality should vary over a "principally blocky" volume.

4.1   Boundary Field Form-Fitting

In the first step, "scaffold triads" are placed along feature curves and at feature vertices. These triads are aligned with the local features as best as is possible, and specify required directionality but not eigenvector ordering or anisotropy. Many geometric features can be captured this way, however, the orthogonality of the triads limits capture of features such as knife edges or a vertex with more than four incident feature curves.

Then, form-fitting is applied independently to each surface. For a given surface, one of its bounding curve's nodes is selected to begin. This node's metric tensor is initialized with the directions from its scaffold triad, and the following procedure is performed:

---

while not all bounding nodes have been visited **do**
    Get the metric tensor of the current node, and **snap its directions** to the closest directions on the node's scaffold.
    Mark the node as visited and as a boundary condition (BC) node.
    Solve an approximate surface Laplacian over the current surface, with the current set of tensor BCs.
    Move to the next unvisited, adjacent node on the bounding node-loop(s) of the surface.
end while

The stencil for the approximate surface Laplacian is:

$$\mathsf{M}^{k+1} = \left(\sum_{j=1}^{N} w_j \right)^{-1} \sum_{j=1}^{N} w_j \mathsf{T}_j \mathsf{M}_j^k \mathsf{T}_j^{\mathsf{T}} , \tag{6}$$

where M^{k+1} is the updated tensor at the current node, w_j is the Floater's mean value coordinate [15] associated with neighbor node $j = 1...N$, T_j is the transformation from the surface normal at neighbor node j to the surface normal at the current node, and M_j^k is the tensor at neighbor node j.

4.2 Interior Form-Fitting

By this point, all surface fields have been established. The fields of adjacent surfaces may not match up in terms of ordered eigenvectors, but they are directionally-compatible due to the shared scaffold triads used while generating them. The tensors on the interior of each surface are snapped to the local surface normal to ensure alignment, and then all surface tensors are converted to scaffold triads. A similar procedure is then employed to march over unvisited boundary nodes and complete the volume field:

Pick a starting node on the boundary and generate a complete metric tensor with directions from its scaffold.
Set this node as visited.
Push node onto boundary-node queue.
while boundary-node queue not empty **do**
    Set current node to the front node of the queue.
    Pop the front of the queue.
    Snap the current node's metric tensor to its scaffold directions.
    Set the current node as a BC node.
    Solve tensor-component Laplacian over volume with BCs.
    for each node adjacent to the current node **do**
        if current adjacent node is on the boundary and is unvisited **then**
            Push onto queue.
            Set visited flag.
        end if
    end for
end while

This concludes the description of the form-fitting procedure. After this, all surface tensors are set as boundary conditions to solve for the exterior volume field. This allows streamlines and surfaces to extend slightly past the boundary as necessary. The nodal tensors of the completed field are modified to represent the desired anisotropy, and the umbilics of the field are extracted using a technique described in [25]. We have modified the second technique presented in that work to extract connected umbilic structures (curves, surfaces, and sub-volumes) as well as discrete points for full- and transverse-isotropy, for the piecewise-linear tensor fields generated here. In the next stage, the tensor field and umbilic information is passed on to the tiling and hex-dominant meshing algorithms.

5   Hex-Dominant Mesh Generation

Individual hex fronts are generated in regions that are free of umbilics and in regions of interest. A hex front is initiated by inserting a seed element on the interior or just within the boundary of the volume. This exposes six hex faces on the "skin," or collection of 2D elements on the boundary of the current hex mesh. The next elements to be inserted will use some of the current skin faces and add new ones, after which the skin is updated. As the skin grows, different combinations of existing skin faces may be used to form new elements. To avoid combinations that might lead to self-intersection or topological invalidity, insertion operators are used to plan and execute insertions that utilize groups of existing skin faces, referred to as insertion face groups (IFG). The IFGs, in turn, are identified by studying their skin nodes. When an operator is used to perform an insertion, depending on the IFG type, quarter-bands are appropriately placed and intersected to obtain the new nodes that define the target element shape. As appropriate, boundary-conformity operations may be invoked to associate proximal mesh entities with boundary features.

Once the fronts have been tiled in ROIs and away from field umbilics, there are two possible strategies for completing the hex-dominant mesh. The first is to perform a Boolean subtraction of the fronts from the volume, generate a hex-dominant mesh in the remaining volume, and then unite the meshes. This is similar to the Hex-Tet algorithm by Meyers et al. [27]. The second is to incorporate the front nodes into the rectangular bubble-packing process, which has been selected for this work. Although it does not guarantee 100% hex recovery, in practice the distribution of front nodes leads to significant recovery. Furthermore, it may allow for better overall element quality and transitions between hex and hex-dominant regions. The potential for non-conformal configurations with hanging-nodes and other complications noted in [24] are also avoided by taking this route.

5.1   Topological Insertion Operators and Face Groups

The rules that determine what types of elements to insert are fundamentally based on the local element-vertex connectivities of the skin vertices and faces. These vertex types can be enumerated by considering a central vertex in a structured hex mesh. The vertex is initially surrounded by eight elements, and by removing different permutations of elements, a unique set of configurations of skin faces using the vertex can be obtained. Figure 3 lists the types considered in this work. Some configurations have been excluded because they naturally do not arise due to the way elements are inserted; for instance, configurations where two hexes only share one edge or where there is a hexahedral through-hole or void.

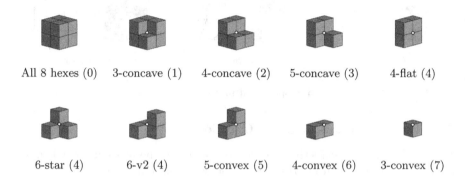

All 8 hexes (0)   3-concave (1)   4-concave (2)   5-concave (3)   4-flat (4)

6-star (4)   6-v2 (4)   5-convex (5)   4-convex (6)   3-convex (7)

Fig. 3. Skin vertex types

The number of elements removed from the original eight is indicated in parentheses next to each vertex type name. In order, these vertex types will be referred to as: **3c, 4c, 5c, 4f, 6*, 6v2, 5v, 4v,** and **3v** from here on.

We will now present the topological insertion operators which process IFGs. The set of insertion operators and their IFGs is given in the figure below. Each possible insertion (left) is shown with its corresponding IFG (right) in Figure 4. Sets of required vertex types that are used to identify IFGs are enumerated in Table 1.

Table 1. Vertex type requirements for IFGs

IFG type	Vertex type requirements
Side-1	Type $(\{v_0, v_1, v_2, v_3\}) \in \{\mathbf{4f, 5v, 4c, 3v}\}$
Bracket-2	Type $(\{v_2, v_5\}) \in \left\{ \begin{array}{c} \mathbf{4c, 5c, 6^*,} \\ \mathbf{6v2, 5v} \end{array} \right\}$, Type $(\{v_0, v_1, v_3, v_4\}) \in \{\mathbf{4f, 5v, 4v, 3v}\}$
Corner-3	Type $(v_6) = \mathbf{3c}$, Type $(\{v_0, v_2, v_4\}) \in \{\mathbf{4c, 5c, 6^*, 6v2, 5v}\}$, Type $(\{v_1, v_3, v_5\}) \in \{\mathbf{3v, 4f, 4v, 4c, 5v}\}$
Cup-3	Type $(\{v_1, v_2, v_5, v_6\}) \in \{\mathbf{4c, 5c, 6^*, 6v2, 5v}\}$, Type $(\{v_0, v_3, v_4, v_7\}) \in \{\mathbf{5v, 4v, 3c}\}$
Scoop-4	Type $(\{v_0, v_1\}) = \mathbf{3c}$, Type $(\{v_2, v_3, v_4, v_5\}) \in \{\mathbf{4c, 5v, 5c, 6v2, 6^*}\}$, Type $(\{v_6, v_7\}) \in \{\mathbf{3v, 4f, 4v, 5v}\}$
Bucket-5	Type $(\{v_0, v_1, v_2, v_3\}) = \mathbf{3c}$, Type $(\{v_4, v_5, v_6, v_7\}) \in \{\mathbf{4c, 5c, 6^*, 6v2, 5v}\}$

Bracket-2*: There is an ambiguous case when both and are of type *5-convex*. To resolve this, an additional requirement is enforced: the edge between them should be used by three hexes

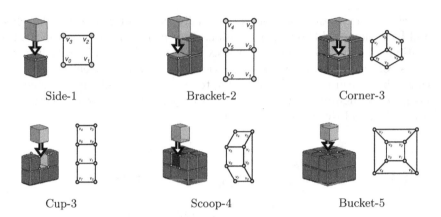

Fig. 4. Topological insertion operators

5.2 Planning and Scheduling Insertions

In this section, we discuss the precedence among the operators at a particular spot and how potential insertions over the entire skin are processed.

Due to the nature of the vertex type requirements, more complex insertions automatically precede simpler ones over an applicable skin region. This prevents some cases in which the mesh propagates into itself or forms sharp cracks. If vertices are instead classified by geometric criteria such as dihedral angles between the faces containing a vertex, then insertions should be searched for starting from the most complex one (one that is not contained in any other). One possible ordering is: *Bucket-5*, *Scoop-4*, *Cup-3*, *Corner-3*, *Bracket-2*, and finally, *Side-1*. This has been adopted as the current insertion precedence.

In order of precedence, IFGs are placed on unclaimed portions of the skin and also pushed onto an insertion queue. Once again, in order of precedence, the top IFG in the queue is popped and then used to place an element. If this fails, the IFG is maintained but marked as *failed* and pushed back onto the queue. Additionally, if an IFG has vertices that fall outside the volume, it is marked as *do_not_insert*. When an insertion is successfully made from an IFG, the skin and vertex types are locally updated and the IFG is deleted. The affected IFGs (e.g. adjacent ones) are deleted and removed from the queue, then local re-planning is performed. One benefit is that a failed insertion (e.g. due to field behavior) may eventually succeed if neighboring insertions succeed and trigger a reevaluation. Also, this scheme naturally prefers uniform front growth. As long as umbilics and highly distorted field regions are avoided, both features lead to avoidance of self-intersection.

Tiling proceeds until no IFGs remaining in the queue can be processed. The goals of avoiding umbilics, achieving boundary conformity, and staying within a ROI use the *do_not_insert* flag to appropriately terminate tiling in particular directions.

5.3   Element Shaping

Consider a single hexahedron: If its faces are aligned with the metric eigenvectors, as should approximately be the case, then the faces are discrete analogues to similarly placed, continuous streamsurfaces. Mesh edges are then analogous to curves of intersections between any two adjacent streamsurfaces, and mesh nodes are analogous to the intersection of any three adjacent streamsurfaces. This is the guiding principle behind the formation of elements in the proposed method. When an element is to be placed, it may use existing skin faces (IFGs) and add new faces. Streamsurfaces are generated to be parallel to the new faces. The intersections of triplets of these streamsurfaces are used to locate newly introduced nodes needed to form a new hex. Figure 5 illustrates the collections of streamsurfaces used to define the insertion types: *Seed*, *Side-1*, *Bracket-2*, and *Corner-3*. The remaining types shown in Figure 4 only use existing nodes and do not require this procedure.

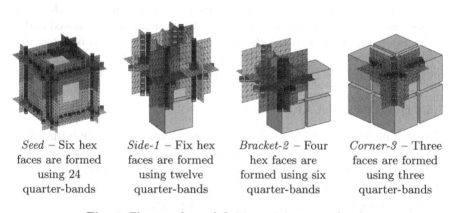

| *Seed* – Six hex faces are formed using 24 quarter-bands | *Side-1* – Fix hex faces are formed using twelve quarter-bands | *Bracket-2* – Four hex faces are formed using six quarter-bands | *Corner-3* – Three faces are formed using three quarter-bands |

Fig. 5. Element shape definition using quarter-bands

To generate a seed element, half unit-length streamlines are integrated along all six eigenvector sign-direction pairs. From the endpoint of each streamline along direction (σ_i, e_i), four quarter-band surfaces are generated from the eigen-directions normal to the local (σ_i, e_i) direction. Triplets of these quarter-bands intersect at a potential node location of the new seed hex. The intersection process is accelerated using a uniform lattice and Möller's presentation of the separating axis theorem [23] to cull the set of triangle-triangle intersection tests that are performed.

For *Side-1*, *Bracket-2*, and *Corner-3* insertions, the eigen-directions for the quarter-bands are determined by looking at the eigen-directions of the metric tensors at the local skin nodes. The eigen-directions which align with skin edges or normals are selected to form the surfaces.

By design, this element formation method primarily aligns newly inserted skin faces with the metric field. Meeting the size requirement is secondary, and thus is only approximately satisfied. Otherwise the mesh would progressively deviate from local field directions, and local eigen-direction selection could not be used.

5.4 Boundary Conformity

Because the proposed method is designed to produce a boundary-aligned field and hence elements that approach boundary-alignment as they reach the surface, a relatively simple boundary capture method suffices for many models.

When elements are inserted in proximity to the boundary, a series of local and greedy snapping moves are attempted to enable capture of feature vertices, feature curves, and surfaces. For a newly inserted hex, nodes are considered for snapping to candidate feature vertices, then edges to candidate feature curves, and finally, faces to boundary surfaces. Non-dimensional fitness scores have been developed for each class to facilitate prioritizing and qualification of moves. To exclude moves that will result in poor quality elements, the shape metric is evaluated for the hexes incident on a moved node. If the minimum metric is below the threshold, then the move is reverted; 0.3 is used in this work. Additionally, to prevent the mesh from spilling out of the volume, the IFGs containing projected faces are appropriately marked.

5.5 Hex-Dominant Mesh Finalization

The packing approach to hex-dominant mesh conversion from an initial constrained Delaunay tet mesh is only slightly modified for the purpose of this work. The following overview includes the necessary alterations, in bold, for incorporating the hex fronts into the process:

1. **Hex front nodes are packed as fixed bubbles.**
2. The rest of curves, surfaces, and interior are then packed.
3. **Trimming is performed on regular bubbles that are within one metric unit of the front node bubbles.**
4. **After the surface is remeshed using the surface bubbles, edge swaps are performed to recover hex edges that lie on the boundary.**
5. A constrained Delaunay tet mesh is obtained from the bubble centers.
6. Topological and geometric quality improvement is performed on the tet elements.
7. Tet to hex conversion is attempted and followed by further quality improvement.

6 Results

To demonstrate the efficacy of the proposed method, a large deformation, elasto-plastic analysis of a lower control arm has been performed using

ABAQUS [26]. Figure 6 shows the control arm model, computed volume tensor field along with umbilic structures, hexahedral mesh fronts, and the resulting hex-dominant mesh. The oriented bounding box of the model has dimensions of 16.77 × 4.61 × 17.60 units and pseudo-isotropy is imposed on the field with a nominal mesh size of 0.2 units. The current implementation is not yet fully optimized for performance, but the overall time for field generation, tiling, and generation of the final hex-dominant mesh is under 15 minutes.

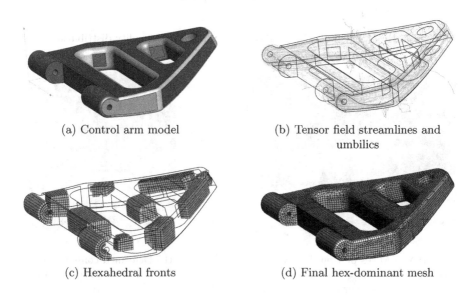

(a) Control arm model   (b) Tensor field streamlines and umbilics

(c) Hexahedral fronts   (d) Final hex-dominant mesh

Fig. 6. Control arm results

A total of 8,046 hexes with an average length of 0.20 were generated during tiling. The final hex-dominant mesh consists of 62,300 elements, with an average edge length of 0.23. By quantity, 20,181 (32%) of the elements are hexes and 42,119 (67%) are tets. The hex elements constitute 73.52% of the total volume, whereas only 26.48% is occupied by tets. The minimum and maximum scaled Jacobian metrics of the hexes are 0.4 and 1.0, and the average is 0.93. The radius-ratio metric for the tet elements ranges from 3.0 (ideal) to 697.69, with an average value of 4.12. The distributions of hex and tet element quality according to these metrics are presented in Figure 7.

For the analysis, an elasto-plastic material model based on isotropic hardening was specified with generic parameters for steel: E=210 GPa, ν=0.3, $\sigma_y = 480$ MPa, ultimate strength $\sigma_u = 550$ MPa, and tangent modulus $E_{p,t} = 349.42$ MPa. The interior cylindrical surfaces on the left portion of the model were displacement constrained, and lateral and vertical loads of 1,080 kN and 540 kN, respectively, were transmitted to the cylindrical surface

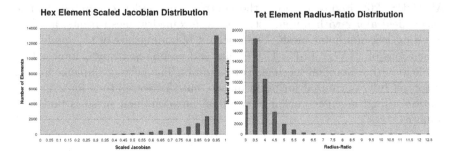

Fig. 7. Hex (Scaled Jacobian) and tet (Radius-Ratio) quality distributions

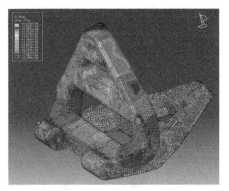

(a) Von-Mises plot on deformed configuration along with undeformed configuration

(b) Deformed configuration with plastic zones shown in gray

Fig. 8. Von-Mises stress distribution with fully plastic zones in deformed configuration

at the apex. The deformed configuration is shown in Figure 8, in which the gray regions denote fully plastic zones.

More examples of this method are shown in Figure 9. Each example shows the hexahedral fronts followed by the resulting hex-dominant mesh with the exterior targeted ROIs circled. In all examples, fronts are seeded at interior positions away from umbilics. Colored faces indicate that a similarly colored surface has been captured by the mesh.

The data for these models are summarized in Table 2. Inspection of the resulting hex-dominant meshes reveals that a significant portion of the hex fronts are recovered at the end of meshing process, and with reasonable quality overall. However, the results also warrant further work on improvement of the worst quality elements (primarily tets) and perhaps a more aggressive strategy for hex recovery.

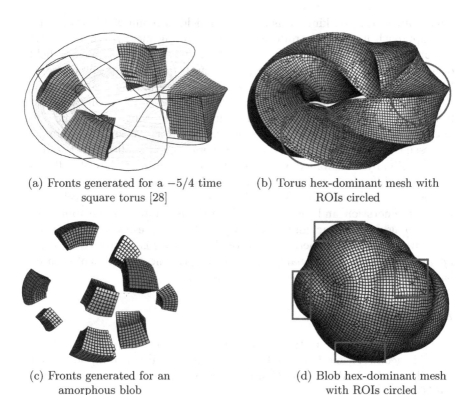

(a) Fronts generated for a $-5/4$ time square torus [28]

(b) Torus hex-dominant mesh with ROIs circled

(c) Fronts generated for an amorphous blob

(d) Blob hex-dominant mesh with ROIs circled

Fig. 9. Additional examples

Table 2. Statistics for additional examples

Model	Square Torus	Blob
Oriented bounding box dims:	$290.11 \times 294.90 \times 100.00$	$25.80 \times 20.51 \times 20.36$
Specified anisotropy:	4.0 pseudo-isotropic	0.5 pseudo-isotropic
# of tiled hexes (avg. edge length):	9,811 (4.11)	5,704 (.46)
# of hex-dom. elements (avg. edge length):	105,052 (4.47)	127,157 (.52)
% hex by volume (by number):	76.90% (36%)	75.32% (33%)
% tet by volume (by number):	23.10% (63%)	24.68% (66%)
(Min / Avg. / Max) Hex Scaled Jacobian:	0.40 / 0.93 / 1.00	0.40 / 0.92 / 1.00
(Min / Avg. / Max) Tet Radius ratio:	3.00 / 3.99 / 87.91	3.00 / 3.82 / 129.26

7   Conclusion

A method for generating quality hex-dominant meshes with targeted all-hex regions has been presented. This is achieved by tiling fronts of hexahedral elements throughout the volume and at ROIs, and then incorporating the front nodes in the rectangular bubble-packing method. To guide element shaping and marching directions during tiling, and bubble directionality and anisotropy during packing, a metric tensor field is required. To this end, a new approach has been introduced for constructing piecewise-linear metric tensor fields over volumes. This approach produces boundary-aligned fields of uniform anisotropy by incrementally "form-fitting" an initially Cartesian field to the directions specified by boundary curves and surfaces. Provided with the field, the tiling process is able to generate topologically valid hex fronts using insertion operators and boundary conformity operations. The nodes of the fronts are set as fixed bubbles, and through other pre- and post-processing steps, the method is able to generate good quality hex-dominant meshes. Finally, the ability of the method to produce meshes of sufficient quality for difficult, non-linear analyses has been demonstrated.

Plans for future extensions to this work include refining the techniques for field generation and interpolation as well as improving the robustness of the tiling process (including boundary conformity), especially with variations in anisotropy. Furthermore, although this work is motivated by *a priori* anisotropy and directionality requirements, the incorporation of solution-based metrics is currently being investigated.

Acknowledgements

The authors would like to thank Dr. Soji Yamakawa for the use of and help with his geometry, meshing, and graphics class library. The authors would also like to thank Mr. Iacopo Gentilini for his guidance and expertise with ABAQUS.

References

1. Lu, Y., Gadh, R., Tautges, T.J.: Volume Decomposition and Feature Recognition for Hexahedral Mesh Generation. In: Proceedings of the 8th International Meshing Roundtable, pp. 269–280 (1999)
2. Schneiders, R., Schindler, R., Weiler, F.: Octree-based Generation of Hexahedral Element Meshes. In: Proceedings of the 5th International Meshing Roundtable, pp. 205–216 (1996)
3. Zhang, Y., Hughes, T.J.R., Bajaj, C.L.: Automatic 3D Mesh Generation for a Domain With Multiple Materials. In: Proceedings of the 16th International Meshing Roundtable, pp. 367–386 (2007)
4. Blacker, T.D., Meyers, R.J.: Seams and Wedges in Plastering: A 3D Hexahedral Mesh Generation Algorithm. Engineering with Computers 2(9), 83–93 (1993)
5. Staten, M.L., Kerr, R.A., Owen, S.J., Blacker, T.D.: Unconstrained Paving and Plastering: Progress Update. In: Proceedings of the 15th International Meshing Roundtable, pp. 469–496 (2006)
6. Blacker, T.: Meeting the Challenge for Automated Conformal Hexahedral Meshing. In: Proceedings of the 9th International Meshing Roundtable, pp. 11–19 (2000)
7. Yamakawa, S., Shimada, K.: High Quality Anisotropic Tetrahedral Mesh Generation Via Ellipsoidal Bubble Packing. In: Proceedings of the 9th International Meshing Roundtable (2000)

8. Shimada, K., Liao, J.-H., Itoh, T.: Quadrilateral Meshing With Directionality Control through the Packing of Square Cells. In: Proceedings of the 7th International Meshing Roundtable, pp. 61–76 (1998)
9. Viswanath, N., Shimada, K., Itoh, T.: Quadrilateral Meshing With Anisotropy and Directionality Control Via Close Packing Of Rectangular Cells. In: Proceedings of the 9th International Meshing Roundtable, pp. 217–225 (2000)
10. Bossen, F.J., Heckbert, P.S.: A Pliant Method for Anisotropic Mesh Generation. In: Proceedings of the 5th International Meshing Roundtable, pp. 63–76 (1996)
11. Li, X.Y., Teng, S.H., Üngör, A.: Biting Ellipses to Generate Anisotropic Mesh. In: Proceedings of the 8th International Meshing Roundtable, pp. 97–108 (1999)
12. Strang, G.: Linear Algebra and Its Applications, 3rd edn. Brooks-Cole (1988)
13. Sirois, Y., Dompierre, J., Vallet, M.G., Guibault, F.: Mesh Smoothing Based On Riemannian Metric Non-Conformity Minimization. In: Proceedings of the 15th International Meshing Roundtable, pp. 271–288 (2006)
14. Zhang, E., Hays, J., Turk, G.: Interactive Tensor Field Design and Visualization on Surfaces. IEEE Trans. Visual Comput. Graphics 13(1), 94–107 (2007)
15. Palacios, J., Zhang, E.: Rotational Symmetry Field Design on Surfaces. In: Proceedings of ACM SIGGRAPH, p. 55 (2007)
16. Alliez, P., Cohen-Steiner, D., Devillers, O., Levy, B., Desbrun, M.: Anisotropic Polygonal Remeshing. In: Proceedings of ACM SIGGRAPH, pp. 485–493 (2003)
17. Tchon, K.F., Dompierre, J., Vallet, M.G., Camarero, R.: Visualizing Mesh Adaptation Metric Tensors. In: Proceedings of the 13th International Meshing Roundtable, pp. 353–363 (2004)
18. Golub, G.H., Van Loan, C.F.: Matrix Computations, 3rd edn. JHU Press, Baltimore (1996)
19. Sundareswara, R., Schrater, P.: Extensible Point Location Algorithm. In: Proceedings of the 2003 International Conference on Geometric Modeling and Graphics, pp. 84–89 (2003)
20. Pennec, X., Fillard, P., Ayache, N.: A Riemannian Framework for Tensor Computing. Int. J. Comput. Vision 66(1), 41–66 (2006)
21. Hesselink, L., Levy, Y., Lavin, Y.: The Topology of Symmetric, Second-Order 3D Tensor Fields. IEEE Trans. Visual Comput. Graphics 3(1) (1997)
22. Press, W.H., Teukolsky, S.A., Vetterling, W.T., Flannery, B.P.: Numerical Recipes in C: The Art of Scientific Computing, 2nd edn. Cambridge Univ. Press, UK (1992)
23. Akenine-Möller, T.: Fast 3D triangle-box overlap testing. In: SIGGRAPH 2005: ACM SIGGRAPH 2005 Courses, Article 8 (2005)
24. Yamakawa, S., Shimada, K.: Fully-automated hex-dominant mesh generation with directionality control via packing rectangular solid cells. International Journal for Numerical Methods in Engineering 57(15), 2099–2129 (2003)
25. Zheng, X., Parlett, B.N., Pang, A.: Topological lines in 3D tensor fields and discriminant Hessian factorization. IEEE Trans. Vis. and Comput. Graphics 11(4) (2005)
26. Simulia Corp., ABAQUS Analysis User's Manual, Version 6.8 (2008)
27. Meyers, R.J., Tautges, T.J., Tuchinsky, P.M.: The 'Hex-Tet' Hex-Dominant Meshing Algorithm as Implemented in CUBIT. In: Proceedings of the 7th International Meshing Roundtable, pp. 151–158 (1998)

28. Edelsbrunner, H.: Square Tori: -5/4 time. 180 Wrapped Tubes,
    http://www.cs.duke.edu/~edels/Tubes/tori/quad/n5.stl
29. Yamakawa, S., Shimada, K.: Subdivision Templates for Converting a Non-
    conformal Hex-Dominant Mesh to a Conformal Hex-Dominant Mesh with-
    out Pyramid Elements. In: Proceedings of the 17th International Meshing
    Roundtable, pp. 497–512 (2008)

Using Parameterization and Springs to Determine Aneurysm Wall Thickness

Erick Johnson, Yongjie Zhang, and Kenji Shimada

Carnegie Mellon University, 5000 Forbes Ave. Pittsburgh, PA 15289
eljohnso@andrew.cmu.edu, jessicaz@andrew.cmu.edu, shimada@cmu.edu

Abstract. Aneurysms are an enlargement of a blood vessel due to a weakened wall and can pose significant health risks. Abdominal aortic aneurysms alone are the 13th leading cause of death in the United States, with 15,000 deaths annually. While there are recommended guidelines for doctors to follow in the treatment of specific aneurysms, they cannot guarantee a satisfactory outcome. Computer simulations of an aneurysm may be able to help doctors in their treatment; however, the results are inaccurate if the vessel wall thickness is poorly measured. In order to provide more accurate, patient-specific simulations, not only does geometry for the fluid domain need to be created from medical images for analysis, but the creation of more accurate models for the wall needs to be accomplished as well. This paper proposes a solution to the latter by deforming the mesh from a healthy vessel into one with an aneurysm through parameterization and the use of a spring model. The thickness of the resulting wall model is empirically valid and fluid-structure interaction simulations show significant improvements when using a variable versus a uniform wall thickness.

Keywords: Aneurysm, Deformation, Parameterization, Spring System.

1 Introduction

Aneurysms are an enlargement of a blood vessel due to a weakened wall, and simulating them is an area of significant research and concern. They can occur anywhere in the vascular system, though are most common along the abdominal aorta and in the brain. The simulation of aneurysms is of critical importance because of the large health risk they pose. Rupturing of an abdominal aortic aneurysm (AAA) occurs in 1–3% of men aged 65 or older and is 70–95% fatal [7]. This paper presents a method to approximate the wall thickness of patient-specific geometry, which will increase the accuracy of simulations.

While there are recommended guidelines for doctors to follow in the treatment of specific aneurysms, they cannot guarantee a satisfactory outcome. Even with the guidelines for AAAs [7], a lot is left to the best judgment of the doctor due to the many variables. Smokers between the ages of 65–75 are encouraged to have a screening, while it is elective for non-smokers. An annual checkup is recommended if the aneurysm is larger than 4 cm in diameter, otherwise it is not a primary health risk. If open surgery is deemed necessary, there is a 4–5% chance of mortality with almost 33% of patients developing other complications. The rupture of cerebral aneurysms are harder to predict and have a much more varied outcome. Some victims may suffer no complications while it is fatal to others. Being able to better predict the growth and failure of the vessel wall or, in the future, the outcome of surgical procedures, could significantly decrease the mortality rate of patients with aneurysms.

Computer simulations are starting to be used to simulate blood flow through aneurysms in order to predict and better understand ruptures. This entails creating geometry of the patient's aneurysm from medical images and running a computational fluid dynamic (CFD) or fluid-structure interaction (FSI) analysis. In addition to all of the other aspects involved in providing doctors with *instantaneous* simulations, a major setback is accurately measuring the thickness of blood vessel walls. A doctor will not rely on an analysis if the results vary by an order-of-magnitude due to poor wall thickness data. In order to provide increasingly precise, patient-specific simulations, not only does the geometry for the fluid domain need to be created, but the creation of a more accurate wall model needs to be accomplished as well.

This paper presents a method to improve the accuracy of FSI simulations for aneurysms by estimating the wall thickness with the blood vessel geometry. The proposed method maps two geometries onto the same domain in order to deform one into the other. This deformation is achieved through two steps. First, parameterization is used to move the mesh from the healthy blood vessel onto the aneurysm, which provides the initial configuration of the spring system. Second, the numerical relaxation of a spring system allows the forces between the nodes to balance in a way that approximates the stretching of the vessel wall. FSI analyses are used to test the results of the variable wall thickness and compare it against the solution for a uniform wall.

The paper continues in the next section with a discussion of previous work. Section 3 provides a description of the method to calculate the aneurysm wall thickness with Section 4 explaining the boundary-layer meshing scheme used to prepare the models for the analyses. The results found in the FSI analyses of the aneurysm models are presented in Section 5. The paper is concluded in Section 6 with a discussion and future work.

2   Previous Work

In computer simulations, the wall thickness plays a key role in determining the shear stress on the vessel wall. In order to provide a patient-specific simulation, accurate modeling of the target area is required. A lot of work is being done to create models from medical images for use in simulations, see the work of Zhang *et al.* for an example of creating smooth geometry with

NURBS (Nonuniform Rational B-Splines) [33]. While there is a push to auto-mate this part of the process, significant user input is still required for proper classification and segmentation. In many cases the wall thickness is not mea-surable from the image data. Even though AAA walls can be seen in medical images, an exact thickness is hard to determine, especially in the thinnest regions. This is because the typical CT scan has a resolution of about one millimeter [15]. Higher doses of X-rays can be used to achieve sub-millimeter dimensioning, but pose a greater risk to the patient. Additionally, vessel wall-thickness cannot be seen for smaller vessels, such as those found in the brain. Intravascular ultrasound (IVUS) can be used to image coronary arteries with 0.2 mm accuracy [5]. However, this is near the limit of successfully captur-ing the thinnest regions in a AAA and, besides not being usable in narrower vessels, i.e. cerebral arteries, it cannot resolve a thinner wall thickness.

In order to perform CFD simulations, a mesh of the fluid domain is re-quired. Two popular methods are the advancing front [10, 16, 22] and octree [34, 13]. An advancing front approach builds a mesh off the surface, in the region of the boundary layer, and then fills the remaining space. Garimella and Shephard give a comprehensive explanation of how to generate an all tetrahedral mesh with anisotropic elements in the boundary layer [10]. A variation of this is given by Khawaja, where the boundary-layer elements re-main wedges, producing a hybrid mesh [16]. Sahni et al. were able to obtain more accurate wall shear stress (WSS) results by using an adaptive approach to improve an existing boundary-layer mesh. Using the solution to generate tensors at the boundary-layer nodes, the anisotropy in the mesh could be changed to improve the solution [22]. Using an octree mesh provides a signif-icantly more structured interior mesh, but is typically thought to have poorer quality near the boundary. Zhang uses pillowing near the boundary to create higher-quality elements that are also aligned with the flow near the surface [34]. Isaksen et al. have used these meshes to run FSI analyses on cerebral aneurysms [13].

CFD and FSI analyses are used to study the blood flow and WSS through localized regions, e.g. aneurysms, and through significantly larger portions of the vascular system [12, 26, 6, 22, 31, 19, 25, 29, 1, 34]. Torii showed the impact of both the cerebral aneurysm shape and the use of FSI on the flow and stress results [29]. Using an assumed 0.3 mm uniform wall thickness, Torii et al. found a 20% decrease in WSS when using compliant, instead of rigid, walls. A similar result was found by Bazilevs through a range of cerebral aneurysm models [1]. These models again assumed a uniform wall thickness because patient values are near impossible to obtain. Scotti et al. showed upwards of a 400% increase in WSS when a variable wall-thickness is used over its uniform wall-thickness counterparts [26, 25]. The thickness of the wall used for their aneurysms is a predetermined range that falls within the measured data from Raghavann [21]. In Raghavan's work AAAs from four cadavers were cut into strips and measured to determine a tensile yield stress for the vessel-wall tissue. It was found that the thickness varied from about

1.5 mm in the healthy regions of the aorta (approximately 15% of the radius) to a minimum of 0.23 mm at a rupture site.

Parameterization is used frequently for model texturing and surface decimation [4, 8, 17, 27, 23, 32]. It takes three-dimensional geometry and projects it onto a planar (or series of planar) domain(s) — this is not completely true as there are some methods that projects the mesh onto the surface of a sphere. A proper map should yield a one-to-one relationship between any point on itself to its parent surface. Sheffer wrote a survey in 2006 describing much of the work on linear mapping [27]. A lot of the mapping schemes will only fit elements onto convex domains and depending on the parameterization scheme different guarantees can be made about the quality. For more complex domains, i.e. closed solids or n-manifold surfaces, it is common to partition the shape into smaller, manifold regions. Kim and Yin showed alternative methods for mapping multiply connected discs. Kim *et al.* used Ricci flow to move a face with holes, for the mouth and eyes, onto an annulus in order to perform expression recognition [17]. Slit mapping was presented by Yin to show a linear solution to the same problem with the ability to set arbitrary boundaries to the inner and outer ring [32].

Model deformation is commonly accomplished through the use of mass-spring systems [28, 11, 2, 3]. Using elasticity theory, a stiffness matrix could be created and solved to model and animate the deformation of rigid bodies [28]. This however can become computationally expensive as the domain increases in complexity. Gudukbay showed solving local spring equations was a good substitute to the older methods and could still provide time-dependent deformations [11]. Both Chen and Cui have recent work for haptic feedback with surgery simulations that utilizes mass-spring systems to provide displacement and force information to the user [2, 3]. It should be noted that the blood-vessel tissue is constantly trying to repair itself and while an aneurysm is growing the walls are slowly becoming stronger as well. Kroon and Holzapfel have an finite element simulation that incorporates the growth and breakdown of collagen in the creation of a cerebral aneurysm [18].

Most of the previous vascular modeling methods have used walls with a uniform thickness when performing analyses. Some methods are beginning to use variable thicknesses, but these tend to be coarse approximations with simple geometry. This paper will present a better approximation to the wall thickness by deforming a healthy blood vessel into one with an aneurysm.

3   Wall-Thickness Estimation

To determine the aneurysm wall thickness a combination of parameterization and spring relaxation is used. The proposed method tries to mimic the process of aneurysm growth by assuming the amount of tissue remains relatively constant while the vessel wall is stretched. If a single surface-mesh can be placed on both the healthy and aneurysm geometries then the change in area of the elements should also represent a local change in wall thickness.

As outlined in Figure 1, the mapping provides a way to move the surface triangulation from the deforming vessel, S_0, onto the surface of the aneurysm, S_1. While the two meshes now lie on the same domain, the deformed surface, S_2, is not a physically meaningful solution. A spring system is used to approximate the elastic nature of the vessel tissue and its relaxation allows the nodes of the deformed mesh to shift their locations along the aneurysm surface. This provides a solution, S_3, that better reflects how the healthy vessel was deformed. By comparing the original and deformed mesh, the wall thickness can be found through a simple volume preserving calculation.

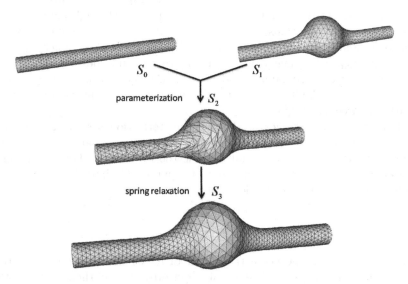

Fig. 1. Deformation of a fusiform aneurysm: S_0 and S_1 are combined in a map to create S_2. S_3 is the spring-relaxed model of S_2. S_0 and S_1 contain 2272 and 1720 triangles respectively.

3.1   Deformation through Parameterization

Moving the mesh from the healthy blood vessel onto the geometry of the aneurysm is achieved in three steps. First, the meshes for the healthy and aneurysm models are mapped onto annuli. Second, both maps are overlaid, which allows nodes from the healthy-surface map to be matched to a unique location in the map of the aneurysm. Third, using this relationship, nodes from the healthy mesh can be moved onto the desired geometry.

In order to create the mappings, the surface being *deformed*, S_0, and the *desired* surface, S_1, will need to be triangulated, t_0 and t_1 respectively. A model is required for both the desired, aneurysm shape and the deforming, healthy, blood vessel. While the aneurysm model can be created from the segmentation of medical images, unless a model already exists for the healthy vessel, an approximation will be required. An advancing front method is used to create a uniform triangulation on S_0 and is recommended to avoid weighting issues while smoothing with the spring system. The triangle mesh creates M nodes on the surface, with m nodes, $n_i, i = 1, ..., m$, being interior to those nodes on the boundary, $n_i, i = m + 1, ..., M$. Every node, n_i, has coordinates on the surface, x_i, and a set of valent nodes, V_i.

The nodes from S_0 and S_1 are mapped onto an annulus using a uniform parameterization in polar coordinates, see Figures 2(a) and 2(c). The annulus is chosen as the working domain primarily because it matches the topology of a cylinder and is still robust for geometries with multiple branches. Each surface has an inlet and outlet that are forced, respectively, onto the outer and inner rings of the annulus. This is accomplished through homogeneous coordinate transformations that translate and rotate each curve onto the same plane and scales them to the proper radii. The radii of the inner, $r_{in} = 1$, and outer, $r_{out} = 10$, rings can be different values, and does not change the final results. All the remaining nodes are then projected normally onto the annulus' plane so the iterative solution can be calculated in two-dimensions instead of three, see Figure 2(b).

Many weighting schemes used for parameterization are meant for convex domains, and as a result, solving the system over an annulus produces triangles that are tangled and fall outside the its bounds. This can be circumvented by moving the calculations of Equation (1) into the polar space. Using a uniform parameterization [30],

$$\acute{x}_i = \sum_{j \in V_i} \frac{\acute{x}_j}{v_i}, \tag{1}$$

places each node in the geometric center of its neighbors, where v_i is the valence number of node n_i. The initial, strict projection of the nodes directly onto the plane may create tangled triangles; however, iterating over the interior nodes with Equation (1) will pull them apart, see Figure 2(c). While most of the triangles will be highly anisotropic, they will also be untangled in polar coordinates. Unfortunately a few inverted triangles may appear when moving back to a Cartesian frame. These can be made positive with a couple iterations of an algorithm presented by Freitag. It is used to maximize the area of the triangles around every node [9]. At each node, a patch is constructed with its neighboring triangles and the minimum area is maximized. This yields a valid triangulation with the same topology in $\Re^2$ as the surfaces in $\Re^3$. The maps only need untangled triangles in order to provide the one-to-one relationship. Higher-quality elements in the maps may improve the initial deformation, but are not critical to the overall method.

The projected nodes now have coordinates $\acute{x}_i$ on the annulus and the projected triangles are $\acute{t}_i$. As can be seen in Figure 2, a uniformly meshed cylinder will appear very regular, with the triangles becoming increasingly skewed as their radial location increases. This is in contrast to the map of S_1, showing a denser concentration of triangles on the side corresponding to the offset aneurysm.

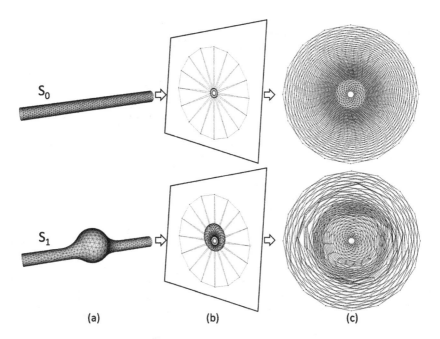

Fig. 2. Mapping of the nodes from a surface to the annulus; (b) shows the initial projection of nodes onto the annuli's plane. (c) shows the final maps after iterating with Equation (1).

The overlay of the two maps provides the relationship needed between the nodes from S_0 and the triangles from S_1, see Figure 3(c). With both maps overlaid, each non-boundary node, $n_{0,i}$, can be found to lie in a triangle, $\acute{t}_{1,j}$, as shown in Figure 3(c) pullout. As a result of the maps covering the same domain and having no tangled elements, *every* node in the map of S_0 will correspond to a unique point within the map of S_1. Further, by knowing every point within $\acute{t}_{1,j}$ also lies uniquely in its surface counterpart, $t_{1,j}$, locating the barycentric coordinates of $\acute{x}_{0,i}$ in $\acute{t}_{1,j}$ also defines its location, $x_{2,i}$, in $t_{1,j}$ on S_1, see Figure 3(d). Finding which triangle each node lies in starts with a search for the coordinate in the map of S_1 that yields the shortest distance to $\acute{x}_{0,i}$. Any triangle connected to that location is chosen as the initial guess for $\acute{t}_{1,j}$, with the surrounding triangles being moved through until all the barycentric coordinates are positive for $\acute{x}_{0,i}$.

Once the barycoordinates of all the nodes, $n_{0,i}$, have been found, they create the resultant surface, S_2. The healthy, blood vessel now matches the *shape* of the vessel with the aneurysm. Unfortunately, this alone does not provide accurate enough information to calculate the wall thickness. As a result of the weighting scheme used in the map, the parametric transformation between two different surfaces is not guaranteed to be the same, *i.e.* if the inner and outer radii of the annulus are 1 and 10 respectively, r = 5 may

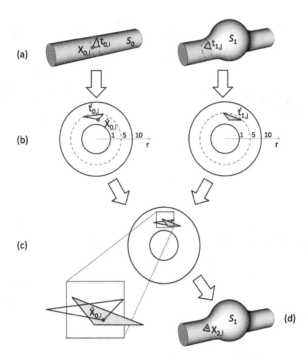

Fig. 3. Overview of the deformation following a node from S_0 (the red dot). (a) The original surface. (b) The map of a surface on its annulus resulting in the mapping of t_i into $\acute{t}_i$. The dashed line represents the location of $r = 5$ on both surfaces and the map. (c) An overlay of the maps allows the barycentric coordinates for $\acute{x}_{0,i}$ within $\acute{t}_{1,j}$ to be found. (d) The barycoordinates found in (c) are the same for $x_{2,i}$ in $t_{1,j}$ and move the nodes from S_0 onto S_1.

not correspond to the same axial coordinate in the original geometry, see Figures 3(a) and 3(b). Consequently, using the mapping technique, *by itself*, will create a bias as to where the wall is thinnest and not represent the true deformation. By converting the mesh into a spring system and allowing the nodes to move, this can easily be corrected.

3.2   Relaxation Using Springs

To make the deformation physically meaningful the deforming geometry should not only match the shape of the desired model, but the elements should stretch in an elastic manner. That is, a blood vessel will have deformed more significantly near the aneurysm than its ends, and the triangulation of S_2 should represent that properly. In order to accomplish this, a linear spring system is created from the mesh, *i.e.* the edges of the triangles become springs attached at the nodes. Through a series of iterations the springs are relaxed until their displacement is below some threshold; here it is less than 5% of the shortest edge. Additionally, the nodes are projected back onto S_1, in a direction normal to the surface, after every iteration to ensure S_2 does not deviate from the desired shape. Equation (2) calculates the force, F_i, acting on every node by comparing the current length of each attached spring against its undeformed length,

$$F_i = k \sum_{j \in V_i} l_{i,j} - ||x_{0,j} - x_{0,i}|| \hat{l_{i,j}}, \tag{2}$$

where $l_{i,j} = (x_{2,j} - x_{2,i})$ is the vector of the current spring between nodes $n_{2,j}$ and $n_{2,i}$ and $\hat{l_{i,j}}$ is the regularized direction. For this work the spring stiffness, $k = 1$, is a constant because of the uniform mesh and the isotropic material properties assumed. Every node is then updated by,

$$x_{2,i} = x_{2,i} + s \frac{F_i}{k}, \tag{3}$$

a fraction of the distance determined by the force acting on it. The step size of $s = 0.1$ prevents the nodes from taking too large a jump and tangling. The process is outlined in Algorithm 1 and the final surface is S_3.

---

while *movement* $> \epsilon$ **do**
    for *every node* $n_{2,i}$ **do**
        calculate force at $n_{2,i}$ using Equation (2)
        move $n_{2,i}$ using Equation (3)
        project $n_{2,i}$ to S_1
    end
end

Algorithm 1. Spring Correction

The deformed, blood vessel now matches the shape of the aneurysm and takes the elasticity of the vessel tissue into account. Even though this spring model can only provide a linear approximation, it is a first guess at trying to represent the complex elasticity and anisotropic nature of the tissue. However, steps can be taken to improve this. If t_0 is not a uniformly generated, triangle mesh, then each spring will need a stiffness relative to its original length in S_0. Additionally, since the axial and circumferential stiffness of the tissue are different, the material properties of tissue should be incorporated through a change in the spring constant as well. A non-linear spring system may also prove to represent the true elastic nature of the tissue more completely.

3.3   Thickness Calculation

The wall thickness at every node in S_3 can now be calculated. The deformed mesh has elements of different sizes than originally and because the topology was never altered, the change in area between $t_{0,i}$ and $t_{3,i}$ can be compared directly. Since the Poisson's ratio of tissue is near 0.5, the change in wall thickness will be inversely proportional to the change in an elements area. At each node $n_{3,i}$, the wall thickness, $d_{3,i}$, is related to the average ratio of the change in area of the neighboring triangles,

$$d_{3,i} = d_{0,i} \frac{\sum_{j \in V_i} \frac{A_{0,j}}{A_{3,j}}}{v_i}, \qquad (4)$$

where $A_{0,j}$ and $A_{3,j}$ are the area of triangles $t_{0,j}$ and $t_{3,j}$ respectively and $d_{0,i}$ is the initial thickness, which is assumed to be uniform.

Geometry and a mesh for the blood vessel wall could be created directly from this information. It should be noted that while S_3 has the wall thickness values associated with the location of its nodes, S_1 has a significantly higher quality mesh because it was never stretched. The barycentric coordinates of $n_{1,i}$ within $t_{3,j}$ should be used to map the wall thicknesses to S_1 since the geometry of S_1 and S_3 match. Figure 4 shows the final wall thickness, the outermost layer, and the fluid mesh, which is discussed in Section 4. As can be seen in the figure, there can be a significant change in the wall thickness through a very small distance.

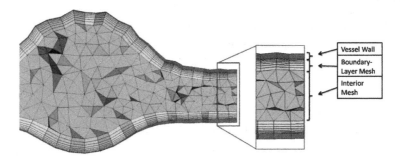

Fig. 4. The vessel wall (3440 wedges) with the boundary-layer mesh (17200 wedges) and interior mesh (6913 tetrahedra) for a fusiform aneurysm.

4   Boundary-Layer Meshing

Once the wall thickness has been found, a mesh of the fluid domain needs to be generated in order to run FSI simulations. To properly capture the rapidly changing velocity within the boundary layer, anisotropic wedges are used along the walls and created in an advancing-front manner [14]. Every surface node has a direction and size that controls the growth of the wedges to prevent a collision of fronts and tangling of elements. After a specified number of layers have been generated, the remaining space is filled with tetrahedra.

The growth of the boundary-layer mesh is dictated by a hair at each node. A hair is a vector connected to a node. The initial direction of the hair is the normal of the surface, while its length is chose to provide near isotropic elements in the outermost layer. The hairs are then smoothed to make sure the wedges do not tangle.

Smoothing the direction is accomplished by moving a hair to the average of its neighbors. If the difference between the hair and the average is larger than 30° the hair is instead moved iteratively to allow its neighbors to move towards it as well. Though not common in biological models, there is a chance a hair could point outside the volume. In these rare instances the hair is marched towards a location where the dot product between it and all the neighboring surfaces is positive, if it exists.

The length of the hair is then smoothed with regards to both the geometry and its neighbors. Concerning the geometry, using a uniform length for every hair would produce increasingly skewed elements near sharp corners. To remedy this, the length of a hair is increased relative to the angle of the corner. This ensures perpendicular corners remain perpendicular. Additionally, a hair's length should also be smoothed with respect to its neighbors. Using the average length would cause a ballooning of the boundary-layer mesh around corners. Instead, the length becomes the average of the normal component of its neighbors' hairs.

Even though the boundary-layer mesh can now grow with well-shaped elements, there is still a chance for fronts from competing surfaces, and even adjacent hairs in high-curvature regions, to intersect. To avoid these collisions, a sizing function based on the radius of the medial-ball is used [20]. The radius of the medial-ball is placed in a background octree-mesh and the values are smoothed with a quadratic, inverse-distance weighted interpolation. If the length of a hair is larger than the local sizing-function, it is shortened. This method for collision avoidance allows a wider range of length scales to be present in the biological models without having to specify proper sizing for each part of the boundary-layer mesh.

5   Results

Models for the fluid region of a fusiform and saccular aneurysm were created to represent both the healthy blood vessel and one with an aneurysm. A fusiform aneurysm occurs when the entire circumference of the blood vessel dilates, though it is commonly not axisymmetric. A saccular aneurysm is where the growth occurs in a small region of the wall and exhibits necking. With these models, the method presented was used to deform the healthy vessel into the unhealthy one in order to yield geometry with a varying thickness for the wall. These were compared against measured values for wall thickness and FSI analyses were performed to validate the method against expected results.

As shown in Figure 1, the healthy geometry for the fusiform set was a cylinder with a diameter of 2 cm and a length of 24 cm. The aneurysm, with a diameter of 6 cm, was placed in the center of the healthy vessel with an offset from the axis of 0.75 cm. The initial wall thickness was calculated to be 1.50 mm. After deformation, the minimum wall thickness was found to be 0.22 mm, with an average of 1.11 mm. The deformation can be seen in Figure 1 and the final wall in Figure 4. With 2272 triangles and 1152 nodes in S_0 and 1720 triangles and 876 nodes in S_1, it took 200 s to create the parameterizations, deform the surface, relax the springs, and calculate the final wall thickness. Surgery would be recommended for an aneurysm of this size and the wall thickness is near the limit measured by [21]. The saccular aneurysm had a maximum diameter of about 4 cm and was placed off-center from the axis of the 2 cm diameter vessel. The minimum thickness of the wall was 0.41 mm with an average of 1.23 mm, see Figures 5 and 6. With 6588 triangles and 3320 nodes in S_0 and 7968 triangles and 4010 nodes in S_1 it took 2771 s to determine the wall thickness. In both cases the majority of the time was taken to relax the spring system and needs to be looked at for efficiency. All the surface were meshed using the advancing front algorithm in CUBIT [24].

The vessel wall and fluid volume of the fusiform geometry were then meshed in order to perform FSI simulations. The wall was meshed with 2 layers for a total of 3440 wedges. The wedges in the wall had a maximum dihedral-angle (MDA) range between 90°–106°. Due to the dimensions of the blood vessel, the boundary layer plays a particularly important role in the development of the flow and the resultant WSS. As a result the mesh in

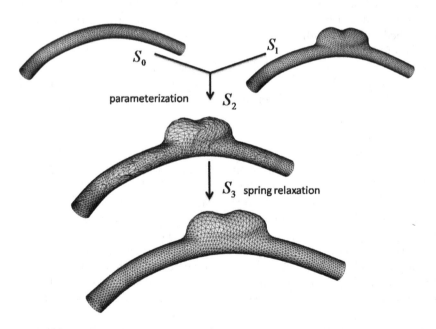

Fig. 5. Deformation process of the saccular aneurysm. S_0 and S_1 contain 6588 and 7968 triangles respectively.

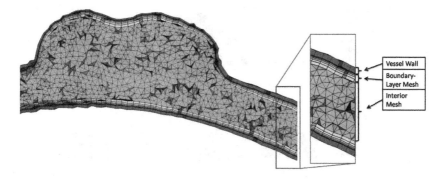

Fig. 6. The vessel wall (15936 wedges) with the boundary-layer mesh (79680 wedges) and interior mesh (57852 tetrahedra) for the saccular aneurysm.

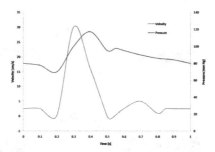

Fig. 7. The cardiac cycle used in the FSI analyses for the inlet, velocity, and outlet, pressure, boundary conditions.

the region of the boundary layer should be sized appropriately in order to capture the steep and changing gradient. The fluid-volume mesh was created using the advancing-front method presented and implemented in CUBIT for boundary-layer meshing and consisted of 17200 wedges (10 layers along the wall) and 6913 tetrahedra. There were 20 wedges with a MDA over 150° and an aspect ratio range of 2-63; the aspect ratio is the ratio between the longest and shortest edge lengths of an element. The fluid-domain tetrahedra had a MDA between 73°–135°.

The FSI analyses were run to compare the WSS and mesh displacement between simulations with uniform and variable wall thicknesses for the fusiform model. They were run with ANSYS® and ANSYS®, CFX™. The blood flow was approximated as a Newtonian fluid with a dynamic viscosity of $\mu = 3.85$ cP and solved using a shear stress transport model. A simplified representation of the cardiac cycle, Figure 7, was used to provide boundary conditions giving an inlet velocity with a peak of 31 cm/s at 0.3 s and an outlet pressure with a peak of 118 mm Hg at 0.4 s. The vessel wall was treated as an

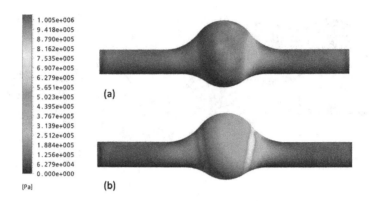

Fig. 8. The shear stress at 0.4 s for the (a) uniform and (b) variable wall thickness.

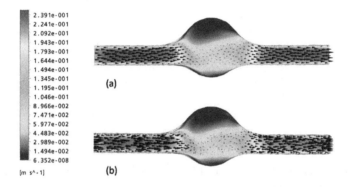

Fig. 9. The velocity with a vector field at 0.4 s for the (a) uniform and (b) variable wall thickness.

isotropic material with a Young's modulus of $E = 2.7$ MPa, a density of $\rho = 2000$ kg/m³, and a Poisson's ratio of $\nu = 0.45$ and the ends were fixed in place.

The analysis was first performed on the geometry with a uniform wall thickness (1.5 mm). With a pulsatile flow the maximum WSS was 0.31 MPa and occurred on the anterior and posterior regions of the aneurysm, Figure 8(a). A second simulation was run using the variable wall-thickness determined by the proposed method. The results match well with what is expected. The max WSS for the variable wall thickness was 1.04 MPa at about 0.4 s and can be seen in Figure 8(b). Using a variable wall thickness produced a 330% increase in the WSS even though the aneurysm geometry was unchanged. This is similar to what was seen in the work of Scotti and is below the yield stress shown by Raghavan (a mean of 1.27 MPa). The velocity profiles, Figure 9, are fairly similar between the two runs. Small differences

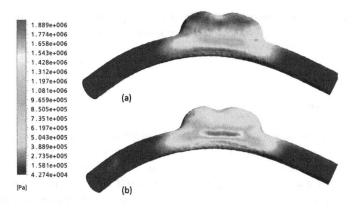

Fig. 10. The wall shear stress at 0.4 s for the saccular geometry with a (a) uniform and (b) variable wall thickness.

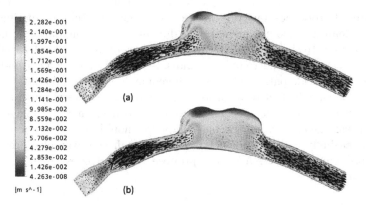

Fig. 11. The velocity with a vector field at 0.4 s for the saccular geometry with a (a) uniform and (b) variable wall thickness.

can be seen between the two though as a result of the variable wall being capable of more deformation in the thinner regions.

The same analyses was also performed on the saccular aneurysm in order to show results for a more complex model. There were 15936 wedges in the wall and 79680 in the blood volume, which also contained 57852 tetrahedra. The MDA for the tetrahedra ranged between 72°–140° in the fluid region. The wedges in the wall had a MDA between 90°–114° and in the fluid domain was between 90°–140° with an aspect ratio of 2.5–110. The analysis again showed a velocity profile that was very similar between the uniform and variable walls, a cross-section can be seen in Figure 11. The WSS, in the variable wall thickness model, had a max value of 1.89 MPa around 0.4 s and was not where failure would be predicted, see Figure 10(b). This is likely a result of the geometry created – it was not medical data – producing a stress concentration and not indicative of the method. The thinnest regions in the variable model show a maximum WSS of 1.31 Mpa and is significantly closer to the expected value and location. There is more than a 250% increase when compared against the WSS in the same region of the uniform model (0.49 Mpa).

6 Conclusion

The proposed method provides a very good start to improve patient-specific geometry for simulations of aneurysms. Having shown how significantly the wall shear stress changes with the addition of variable wall thicknesses, it can only be argued that better predicting a patient's unique geometry is crucial. That being said, the growth of an aneurysm is more complicated than deforming an isotropic material with a uniform wall thickness. To start, the blood-vessel wall is not isotropic and repairs itself while being stretched. Moreover, as the patient ages, calcification and plaquing will locally alter the material properties and wall thickness in a way that fully predicting how the wall deforms becomes increasingly difficult. More boundary conditions should be used to simulate the natural, axial tension in the blood vessel and external pressure from surrounding organs. Regardless, having a method that can start to provide patient-specific wall geometry is a needed step.

Future work will start to incorporate some of these ideas through the use of non-linear, anisotropic springs. In addition to using medical image data for aneurysm geometry, more robust mappers should be used to accommodate increasingly complex geometry. Even though the examples presented were approximations of AAAs, the proposed method is extendable to any aneurysm.

References

1. Bazilevs, Y., Hsu, M.C., Zhang, Y., Wang, W., Liang, X., Kvamsdal, T., Brekken, R., Isaksen, J.: Computational vascular fluid-structure interaction: Methodology and application to cerebral aneurysms. Submitted to Biomechanics and Modeling in Mechanobiology (2009)
2. Chen, P., Barner, K.E., Steiner, K.V.: A displacement driven real-time deformable model for haptic surgery simulation. In: HAPTICS 2006: Proceedings of the Symposium on Haptic Interfaces for Virtual Environment and Teleoperator Systems, Washington, DC, USA, pp. 499–505 (2006)
3. Cui, T., Song, A., Wu, J.: Simulation of a mass-spring model for global deformation. Frontiers of Electrical and Electronic Engineering in China 4(1), 78–82 (2009)
4. Eck, M., DeRose, T., Duchamp, T., Hoppe, H., Lounsbery, M., Stuetzle, W.: Multiresolution analysis of arbitrary meshes. In: SIGGRAPH 1995: Proceedings of the 22nd annual conference on Computer graphics and interactive techniques, pp. 173–182. ACM Press, New York (1995)
5. Elliott, M.R., Thrush, A.J.: Measurement of resolution in intravascular ultrasound images. Physiological Measurement 17(4), 259–265 (1996)

6. Figueroa, C.A., Vignon-Clementel, I.E., Jansen, K.E., Hughes, T.J.R., Taylor, C.A.: A coupled momentum method for modeling blood flow in three-dimensional deformable arteries. Computer Methods in Applied Mechanics and Engineering 195(41–43), 5685–5706 (2006)
7. Fleming, C., Whitlock, E.P., Beil, T.L., Lederle, F.A.: Screening for abdominal aortic aneurysm: A best-evidence systematic review for the u.s. preventive services task force. Annals of Internal Medicine 142(3), 203–211 (2005)
8. Floater, M.S.: Parametrization and smooth approximation of surface triangulations. Computer Aided Geometric Design 14(3), 231–250 (1997)
9. Freitag, L.A., Plassmann, P.: Local optimization-based simplicial mesh untangling and improvement. International Journal for Numerical Methods in Engineering 49, 109–125 (2000)
10. Garimella, R.V., Shephard, M.S.: Boundary layer mesh generation for viscous flow simulations. International Journal for Numerical Methods in Engineering 49, 193–218 (2000)
11. Güdükbay, U., Özgüç, B., Tokad, Y.: A spring force formulation for elastically deformable models. Computers & Graphics 21(3), 335–346 (1997)
12. Hose, D.R., Lawford, P.V., Narracott, A.J., Penrose, J.M.T., Jones, I.P.: Fluid-solid interaction: Benchmarking of an external coupling of ANSYS with CFX for cardiovascular applications. Journal of Medical Engineering & Technology 27(1), 23–31 (2003)
13. Isaksen, J., Bazilevs, Y., Kvamsdal, T., Zhang, Y., Kaspersen, J.H., Waterloo, K., Romner, B., Ingebrigsten, T.: Determination of wall tension in cerebral artery aneurysms by numerical simulation. Stroke 39, 3172–3178 (2008)
14. Johnson, E., Yamakawa, S., Shimada, K., Brewer, M.L., Owen, S.J.: Generating hexahedral boundary-layer meshes for CFD simulations (Manuscript in preparation) (2009)
15. Ketcham, R.A., Carlson, W.D.: Acquisition, optimization and interpretation of x-ray computed tomographic imagery: Applications to the geosciences. Computers and Geosciences 27(4), 381–400 (2001)
16. Khawaja, A., Kallinderis, Y.: Hybrid grid generation for turbomachinery and aerospace applications. International Journal for Numerical Methods in Engineering 49, 145–166 (2000)
17. Kim, J., Wang, S., Zeng, Y., Wang, Y., Gu, X., Qin, H., Samaras, D.: Ricci flow for 3D shape analysis. In: Proceedings of IEEE International Conference on Computer Vision 2007, Rio de Janeiro, Brazil (October 2007)
18. Kroon, M., Holzapfel, G.A.: Modeling of saccular aneurysm growth in a human middle cerebral artery. Journal of Biomechanical Engineering 130(5), 051012.1–051012.10 (2008)
19. Mabotuwana, T.D.S., Cheng, L.K., Pullan, A.J.: A model of blood flow in the mesenteric arterial system. BioMedical Engineering OnLine 6, 17 (2007)
20. Quadros, W.R., Owen, S.J., Brewer, M.L., Shimada, K.: Finite element mesh sizing for surfaces using skeleton. In: Proceedings, 13th International Meshing Roundtable, Williamsburg, VA, USA, pp. 389–400 (2004)
21. Raghavan, M.L., Kratzberg, J., de Tolosa, E.M.C., Hanaoka, M.M., Walker, P., da Silva, E.S.: Regional distribution of wall thickness and failure properties of human abdominal aortic aneurysm. Journal of Biomechanics 39(12), 3010–3016 (2006)

22. Sahni, O., Müller, J., Jansen, K.E., Shephard, M.S., Taylor, C.A.: Efficient anisotropic adaptive discretization of the cardiovascular system. Computer Methods in Applied Mechanics and Engineering 195(41–43), 5634–5655 (2006)
23. Sander, P.V.: Sampling-efficient mesh parametrization. PhD thesis, Harvard University (2003)
24. Sandia National Laboratory. Cubit: Geometry and mesh generation toolkit
25. Scotti, C.M., Jimenez, J., Muluk, S.C., Finol, E.A.: Wall stress and flow dynamics in abdominal aortic aneurysms: Finite element analysis vs. fluid-structure interaction. Computer Methods in Biomechanics and Biomedical Engineering 11(3), 301–322 (2008)
26. Scotti, C.M., Shkolnik, A.D., Muluk, S.C., Finol, E.A.: Fluid-structure interaction in abdominal aortic aneurysms: Effects of asymmetry and wall thickness. BioMedical Engineering OnLine 4, 64 (2005)
27. Sheffer, A., Praun, E., Rose, K.: Mesh parameterization methods and their applications. Foundations and Trends in Computer Graphics and Vision 2(2), 105–171 (2006)
28. Terzopoulos, D., Platt, J., Barr, A., Fleischer, K.: Elastically deformable models. Computer Graphics 21(4), 205–214 (1987)
29. Torii, R., Oshima, M., Kobayashi, T., Takagi, K., Tezduyar, T.E.: Fluid-structure interaction modeling of blood flow and cerebral aneurysm: Significance of artery and aneurysm shapes. Computer Methods in Applied Mechanics and Engineering (2008) doi:10.1016/j.cma.2008.08.020
30. Tutte, W.T.: How to draw a graph. London Mathematical Society 13, 743–768 (1963)
31. Valencia, A.A., Guzmán, A.M., Finol, E.A., Amon, C.H.: Blood flow dynamics in saccular aneurysm models of the basilar artery. Journal of Biomechanical Engineering 128, 516–526 (2006)
32. Yin, X., Dai, J., Yau, S.-T., Gu, X.: Slit map: Conformal parameterization for multiply connected surfaces. In: Chen, F., Jüttler, B. (eds.) GMP 2008. LNCS, vol. 4975, pp. 410–422. Springer, Heidelberg (2008)
33. Zhang, Y., Bazilevs, Y., Goswami, S., Bajaj, C., Hughes, T.J.R.: Patient-specific vascular NURBS modeling for isogeometric analysis of blood flow. Computer Methods in Applied Mechanics and Engineering 196(29–30), 2943–2959 (2007)
34. Zhang, Y., Wang, W., Liang, X., Bazilevs, Y., Hsu, M.C., Kvamsdal, T., Brekken, R., Isaksen, J.: High fidelity tetrahedral mesh generation from medical imaging data for fluid-structure interaction analysis of cerebral aneurysms. Computer Modeling in Engineering & Science 42(2), 131–150 (2009)

Hybrid Mesh Generation for Reservoir Flow Simulation in CPG Grids

T. Mouton[1], C. Bennis[1], and H. Borouchaki[2]

[1] Institut Français du Pétrole,
1 et 4 avenue de Bois Préau,
92852 Rueil-Malmaison Cedex, France
thibaud.mouton@ifp.fr, chakib.bennis@ifp.fr
[2] Université de Technologie de Troyes,
12 rue Marie Curie,
BP 2060 10010 Troyes Cedex, France
houman.borouchaki@utt.fr

Abstract. In this paper, we introduce a new method to generate a hybrid mesh from a CPG (Corner Point Geometry) reservoir grid and a radial mesh around a well. The method is an extension of the approach proposed in [1] to the case of CPG grids with high level of deformation. This ensures a fully functional mesh generation for realistic cases. The main idea is first to construct a mapping between the real space containing the CPG grid along with the radial mesh of the well and a virtual space where the CPG reservoir grid becomes a Cartesian grid. Then, because this mapping damages the circular property of the radial mesh, an appropriate radial mesh is built in the virtual space and the initial mapping is modified by taking into account the new radial mesh in the virtual space. To this end, an optimization technique using mesh refinement procedures is applied. The mapping combined with the mentioned deformation allows us to generate an unstructured polyhedral transition mesh (between the reservoir grid and the radial mesh) in the virtual space using the algorithm proposed in [1]. Finally, coming back to the real space, the obtained hybrid mesh may require a post processing step to recover the requested finite volume properties.

Introduction

The new technological improvements in 3D seismic imagery and drilling/production enable today to obtain a realistic and faithful image of the internal architecture of the reservoir and to drill deviated and complex 3D wells with several levels of ramification. Well trajectories can be well adapted to the geometry of the reservoir in order to optimize its production. In this new technological context, the mesh generation becomes a crucial step in the reservoir flow simulation of new generation. Meshes allow us to represent the geometry of the geological structures with discrete elements on which the simulation is processed. A better comprehension of the physical phenomena requires us to simulate 3D multiphase flows in increasingly complex geological structures, in the vicinity of several types of singularities such as complex wells. All these complexities must initially be taken into account within the mesh construction and the mesh must faithfully represent all this heterogeneous information.

Whereas the classical meshes are totally structured or unstructured, in [5], a hybrid mesh model was proposed in 2D to capture the radial characteristics of the flow around the wells. It combines the advantages of the structured and unstructured approaches, while limiting their disadvantages. In [3], the generation methodology was extended to 3D case where the reservoir grid is Cartesian. The hybrid mesh is composed of a structured hexahedral mesh describing the reservoir field, structured radial meshes adapted locally to flow directions around each well and unstructured polyhedral meshes (based on power diagrams) connecting the two structured meshes.

In this paper, the generation of the hybrid mesh is extended to the 3D highly deformed cases. In Section 1, we present the numerical constraints imposed by finite volume schemes that will govern the mesh construction. In Section 2, we quickly recall the methodology used to generate a hybrid mesh in 3D and introduce the problems induced by the cartesian methodology in the real reservoirs grids. A solution to generate a hybrid mesh in such a grid is given in section 3.

1   Problem Statement

We present here a quick overview of the problem. In order to achieve numerical simulations of a phenomenon, we need first to establish a mathematical model of the phenomenon. This model is then discretized in order to define a numerical scheme. Finally, this numerical scheme is applied to a grid which is the discretization of the space where the phenomenon occurs.

1.1   Mathematical Model of Flow Simulation

The reservoir simulation is the whole set of operation allowing the modelisation behaviour of a petroleum reservoir. The aim of this simulation is to drive the reservoir exploitation and to argue the different technical choices to make. In a petroleum reservoir, flows are the consequences of variations in space and time of pressure and saturation in the water, oil and gas phases. These variations are induced by the production or the injection of one of this fluids into the different wells. The mathematical modelisation of the flow allows to take into account simultaneously the whole set of physical relationships describing the interactions inside the medium. Let us consider the case of a compressible, diphasic flow without capillary pressure (the different phases have the same pressure) at a constant temperature and without chemical reaction. The global equation can be written as follows:

$$\rho \frac{\mathrm{d}P}{\mathrm{d}t} + div(\rho \frac{\mathsf{K}}{\mu} (-\mathbf{grad\ P} + \rho\ \mathbf{g})) = 0\,. \tag{1}$$

where P is the pressure, ρ the density of the fluid, $\mathbf{g}$ the gravity, K the permeability tensor and μ the viscosity of the fluid.

1.2 Numerical Schemes

Most common numerical schemes used in reservoir flow simulation are volume finite schemes. They use the mass conservation law for two adjacent cells. Each discretization point is associated with a control volume. This control volume is defined by a set of bounding faces.

Considering that the pressure in a cell is constant, for a face $\partial C_1 \cap \partial C_2$ separating a cell C_1 to a cell C_2, if the face normal $\mathbf{n} \simeq \frac{\mathbf{o_1 o_2}}{o_1 o_2}$, then we can approximate the flow between both cells through the face with area A_i by

$$- \int_{\partial C_1 \cap \partial C_2} \mathsf{K} \, \mathbf{grad} \, \mathbf{p} \simeq A_i \, \mathsf{K} \, \frac{p_{C_2} - p_{C_1}}{o_1 o_2} \, \mathbf{n}. \qquad (2)$$

This way to approximate the flow requires meshes to respect some conditions:

1. **Mesh conformity**, ensuring a face to face connection.
2. **Mesh orthogonality**, ensuring that the normal vector of the face is colinear to the line joining the centers of the adjacent cells.
3. **Face planarity**, so that the flow approximation is more precise.
4. **Auto centering**, ensuring that the cell center lies inside the cell.

1.3 Meshes Overview

In order to represent the reservoir using discrete elements, the hybrid mesh is composed of three kinds of elementary meshes:

- A structured CPG grid, respecting the geological features, is used to describe the reservoir field.
- To gain accuracy at the drainage areas, a structured radial circular mesh adapts locally to the radial nature of the flows around the wells.
- Finally, these structured meshes are connected together by the use of unstructured polyhedral mesh with respect to conformity and finite volume properties.

While the structured grid generation is a well known process, the construction of the unstructured transition mesh in 3D constitutes a major issue. The structured CPG mesh of the reservoir grid is constructed through the use of transfinite interpolations, projections onto the geological interfaces (horizons and faults) and a relaxation procedure [2]. The structured radial mesh is computed given the well's trajectory, the drainage area radius and the progression of cells' size. The separate construction of these grids leads to incompatibilities due to a lack of common structure and a transition mesh is needed to carry out a correct connection.

Then, to generate such a transition mesh, a method using *power diagrams* was introduced in [5]. As a generalisation of Voronoï diagrams, power diagrams provide polyhedral, convex, and orthogonal cells. In addition, these allow here to reach the mesh conformity which would not be possible using Voronoï diagrams. In [3], the generation methodology was extended to 3D case where the reservoir grid is Cartesian.

2 Hybrid Meshes and Non Cartesian Grids

2.1 General Methodology and Its Restriction

The general methodology used to build an hybrid mesh from a reservoir grid and a radial mesh has been detailled in [1]. This methodology consists of the following steps:

1. the definition of the volume (called *cavity*) inside which will be built the polyhedral mesh and the boundaries of this volume,
2. the definition of the centers (called *sites*) of the polyhedral cells,
3. the weighted Delaunay triangulation of the sites previously defined and the polyhedral mesh computation.
4. the adjustment of the polyhedral mesh using topological considerations.

This methodology especially details how to link a radial mesh to a cartesian grid modelling the reservoir by the use of a polyhedral transition mesh. It is important to note that using this methodology must be restricted to the Cartesian reservoir grids. Unfortunately, the Cartesian modelling of a reservoir is not enough to take into account all its geological complexities. On the other hand, the Corner Point Geometry grids are able to handle non geological caracteristics like faults, sedimentation and erosion much more realistically. The problem occurring with such grids is that they do not respect some essential conditions for the hybrid mesh generation, in particular the planarity of the cavity faces, the cocyclicity of the cavity faces and the Delaunay admissibility of the cavity edges.

The methodology used to build the radial mesh is such that it ensures this properties are respected. That is why, cavity faces resulting from the radial mesh are always valid. In contrary, CPG grids are composed of quadrilateral faces whose vertices are not planar and thus non co-circular. Furthermore, this kind of grid does not ensure the Delaunay admissibility of the cavity edges, inherent properties of cartesian grids. Consequently, the methodology developped for the cartesian reservoir grid is not applicable to the CPG reservoirs.

2.2 A Solution Using Grid Deformation

At first, (because we will often talk about it), it is important to explain what we call *grid deformation*. The undeformed grid is the cartesian grid. In a

CPG grid, the cell faces are aligned with the geological structures of the real reservoir. By doing such thing, the cells of the reservoir are modified against the cartesian position, so that they are deformed. Quantification of such a deformation is not what we care about, whereas we need to ensure that all edges of the mesh are Delaunay admissible.

A grid deformation approach was proposed in [1]. This previous work consisted in deforming the CPG grid of the reservoir in order to make it cartesian. The aim was to restore the co-cyclic properties of the external faces of the *core*. This approach is based on two different spaces : a space called *real* ($\mathcal{E}_{real}$) in which the reservoir mesh is CPG and a space called *reference* ($\mathcal{E}_{reference}$), image of $\mathcal{E}_{real}$ where this one has a cartesian geometry. A bijective function ϕ ensures the definition of the image of one point from $\mathcal{E}_{real}$ to $\mathcal{E}_{reference}$.

When the reservoir grids are highly deformed, the radial mesh image computed using ϕ is too deformed and the essential conditions are not fulfilled anymore. The polyhedral mesh generation can not be achieved directly using the cartesian methodology, or the resulting mesh will not correspond to the cavity boundaries topology. The aim of the work we are talking in this paper is to develop a methodology which restore the necessary geometrical properties of the radial mesh surface. The recovery is achieved by an extension of the function ϕ computation.

3   Hybrid Meshes for Real Grids with High Deformation

As we explained in the previous section, the solution based on the grid deformation improve the external boundary of the cavity. Unfortunately, this solution impacts the radial grid which suffers the reservoir grid deformation. We present here an extension of this method which ensures that both radial grid and reservoir grid remain optimal for power diagram generation. To do so, we first modify the way defining the interpolation space, and we develop a set of tools allowing us to compute the intended mapping.

3.1   Mapping the Cavity Space

In the following, we define the *core* as the whole set of simplices covering the cavity intended to be filled by the polyhedral mesh.

The global definition of the function ϕ is the sum of the elementary functions ϕ_i defined inside each element of the *core*. An elementary function ϕ_i is defined as follows : Let P be a point contained by a tetrahedron T ($volume(T) \neq 0$). A, B, C and D are the tetrahedron vertices. P can be written as a linear combination of the four vertices:

$$P = \alpha A + \beta B + \gamma C + \delta D \qquad (3)$$

where α, β, γ et δ are the barycentric coordinates of point P associated with A, B, C and D. The global function ϕ is the combination of all the elementary functions. Ensuring that the different tetrahedra do not overlap, the global function is also bijective on the whole *core*. Knowing the barycentric coordinates of a point in a tetrahedron T, the computation of its new position is obvious. Thus, we have a simple piecewise linear interpolation function that can be defined on any arbitrary volume.

The aim of our methodology is to find a bijective mapping between both $\mathcal{E}_{real}$ and $\mathcal{E}_{reference}$ spaces. In the $\mathcal{E}_{reference}$ space, both surfaces on which will fit the transition mesh have to fulfill the geometrical requirements. The whole methodology is illustrated in Figure 1.

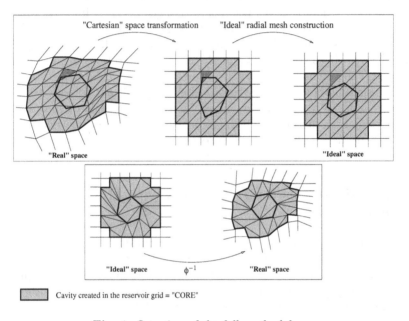

Cavity created in the reservoir grid = "CORE"

Fig. 1. Overview of the full methodology.

The first step was done by mapping the real reservoir grid to a non uniform reservoir grid which provide the $\mathcal{E}_{reference}$ space. This mapping is a temporary snapshot of the function ϕ, but properties of radial mesh are not preserved. The problem is that using this mapping, the obtained radial mesh is not suitable to apply the polyhedral mesh generation algorithm. Thus, we start defining a way to obtain an ideal radial mesh in the $\mathcal{E}_{reference}$ space. In reservoir modelling, physical wells are simply modelled by a curve following the trajectory of the well. In order to obtain a surface with the good properties, a simple idea is to build another radial mesh from the update trajectory ($= \phi$(initial trajectory)) of the well in the reference space. This process allows us to obtain a new external surface. The reconstruction method is the standard methodology used to build a radial mesh. This methodology is a simple *sweep* of a circular section along the well trajectory. We now call

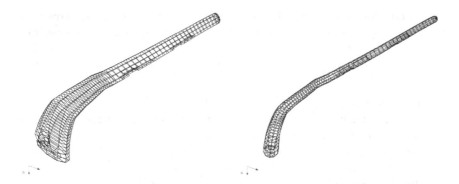

Fig. 2. Deformed radial mesh surface and its image using reconstruction.

$\mathcal{S}_{radial}$ the ideal surface obtained after radial mesh reconstruction. Both deformed and ideal surfaces (see Figure 2) have the same topology, i.e. the same connectivity between vertices.

At this point, we have the two different geometries which correspond to the surfaces in the real and the ideal spaces. The main objective is then to map the cavity volume using a function giving us the position of a point in one of the two spaces, knowing the position in the other space. It is obvious that vertices of $\mathcal{S}_{radial}$ do not conform to the initial mapping (see Figure 1, a specified vertex of the deformed $\mathcal{S}_{radial}$ is inside the horizontally hatched triangle whereas the same vertex of the ideal $\mathcal{S}_{radial}$ is inside the vertically hatched triangle). To correct the mapping, the simplest idea is to replace the tetrahedra obtained from the reservoir cells by an other set of tetrahedra obtained from the Delaunay triangulation of the vertices of the cavity boundaries. This ensures that the space is mapped with a set of tetrahedra without self intersection neither void and provides us with a piecewise representation of the cavity. In order to ensure the consistency of the mapping function, the cavity boundary must be preserved in the triangulation. This can be realized by generating the Delaunay triangulation in the ideal space. In the case where the deformation of the reservoir grid is small, the Delaunay triangulation remains valid in the real space. We thus have the space mapped with a bijective piecewise function. Unfortunately, this is not the case if we apply this technique on a highly deformed reservoir grid. In such a case, the bijectivity of the function will not be ensured and an additional step is necessary to restore the bijectivity of the mapping function.

3.2   Correction of the Mapping

Point insertion algorithm

In the case of highly deformed reservoir grids, the triangulation built in the ideal space will not be valid in the real space. Indeed, some of the elements can be invalid which creates locally self intersections. An optimisation procedure has been proposed to restore the validity of the triangulation. The basic idea is to insert new vertices in the triangulation in order to catch the bending induced by the mapping between the two spaces. It consists on the following steps:

1. mark the invalid tetrahedra in the real space,
2. insert the barycenter of the marked tetrahedron in the delaunay triangulation, thus in the ideal space (step 1 in Figure 3),
3. compute the best position of the inserted vertices so that their associated ball of tetrahedra is valid (step 2 in Figure 3),
4. stop if no invalid tetrahedra remain, else loop again.

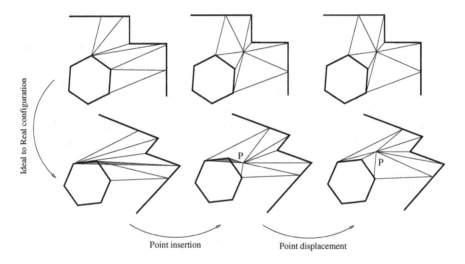

Fig. 3. Point insertion algorithm.

The insertion of vertices in the triangulation is done in the ideal space. By consequence, all tetrahedra are valid in the ideal space. On the other hand, this can not be ensured in the real space, and inverted tetrahedra can be produced. We thus have to find a position for the inserted vertex that make simultaneously all its adjacent tetrahedra non inverted when it is possible. The methodology used to compute the best position of the free vertex is discussed below.

Untangling algorithm using an objective function

The optimization loop we described involves to find an optimal position for the vertex of a ball of tetrahedra. An efficient algorithm was published in [4] which consists in minimizing the following objective function:

$$f(p_i) = \sum_{j=0}^{N_j} (|V_j| - V_j)^2 \, . \tag{4}$$

where V_j is the volume of the jth tetrahedron adjacent to vertex p_i. The convexity of such a function is provable, so that a local minimum is the global minimum of the function. Due to the absolute value, the function is not differentiable. Consequently, the function is not "smooth" and there is not an unique minimum but a *region* where the function has the same value, corresponding to the location of p_i where no tetrahedra has a null volume. If such a region does not exist, then there is a single minimum which minimizes the sum of the square of negative volumes. Nevertheless, negative volumes remain in this case.

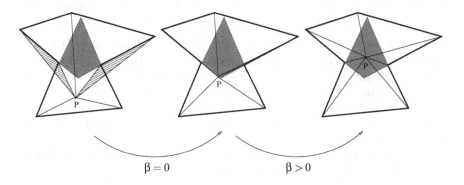

$\beta = 0$ $\beta > 0$

Fig. 4. Point displacement algorithm.

Minimizing the above function using a steepest descent algorithm leads to find a point on the boundary of the stationnary region. In this case, at least one tetrahedron has a null volume. This is not what was initially wanted, because tetrahedra with a null volume create a singularity in the deformation function. Our aim is thus to find the best location for the vertex p_i inside the "stationnary region" (grey area in Figure 4). To achieve such an objective, an idea is to cheat about the computed volume of the tetrahedra. Whereas the volume of a considered tetrahedron is null, a small value is subtracted to the computed volume, so that the declared volume is still negative. That is why a parameter β is introduced, which has the effect of reducing the area of the stationnary region and thus allowing the inserted vertex to "enter" in the "stationnary region". Thanks to this parameter, at convergence, there is not anymore tetrahedra with a null volume. The objective function becomes:

$$f(p_i) = \sum_{j=0}^{N_j} (|V_j - \beta| - (V_j - \beta))^2 \, . \tag{5}$$

The main difference with the algorithm presented in [4] is that our minimization includes two different steps. First, we do the computation with $\beta = 0$, then we compute the parameter β considering the local configuration. Finally, we launch again the minimization of the function with the new value of β, starting from the vertex position found at the first step.

Remark: An important point is the determination of the value of the parameter β. This parameter must be computed so that the smallest volume is maximized. If β is too small, at least one tetrahedron would keep a volume very small, whereas an other position of p_i would give a higher volume. If β is too high, the computed position can be outside the "stationnary region". The parameter β cannot be a priori defined because it depends on the configuration encountered. In practice, we compute the volume of the smallest tetrahedron which has a non null volume and associate this value (divided by the number of null volume plus one) to β. The aim is to distribute this smallest non null volume between the tetrahedra having a null volume.

As we said, the "stationnary region" we discussed could also not exist for the given configuration. In such a case, the final value of the objective function is more than zero and the point is just not moved. Thus, we loop over all free vertices and we wait until one or more vertices of a given configuration is moved. If no vertex move occurs within an entire loop, the configuration is blocked and new free vertices have to be inserted in order to allow the displacement of the blocked vertices.

The final triangulation provides us a way to transpose the polyhedral mesh (built in the ideal space) in the real space.

4   Numerical Example

We give here an example of an hybrid mesh built inside a deformed reservoir grid. The reservoir grid (see Figure 6) is made of 18000 cells and is considered to be deformed enough to be a good example of what could be encountered in general. The radial mesh is a classical grid built around the well trajectory containing 2200 cells.

Figure 7 shows in detail the radial mesh in the real space and in the intermediate space. The intermediate space generated before the ϕ correction changes the reservoir grid into a cartesian one (step O_1 in Figure 5). The right picture shows the radial mesh in the intermediate space. The initial radial characteristic of the mesh is lost.

The radial mesh is then rebuilt (step O_2) and the Delaunay triangulation of the vertices of the cavity is processed. The triangulation obtained is not valid in the real space (143 invalid tetrahedra). The vertex insertion algorithm is thus launch and after 6 insertion loops and 101 vertices finally inserted, no invalid tetrahedron remains.

The final polyhedral mesh is obtained using the cartesian methodology (step O_3) and the deformation defined by ϕ^{-1} (step O_4). Graphical results of

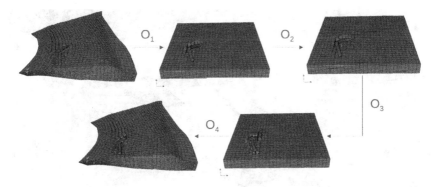

Fig. 5. Detailled description of the generation methodology.

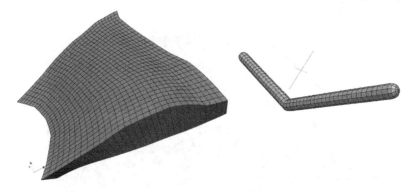

Fig. 6. Initial reservoir grid (30 × 30 × 20 cells) and radial grid around the well (5 × 10 × 44 cells).

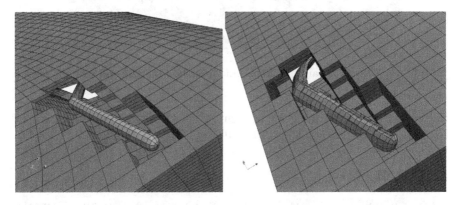

Fig. 7. Detail of the radial in the real space and in the intermediate space (cartesian configuration of the reservoir grid).

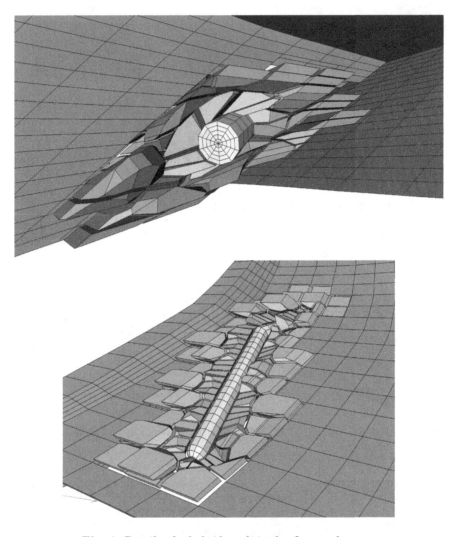

Fig. 8. Details of a hybrid mesh in the $\mathcal{E}_{real}$ real space.

the whole hybrid mesh are given in Figure 8. The generated polyhedral mesh contains 4887 vertices, 6432 faces and 950 cells.

5   Conclusions

The results obtained by the second methodology are encouraging and the hybrid mesh generation is now possible on realistic cases. However, some additional work needs to be done. The mapping could be well adapted by using, after the untangler, a smoother rather than setting a parameter depending of the configuration. Then, this method induces a severe deformation to the polyhedral mesh which makes it invalid for simulations. A correction step by means of optimization is required. Two kinds of optimization are used:

- A priori optimization: the deformation can be quite severe locally. The aim of this optimization is to refine locally the polyhedral mesh so that the deformation is distributed over several cells. The refinement process is applied while the deformation per cell remains too high.
- A posteriori optimization: due to the fact that the faces and edges of the mesh are not suited to the simulations, a classical mesh optimization based on geometrical criterions will be applied in the $\mathcal{E}_{real}$ space.

Finally, let's not forget that we are not looking for a polyhedral mesh that has perfect faces and edges. Indeed, a power diagram owning all the given cavity faces does not exist. That is the reason why we are looking for a compromise as close as possible to a power diagram. We now have a methodology to build a conformal polyhedral mesh from surfaces which do not suit the power diagram generation methodology requirements. This new approach is very promising an simulation tests on real data are planned in order to validate the efficiency of this technology.

References

1. Bennis, C., Borouchaki, H., Flandrin, N.: 3d conforming power diagrams for radial lgr in cpg reservoir grids. Eng. with Comput. 24(3), 253–265 (2008)
2. Bennis, C., Sassi, W.: Method to generate a 3d mesh conforming to the geometry of a specified body in order to build a representative model of this body. IFP patent, national registration number 96/04.567 (avril 1996)
3. Flandrin, N., Borouchaki, H., Bennis, C.: 3d hybrid mesh generation for reservoir flow simulation. In: International Meshing Roundtable, pp. 133–144 (2004)
4. Knupp, P.M.: Hexahedral and tetrahedral mesh untangling. Engineering with Computers 17, 261–268 (2001)
5. Boissonnat, J.-D., Balaven, S., Bennis, C., Yvinec, M.: Conforming orthogonal meshes. In: International Meshing Roundtable, pp. 219–227 (2002)

VECTIS Mesher – A 3D Cartesian Approach Employing Marching Cubes

Lukáš Plaček

Ricardo Prague, s.r.o.
Thámova 11-13
186 00 Prague
Czech Republic
Lukas.Placek@ricardo.com

Abstract. This paper describes the main principles used in the development of the VECTIS mesher. The mesher produces unstructured 3D meshes suitable for Finite Volume Methods. It is based on the Cartesian approach. In contrary to the traditional approaches which use exact shape of boundary faces of cut cells, this mesher employs Marching Cubes method for generation of majority of boundary faces. Only in problematic parts of the geometry, when the danger of chamfering of sharp features occurs or when watertightness of the cell might not be ensured, the Exact Fit method is used to produce the patches. Because two different methods are used for generation of patches, additional effort needs to be made to tie the boundary polygons to prevent gaps. A new algorithm for determining the most suitable configuration of triangles of Marching Cubes patterns is proposed. In cartesian meshers, a problematic situation occurs whenever triangles of the surface lay exactly on a side of the intersecting box. In order to prevent these collisions, an approach called Dual Levels has been introduced. The implemented method of cell refinement is presented. The paper also explains the way how the problem of cells that are too concave was resolved. The algorithm of the whole meshing task is described in detail. The new mesher has significantly lower time and memory demands in comparison with its predecessor. The main approaches responsible for this improvement are discussed.

1 Introduction

For many applications, it is advantageous to use meshing based on the Cartesian approach. Although this type of meshing have a drawback in lower quality of cells near boundary, there are also significant advantages. Cartesian meshers are robust, so when the input surface geometry is prepared in a reasonable quality (it is closed and there are no self-intersections) the meshing process does not require further interaction with the user. Because generation of inner parts of meshes is very easy, Cartesian meshers can quickly produce millions of cells.

There are two basic ways how Cartesian meshes can be generated. The **first approach** uses so called cut cells, which means that the boundary polygons have arbitrary shape and size defined by the intersection of the input geometry and the cutting box. When this approach is used, the solver that uses these meshes needs to be able to cope with the fact that neighbourhood of cells with significantly different size is possible. A finite volume scheme like this has been presented in [5]. Another possibility may be merging of cells (like in [3]). There is a lot of literature dealing with this kind of Cartesian meshing, for example [1], [2], [3] or [4]. The basic knowledge necessary to start working on the development of a Cartesian mesher is well summarised in [3]. In order to try to increase the quality of surface cells, the **second approach** is used by several development groups. In this case, the cartesian cells are generated only inside and then, the gap between the inner cells and the boundary is tried to be filled by cells with as high quality as possible. This approach is sometimes called I2B (interior to boundary). When using this way of mesh generation, the preservation of sharp features is the most challenging part of the task. This approach can be represented for example by these papers: [6], [7] or [8].

The mesher described in this paper uses the first approach (cut cells) and the generated 3D meshes are used by VECTIS solver, which is based on principles described in [5]. The mesher uses an unique approach of generation of patches. The exact shape of patches of cut cells (generated by a method in further text called Exact Fit) is used only in these boxes which are intersected by a sharp feature. All the other patches (the majority) are generated by Marching Cubes method (see [10]), which is very straightforward and quickly generates simple patches. Usage of Marching Cubes has a great advantage also in its ability to overcome problems with small flaws in the input geometry. The real world geometries often contain problems when healthy triangles are connected by a thin triangle with opposite orientation (this flaw is called "folded geometry"). When boundary faces are produced by Exact Fit, this kind of flaw leads to problematic patches and an additional cleaning algorithm needs to be employed. However, these flaws are invisible for Marching Cubes method; therefore, the surface is naturally cleaned up. Also, a new method for choice of the proper pattern of Marching Cubes is proposed (see section 4.7). Usage of Marching Cubes and the mixture of two different methods of patch generation creates a new problem which needs to be overcome: watertightness of the surface polygons need to be ensured. It is described in this paper how to do it (see section 4.8). There is another unique approach used in the mesher which ensures that collisions of the input surface with sides of cutting boxes are avoided. This technique is referred as Dual Levels in this paper. The approach is based on slight shifting of nodes of the input geometry to discrete levels and cutting planes are shifted to another discrete levels (shifted by half step), so as it can never happen that a triangle of input surface lies exactly on a cutting plane. The step of the discrete levels is far smaller than the manufacturing tolerance of the real component. However, the step is big enough to ensure robustness of the routines for generation of faces of cells. The technique is described in section 4.2. In order to cope with concave cells, the mesher uses two techniques to overcome problematic situations. The first approach uses splitting of cells to convex features (called Cell Splitting in this text and described in section 4.11). For those situations when Cell Splitting fails (this can happen if the input surface contains flaws), a less exact but robust method is used (called IP patches in this paper and described in section 4.11). This second method is based on the maximal simplification of the inner parts of patch structure. The algorithm of the whole meshing task is described in detail in section 6. In comparison with its predecessor (the main principles on which the previous mesher was based were presented on 3^{rd}International Meshing Roundtable [9]), the new mesher produces cells with much higher quality. This is caused mainly by better choice of Marching Cubes patterns, better strategy of decisions which method should be used (Exact Fit or Marching Cubes) when generating patches in a box and the new Dual Levels technique. Also, the new mesher has significantly lower time and memory demands. The main approaches responsible for the improvement in speed and memory consumption are discussed.

2   Context of the Mesher in VECTIS-MAX System

The mesher described in this paper is part of the new version of VECTIS program (the new program is called VECTIS-MAX). VECTIS is a three-dimensional computational fluid dynamics program that has been developed specifically to address fluid flow simulations in the vehicle and engine industries. VECTIS allows the simulation of a number of applications: in-cylinder air motion and mixture preparation, spray dynamics, combustion modelling, intake system design and optimisation such as exhaust gas re-circulation or air/fuel ratio distribution, exhaust system development such as catalyst optimisation and thermal analysis, coolant jacket design and development, under-bonnet (under-hood) thermal simulations.

The whole VECTIS system consists of preprocessor, mesher, solver and postprocessor. In the preprocessor, user can load triangulated geometry in STL format and ensure (with help of implemented tools) that the triangulated surfaces are clean (i.e. there are no open edges nor mutual intersections of the triangles). Additionally, groups of triangles forming different boundary regions can be identified. In the preprocessor, the user also specifies how the mesher should create cells (regions with different density of cells can be specified). When this information is prepared, the mesher can be run. The meshing process is fully automatic and no further interaction with the user is required. Then, the graphical user interface of the solver can be used to define the input file for the solver. Here, boundary conditions are specified on the previously identified boundaries and parameters of the simulation can be set. Then, solver can be run and the results are visualised by the postprocessor.

3 Requirements for Mesh Quality

VECTIS-MAX solver uses an unstructured mesh. Cells can be formed by any number of polygons. They can have a nearly arbitrary shape; however, there are several criteria that each cell needs to fulfil.

Shape of polygons: It must be possible to split each polygon forming the cell into triangles starting from its centre (so called *star triangulation*). The polygon may be slightly concave; however, there must be direct visibility of each of its nodes from the face centre.

Watertight cells: Each cell needs to be properly enclosed by its faces. When polygons are split into triangles by the star triangulation (see the previous condition) and the *face surface vector* $\mathbf{A_j}$ is calculated for each triangle

$$\mathbf{A}_j = \sum_{i=1}^{N} \mathbf{A}_i = \frac{1}{2} \sum_{i=1}^{N} [(\mathbf{r}_{i-1} - \mathbf{r}_c) \times (\mathbf{r}_i - \mathbf{r}_c)] \tag{1}$$

(where N is the number of nodes of the polygon, which is equal to number of triangles used for the star triangulation; $\mathbf{r_i}$ is the position vector of i-th node; $\mathbf{r_c}$ is the position vector of the centre of j-th face), the geometric conservation law has to be ensured:

$$\oint_A d\mathbf{A} = \sum_{j=1}^{N_f} \mathbf{A}_j = 0 \tag{2}$$

(where N_f is the number of faces in the cell)

Angle condition for boundary faces: Angle between the vector *cell_centre* $\longrightarrow$ *face_centre* and the normal vector of the boundary face needs to be less than 90°(the scalar product of these vectors needs to be positive). In figure 1, there are examples of cells which do and do not meet the criterion. The cell on the left side satisfies the criterion, even though it is slightly concave. The cell on the right side does not fulfil the criterion, because the angle α is greater than 90°.

Angle condition for inner faces: A similar condition as for boundary faces needs to be fulfilled for inner faces. In this case, the angle of the two characteristic vectors needs to be less than 75°(the condition is more strict).

4 General Approach

The approach is based on cutting the whole domain according to a global mesh (defined by user) into boxes. When a box is intersected by the input geometry, it can be further refined. For each hexahedral box, input/output statuses of the eight vertices are remembered. Boundary faces (patches) are

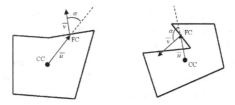

Fig. 1. A cell which does fulfil the angle criterion (left) and a cell which does not (right); CC represents the cell centre, FC is the centre of the tested face, **v** is the normal vector of the face; α is the tested angle

generated in all boxes intersected by the surface. Then, sides of the box are tested. If patches were already generated in both adjacent boxes, they are tied in order to close gaps; also, inner faces are generated on the common rectangular face of the two boxes. Then, all fully inner rectangular faces are generated on the rest of common sides of boxes. Finally, the output file is generated. In the next subsections, this general approach is described in detail.

4.1   Scaling

In order to eliminate the influence of dimensions of the input geometry (a fuel injector nozzle in millimetres, a boat in dozens of meters), the input geometry is proportionally scaled so as the longest dimension fits between 0.0 and 10.0.

4.2   Dual Levels

The problematic situation when the input geometry intersects a box exactly on one of its sides (the triangle lies exactly on the cutting plane) needs to be avoided. In order to cope with this situation, the technique Dual Levels has been proposed. The approach is based on slight shifting of nodes of the input geometry to discrete levels and cutting planes are shifted to another discrete levels (shifted by half step), so as it can never happen that a triangle of input surface lies exactly on a cutting plane. In order to get the discrete levels of the two grids, the whole working space is confined to a cube that is divided to 4200000000 levels in each direction (x,y,z), so as the position of each node can be described by three unsigned integer values. Then, odd levels are used to find positions of vertices of surface triangles and even levels help to find new positions of the cutting planes. The technique of describing x,y and z positions of vertices by three unsigned integers is used also for storage of nodes and it is described in detail in section 7. Usage of this technique does not mean that the input geometry is changed. In fact there is no error introduced, because the step of the discrete levels is far smaller than the manufacturing tolerance of the real component. However, the step is big enough to ensure robustness of the routines for generation of faces of cells. For example, even if the calculated domain has 10 meters in its longest dimension, the step defining the fine grid is 2.4×10^{-9} m. Then, the odd grid defining positions of vertices of input triangles has step 4.8×10^{-9} m, which is far below any used manufacturing tolerances. Usage of this technique makes the task of generation of polygons of cells much easier.

4.3   Shoeboxes

Shoeboxing is a system that helps to quickly limit number of elements that need to be taken into account when intersection tests are performed. The 3D space is divided to NI x NJ x NK boxes (so called shoeboxes). Then, for each input triangle all shoeboxes that are intersected by it are found. The index of the triangle is remembered by all affected shoeboxes. In the mesher, the shoeboxes are identical to the global boxes (that are defined by the user). Usually, users define higher density of meshlines in those regions where fine details in the geometry (modelled by many small triangles) occur, which naturally makes the searching system sufficiently balanced.

4.4   In/Out Status

In the stage of box generation, it is necessary to find the in/out statuses of the vertices of all potential boxes. In order to find the in/out status of a point, a ray-casting method (described for example in [3]) is used. Six rays are released in directions -x, +x, -y, +y, -z and +z. All intersections of the rays with the triangles of the surface are found. The in/out status is then evaluated according to the number of the intersections with the surface (odd number indicates inside status, even number means outside). Those rays containing surface intersections that are too close to each other are not taken into account to avoid further analysing whether the status should be reversed or not. Two different situations are possible when two very close intersections are found: the ray just hit edge of two triangles with similar unit normal vectors (the in/out status needs to be reversed) or it just touched a sharp feature (the in/out status needs to stay unchanged). If there is not any reliable ray, new set of rays needs to be released under different angles.

The described algorithm is used when the in/out status of a node needs to be found after the boxes are generated. However, for determining the in/out statuses of the initial box vertices, a slightly modified approach also based on the described principles is used: the released rays are used for whole row of vertices. This method is described for example in [11].

4.5   Box Generation

First, so called *global boxes* are constructed according to the given meshlines; in/out statuses of their vertices are found. Each global box is marked with

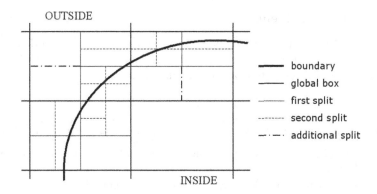

Fig. 2. 2D analogy of refinement of global boxes

one of the three statuses: *completely_inside, completely_outside* or *intersected*. Then, all completely inside and completely outside global boxes which neighbour to an intersected global box need to be tested for intersections on their twelve edges and intersections on their six sides. This test is necessary because it is possible that the surface penetrates the global box through an edge (there is even number of intersections on the edge) or through a side (there is a closed polygon forming the intersection on the side). If one of these cases is detected, the status of the global box needs to be changed to *intersected*.

Then, all the boxes marked as *intersected* are processed. For each the processed global box, the maximum possible refinement is found. According to the maximum refinement, a 3D array of the in/out statuses of the vertices of all the potential boxes is found [dimensions of the array depend on the maximum refinement depth D_{max}: $(2^{D_{max}} + 1) \times (2^{D_{max}} + 1) \times (2^{D_{max}} + 1)$]. Also, a 3D array of statuses of the potential boxes is assembled (dimensions of the 3D array are $2^{D_{max}} \times 2^{D_{max}} \times 2^{D_{max}}$). The statuses are the same as for the global boxes: *completely_inside, completely_outside* or *intersected*. Then, $1 \times 1 \times 1$ boxes need to be combined in order to find the minimum number of boxes which can completely cover the space defined by $1 \times 1 \times 1$ boxes with inside or intersected status. This task is solved by searching for the minimum number of splits. The process of box generation is illustrated in figure 2. The algorithm of box generation does not allow neighbourhood of boxes with too different levels of refinement. When a big box is touched by more than one split line on one of its side (when 2D analogy is considered), it needs to be refined in the same direction. This situation can also be seen on the figure 2. The big box on the left part of upper-left global box needs to be split horizontally; otherwise, it would be touched by three horizontal splits from two different levels of refinement. Also, the box forming the lower part of upper-right global box needs to be split vertically because of the two split lines from two different levels of refinement.

4.6 Types of Refinement

As explained in the previous section (4.5), refinement depth (D) defines the maximum possible divisions of the global box. The maximal number of $1 \times 1 \times 1$ boxes that can be generated can be calculated as

$$N_{1 \times 1 \times 1} = 2^D \times 2^D \times 2^D \tag{3}$$

For the user, there are three ways how to affect the refinement of global boxes:

1) **Global refinement depth** defines the default refinement depth valid for all boxes with no other specification
2) **IJK refinement block** allows to set a refinement level to a rectangular block of global boxes. Each global box is defined by its I,J,K integer coordinates in the system of meshlines; therefore, the block can be specified by six integer values (IS, IE, JS, JE, KS, KE – where S stands for start and E stands for end) and values specifying the refinement depth: allowed *refinement level* and *forced refinement*. If the forced refinement is specified, the global box will be split regardless of whether the input surface intersects it or not.
3) **Boundary refinement** specifies the refinement depth which is to be used in the global boxes that are intersected by a particular boundary. With this type of refinement, three variables can be set for a boundary:
   - Refinement depth at the boundary
   - Refinement blending distance (This parameter specifies an integer value which is used to control how the refinement at the boundary blends into the refinement level of the surrounding cells. Blending is achieved by giving the cells at the boundary a forced refinement level which is less than or equal to the specified refinement depth, and propagating away from the boundary in layers of successively lower forced refinement. The blend distance defines how many layers of cells should be at each forced refinement level.)
   - Blend to boundary depth -1 (This is yes/no information specifying whether the blending should start from "refinement depth–1" instead from "refinement depth")

Boundary refinement is applied after IJK refinement blocks. Therefore, the allowed refinement depth can be changed in global boxes that has previously been affected by an IJK refinement block. However, the forced refinement level and refinement depth are always only increased by boundary refinement specifications - never decreased.

4.7 Generation of Patches

Inside each box intersected by the surface, it is necessary to generate boundary faces (so called patches). First, patches are generated as triangles and later, triangles with similar unit normal vectors are combined to convex polygons. For generation of the triangular patches the concept of combination of methods Marching Cubes and Exact Fit is used.

Marching Cubes

Marching Cubes is a well known method used in computer graphics (described in [10]). The method defines 14 basic patterns to create triangles on the intersections of edges of a cube (*box* in the terminology of VECTIS-MAX). The patterns differ according to the in/out statuses of vertices of the cube. The majority of the patterns have more than one possibility of how to create triangles (e.g. if there are four intersected edges on the cube in pattern no. 2, two configurations of triangles are possible: "[4,2,9]+[9,2,10]" or "[4,10,9] + [4,2,10]".

During the development of the new mesher, several ways to find the optimal configuration of triangles have been tried. Based on these experiments, the method used in the older 3D Cartesian mesher was rejected. The method was based on the comparison of the in/out statuses of selected nodes in the box when intersected by the original surface and in the box intersected by the tested configuration of triangles. There are also some other possibilities described in [13].

However, during the development of the mesher, a new simple method was found which seems to be sufficient and much quicker than the other tested methods. This method is based on evaluation of a criterion calculated from the scalar product of unit normal vectors of the proposed triangular patch ($\mathbf{n_p}$) and the triangle of the surface intersecting the edge of the box ($\mathbf{n_s}$). The criterion C can be calculated as

$$C = \frac{(C_{max} - 1)b^{1-s} + b^2 - C_{max}}{(b^2 - 1)} \tag{4}$$

where b is the base of the used logarithm, s is the scalar product $s = \mathbf{n_p} \cdot \mathbf{n_s}$ and C_{max} is the chosen maximum value of the criterion (the value for the worst case when the unit normal vectors are exactly opposite). This criterion is calculated for each node of each triangle and the average is taken as the value evaluating the configuration of triangles. The configuration with the minimum value of the criterion is chosen. The shape of the function of the criterion is visualised in figure 3. The criterion is designed to strictly refuse those configurations where unit normal vectors point to opposite half-spaces. Values $b = 7$ and $C_{max} = 4$ were found as reasonable for this criterion, so they are used in the mesher.

Exact Fit

If it is not possible to use the Marching Cubes method, the Exact Fit algorithm is applied. This approach of patch generation is based on the splitting

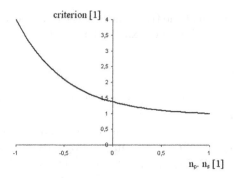

Fig. 3. Criterion for evaluation of different possibilities in Marching Cubes patterns

of surface triangles according to the sides of the box; only the triangles lying inside the box are kept as patches.

Choice of the patching method

Marching Cubes method can be used only when it is sure that the simplification would not cause chamfering of sharp features, problems with watertightness or damage of the border between different boundaries. When one of the following conditions is true, the Exact Fit must be used:

- one of the edges of the box is intersected more than once (in this case, tying of patches may not be feasible)
- the intersection polyline forms a closed polygon on one of the sides of the box (the geometry would be invisible for Marching Cubes)
- there is a sharp feature detected in the box (Marching Cubes would chamfer the feature)
- there is more than one boundary index detected among the triangles intersecting the box (Marching Cubes would damage the border between two boundaries)

4.8   Tying of Patches

Whenever patches are generated in a box, all rectangular faces of the box are tested to determine whether boxes on its both sides have already been processed. When a rectangular side is found, whose both adjacent boxes have been processed, the patches in the two boxes need to be tested for watertightness. If there is a gap between patches, they need to be tied so as the non-conformance is avoided. Two main cases when gaps appear are illustrated in figure 4. The picture on the left shows the situation when there is one box neighbouring with two boxes. Different simplifications from using the Marching Cubes method from both sides gives patches that are not properly tied. The picture on the right illustrates the situation when the Exact Fit is

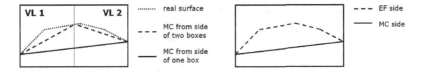

Fig. 4. Gap between patches caused by different refinement on two sides of common rectangular side of boxes (left side) and gap between patches caused by usage of different methods of patch generation (right side)

used from one side of the common rectangular side of boxes and Marching Cubes is used from the other side.

When tying of patches is used in the mesher, the corresponding nodes need to be found first. This action divides the problem to several simpler parts when several segments need to be tied to one segment. The single segment is divided so as the lengths of corresponding segments respect ratio of total lengths of the polylines:

$$\frac{L_j}{L'_j} = \left(\sum_{i=1}^{N} L_i\right) / \left(\sum_{i=1}^{N} L'_i\right) \tag{5}$$

where L_j is length of j-th segment of the polyline before tying of patches and L'_j is its length after movement of the nodes. Then, the nodes from the more complex side are moved to the new positions.

4.9   Generation of Inner Faces

Each generated inner face is tested to determine whether it can be divided into triangles by the star triangulation (see section 3). Those faces that are so concave that correct triangles cannot be formed, need to be split to more convex parts. The algorithm of searching the optimal cutting edge from the most concave angle (described in [12]) is used. Boundary faces do not require this treatment, because their convexity is ensured by the process of combination of triangular patches to polygons.

4.10   Removal of Small Cells

Volume of each generated cell is tested and compared with the size of its box. If the ratio V_{cell}/V_{box} is less than a defined constant R, the cell needs to be deleted. When a cell is removed, its inner faces need to become boundary faces of its neighbours. The constant R can be chosen by the user for each boundary region. Defaultly, $R = 1.0 \times 10^{-3}$ is set; for boundaries representing input/output or cyclic boundary, the value 1.0×10^{-100} is used.

4.11   Problem of Concave Cells

In order to overcome problems with cells that are so concave that the *angle condition for boundary faces* mentioned in section 3 is not fulfilled, two techniques are applied: *Cell splitting* and *IP patches*. Cell splitting inheres in searching for a cutting polygon which would divide the concave cell into two parts with better properties.

When this method fails (usually due to a flaw in the input surface), IP patches can be generated instead. This approach inheres in replacing the patch structure by polygons forming the intersection of the box with the surface (IP means intersection polygons). When IP patch method is applied, the shape of the geometry is not covered as well as if Cell splitting were used. Therefore, IP patches should be used only as a last chance mechanism when Cell splitting algorithm fails, just to avert failure of the whole meshing task.

Cell Splitting

First, the optimal cutting plane is found. In order to do this, the following procedure is performed. The edge representing the worst concave feature needs to be identified. Then, the adjacent edges are tested; if some of them are also not convex, the *concave feature polyline* will grow. Then, the cutting plane angle α_c is calculated as half of the average angles α_i adjacent to N edges of the concave feature polyline:

$$\alpha_c = \frac{1}{2}\left[\left(\sum_{i=1}^{N}\alpha_i l_i\right)\Big/\left(\sum_{i=1}^{N}l_i\right)\right] \tag{6}$$

The average is weighted by lengths of the edges (l_i).

Then, the intersections of edges of the cell with the cutting plane are found and the cutting polygon (the closed polyline of the intersection) can be finished. Whenever an edge is intersected very close to one of its end vertices (the deviation of the two adjacent edges of the cutting polygon from the cutting plane is less than 5°), the end vertex is used instead.

IP patches

The unit normal vector and the face centre are assigned to each IP patch. These properties are calculated in the mesher and they are passed to the solver (in the contrary with other types of faces whose properties are found during the run of the solver). The normal vector of the IP patch is calculated from the condition described by equation (2). Sum of face surface vectors $\mathbf{A_j}$ of all other faces determines the face surface vector of the IP patch. A face centre needs to be found so as the angle criterion (see the section 3) is fulfilled. The process needs to be done iteratively, since each movement of the face centre affects the cell centre.

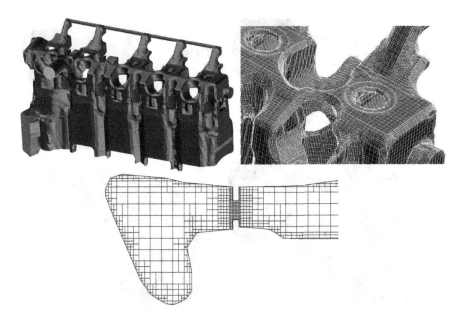

Fig. 5. Coolant jacket (input geometry, part of the mesh and a slice view of the gasket hole)

5   Examples of Generated Meshes

In order to show some examples of generated meshes, geometries of two typical problems often solved by VECTIS were chosen.

1) **Exhaust manifold with turbine:** The input geometries, and parts of generated meshes are illustrated in figure 6. The pictures represent exhaust manifold with a turbine housing. Images on the left side are linked to the solid part of the multi-domain simulation. On the right side, there is the fluid part of the simulation. When both geometries are meshed, the common interface needs to be made conformal in order to prepare them for the multi-domain simulation in the solver. Even distribution of meshlines with size of cells 3 mm and refinement depth 1 were chosen.

2) **Coolant jacket:** In figure 5 a typical example of geometry for modelling of flow of water in cooling channels of the engine is shown. The size of the global cells was chosen as 3 mm; together with refinement depth 2. The slice of the mesh in the lower part of the figure shows the critical part between the head and the block of the engine, called the gasket hole. IJK refinement block was used here to enforce refinement depth 3 in order to ensure sufficient number of cells to realistically simulate the flow in this narrow channel.

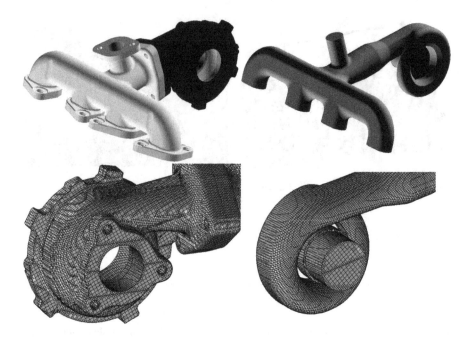

Fig. 6. Exhaust manifold with turbine (input geometries and parts of the meshes)

6   Description of the Meshing Algorithm

When running in the meshing mode, the scheme of the work of the mesher is this:

1) **READ INPUT FILE FOR MESHING TASK:** The input ASCII file (containing information about positions of meshlines and some other specifications for the meshing task) is read.

2) **READ SURFACE TRIANGLES:** The input file containing information about surface triangles (trifile) is read. Coordinates of the triangles are scaled (see the subsection 4.1), the geometrical extents of triangles are found and connectivity information is assembled.

3) **APPLY DUAL LEVELS:** The x,y and z positions of meshlines (defining the cutting planes) are shifted to discrete positions on a very fine grid. At the same time, nodes defining triangles of the input surface are moved to lay on a different grid (with its levels shifted by half-step). This averts collisions of cutting planes with with triangles perpendicular to the principal axes. (see the subsection 4.2)

4) **DETECTION OF OVERLAPPED TRIANGLES:** The algorithm for detection of overlapped triangles is run here. If there are some problematic triangles, a warning is printed together with the list of indexes of the triangles. If problematic triangles are detected, the mesher will continue its work. Usually, the mesher automatically overcomes small

problems in the input triangulated surfaces. However, if it happens that the final gridfile contains cells with low quality, the user should try to improve the flaws in the surface detected in this step and run the mesher again.

5) **PREPARE SHOEBOXES:** The system of shoeboxes is established here (see the subsection 4.3).

6) **CONSISTENTLY ORIENT TRIANGLES:** Since the orientation of the input triangulated surfaces is random in the input file, it is necessary to orient them so as normal vectors of the triangles always point into the flow domain.

7) **PREPARE IN/OUT STATUSES:** Here, in/out statuses of the vertices of the global boxes (that are defined by meshlines) are found by ray-casting method (see the subsection 4.4).

8) **GENERATION OF BOXES:** The boxes are generated in the global boxes (see the subsection 4.5) according to the prescribed refinement specification (see the subsection 4.6).

9) **PREPARE COMMON RECTANGULAR SIDES OF BOXES:** In this part of the algorithm, the boxes generated in the previous step are indexed first. Then, the common rectangular sides of the generated boxes are generated. According to which direction they are perpendicular, U, V and W common sides are distinguished (perpendicular to x, y and z, respectively). They serve for navigation through boxes and for generation of inner faces on them.

10) **GENERATE CELLS IN BOUNDARY BOXES:** For all boxes, it is determined whether the box is intersected by the surface or fully inner or fully outer. The optimal order of processing boundary boxes is found (smaller boxes need to be processed first to be sure that when a box is processed which has more than one neighbour in a direction, all the adjacent boxes are already done). Then, all boxes intersected by the surface are looped and their faces are generated in these steps:

— POLYGON GENERATION PART —

A1) Generate boundary faces (patches): Marching Cubes or Exact Fit method is used for the generation of boundary faces (see the subsection 4.7).

A2) Polygon Simplification: This technique simplifies the patch structure where possible. The algorithm preserves sharp features and borders between different boundaries.

A3) Generate inner faces if neighbours are processed: All rectangular sides common with the adjacent boxes are looped. On those locations where patches of the two adjacent boxes have already been generated from both sides, the patches are tied (described in the subsection 4.8) and polygons of inner faces are generated. If any generated inner face is concave, it is split to convex parts here.

— CELL ASSEMBLING PART —

B1) Find a complete cell: It is found whether there is a cell that has all its polygons already generated. If there is no such a cell, continue with the next box (go to A1).

B2) Distinguish separated volumes: The connectivity of faces is found and the polygons are painted in order to find separated volumes in the box.

B3) Cell splitting: On each separated volume, it is tested whether there are concave features. If there are, split the cell to convex parts.

B4) Delete small cell: The volume of the cell is found. If it is too small, the cell needs to be removed (described in subsection 4.10).

B5) Save faces: The generated boundary and inner faces are saved to auxiliary files so as they can be retrieved back later when assembling the final grid (this concept is described below in subsection 7.2). The algorithm continues with B1.

11) **FINISH COMMON FACES:** Rectangular faces of those cells that are fully inner are generated here.

12) **PRINT STATISTICS OF GENERATED MESH:** The statistics of the generated grid is printed. The report contains information about number of generated cells, numbers of boxes processed by Exact Fit and by Marching Cubes methods, number of cells that needed to be split to convex parts. If problems occur, number of cells with negative volume, gaps or angle problems is reported here (see the section 3). Problems like this are usually linked to topological problems of the input surface.

13) **WRITING THE MESH FILE:** The polygons stored in the auxiliary files (saved in the step B5) are subsequently read while the output arrays are assembled at the same time. Then, the output gridfile is written out. Finally, the auxiliary files are deleted.

7 Tools Helping to Decrease Time and Memory Demands

It is hard to compare the previous VECTIS mesher with the new approach since both are doing different tasks. The new VECTIS-MAX mesher needs to perform more actions in order to meet the higher requirements for the mesh quality. Despite this fact, measurement on several typical cases has shown that the new program consumes 74 % of memory and 64 % of time in comparison with the previous system. In the following text, the main features are described that are believed to be responsible for the significant improvement.

7.1 Features Improving Time Efficiency

Integer storage of coordinates of vertices

In order to ensure the test for existence of a node to be efficient, the approach recommended in [3] and [12] was used for storage of nodes. The whole working space is confined to a cube that is divided to 4200000000 levels in each direction (x,y,z), so as the position of each node can be described by three unsigned integer values. The choice of the number of levels is linked to the capacity of unsigned int type on 32-bit computers, which is 4294967295. All existing nodes are kept sorted according to their coordinates. When comparing two nodes, x determines which node is less; if both nodes have the same x, the decision is done according y, etc. Whenever a new node needs to be created, it is very quick to find out whether the node already exists or not, because of the binary search in the sorted array.

This concept turned out to be very efficient. Of course, it saves certain amount of memory (*unsigned int* consumes half the number of bytes than the *double* type on 32-bit computers). However, more important is its time efficiency. It is believed that this concept is mainly responsible for the higher time efficiency of the new mesher.

This usage of integer values is limited only to storage of coordinates of vertices. The full integer arithmetic (described in [3]) was not implemented.

7.2   Features Improving Memory Efficiency

Avoidance of STL containers

During the development of the mesher, it was found that when containers from the Standard Template Library (set, map and list) had been replaced by simple classes using "malloc" allocation, the time and memory requirements were significantly lower. A comparison of STL and non-STL approach was done on storage of one million nodes to a map. Memory requirements fell down to 16 % when std::map was avoided. Time necessary for storage of the nodes was decreased to 23 %; time needed to retrieve the nodes decreased to 56 %. The development team believes that the savings are caused by allocations and reallocations with a reasonable step. If an STL container is to be extensively used in a program, its allocator should be changed to avoid too many allocations by small chunks.

Temporary storage of polygons

Whenever a cell is generated (all its polygons are prepared), those polygons that are no longer needed can be stored in an auxiliary file. The memory occupied by the polygons can be reused for polygons of another cell. In order to make this technique efficient, the optimal order in which boxes are processed needs to be found so as neighbours of already done cells are processed as soon as possible.

Arrays of low-bit information

During the run of the mesher, it is often needed to store long arrays of information that requires only low number of bits (e.g. in/out status of nodes, done status of a box, ...). However, allocation of an array of *bool* type consumes eight bits for each entry. Therefore, a tool that can store 1D, 2D or 3D array of entries of arbitrary bit-length in a chunk of memory (unsigned char type is used) has been prepared. For storage or retrieval of each entry some additional time is consumed. This is caused by necessity to find the proper unsigned char(s) containing the information and perform appropriate bit operations in order to find the bits that should be used. However, the additional time disadvantage seems to be low price for the memory reduction benefit.

References

1. Aftosmis, M.J., Berger, M., Melton, J.: Robust and Efficient Cartesian Mesh Generation for Component-Based Geometry. In: 35th AIAA Aerospace Sciences Meeting, Reno NV (January 1997)
2. Berger, M., Aftosmis, M.J.: Aspects (and Aspect Ratios) of Cartesian Mesh Methods. In: 16th International Conf. on Num. Meth. in Fluid Dynamics (July 1998)
3. Aftosmis, M.J.: Solution Adaptive Cartesian Grid Methods for Aerodynamic Flows with Complex Geometries. In: 28th Computational Fluid Dynamics Lecture Series. Von Karman Institute for Fluid Dynamics, Belgium (1997)
4. Ingram, D.M., Causon, D.M., Mingham, C.G.: Developments in Cartesian cut cell methods. Mathematics and Computers in Simulation 61, 561–572 (2003)
5. Przulj, V., Birkby, P., Mason, P.: Finite Volume Method for Conjugate Heat Transfer in Complex Geometries Using Cartesian Cut-cell Grids. CHT-08: Advance. In: Computational Heat Transfer, Marrakech (May 2008)
6. Kini, S., Thoms, R., Zhu, F.: A Fast and Fully Automated Cartesian Meshing Solution for Dirty CAD Geometries. SAE International paper 2008-01-2998 (December 2008)
7. Kanade, K., Lietz, R., et al.: Rapid Meshing for CFD Simulations of Vehicle Aerodynamics: SAE International paper 2009-01-0335 (April 2009)
8. Kini, S., Thoms, R.: Multi-domain Meshes for Automobile Underhood Applications. SAE International paper 2009-01-1149 (April 2009)
9. Bardsley, M.: Automatic Mesh Generation for CFD. In: 3rd International Meshing Roundtable, Albuquerque, New Mexico (1994)
10. Lorensen, W.E., Cline, H.E.: Marching Cubes: a High Resolution 3D Surface Construction Algorithm. Computer Graphics 21, 163–169 (1987)
11. Tarini, M., Callieri, M., Montani, C., Rocchini, C.: Marching Intersections: An Efficient Approach to Shape-from-Silhouette. In: Vision Modeling and Visualization Conference, Erlangen, Germany (2002)
12. O'Rourke, J.: Computational Geometry in C. Cambridge University Press, Cambridge (1994)
13. Frey, P.J., George, P.L.: Mesh Generation. HERMES Science Publishing, Oxford (2000)

Shape Operator Metric for Surface Normal Approximation

Guillermo D. Canas and Steven J. Gortler

School of Engineering and Applied Sciences
Harvard University
33 Oxford St. Cambridge, MA
{gdiez,sjg}@seas.harvard.edu

Abstract. This work deals with the problem of practical mesh generation for surface normal approximation. Part of its contribution is in presenting previous work in a unified framework. A new algorithm for surface normal approximation is then introduced which improves upon existing ones in a number of aspects. In particular, it produces better approximations of surfaces both in practice and in the theoretical limit regime. Additionally, it resolves in a simple way some of the problems that previous methods for surface approximation suffered from.

1 Introduction

Computing high-quality approximating meshes from surfaces is an important problem in computational geometry, with many practical implications. Although the approximation criteria can vary greatly, often, approximating either surface position, or a surface's normal field can be a good criteria in practice. As has been argued elsewhere [7, 14], approximating a surface while minimizing normal approximation error can be useful in many applications.

There is a considerable body of previous work that deals with the surface approximation problem. Some notable examples include ϵ-nets [6], for surface and normal approximation, the Quadric Error Metric algorithm (QEM) [9] for surface approximation, and Variational Shape Approximation (VSA) [14], for surface and normal approximation.

In this paper, it is first discussed how the above three methods can be interpreted from within a unified framework. In this interpretation, they are essentially all minimizing a k-means like energy, where only the distance metrics are different. Interestingly enough, in the limit, for smooth surfaces, these distance measures converge to each other. The other key difference that is explored is how these means are used by the different algorithms to produce the output triangulation.

Next, a novel distance measure is proposed (Shape Operator Metric, or SOM), with a corresponding algorithm, that fills a natural gap in this framework. In particular, like VSA, it is designed for normal approximation. But, like QEM, it does not require a region-triangulation step. Such a step can complicate the implementation and, as it is discussed, it introduces a constant factor of inefficiency close to 2, in the limit of approximation.

2   Framework

Considered here are meshing algorithms for surface approximation that try to either meet a uniform error bound, or minimize the average approximation error over a surface M (minimizing error in the $\mathcal{L}^\infty$ or $\mathcal{L}^2$ sense respectively). These kinds of algorithms can be naturally described as an optimization problem:

$$\operatorname*{argmin}_{\{p_j\}, \mathcal{V}_j} E_{\mathrm{x}}^\infty = \operatorname*{argmin}_{\{p_j\}, \mathcal{V}_j} \max_j \max_{p \in \mathcal{V}_j} D_{\mathrm{x}}(p_j, p) \tag{1}$$

$$\operatorname*{argmin}_{\{p_j\}, \mathcal{V}_j} E_{\mathrm{x}}^2 = \operatorname*{argmin}_{\{p_j\}, \mathcal{V}_j} \sum_j \int_{\mathcal{V}_j} D_{\mathrm{x}}(p_j, p) dp \tag{2}$$

over both a set of *means* $\{p_j\}$ (points on the surface), and a corresponding partition $\{\mathcal{V}_j\}$ of M composed of the Voronoi cells of $\{p_j\}$ with respect to a chosen distance function D_{x}.

Optimal Voronoi partitions have in all (except perhaps the rarest) cases neither the shape nor the topology of a triangle mesh. Some further step is generally necessary before producing a triangle mesh as output. In the sequel, a meshing algorithm is referred to as a *primal algorithm* if it discretizes the boundaries of Voronoi cells, triangulates their interior, and outputs this set of triangles, as in [14]; while an algorithm in which the means are instead directly connected using the dual topology of the partition $\{\mathcal{V}_j\}$ to produce a triangle mesh, as in [6], will be denoted as a *dual algorithm*.

Apart from the added algorithmic complexity, primal algorithms have an inherent approximation inefficiency in the limit. Roughly speaking, in smooth surface regions, in the limit, the relative sizes and aspect ratios of the Voronoi regions are optimized by minimizing the above energies. These relative sizing and aspect ratios will be maintained under mesh duality. But these sizes and aspect ratios are altered (within a constant factor) when the Voronoi regions are triangulated. The limit regime is explored in more detail in Appendix B, while the non-limit case is discussed experimentally in Section 5.

The algorithms of [6, 9, 14], as well as the one introduced in Sec. 3, all fit into this framework. In particular, the method in [14] introduced the idea of directly optimizing energies with the above form using a k-means/Lloyd-Max type algorithm. It then applies a primal meshing step to the resulting partition.

In the work of [6] one finds a set of means $\{p_j\}$ with bounds on the energy of Eq. 1. The means are then connected in a dual triangulation.

The QEM method of [9] applies a sequence of edge collapses to the input mesh, which can essentially be interpreted as an attempt to minimize 2. In particular, upon completion of the QEM algorithm, each vertex of the output triangulation can be thought of as a mean with a region of the surface associated to it: the portion of the surface that it uses to evaluate its associated (quadric) error (with adjacent regions slightly overlapping). In this sense QEM can be considered a dual algorithm. Even though the connectivity of the triangle mesh is not directly related to the Voronoi regions of Eq. 1, its limit behavior is analogous.

2.1 Surface Approximation

The algorithms of [6, 9, 14] use the following distances when optimizing Eqs. 1 or 2:

$$D_{\mathrm{II}}(p_j, p) = \min_{\gamma \in P(p_j, p)} \int_\gamma q_{\mathrm{II}}^c(\gamma'(t); \gamma(t))^{\frac{1}{2}} dt \tag{3}$$

$$D_{\mathrm{QEM}}(p_j, p) = <p - p_j, n(p)>^2 \tag{4}$$

$$D_{\mathrm{sVSA}}(p_j, p) = <p - p_j, n(p_j)>^2 \tag{5}$$

where $q_{\mathrm{II}}^c(\gamma'(t); \gamma(t))$ is the "convexified" (using the absolute value of the eigenvalues) second fundamental form at point $\gamma(t)$ and applied in direction $\gamma'(t) \in T_{\gamma(t)}M$, and $P(p_j, p)$ is the set of all paths that connect p_j to p on the surface.

As described in Appendix A, for smooth surfaces, it is possible to write: for all $\lambda > 0$, for all non-parabolic $p_j \in M$ there is an open neighborhood V of p_j such that $\forall p \in V$:

$$D_{\mathrm{QEM}}(p_j, p) \simeq_\lambda D_{\mathrm{II}}(p_j, p)^4 \simeq_\lambda D_{\mathrm{sVSA}}(p_j, p) \tag{6}$$

where the notation $\simeq_\lambda$, borrowed from [6], implies tight approximation to any desired degree of accuracy. Note that the exponent 4 above arises from the fact that ϵ-nets minimize a form of Euclidean distance between the surface and the approximation, while QEM and $sVSA$ minimize *squared* Euclidean distance. Equation 6 is valid only for elliptic points p_j. For hyperbolic points, D_{QEM} and D_{sVSA} still converge to the same value, but $(D_{\mathrm{II}})^4$ is only an upper bound of D_{QEM} and D_{sVSA}. [Note that one could have defined D_{II} using $|q_{\mathrm{II}}|^{1/2}$ in the integrand of 3, where q_{II} is the second fundamental form. This would make $(D_{\mathrm{II}})^4$ be a tight approximation of D_{QEM} and D_{sVSA} everywhere non-parabolic on M, but would no longer be a Riemannian metric.]

The distance D_{II}, is too expensive to compute in practice (since each evaluation involves computing a shortest path under the q_{II}^c surface metric). In contrast, both D_{QEM} and D_{sVSA} are efficiently computed using only local information at the arguments p_j and p.

2.2 Normal Approximation

The problem of computing a mesh that approximates the normal field of a surface is considered next. It is noted that for this problem, the normals of the approximating mesh are piecewise constant. However, instead of being inferred from the vertex positions, the normals of the output mesh are optimally *assigned* to triangles. This distinction is necessary to avoid difficulties like those described in [5, 7] that can occur when triangles have large internal angles, even if they have the right limit shape and size.

A similar analysis to that of Sec. 2.1 can be made in this case. Here, the two relevant algorithms that are considered are [6, 14]. They use the following distances to optimize Eqs. 1 and 2 respectively:

$$D_{\mathrm{III}}(p_j, p) = \min_{\gamma \in P(p_j, p)} \int_\gamma q_{\mathrm{III}}(\gamma'(t); \gamma(t))^{\frac{1}{2}} dt \tag{7}$$

$$D_{\mathrm{nVSA}}(p_j, p) = \|n(p_j) - n(p)\|^2 \tag{8}$$

where q_{III} is the surface's third fundamental form, and $n : M \to \mathbb{S}^2$ is the Gauss map.

Analogously as proven in Appendix A, it is, for p in an appropriate, small enough neighborhood of a non-parabolic p_j:

$$D_{\mathrm{III}}(p_j, p)^2 \simeq_\lambda D_{\mathrm{nVSA}}(p_j, p) \tag{9}$$

2.3 Behavior

To aid in our discussion, three different kinds of regions on a surface will be considered, and the algorithms under consideration evaluated separately for each. The following distinct types of regions on surfaces are considered:

In smooth and non-parabolic regions, it can be shown that, in the limit, the regions of the partitions generated by optimizing D_{sVSA} D_{nVSA} and D_{QEM} have the proper aspect ratio [10, 14], which is a necessary condition for optimality for their respective surface or normal approximation problem. It can also be shown that, for everywhere-elliptical surfaces, and in the limit, the method of ϵ-nets [6] using D_{II} produces results that are within a constant factor of the globally optimal $\mathcal{L}^\infty$ minimizer for the surface surface approximation problem. As discussed in Appendix B, in the limit, primal algorithms such as VSA will need roughly twice as many triangles as compared to dual algorithms such as QEM and ϵ-nets.

Near sharp features, these algorithms behave quite differently. In particular, one can see that D_{QEM} measures error "from the viewpoint" of the variable of integration $p \in M$, while D_{sVSA} does so from the viewpoint of the mean p_j. As a result, QEM places means at high-curvature points, and thus is suited as a dual algorithm, while sVSA (the surface approximation version of VSA [14]) places them at *low*-curvature points, making it better suited as a primal algorithm. nVSA also tends to place means at low-curvature points. It is less clear the authors how ϵ-nets behaves in this regard.

Parabolic/curved regions are places on a smooth surface near, or at parabolic points where there is significant higher-than-second-order bending (i.e. the surface is not locally well approximated by a quadratic patch), such as near the parabolic line on a torus. Algorithms often need special care in this case. Near parabolic/curved points, an optimization using D_{QEM} undersamples regions near curved parabolic lines. The original QEM algorithm [9] deals with this case by introducing special rules to prevent flips before edge collapses (which strictly-speaking breaks the energy-minimization formulation of Eq. 2). For ϵ-nets, an additional *isotropic* term is added to the distance to cope with such regions. VSA deals with this case, in which the Voronoi cell boundaries are highly curved, by discretizing these boundaries and triangulating the cells finely enough as to avoid undersampling.

3   Shape Operator Metric for Normal Approximation

An obvious missing piece in this description is an algorithm that converges to Eq. 9 in the limit, but places means at high curvature points away from it, making it most suitable as a dual algorithm. In some sense this algorithm would be to nVSA what QEM is to sVSA. Moreover it can be efficiently computed an optimized (as in Eq. 1 or 2), has high approximation efficiency in the sense of Appendix B, and it avoids heavy undersampling near curved parabolic lines.

To begin, consider the definition $D_{\mathrm{nVSA}}(p_j, p) = \|n(p_j) - n(p)\|^2$, which measures normal error from either p_j or p, and, similarly as QEM, construct an approximation that only depends on p_j but not on any higher-order local information at p_j. To do this, a second-order Taylor expansion of $D_{\mathrm{nVSA}}(p_j, p)$ around p is constructed (note that the zero-th and first order terms vanish):

$$D_{\mathrm{SOM}}(p_j, p) \equiv (p_j - p)^T \frac{\partial D_{nVSA}(p'_j, p)}{\partial^2 p'_j}(p_j - p) \qquad (10)$$

$$= (p_j - p)^T S(p)^2 (p_j - p) \qquad (11)$$

where $S(p)$ is a $\mathbb{R}^{3 \times 3}$ shape operator matrix $S(p) = k_1(p)e_1(p)e_1(p)^T + k_2(p)e_2(p)e_2(p)^T$, $\{k_1, k_2\}$ are the principal curvatures, and $\{e_1, e_2\}$ the principal directions.

Note that D_{SOM}, like D_{QEM} and D_{VSA}, can be efficiently computed only from local information at the endpoints, and, as will be shown in Sec. 4, results in an energy of the type of Eq. 1 or 2 that can be efficiently minimized using standard algorithms [13, 12].

The SOM algorithm then simply outputs the dual trianglulation of this computed surface partition.

It follows from the fact that this is a dual algorithm, whose distance converges in the limit to that of D_{nVSA}, and from the discussion of Appendix B, that this algorithm has the desired favorable (limit) efficient approximation characteristics when compared to the primal algorithm of [14]. It is also shown in Appendix C that this algorithm produces elements that conform to the theoretically optimal limit shape and orientation.

Unlike [6, 9, 14], curved parabolic regions are dealt with in a natural way, without special consideration, which adds to the simplicity of the algorithm. The side figure illustrates this point, where a mean p_j is placed at a parabolic line (red). Because the parabolic line is curved, p_j does not lie along the flat direction when viewed from the point of view of nearby points p. An SOM primal-region centered around p_j thus cannot grow too much along the parabolic line if the parabolic line curves.

It is possible to see that minimizing Eq. 2 using Eq. 10 has the effect of placing means at points of high-curvature. Consider the closely-related problem of gradient approximation of a scalar function f defined on the plane, and an analogous distance $D_{\text{fSOM}} = (p_j - p)^T H_f^2(p)(p_j - p)$ with $p_j, p \in \mathbb{R}^2$, where H_f is the Hessian of f. In an everywhere-isotropic region, $D_{\text{fSOM}} = k(p)\|p_j - p\|^2$, which, used for $\mathcal{L}^2$ minimization in a form analogous to Eq. 2 over the plane, is an instance of the *weighted k-means* problem, which is well-known to place means at points with high weight [1] (high-curvature in this case). The case where H is not isotropic behaves similarly, but the weight can be thought of as directionally-varying.

4   Implementation of SOM

The energy of Eq. 10 is minimized in a way very similar to the algorithm of [12], which uses a probabilistic seeding of means, followed by a Lloyd relaxation [13] and has theoretical guarantees of closeness to the global optimum. In this work, the probabilistic seeding is simply replaced by iteratively placing means at the surface point with maximum minimum-distance to the current set of means, similarly as the greedy algorithm for computing ϵ-nets of [11]. This is also similar to the optimization method of [6], except that the seeding is followed by a Lloyd relaxation, and is also similar to [14]. The shape operator matrix S of Eq. 10 is estimated using the algorithm of [3].

Once the seed means have been placed, the Lloyd relaxation has two stages. The first creates a distance-dependent Voronoi partition of the surface, and the second computes the new means' locations from the current partition.

To compute a Voronoi partition, all vertices (as opposed to input triangles) are tagged as belonging to some primal Voronoi region, and Voronoi region boundaries are computed by splitting input triangles that have vertices in different regions, as in the side figure below. A Voronoi region is thus not constrained to be a collection of faces, but can have a boundary that cuts across triangles, which may slightly improve accuracy in practice. Also, in this way, Voronoi regions can meet at most at 3-way junctions. These 3-way junctions naturally dualize into triangles. Note that this generalizes to higher-dimensions, so that, by construction, it will only output simplicies.

Given a Voronoi partition of the surface, the new means' locations are computed. First, note that the energy of Eq. 2 for the distance D_{SOM} can be written as

$$E_{\text{SOM}}^2 = \sum_j p_j^T (\int_{\mathcal{V}_j} S(p)^2 dp) p_j - \tag{12}$$

$$- 2p_j^T (\int_{\mathcal{V}_j} S(p)^2 \cdot p \, dp) + (\int_{\mathcal{V}_j} p^T \cdot S(p)^2 \cdot p \, dp) \tag{13}$$

and so it is quadratic in p_j.

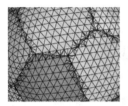

It is possible to compute the minimizer p_j of Eq. 12 by solving a small linear system, but this would return a mean p_j which is not constrained to be on the surface. Instead, the constants in equation 12 are computed in a first pass: $A_j = \int_{\mathcal{V}_j} S(p)^2 dp$ and $b_j = \int_{\mathcal{V}_j} S(p)^2 \cdot p \, dp$, for each Voronoi region $\mathcal{V}_j$. Then, for each input triangle (or split triangle) inside region $\mathcal{V}_j$ the barycentric coordinates (u, v) of the minimizer p_j of Eq. 12 can be found by solving $R^T A_j R \begin{pmatrix} u \\ v \end{pmatrix} = R^T b_j$ where $R \in \mathbb{R}^{3 \times 2}$ is some basis of the supporting plane of the triangle. The minimizer may fall outside the triangle, so it is necessary to look for it along triangle edges and vertices as well. The final mean is the minimum over all the minimizers on each triangle, guaranteeing that p_j is a point on the surface. Finally, instead of outputting p_j directly as a (dual) vertex, a quadric error metric [9] for its associated Voronoi region $\mathcal{V}_j$ is first computed, and its minimizer along the line passing through p_j in direction $n(p_j)$ is output. This small perturbation slightly improves the approximation.

5 Results

Some surfaces processed by the SOM algorithm are shown in figures 1 through 3. These meshes are compared with those produced by VSA, which are computed by exactly following [14]. Note that, unlike SOM, VSA has a free parameter (the precision used to discretize the partition regions's boundaries), which has been tuned to improve VSA's output. These results are also compared with the output of QEM [9]. Note that QEM optimizes (RMS) distance from the surface to the approximation, instead of normal error, and therefore the comparison is not strictly relevant; but it is included it as reference. Runtimes for SOM range from 5 sec. (bunny, input: 70k tris.) to 40 sec. (statue, input: 512k tris.), on a single core 2.0GHz CPU.

Even though it is not necessarily what is being optimized for in this work, it is possible to consider ($\mathcal{L}^\infty$) Hausdorff, and RMS error in the sense of surface approximation. Note that, in most cases, QEM produces slightly better approximation of the surface than SOM, and significantly better than VSA.

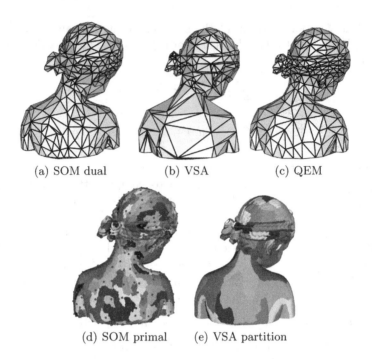

(a) SOM dual   (b) VSA   (c) QEM

(d) SOM primal   (e) VSA partition

Fig. 1. SOM: 500 vert., 996 tris. (Hausdorff error = 1.79e-2, RMS error = 2.24e-3). **VSA**: 528 verts., 1076 tris. (Hausdorff error = 2.31e-2, RMS error = 5.06e-3). — **QEM**: 502 verts., 1000 tris. (Hausdorff error = 1.93e-2, RMS error = 1.93e-3).

This is expected, as QEM optimizes surface approximation error (RMS error in the figures), while VSA and SOM both optimize normal error instead. Notice that, for smooth surfaces, and using (almost) the same number of triangles, SOM's approximation is appreciably finer than VSA's. On smooth surfaces, the approximation is significantly better for SOM at a given sampling rate. As can be seen in the primal partitions in figures 1 and 2, with an equal triangle budget, SOM is able to partition the surface into smaller regions that capture detail better. Note that the bunny is particularly troublesome for VSA, when compared with SOM, because its bumpy surface produces very curved regions that can output many triangles when their boundaries are discretized by the VSA algorithm. In general, in the above figures, triangles are elongated along the directions of minimum curvature, and tend to show very high anisotropy in places where this is possible: like the ears of the bunny or the statue's arms. Note that our algorithm offers no guarantees in terms of normal flips in the triangulation, which could show up occasionally in sparsely sampled regions. This behavior is similar to VSA and ϵ-nets, which also cannot guarantee to be free of flips.

Figure 3(d-f) shows a surface composed of roughly flat parts separated by sharp features. On these kinds of surfaces VSA does particularly well, since

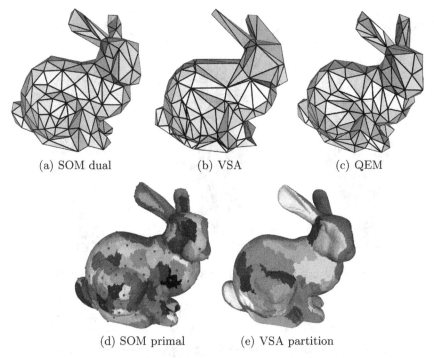

(a) SOM dual   (b) VSA   (c) QEM

(d) SOM primal   (e) VSA partition

Fig. 2. SOM: 200 verts., 396 tris. (Hausdorff error = 3.32e-1, RMS = 5.02e-2). **VSA**: 199 verts., 409 tris. (Hausdorff error = 7.03e-1, RMS error = 9.93e-2). — **QEM**: 202 verts., 400 tris. (Hausdorff error = 2.50e-1, RMS error = 4.31e-2).

it operates by locating roughly-flat patches and triangulating them. In particular, the region-triangulation phase of VSA is well-tuned to this problem, since the desired behavior in this case is to triangulate the flat polygons. SOM in this case naturally places means at sharp corners. But its connectivity is guided by the shape operator, which is almost everywhere degenerate here. This case is dealt with by computing the final mesh connectivity using a modified shape operator, which is set to a very high (isotropic) value in flat regions, effectively simulating a flat-polygon triangulation step (similar to the constrained Delaunay triangulation used in [14]).

5.1 Numerical Validation

Unlike for surface approximation, there is, as far as the authors are aware, no standard way of measuring the surface *normal* approximation on a surface. If there was, away from the limit regime, a well-defined one-to-one correspondence between points on the surface and points on the approximation, then it would be possible to compute the $(\mathcal{L}^2)$ approximation error by integrating the distance between corresponding normals over the surface. However, this correspondence is not available. To analyze approximation error, the very

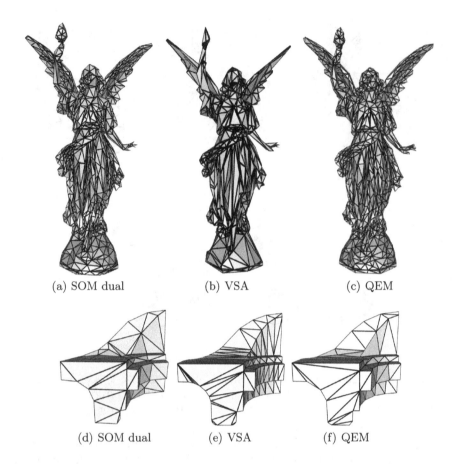

(a) SOM dual   (b) VSA   (c) QEM

(d) SOM dual   (e) VSA   (f) QEM

Fig. 3. Lucy SOM: 1500 verts., 2988 tris. (Hausdorff = 14.598, RMS = 1.866).
Lucy VSA: 1456 verts., 2990 tris. (Hausdorff error = 44.688, RMS error = 5.911).
Lucy QEM: 1496 verts., 2988 tris. (Hausdorff error = 11.834, RMS error = 1.472).
Fandisk SOM: 80 verts., 156 tris. (Hausdorff error = 0.118, RMS error = 0.0157).
Fandisk VSA: 80 verts., 156 tris. (Hausdorff error = 0.0596, RMS error = 0.0131).
Fandisk QEM: 80 verts., 156 tris. (Hausdorff error = 0.264, RMS error = 0.0152).

closely-related problem of approximation of the gradient of a height field over the plane is considered [it has optimal limit aspect ratio ξ_1/ξ_2, where ξ_i are the eigenvalues of the heigh field Hessian [4, 5]]. Because both VSA and SOM only look at normals and shape operators, it is possible to naturally adapt both to the gradient approximation case by measuring distances between gradients, as opposed to normals, by computing a Hessian of the height field at each point, instead of a shape operator. Both algorithms must also be extended to force them to conform to the boundary of the domain. There is however, to our knowledge, no equivalent natural generalization of QEM [9] to the heigh field approximation case. Once again, the tunable parameter in

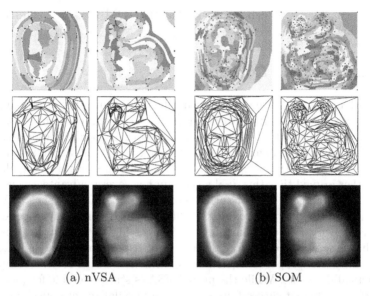

(a) nVSA   (b) SOM

Fig. 4. Gradient approximation using nVSA and our algorithm. Primal (top), algorithm output (middle), height field approximation (bottom). Red marks in the primal are locations of vertices in the dual. In both (a) and (b), the mask (left side) is approximated with 356 triangles, and the bunny (right side) with 468 triangles.

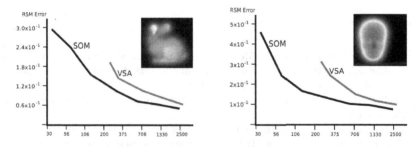

Fig. 5. RMS ($\mathcal{L}^2$) gradient error plots for the mask and bunny height fields (top-right corners.)

VSA has been adjusted to the best results obtained. The input is a surface that is finely scan-converted on a squared piece of the plane (figure 5 top-right corners.) Planar meshes obtained this way are shown in figure 4, while figure 5 shows the corresponding error plots for these two inputs, at several approximation levels. Notice how the mesh approximating the mask in 4.b more closely matches the features of the input than 4.a, even though both have the same number of triangles. The difference in RMS error is not always large for a fixed number of triangles, though it is significant. If, alternatively, an RMS error level is fixed, and the VSA and SOM approximations with that error are considered, it can be noted that the SOM mesh has significantly fewer triangles.

6 Summary and Conclusion

This work begins by placing some established algorithms for surface approximation into a common framework. From this analysis, it becomes apparent that a *dual* variational algorithm for surface normal approximation was previously missing. Such algorithm is introduced next, and its limit behavior shown to conform with the theoretical asymptotic aspect ratio (Appendix C). It is further argued that this dual algorithm has several advantages over primal variational algorithms for surface normal approximation (such as VSA). In particular, the limit approximation efficiency is discussed in Appendix B, which is shown to be approximately 1.75 times higher for a dual algorithm with the same (asymptotically optimal) limit aspect ratio. The approximation results of the proposed algorithm and established ones are also compared on practical data sets. While the primal VSA is still preferable for piecewise flat surfaces, where it successfully splits them into flat regions which are then triangulated, for general curved surfaces, the algorithm introduced in this paper is shown to perform better. This is further shown on quantitatively experiments, which are carried out on the very closely related problem of gradient approximation over the plane.

References

1. Morii, F.: A Generalized K-Means Algorithm with Semi-Supervised Weight Coefficients. In: ICPR 2006: Proceedings of the 18th International Conference on Pattern Recognition, pp. 198–201 (2006)
2. Gruber, P.M.: Optimum quantization and its applications. Adv. Math. 186, 456–497 (2004)
3. Rusinkiewicz, S.: Estimating Curvatures and Their Derivatives on Triangle Meshes. In: Symposium on 3D Data Processing, Visualization, and Transmission (2004)
4. D'Azevedo, E.F., Simpson, R.B.: On Optimal Triangular Meshes for Minimizing the Gradient Error. Numerische Mathematik 59, 321–348 (1991)
5. Canas, G.D., Gortler, S.J.: On Asymptotically Optimal Meshes by Coordinate Transformation. In: 15th International Meshing Roundtable (2006)
6. Clarkson, K.L.: Building Triangulations Using Epsilon-Nets. In: STOC: Proceedings of the Thirty-eighth Annual SIGACT Symposium (2006)
7. Shewchuk, J.R.: What Is a Good Linear Finite Element? - Interpolation, Conditioning, Anisotropy, and Quality Measures. In: Eleventh International Meshing Roundtable, Ithaca, New York, pp. 115–126 (2002)
8. Nadler, E.: Piecewise linear best l2 approximation on triangulations. Approximation Theory V, 499–502 (1986)

9. Garland, M., Heckbert, P.: Surface Simplification Using Quadric Error Metrics. In: ACM SIGGRAPH (1997)
10. Heckbert, P.S., Garland, M.: Optimal triangulation and quadric-based surface simplification. Comput. Geom. Theory Appl. 14(1-3), 49–65 (1999)
11. Gonzalez, T.: Clustering to Minimize the Maximum Inter-Cluster Distance. Theoretical Computer Science 38, 293–306 (1985)
12. Arthur, D., Vassilvitskii, S.: K-means++: the advantages of careful seeding. In: SODA: Symposium On Discreet Algorithms, pp. 1027–1035 (2007)
13. Lloyd, S.P.: Least squares quantization in PCM. IEEE Transactions on Information Theory 28(2), 129–136 (1982)
14. Cohen-Steiner, D., Alliez, P., Desbrun, M.: Variational shape approximation. In: ACM SIGGRAPH (2004)

Appendix

Appendix A: Distance Tightness Bounds

As in [6], it is said that $a \leq_\lambda b \Leftrightarrow a \leq (1+\lambda)b$, and that $a \simeq_\lambda b \Leftrightarrow a \leq_\lambda b \wedge b \leq_\lambda a$. $q_{II}(t,p)$ is the second fundamental form at point p in direction t, and q_{II}^c is its "convexified" version (taking the absolute value of the eigenvalues). The definitions of D_{II} and D_{III} are in Eqs. 3 and 7.

Lemma 1. *For all $\lambda > 0$, for all non-parabolic $p_j \in M$, there's an open neighborhood $V \ni p_j$ of M such that $\forall p \in V$, $D_{QEM}(p_j, p) \leq_\lambda D_{II}(p_j, p)^4$, and $D_{sVSA}(p_j, p) \leq_\lambda D_{II}(p_j, p)^4$.*

Proof. Lemma 4.1 of [6] shows that for all $\lambda > 0$, for all non-parabolic $p_j \in M$, there's an open neighborhood $V \ni p_j$ of M such that $\forall p \in V$, $D_E(p_j, T_p M) \leq_\lambda D_{II}(p_j, p)^2$, where $D_E(p_j, T_p M)$ is the Euclidean shortest distance from p_j to the plane tangent to the surface at p. Then, by the symmetry of D_{II}:

$$D_{QEM}(p, p') = D_E(p_j, p)^2 \leq_\lambda D_{II}(p_j, p)^4 =$$
$$= D_{II}(p, p_j)^4 \geq_\lambda D_E(p, p_j)^2 = D_{sVSA}(p_j, p)$$

where V is chosen small enough such that the neighborhood $V' \ni p$ of the last approximate inequality above contains p_j as well.

Note that the other direction of the inequality is not true in general in neighborhoods that are not elliptic. If D_{II} had been defined using $|q_{II}|^{1/2}$ instead, then it would've been possible to write: $D_{QEM}(p, p') \simeq_\lambda D_{II}(p_j, p)^4 \simeq_\lambda D_{sVSA}(p, p')$ at every non-parabolic point.

Lemma 2. *For all $\lambda > 0$, for all non-parabolic $p_j \in M$, there's an open neighborhood $V \ni p_j$ of M such that $\forall p \in V$, $D_{SOM}(p_j, p) \simeq_\lambda D_{III}(p_j, p)^2$, and $D_{nVSA}(p_j, p) \simeq_\lambda D_{III}(p_j, p)^2$.*

Proof. From the fact that $D_{\text{SOM}}(p_j, p)$ is a second-order Taylor approximation of $D_{\text{nVSA}}(p_j, p)$ around p, and that p_j is not parabolic, with V chosen small enough not to contain parabolic points (which is possible since the set of non-parabolic point is open) follows that $\forall \beta > 0$ there's a neighborhood $V \ni p_j$ such that $\forall p \in V$, $D_{\text{nVSA}}(p_j, p) \simeq_\beta D_{\text{SOM}}(p_j, p)^2$ for $0 < \beta < \lambda, (1 + \beta)^2 = 1 + \lambda$. It is also possible to choose a neighborhood $V' \ni p_j$ small enough such that [2]:

$$D_{\text{SOM}}(p_j, p) = (p_j - p)^T S^2(p)(p_j - p) \simeq_\beta D_{\text{III}}(p_j, p)^2$$

In particular, because $\beta < \lambda$, then also $D_{\text{SOM}}(p_j, p) \simeq_\lambda D_{\text{III}}(p_j, p)^2$. Finally, inside the intersection of the two neighborhoods from the two claims, the transitivity property $x \leq_\beta y \leq_\beta z \Rightarrow x \leq (1 + \beta)^2 z$ yields $D_{\text{nVSA}}(p_j, p) \simeq_\lambda D_{\text{III}}(p_j, p)^2$.

Appendix B: Limit Approximation Efficiency

As pointed out in [6], an optimal solution of Eq. 1 (or 2), in the limit regime, for a small enough, regular (everywhere elliptical or hyperbolic) neighborhood of a surface point, looks like a (stretched) regular hexagonal tiling. A dual algorithm outputs the dual of this tiling (blue), which locally is a regular (valence 6) triangulation. A primal algorithm instead triangulates the hexagons directly (green). The limit efficiencies of these dual and primal triangulations are compared next.

The uniform stretching is first undone to obtain an isotropic hexahedral decomposition, which can be shown not to affect the analysis. Note that, although there are multiple ways of triangulating a regular hexagon, all produce the same set of triangles if symmetry and rotation are discounted. In the $\mathcal{L}^\infty$, normal approximation case, the larger triangles of the primal (green) have error equal to $D_{\text{III}}(p_j, v_i)$, same as the error of the dual triangles (and analogously for surface approximation by using D_{II} instead). But there are four primal triangles per hexagon, and only two dual triangles per hexagon, resulting in a factor of two inefficiency of the primal.

The $\mathcal{L}^2$ case is more involved, and it is only analyzed for the normal approximation case that concerns us most here. Optimal normals are *assigned* to each triangle in both triangulations, which can be computed in closed-form. The $\mathcal{L}^2$ normal error over the triangles is then numerically integrated. Starting from the same regular hexagonal tiling, here the error per unit area in the primal and the dual triangulations is different. Using the fact that $\mathcal{L}^2$ error grows as s^4 where s is a uniform scale factor applied to the triangulation, it is possible to scale the dual triangulation until its error per unit area matches that of the primal. Now the average triangle areas can be compared, yielding the inefficiency factor between primal and dual. All computations (including integration and scaling) use conservative bounds. The limit inefficiency factor is $\gamma \in (1.7635, 1.7642)$ (where lower and upper bounds are rounded down and up, respectively). Hence a primal triangulation in the limit uses approx. 75% more triangles to obtain the same $\mathcal{L}^2$ normal error as the dual.

Appendix C: Shape Operator Metric and Aspect Ratio

For a regular (non parabolic) point p on a smooth surface M, and for very fine approximations, it is possible to consider the shape of a neighborhood $\mathcal{N}_p$ of fixed area that locally minimizes Eq. 2 using D_{SOM}. Since the neighborhood is very small and the surface is smooth, the shape operator is approximately constant inside. Therefore, to any desired degree of accuracy, $\mathcal{N}_p$ is the minimizer of $\int_{\mathcal{N}_p} (p' - p)^T \cdot S(p)^2 \cdot (p' - p)dp'$. If this expression is written in a frame centered at p and oriented so that $\hat{z} = n(p)$ and $\{\hat{x}, \hat{y}\}$ are the principal directions of $S(p)$, then this energy is $\int_{\mathcal{N}_p} k_1^2 x^2 + k_2^2 y^2 dxdy$, where k_1, k_2 are the principal curvatures at p. It is easy to show that a neighborhood $\mathcal{N}_p$ of fixed area minimizing this energy is an ellipse of aspect ratio (ratio of major to minor axis) k_1/k_2, which matches the asymptotically optimal aspect ratio for normal approximation of [4]. Note that (around elliptic points) this ratio would have been $|k_1/k_2|^{1/2}$ for the surface approximation energy $\int_{\mathcal{N}_p} |k_1| x^2 + |k_2| y^2 dxdy$, in accordance with [8]. The dual triangulation inherits these properties: in the limit regime, dual triangles have circumscribing ellipses with same orientation and aspect ratio as the primal regions.

The Meccano Method for Automatic Tetrahedral Mesh Generation of Complex Genus-Zero Solids

J.M. Cascón[1], R. Montenegro[2], J.M. Escobar[2], E. Rodríguez[2], and G. Montero[2]

[1] Department of Economics and Economic History, Faculty of Economics and Management, University of Salamanca, Spain
   casbar@usal.es
[2] Institute for Intelligent Systems and Numerical Applications in Engineering, University of Las Palmas de Gran Canaria, Campus Universitario de Tafira, Las Palmas de Gran Canaria, Spain
   {rmontenegro,jmescobar,erodriguez,gmontero}@siani.es
   http://www.dca.iusiani.ulpgc.es/proyecto2008-2011

Abstract. In this paper we introduce an automatic tetrahedral mesh generator for complex genus-zero solids, based on the novel meccano technique. Our method only demands a surface triangulation of the solid, and a coarse approximation of the solid, called meccano, that is just a cube in this case. The procedure builds a 3-D triangulation of the solid as a deformation of an appropriate tetrahedral mesh of the meccano. For this purpose, the method combines several procedures: an automatic mapping from the meccano boundary to the solid surface, a 3-D local refinement algorithm and a simultaneous mesh untangling and smoothing. A volume parametrization of the genus-zero solid to a cube (meccano) is a direct consequence. The efficiency of the proposed technique is shown with several applications.

Keywords: Tetrahedral mesh generation, local refinement, nested meshes, mesh untangling and smoothing, surface and volume parametrization.

1 Introduction

Many authors have devoted great effort to solving the automatic mesh generation problem in different ways [3, 13, 14, 26], but the 3-D problem is still open [1]. In the past, the main objective has been to achieve high quality adaptive meshes of complex solids with minimal user intervention and low computational cost. At present, it is well known that most mesh generators are based on Delaunay triangulation and advancing front technique. However, problems related to mesh quality, mesh adaption and mesh conformity with the solid boundary, still remain.

We have recently introduced the meccano technique in [21, 2, 22] for constructing adaptive tetrahedral meshes of solids. The method requires a surface triangulation of the solid, a meccano and a tolerance that fixes the desired approximation of the solid surface. The name of the method stems from the fact that the process starts from an outline of the solid, i.e. a meccano composed by connected polyhedral pieces. A particular case is a meccano consisting only of connected cubes, i.e. a polycube [25, 19, 27]. The method generates the solid mesh as a deformation of an appropriate tetrahedral mesh of the meccano. The main idea of the new mesh generator is to combine an automatic parametrization of surface triangulations [6], a local refinement algorithm for 3-D nested triangulations [17] and a simultaneous untangling and smoothing procedure [4].

In this paper, we present significant advances in the method. We define an automatic parametrization of a solid surface triangulation to the meccano boundary. For this purpose, we first divide the surface triangulation into patches with the same topological connection as the meccano faces. Then, a discrete mapping from each surface patch to the corresponding meccano face is constructed by using the parameterization of surface triangulations proposed by M. Floater in [6, 7, 8, 9]. Specifically, we describe the procedure for a solid whose boundary is a surface of genus 0; i.e. a surface that is homeomorphic to the surface of a sphere. In this case, the meccano is a single cube, and the global mapping is the combination of six patch-mapping. The solution to several compatibility problems on the cube edges will be discussed.

The extension to more general solids is possible if the construction of an appropriate meccano is assumed. In the near future, more effort should be made in an automatic construction of the meccano when the genus of the solid surface is greater than zero. Currently, several authors are working on this aspect in the context of polycube-maps, see for example [25, 19, 27]. They are analyzing how to construct a polycube for a generic solid and, simultaneously, how to define a conformal mapping between the polycube boundary and the solid surface. Although surface parametrization has been extensively studied in the literature, only a few works deal with volume parametrization and this problem is still open. A meshless procedure is presented in [18] as one of the first tentative to solve the problem. In addition, Floater et al [10] give a simple counterexample to show that convex combination mappings over tetrahedral meshes are not necessarily one-to-one.

In the following Section we present a brief description of the main stages of the method for a generic meccano composed of polyhedral pieces. In Section 3 we analyze the algorithm in the case that the meccano is formed by a simple cube. In Section 4 we show test problems and practical applications which illustrate the efficiency of this strategy. Finally, the conclusions and future research are presented in Section 5.

2   Meccano Technique Algorithm

The main steps of the *meccano tetrahedral mesh generation algorithm* are summarized in this Section. A first approach of this method can be found in [21, 2, 22]. The input data of the algorithm are the definition of the solid boundary (for example a surface triangulation) and a given precision (corresponding to the approximation of the solid boundary). The following algorithm describes the mesh generation approach.

Meccano tetrahedral mesh generation algorithm

1. Construct a meccano approximation of the 3-D solid formed by polyhedral pieces.
2. Define an admissible mapping between meccano and solid boundaries.
3. Construct a coarse tetrahedral mesh of the meccano.
4. Generate a local refined tetrahedral mesh of the meccano, such that the mapping (according step 2) of the meccano boundary triangulation approximates the solid boundary for a given precision.
5. Move the boundary nodes of the meccano to the solid surface according to the mapping defined in 2.
6. Relocate the inner nodes of the meccano.
7. Optimize the tetrahedral mesh by applying the simultaneous untangling and smoothing procedure.

The first step of the procedure is to construct a meccano approximation by connecting different polyhedral pieces. The meccano and the solid must be equivalent from a topological point of view, i.e., their surfaces must have the same genus. Once the meccano is assembled, we have to define an *admissible* one-to-one mapping between the boundary faces of the meccano and the boundary of the solid. In step 3, the meccano is decomposed into a coarse tetrahedral mesh by an appropriate subdivision of its initial polyhedral pieces. This mesh is locally refined and its boundary nodes are *virtually* mapped to the solid surface until it is approximated to within a given precision. Then, we construct a mesh of the domain by mapping the boundary nodes from the meccano plane faces to the true boundary surface and by relocating the inner nodes at a *reasonable* position. After those two steps, the resulting mesh is generally tangled, but it has an admissible topology. Finally, a simultaneous untangling and smoothing procedure is applied and a valid adaptive tetrahedral mesh of the object is obtained.

3 Meccano Technique for a Complex Genus-Zero Solid

In this Section, we present the application of the meccano algorithm in the case of the solid surface being genus-zero and the meccano being formed by one cube. We assume as datum a triangulation of the solid surface. We introduce an automatic parametrization between the surface triangulation of the solid and the cube boundary. To that end, we divide the surface triangulation into six patches, with the same topological connection than cube faces, so that each patch is mapped to a cube face.

We note that even being poor the quality of this initial triangulation, the meccano method can reach a high quality surface and volume triangulation.

3.1 Meccano

A simple cube, $\mathcal{C}$, is defined as meccano. We associate a planar graph, $\mathcal{G}_\mathcal{C}$ to the meccano in the following way:

- Each face of the meccano corresponds to a vertex of the graph.
- Two vertices of the graph are connected if their corresponding meccano faces share an edge.

Figure 1 shows the numbering of cube faces and their connectivities, and Figure 2 represents the corresponding planar graph.

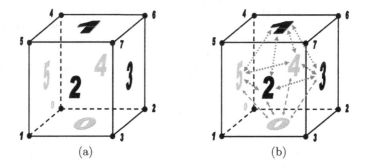

(a)   (b)

Fig. 1. Meccano formed by one cube: (a) notation of nodes and faces of the cube and (b) connectivities of faces

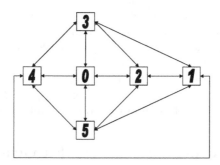

Fig. 2. Planar graph $\mathcal{G}_\mathcal{C}$ associated to the cube

The position of the cube is crucial to define an *admissible* mapping between the cube and solid boundary, as we analyze later. However, its size is less important, because it only affects the efficiency of the mesh optimization step. For a genus-zero solid, if the center of the cube is placed inside the solid, the existence of an admissible mapping is ensured.

3.2   Mapping from Cube Faces to Solid Surface Patches

Once the cube is fixed, we have to determine a mapping between the cube faces and the solid surface triangulation. First, we define the concept of admissible mapping for a cube. Let $\Sigma_\mathcal{C}$ be the boundary of the cube and $\Sigma_\mathcal{S}$ the boundary of the solid, given by a surface triangulation $\mathcal{T}_\mathcal{S}$. We denote by $\Sigma_\mathcal{C}^i$ the i-th face of the cube, i.e. $\Sigma_\mathcal{C} = \bigcup_{i=0}^{5} \Sigma_\mathcal{C}^i$. Let $\Pi : \Sigma_\mathcal{C} \to \Sigma_\mathcal{S}$ be a piecewise function, such that $\Pi_{|\Sigma_\mathcal{C}^i} = \Pi^i$ where $\Pi^i : \Sigma_\mathcal{C}^i \to \Pi^i(\Sigma_\mathcal{C}^i) \subset \Sigma_\mathcal{S}$. Then, Π is called an *admissible* mapping if it satisfies:

a) Functions $\{\Pi^i\}_{i=0}^{5}$ are compatible on $\Sigma_\mathcal{C}$. That is $\Pi^i_{|\Sigma_\mathcal{C}^i \cap \Sigma_\mathcal{C}^j} = \Pi^j_{|\Sigma_\mathcal{C}^j \cap \Sigma_\mathcal{C}^i}$,
   $\forall i, j = 0, \ldots, 5$, with $i \neq j$ and $\Sigma_\mathcal{C}^i \cap \Sigma_\mathcal{C}^j \neq \emptyset$.
b) Global mapping Π is continuous and bijective between $\Sigma_\mathcal{C}$ and $\Sigma_\mathcal{S}$.

We define an automatic admissible mapping in the following Sections. For this purpose, we first construct a partition of the solid surface triangulation into six patches, maintaining the topology of the graph in Figure 2, then we parametrize each patch to a cube face.

Partition of the Solid Surface Triangulation

In the following we call *connected subtriangulation* to a set of triangles of $\mathcal{T}_\mathcal{S}$ whose interior is a connected set. Given a decomposition of the surface triangulation $\mathcal{T}_\mathcal{S}$ in any set of connected subtriangulations, we can associate a planar graph, $\mathcal{G}_\mathcal{S}$, to this partition in the following way:

- Each subtriangulation corresponds to a vertex of the graph.
- Two vertices of the graph are connected if their corresponding subtriangulations have at least one common edge.

We say that a solid surface partition and the meccano are *compatible* if their graphs are isomorphic, $\mathcal{G}_\mathcal{S} = \mathcal{G}_\mathcal{C}$. In our case, since the solid surface is isomorphic to a sphere, it is clear that a compatible partition exists. We now propose an algorithm to obtain a decomposition of the given solid surface triangulation $\mathcal{T}_\mathcal{S}$ into six subtriangulations $\{\mathcal{T}_\mathcal{S}^i\}_{i=0}^{5}$. We distinguish three steps:

a) *Subdivision in connected subtriangulations.* We construct the Voronoi diagram associated to the centers of the six cube faces. We consider that a triangle $F \in \mathcal{T}_\mathcal{S}$ belongs to the i-th Voronoi cell if its barycenter is inside this cell. We generate a partition of $\mathcal{T}_\mathcal{S}$ in maximal connected subtriangulations with this criterion, i.e. two subtriangulations belonging to the same cell can not be connected. We denote as $\mathcal{T}_\mathcal{S}^{ij}$ the j-th connected subtriangulation belonging to the i-th Voronoi cell, and n_i is the total number of subtriangulation in the i-th cell.

b) *Construction of the graph.* We associate a planar graph, $\mathcal{G}_{\mathcal{S}}$ to the partition generated in the previous step. If the center of the cube is inside the solid and the surface triangulation is fine enough, there is one compatible subtriangulation for each Voronoi cell, i.e. there is one *head* subtriangulation $T_{\mathcal{S}}^{i0}$, vertex of the graph $\mathcal{G}_{\mathcal{S}}$, with the same connection as the vertex associated to the *i-th* cube face in $\mathcal{G}_{\mathcal{C}}$. Otherwise, the subtriangulation with the greatest number of elements is chosen as $T_{\mathcal{S}}^{i0}$.

c) *Reduction of the graph.* In order to achieve a decomposition of $T_{\mathcal{S}}$, we propose an iterative procedure to reduce the current graph $\mathcal{G}_{\mathcal{S}}$. In each step all triangles of $T_{\mathcal{S}}^{jk}$ are included in the head subtriangulation $T_{\mathcal{S}}^{i0}$ if:

 - $T_{\mathcal{S}}^{i0}$ is the head subtriangulation with the fewest number of triangles.
 - $T_{\mathcal{S}}^{jk}$ and $T_{\mathcal{S}}^{i0}$ are connected.
 - k is higher than zero.

Then, the vertex $T_{\mathcal{S}}^{jk}$ is removed from the graph and its connectivities are inherited by $T_{\mathcal{S}}^{i0}$. The connectivity of the graph is updated.

After this process, $T_{\mathcal{S}}^{i0}$ could be connected to other subtriangulations $T_{\mathcal{S}}^{il}$ of the same *i-th* cell. In this case, the triangles of all $T_{\mathcal{S}}^{il}$ are included in $T_{\mathcal{S}}^{i0}$, the graph vertices $T_{\mathcal{S}}^{il}$ are removed from the graph, their connectivities are inherited and the graph connectivities are updated. Therefore, the connected subtriangulations are always maximal in all algorithm steps.

This procedure continues iteratively until the graph $\mathcal{G}_{\mathcal{S}}$ is comprised only six head vertices, but the compatibility of $\mathcal{G}_{\mathcal{S}}$ with $\mathcal{G}_{\mathcal{C}}$ can not be ensured. As the computational cost of this algorithm is low, a movement in the cube center, in order to obtain a compatible partition $\{T_{\mathcal{S}}^i\}_{i=0}^5$, does not affect the efficiency of the meccano technique. In what follows we denote $\Sigma_{\mathcal{S}}^i$ the solid surface patch defined by the triangles of $T_{\mathcal{S}}^i$.

Parametrization of the Solid Surface Triangulation

Once the given solid surface $\Sigma_{\mathcal{S}}$ is decomposed into six patches $\Sigma_{\mathcal{S}}^0, \ldots, \Sigma_{\mathcal{S}}^5$, we map each surface patch $\Sigma_{\mathcal{S}}^i$ to the corresponding cube face $\Sigma_{\mathcal{C}}^i$ by using the parametrization of the surface triangulations $T_{\mathcal{S}}^i$ proposed by M. Floater [6]. So, we define $\left(\Pi^i\right)^{-1} : \Sigma_{\mathcal{S}}^i \rightarrow \Sigma_{\mathcal{C}}^i$ and we denote $\tau_F^i = \left(\Pi^i\right)^{-1}(T_{\mathcal{S}}^i)$ as the planar triangulation of $\Sigma_{\mathcal{C}}^i$ associated to $T_{\mathcal{S}}^i$. To obtain τ_F^i, Floater parametrization fixes their boundary nodes and the position of their inner nodes is given by the solution of a linear system based on convex combinations. Let $\{P_1^i, \ldots, P_n^i\}$ be the inner nodes and $\{P_{n+1}^i, \ldots, P_N^i\}$ be the boundary nodes of $T_{\mathcal{S}}^i$, respectively, where N denotes the total number of nodes of $T_{\mathcal{S}}^i$. Fixed the position of boundary nodes $\{Q_{n+1}, \ldots, Q_N\}$ of τ_F^i, the position of the inner nodes $\{Q_1^i, \ldots Q_n^i\}$ is given by the solution of the system:

$$Q_k^i = \sum_{l=1}^N \lambda_{kl} Q_l^i, \qquad k = 1, \ldots, n.$$

The values of the weights of the convex combinations $\{\lambda_{kl}\}_{k=1,\ldots,n}^{l=1,\ldots,N}$ verify

$$\lambda_{kl} = 0, \qquad \text{if } P_k \text{ and } P_l \text{ are not connected}$$

$$\lambda_{kl} > 0, \qquad \text{if } P_k \text{ and } P_l \text{ are connected}$$

$$\sum_{l=1}^{N} \lambda_{kl} = 1, \qquad \text{for } k = 1, \ldots n.$$

In [6] three alternatives are analyzed: uniform parametrization, weighted least squares of edge lengths and shape preserving parametrization. Another choice, called mean value coordinate, is presented in [8]. The goal is to obtain an approximation of a conformal mapping.

In order to ensure the compatibility of $\{\Pi^i\}_{i=0}^5$, the boundary nodes of $\{\tau_F^i\}_{i=0}^5$ must coincide on their common cube edges. The six transformations $\{\Pi^i\}_{i=0}^5$ define an admissible mapping between Σ_C and Σ_S, i.e. the cube boundary triangulation $\tau_F = \bigcup_{i=0}^5 \tau_F^i$ is a global parametrization of the solid surface triangulation $\mathcal{T}_S$.

Two important properties of mapping Π are:

(a) the triangulations τ_F and $\mathcal{T}_S$ have the same topology,
(b) each triangle of τ_F is completely contained in one face of the cube.

We note that usual polycube-maps [25, 19] verify property (a), but they do not verify property (b), i.e., a triangle belonging to $\mathcal{T}_S$ can be transformed by a polycube-map into a *triangle* whose vertices are placed on different faces of the polycube.

The proposed mapping Π is used in a following step of the meccano algorithm to map a new triangulations τ_K (obtained on Σ_C by application of the refinement algorithm of Kossaczky [17]) to the solid boundary. Several problems can appear in the application of this transformation due to the fact that a valid triangulation $\tau_K \neq \tau_F$ on Σ_C can be transformed by Π into a non-valid one on the solid surface.

3.3   Coarse Tetrahedral Mesh of the Meccano

We build a coarse and high quality tetrahedral mesh by splitting the cube into six tetrahedra [17], see Figure 3(a). The resulting mesh can be recursively and globally bisected to fix a uniform element size in the whole mesh. Three consecutive global bisections for a cube are presented in Figures 3 (b), (c) and (d). The resulting mesh of Figure 3(d) contains 8 cubes similar to the one shown in Figure 3(a). Therefore, the recursive refinement of the cube mesh produces similar tetrahedra to the initial ones.

3.4   Local Refined Tetrahedral Mesh of the Meccano

The next step in the meccano mesh generator includes a recursive adaptive local refinement strategy, by using Kossaczky's algorithm [17], of those tetrahedra with a face placed on a boundary face of the initial coarse tetrahedral

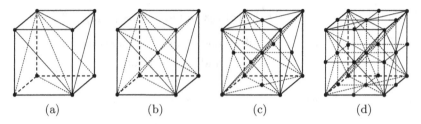

Fig. 3. Refinement of a cube by using Kossaczky's algorithm: (a) cube subdivision into six tetrahedra, (b) bisection of all tetrahedra by inserting a new node in the cube main diagonal, (c) new nodes in diagonals of cube faces and (d) global refinement with new nodes in cube edges

mesh of the cube. The refinement process is done in such a way that the given solid surface triangulation $\mathcal{T}_\mathcal{S}$ is approximated by a new triangulation within a given precision. That is, we seek an adaptive triangulation τ_K on the cube boundary $\Sigma_\mathcal{C}$, so that the resulting triangulation after node mapping $\Pi(\tau_K)$ is a good approximation of the solid boundary. The user has to introduce as input data a parameter ε, which is a tolerance to measure the separation allowed between the linear piecewise approximation $\Pi(\tau_K)$ and the solid surface defined by the triangulation $\mathcal{T}_\mathcal{S}$. At present, we have considered two criteria: the first related to the Euclidean distance between both surfaces and the second attending to the difference in terms of volume.

To illustrate these criteria, let abc be a triangle of τ_K placed on the *meccano* boundary, and $a'b'c'$ the resulting triangle of $\Pi(\tau_K)$ after mapping the nodes a, b and c on the given solid surface $\Sigma_\mathcal{S}$, see Figure 4. We define two different criteria to decide whether it is necessary to refine the triangle (and consequently the tetrahedron containing it) in order to improve the approximation.

For any point Q in the triangle abc we define $d_1(Q)$ as the euclidean distance between the mapping of Q on $\Sigma_\mathcal{S}$, Q', and the plane defined by $a'b'c'$. This definition is an estimate of the distance between the surface of the solid and the current piecewise approximation $\Pi(\tau_k)$.

We also introduce a measure in terms of volume and then, for any Q in the triangle abc, we define $d_2(Q)$ as the volume of the *virtual* tetrahedron $a'b'c'Q'$. In this case, $d_2(Q)$ is an estimate of the lost volume in the linear approximation by the face $a'b'c'$ of the solid surface.

The threshold of whether to refine the triangle or not is given by a tolerance ε_i fixed by the user. We note that other measures could be introduced in line with the desired approximation type (curvature, points properties, etc.).

The *refinement criterion* decides whether a tetrahedron should be refined attending to the current node distribution of triangulation τ_K on the cube boundary $\Sigma_\mathcal{C}$ and their *virtual* mapping $\Pi(\tau_K)$ on the solid boundary $\Sigma_\mathcal{S}$. The separation between triangulations $\Pi(\tau_K)$ and $\Pi(\tau_F) = \mathcal{T}_\mathcal{S}$ is used in the refinement criterion for tetrahedron T:

Refinement criterion

Tetrahedron T is marked to be refined if it satisfies the following two conditions:

1. T has a face $F \in \tau_K$ on the cube boundary.
2. $d_i(Q) \geq \varepsilon_i$ for some node $Q \in \tau_F$ located on face F of T.

From a numerical point of view, the number of points Q (analyzed in this strategy) is reduced to the set of nodes of the triangulation τ_F (defined by the parametrization of Floater) that are contained in face F. We use the nested mesh *genealogy* to implement the refinement criterion efficiently.

Finally, the *refinement procedure* for constructing a local refined tetrahedral mesh of the meccano is summarized in the following algorithm:

Refinement procedure

1. Given the coarse tetrahedral mesh of the meccano.
2. Set a tolerance ε_i.
3. Do
   a) Mark for refinement all tetrahedra that satisfy the *refinement criterion* for a distance d_i and a tolerance ε_i.
   b) Refine the mesh.
   While any tetrahedron T is marked.

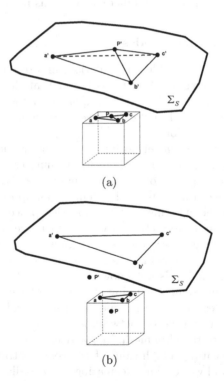

Fig. 4. Node mapping from meccano to real domain: (a) mapping Π from external nodes a, b, c, P to a', b', c', P', and (b) relocation of an inner node P in P'

We denote n_b the number of levels of the nested tetrahedral mesh sequence and τ_K the resulting triangulation of the cube boundary associated to the finest level of the sequence. We note that the refinement procedure automatically concludes according to a single parameter, i.e. ε_i.

3.5   External Node Mapping on Solid Boundary

Once we have defined the local refined tetrahedral mesh by using the method proposed in the previous Section, the nodes of the triangulation τ_K are mapped to the solid surface. Therefore, the triangulation $\Pi(\tau_K)$ is the new approximation of the solid surface.

After this process, due to the properties of Floater's parametrization, $\Pi(\tau_K)$ is generally a valid triangulation. However, unacceptable triangulations can appear. We have checked that this problem only appear when the mesh size of surface approximation $\Pi(\tau_K)$ is the same order than the mesh size of $\mathcal{T}_S$. So, if a more precise approximation of the solid surface is demanded to the meccano approximation, a simple solution is to refine the given solid surface triangulation $\mathcal{T}_S$.

In addition, a tangled tetrahedral mesh is generated because the position of the inner nodes of the cube tetrahedral mesh has not changed.

3.6   Relocation of Inner Nodes

Even if $\Pi(\tau_K)$ is an acceptable triangulation, an optimization of the solid tetrahedral mesh is necessary. Since it is better that the optimization algorithm starts from a mesh with as good a quality as possible, we propose to relocate the inner nodes of the cube tetrahedral mesh in a reasonable position before the mesh optimization.

Although this node movement does not solve the tangle mesh problem, it normally reduces it. In other words, the resulting number of inverted elements is lower and the mean quality of valid elements is greater.

There would be several strategies for defining an appropriate position for each inner node of the cube mesh. The relocation procedure should modify their relative position as a function of the solid surface triangulation before and after their mapping Π, see Figure 4(b). However, an ideal relocation of inner nodes requires a volume mapping from the cube to the complex solid. Obviously, this information is not known *a priori*. In fact, we will reach an approximation of this volume mapping at the end of the mesh generation.

An interesting idea is to use an specific discrete volume mapping that is defined by the transformation between a cube tetrahedral mesh and the corresponding solid tetrahedral mesh. In practice, a good strategy is: we start meshing the solid by using a high value of ε (a coarse tetrahedral mesh of the solid is obtained) and we continue decreasing it gradually. In the first step of this strategy, no relocation is applied. In this case, the number of nodes of the resulting mesh is low and the mesh optimization algorithm is fast. In the following steps a relocation of inner nodes is applied by using the mapping that is defined by the previous iteration.

3.7 Solid Mesh Optimization: Untangling and Smoothing

The proposed relocation procedure, based on volumetric parametrization, is efficient but does not solve the tangling problem completely. Therefore, it is necessary to optimize the current mesh. This process must be able to smooth and untangle the mesh and is crucial in the proposed mesh generator.

The most usual techniques to improve the quality of a *valid* mesh, that is, a mesh with no inverted elements, are based upon local smoothing. In short, these techniques consist of finding the new positions that the mesh nodes must hold, in such a way that they optimize an objective function. Such a function is based on a certain measurement of the quality of the *local submesh*, $N(v)$, formed by the set of elements connected to the *free node* v, whose coordinates are given by $\mathbf{x}$. We have considered the following objective function derived from an *algebraic mesh quality metric* studied in [16],

$$K(\mathbf{x}) = \left[\sum_{m=1}^{M} \left(\frac{1}{q_{\eta_m}} \right)^p (\mathbf{x}) \right]^{\frac{1}{p}}$$

where M is the number of elements in $N(v)$, q_{η_m} is an algebraic quality measure of the m-th element of $N(v)$ and p is usually chosen as 1 or 2. Specifically, we have considered the mean ratio quality measure, which for a tetrahedron is $q_\eta = \frac{3\sigma^{\frac{2}{3}}}{|S|^2}$ and for a triangle is $q_\eta = \frac{2\sigma}{|S|^2}$, $|S|$ being the Frobenius norm of matrix S associated to the affine map from the *ideal* element (usually equilateral tetrahedron or triangle) to the physical one, and $\sigma = \det(S)$. Other algebraic quality measures can be used as, for example, the metrics based on the condition number of matrix S, $q_\kappa = \frac{\rho}{|S||S^{-1}|}$, where $\rho = 2$ for triangles and $\rho = 3$ for tetrahedra. It would also be possible to use other objective functions that have barriers like those presented in [15].

We have proposed in [4] an alternative to the procedure of [12, 11], so the untangling and smoothing are carried out in the same stage. For this purpose, we use a suitable modification of the objective function such that it is regular all over $\mathbb{R}^3$. It consists of substituting the term σ in the quality metrics with the positive and increasing function $h(\sigma) = \frac{1}{2}(\sigma + \sqrt{\sigma^2 + 4\delta^2})$. When a feasible region (subset of $\mathbb{R}^3$ where v could be placed, $N(v)$ being a valid submesh) exists, the minima of the original and modified objective functions are very close and, when this region does not exist, the minimum of the modified objective function is located in such a way that it tends to untangle $N(v)$. With this approach, we can use any standard and efficient unconstrained optimization method to find the minimum of the modified objective function.

In addition, a smoothing of the boundary surface triangulation could be applied before the movement of inner nodes of the domain by using the procedure presented in [5, 20].

4 Test Examples

We have implemented the meccano technique using:

- The parametrization toolbox of the geometry group at SINTEF ICT, Department of Applied Mathematics.
- The module of 3D refinement of ALBERTA code.
- Our optimization mesh procedure describes in Section 3.7.

The parametrization of a surface triangulation patch $\mathcal{T}_\mathcal{S}^i$ to a cube face $\Sigma_\mathcal{C}^i$ is done with GoTools core and parametrization modules from SINTEF ICT, available on the website http://www.sintef.no/math_software. This code implements Floater's parametrization in C++. Specifically, in the following applications we have used the mean value method for the parametrization of the inner nodes of triangulation, and the boundary nodes are fixed with chord length parametrization [6, 8].

ALBERTA is an adaptive multilevel finite element toolbox [24] developed in C. This software can be used to solve several types of 1-D, 2-D or 3-D problems. ALBERTA uses the Kossaczky refinement algorithm [17] and requires an initial mesh topology [23]. The recursive refinement algorithm could not terminate for general meshes. The meccano technique constructs meshes that verify the imposed restrictions of ALBERTA relative to topology and structure. In addition, the minimum quality of refined meshes is function of the initial mesh quality.

The performance of our novel tetrahedral mesh generator is shown in the following applications. The first corresponds to a Bust and the second to the Stanford Bunny. We have obtained a surface triangulation of these objects from internet. For both examples, the *meccano* is just a cube.

4.1 Example 1: Bust

The original surface triangulation of the Bust has been obtained from the website *http://shapes.aimatshape.net*, i.e. AIM@SHAPE Shape Repository, and it is shown in Figure 5(a). It has 64000 triangles and 32002 nodes. The bounding box of the solid is defined by the points $(x, y, z)_{min} = (-120, -30.5, -44)$ and $(x, y, z)_{max} = (106, 50, 46)$.

We consider a cube, with an edge length equal to 20, as meccano. Its center is placed inside the solid at the point $(5, -3, 4)$. We obtain an initial subdivision of Bust surface in seven maximal connected subtriangulations. In order to get a compatible decomposition of the surface triangulation, we use the proposed iterative procedure to reduce the current seven vertices of the graph $\mathcal{G}_\mathcal{S}$ to six. Figure 5(a) shows the resulting compatible partition $\{\mathcal{T}_\mathcal{S}^i\}_{i=0}^5$.

(a)   (b)

Fig. 5. (a) Original surface triangulation of the Bust with a compatible partition $\{\mathcal{T}_\mathcal{S}^i\}_{i=0}^5$ after applying our reduction algorithm and (b) the resulting valid tetrahedral mesh generated by the meccano method

We map each surface patch $\Sigma_\mathcal{S}^i$ to the cube face $\Sigma_\mathcal{C}^i$ by using the Floater parametrization [6]. Once the global parametrization of the Bust surface triangulation is built, see Figure 6(a), the definition of the one-to-one mapping between the cube and Bust boundaries is straightforward.

Fixing a tolerance $\varepsilon_2 = 0.1$, the meccano method generates a tetrahedral mesh of the cube with 147352 tetrahedra and 34524 nodes; see Figures 6(b) and 7(a). This mesh has 32254 triangles and 16129 nodes on its boundary and it has been reached after 42 Kossaczky refinements from the initial subdivision of the cube into six tetrahedra. The mapping of the cube external nodes to the Bust surface produces a 3-D tangled mesh with 8947 inverted elements; see Figure 7(b). The relocation of inner nodes by using volume parametrizations reduces the number of inverted tetrahedra to 285. We apply the mesh optimization procedure [4] and the mesh is untangled in 2

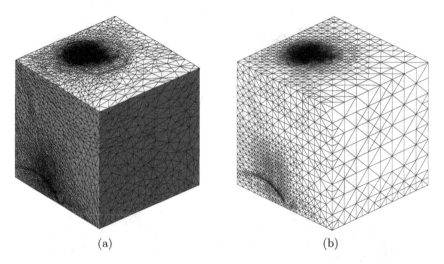

(a)   (b)

Fig. 6. (a) Floater's parametrization of $\{\mathcal{T}_{\mathcal{S}}^i\}_{i=0}^5$ on corresponding cube faces for the bust application, (b) cube tetrahedral mesh obtained by the meccano method

iterations. The mesh quality is improved to a minimum value of 0.07 and an average $\overline{q}_\kappa = 0.73$ after 10 smoothing iterations. We note that the meccano technique generates a high quality tetrahedral mesh (see Figure 5(b)): only 1 tetrahedron has a quality less than 0.1, 13 less than 0.2 and 405 lees than 0.3. In Figure 7, we display two cross sections of the cube and Bust meshes before and after the mesh optimization. The location of the cube is shown in Figure 7(b).

The CPU time for constructing the final mesh of the Bust is 93.27 seconds on a Dell precision 690, 2 Dual Core Xeon processor and 8 Gb RAM memory. More precisely, the CPU time of each step of the meccano algorithm is: 1.83 seconds for the subdivision of the initial surface triangulation into six patches, 3.03 seconds for the Floater parametrization, 44.50 seconds for the Kossaczky recursive bisections, 2.31 seconds for the external node mapping and inner node relocation, and 41.60 seconds for the mesh optimization.

4.2   Example 2: Bunny

The original surface triangulation of the Stanford Bunny has been obtained from the website *http://graphics.stanford.edu/data/3Dscanrep/* , i.e. the Stanford Computer Graphics Laboratory, and it is shown in Figure 8(a). It has 12654 triangles and 7502 nodes. The bounding box of the solid is defined by the points $(x, y, z)_{min} = (-10, 3.5, -6)$ and $(x, y, z)_{max} = (6, 2, 6)$.

We consider a unit cube as meccano. Its center is placed inside the solid at the point $(-4.5, 10.5, 0.5)$. We obtain an initial subdivision of the Bunny surface in eight maximal connected subtriangulations using Voronoi diagram. We reduce the surface partition to six patches and we construct the Floater

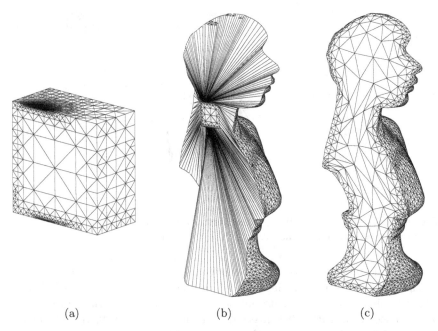

(a)   (b)   (c)

Fig. 7. Cross sections of cube (a) and Bust tetrahedral meshes before (b) and after (c) the application of the mesh optimization procedure

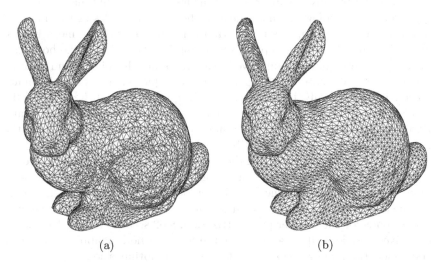

(a)   (b)

Fig. 8. (a) Original surface triangulation of the Stanford Bunny and (b) the resulting valid tetrahedral mesh generated by the meccano method

parametrization from each surface patch Σ_S^i to the corresponding cube face Σ_C^i. Fixing a tolerance $\varepsilon_2 = 0.0005$, the meccano method generates a tetrahedral mesh with 54496 tetrahedra and 13015 nodes. This mesh has 11530

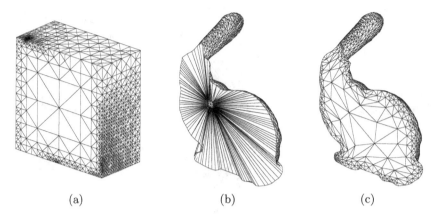

(a)   (b)   (c)

Fig. 9. Cross sections of cube (a) and Bunny tetrahedral meshes before (b) and after (c) the application of the mesh optimization process

triangles and 6329 nodes on its boundary and has been reached after 44 Kossaczky refinements from the initial subdivision of the cube into six tetrahedra. The mapping of the cube external nodes to the Bunny surface produces a 3-D tangled mesh with 2384 inverted elements, see Figure 9(b). The relocation of inner nodes by using volume parametrizations reduces the number of inverted tetrahedra to 42. We apply 8 iterations of the tetrahedral mesh optimization and only one inverted tetrahedra can not be untangled. To solve this problem, we allow the movement of the external nodes of this inverted tetrahedron and we apply 8 new optimization iterations. The mesh is then untangled and, finally, we apply 8 smoothing iterations fixing the boundary nodes. The mesh quality is improved to a minimum value of 0.08 and an average $\overline{q}_\kappa = 0.68$. We note that the meccano technique generates a high quality tetrahedral mesh: only 1 tetrahedron has a quality below 0.1, 41 below 0.2 and 391 below 0.3. In Figure 9, we display two cross sections of the cube and Bunny meshes before and after the mesh optimization. The location of the cube can be observed in Figure 9(b).

The CPU time for constructing the final mesh of the Bunny is 40.28 seconds on a Dell precision 690, 2 Dual Core Xeon processor and 8 Gb RAM memory. More precisely, the CPU time of each step of the meccano algorithm is: 0.24 seconds for the subdivision of the initial surface triangulation into six patches, 0.37 seconds for the Floater parametrization, 8.62 seconds for the Kossaczky recursive bisections, 0.70 seconds for the external node mapping and inner node relocation, and 30.35 seconds for the mesh optimization.

5   Conclusions and Future Research

The meccano technique is a very efficient mesh generation method for creating adaptive tetrahedral meshes of a solid whose boundary is a surface of genus 0. We highlight the fact that the method requires minimum user intervention and has a low computational cost. The procedure is fully automatic and it is only defined by a surface triangulation of the solid, a cube and a tolerance ε that fixes the desired approximation of the solid surface. In addition, the quality of the resulting meshes is high.

The definition of an automatic parametrization of a solid surface triangulation to the meccano boundary is a significant advance for the method. To that end, we have introduced an automatic partition of the given solid surface triangulation for fixing an admissible mapping between the cube faces and the solid surface patches.

In future works, the meccano technique can be extended for meshing a complex solid whose boundary is a surface of genus greater than zero. In this case, the meccano can be a polycube or a set of polyhedral pieces with compatible connections.

Acknowledgments

This work has been supported by the *Secretaría de Estado de Universidades e Investigación* of the *Ministerio de Ciencia e Innovación* of the Spanish Government and FEDER, grant contracts: CGL2008-06003-C03.

References

1. Bazilevs, Y., Calo, V.M., Cottrell, J.A., Evans, J., Hughes, T.J.R., Lipton, S., Scott, M.A., Sederberg, T.W.: Isogeometric analysis: Toward unification of computer aided design and finite element analysis. In: Trends in Engineering Computational Technology, pp. 1–16. Saxe-Coburg Publications, Stirling (2008)
2. Cascón, J.M., Montenegro, R., Escobar, J.M., Rodríguez, E., Montero, G.: A new meccano technique for adaptive 3-D triangulations. In: Proc. 16th Int. Meshing Roundtable, pp. 103–120. Springer, Berlin (2007)
3. Carey, G.F.: Computational grids: generation, adaptation, and solution strategies. Taylor & Francis, Washington (1997)
4. Escobar, J.M., Rodríguez, E., Montenegro, R., Montero, G., González-Yuste, J.M.: Simultaneous untangling and smoothing of tetrahedral meshes. Comput. Meth. Appl. Mech. Eng. 192, 2775–2787 (2003)
5. Escobar, J.M., Montero, G., Montenegro, R., Rodríguez, E.: An algebraic method for smoothing surface triangulations on a local parametric space. Int. J. Num. Meth. Eng. 66, 740–760 (2006)
6. Floater, M.S.: Parametrization and smooth approximation of surface triangulations. Comp. Aid. Geom. Design 14, 231–250 (1997)
7. Floater, M.S.: One-to-one Piece Linear Mappings over Triangulations. Mathematics of Computation 72, 685–696 (2002)
8. Floater, M.S.: Mean Value Coordinates. Comp. Aid. Geom. Design 20, 19–27 (2003)
9. Floater, M.S., Hormann, K.: Surface parameterization: a tutorial and survey. In: Advances in Multiresolution for Geometric Modelling, Mathematics and Visualization, pp. 157–186. Springer, Berlin (2005)

10. Floater, M.S., Pham-Trong, V.: Convex Combination Maps over Triangulations, Tilings, and Tetrahedral Meshes. Advances in Computational Mathematics 25, 347–356 (2006)
11. Freitag, L.A., Knupp, P.M.: Tetrahedral mesh improvement via optimization of the element condition number. Int. J. Num. Meth. Eng. 53, 1377–1391 (2002)
12. Freitag, L.A., Plassmann, P.: Local optimization-based simplicial mesh untangling and improvement. Int. J. Num. Meth. Eng. 49, 109–125 (2000)
13. Frey, P.J., George, P.L.: Mesh generation. Hermes Sci. Publishing, Oxford (2000)
14. George, P.L., Borouchaki, H.: Delaunay triangulation and meshing: application to finite elements. Editions Hermes, Paris (1998)
15. Knupp, P.M.: Achieving finite element mesh quality via optimization of the jacobian matrix norm and associated quantities. Part II-A frame work for volume mesh optimization and the condition number of the jacobian matrix. Int. J. Num. Meth. Eng. 48, 1165–1185 (2000)
16. Knupp, P.M.: Algebraic mesh quality metrics. SIAM J. Sci. Comp. 23, 193–218 (2001)
17. Kossaczky, I.: A recursive approach to local mesh refinement in two and three dimensions. J. Comput. Appl. Math. 55, 275–288 (1994)
18. Li, X., Guo, X., Wang, H., He, Y., Gu, X., Qin, H.: Harmonic Volumetric Mapping for Solid Modeling Applications. In: Proc. of ACM Solid and Physical Modeling Symposium, pp. 109–120. Association for Computing Machinery, Inc. (2007)
19. Lin, J., Jin, X., Fan, Z., Wang, C.C.L.: Automatic polyCube-maps. In: Chen, F., Jüttler, B. (eds.) GMP 2008. LNCS, vol. 4975, pp. 3–16. Springer, Heidelberg (2008)
20. Montenegro, R., Escobar, J.M., Montero, G., Rodríguez, E.: Quality improvement of surface triangulations. In: Proc. 14th Int. Meshing Roundtable, pp. 469–484. Springer, Berlin (2005)
21. Montenegro, R., Cascón, J.M., Escobar, J.M., Rodríguez, E., Montero, G.: Implementation in ALBERTA of an automatic tetrahedral mesh generator. In: Proc. 15th Int. Meshing Roundtable, pp. 325–338. Springer, Berlin (2006)
22. Montenegro, R., Cascón, J.M., Escobar, J.M., Rodríguez, E., Montero, G.: An Automatic Strategy for Adaptive Tetrahedral Mesh Generation. Appl. Num. Math. 59, 2203–2217 (2009)
23. Schmidt, A., Siebert, K.G.: Design of adaptive finite element software: the finite element toolbox ALBERTA. Lecture Notes in Computer Science and Engineering, vol. 42. Springer, Berlin (2005)
24. Schmidt, A., Siebert, K.: ALBERTA - An Adaptive Hierarchical Finite Element Toolbox, http://www.alberta-fem.de/
25. Tarini, M., Hormann, K., Cignoni, P., Montani, C.: Polycube-maps. ACM Trans. Graph 23, 853–860 (2004)
26. Thompson, J.F., Soni, B., Weatherill, N.: Handbook of grid generation. CRC Press, London (1999)
27. Wang, H., He, Y., Li, X., Gu, X., Qin, H.: Polycube Splines. Comp. Aid. Geom. Design 40, 721–733 (2008)

Collars and Intestines:
Practical Conforming Delaunay Refinement

Alexander Rand and Noel Walkington

Carnegie Mellon University

Abstract. While several existing Delaunay refinement algorithms allow acute 3D piecewise linear complexes as input, algorithms producing conforming Delaunay tetrahedralizations (as opposed to constrained or weighted Delaunay tetrahedralizations) often involve cumbersome constructions and are rarely implemented. We describe a practical construction for both "collar" and "intestine"-based approaches to this problem. Some of the key ideas are illustrated by the inclusion of the analogous 2D Delaunay refinement algorithms, each of which differs slightly from the standard approach. We have implemented the 3D algorithms and provide some practical examples.

1 Introduction

Acute input angles pose significant challenges to Delaunay refinement algorithms for quality mesh generation in both two and three dimensions. In 2D, the formation of a conforming mesh is relatively simple: acutely adjacent input segments must be split at equal lengths. Research extending Ruppert's algorithm [16] to accept small input angles [17, 8] focused on finding algorithms which involve simple modifications of Ruppert's algorithm and produce the "best" output meshes in practice. In 3D, producing a conforming tetrahedralization of an arbitrary piecewise linear complex (PLC) involves a substantial construction [10, 6] and quality refinement algorithms have been developed in the context of this construction [5, 11]. Alternative algorithms involving weighted [4, 3] and constrained [17, 19, 18] Delaunay tetrahedralization have also been developed. Due to these different challenges, the algorithms for 3D Delaunay refinement of acute domains are markedly different than those in 2D.

In this paper, we describe two related strategies for the protection of acute angles during 3D Delaunay refinement: collars and intestines. The collar approach generalizes the construction of Murphy, Mount, and Gable [10] and that of Cohen-Steiner, Colin de Verdière, and Yvinec [6] and produces a quality mesh following the ideas of Pav and Walkington [11]. The intestine approach is closely related to the quality refinement algorithm of Cheng and Poon [5]. Unlike these previous algorithms for 3D Delaunay refinement of acute input, our algorithms are motivated by analogous 2D versions and, more notably, have been implemented.

Algorithm 1. Quality Refinement of Acute Input

(PROTECT) Protect acute input angles.
(REFINE) Perform a protected version of Ruppert's algorithm.

Algorithm 1 is the template for both the collar and intestine based refinement algorithms. The (PROTECT) step requires information about the $(d-2)$-dimensional features of the input complex. We note that a brute force computation of this information can be avoided using estimates resulting from certain Delaunay refinement algorithms [13, 14] or an exact computation during sparse Voronoi refinement [9].

Section 2 contains necessary preliminaries for our analysis. Section 3 describes both collar and intestine based Delaunay refinement algorithms in two dimensions, and the three-dimensional algorithms are given in Section 4. Finally, some examples and practical issues are discussed in Section 5.

2 Preliminaries

2.1 Definitions

Our algorithms accept an arbitrary PLC as input and involve an intermediate piecewise smooth complex (PSC), defined below.

Definition 1. *In three dimensions [or two dimensions]:*

- *A **piecewise linear complex** (PLC), $C = (\mathcal{P}, \mathcal{S}, \mathcal{F})$ [$(\mathcal{P}, \mathcal{S})$], is a triple [duple] of sets of input vertices $\mathcal{P}$, input segments $\mathcal{S}$, and polygonal input faces $\mathcal{F}$ such that the boundary of any feature or the intersection of any two features is the union of lower-dimensional features in the complex.*
- *A PLC $C' = (\mathcal{P}', \mathcal{S}', \mathcal{F}')$ [$(\mathcal{P}', \mathcal{S}')$] is a **refinement** of the PLC $C = (\mathcal{P}, \mathcal{S}, \mathcal{F})$ if $\mathcal{P} \subset \mathcal{P}'$, each segment in $\mathcal{S}$ is the union of segments in $\mathcal{S}'$, and every face in $\mathcal{F}$ is the union of faces in $\mathcal{F}'$.*
- *A **piecewise smooth complex** (PSC), $C = (\mathcal{P}, \mathcal{S}, \mathcal{F})$ [$C = (\mathcal{P}, \mathcal{S})$], is a triple [duple] of sets of input vertices $\mathcal{P}$, non-self-intersecting smooth input curves $\mathcal{S}$, and non-self-intersecting smooth input faces $\mathcal{F}$ such that the boundary of any feature or the intersection of any two features is the union of lower-dimensional features in the complex.*

Definition 2. *Let C be a PLC.*

- *The i-**local feature size** at point x with respect to C, $\mathrm{lfs}_i(x, C)$, is the radius of the smallest closed ball centered at x which intersects two disjoint features of C of dimension no greater than i.*
- *The 1-**feature size** of segment s with respect to C, $\mathrm{fs}_1(s, C)$, is the radius of the smallest closed ball centered at a point $x \in s$ intersecting a segment or input vertex of C which does not intersect s.*

If the argument supplied to the local feature size function is a set of points, rather than a single point, then the result is defined to be the infimum of the function over the set, *i.e.* $\mathrm{lfs}_i(s, C) := \inf_{x \in s} \mathrm{lfs}_i(x, C)$. Often the PLC argument supplied to the local feature size is that of the input complex and in this case the argument will be omitted. The subscript will be omitted when it is equal to $(d-1)$ where d is the dimension, i.e. $\mathrm{lfs}(x) := \mathrm{lfs}_2(x)$ in 3D.

For a PSC we will use the same definition of local feature size as for a PLC. Typically the definition of local feature size for a PSC also involves the radius of curvature or the distance to the medial axis. Since we will only use a very restricted class of PSCs the simpler definition is sufficient. In our particular constructions the radius of curvature is proportional to (and often equal to) the local feature size defined above.

2.2   Generic Delaunay Refinement Algorithm

The Delaunay refinement algorithms which we will consider have the form of Algorithm 2. Additionally, we require that each of the operations involve only local computations in the Delaunay triangulation of the current vertex set. To specify an algorithm from Algorithm 2, it is necessary to carefully describe the following four statements.

Algorithm 2. Delaunay Refinement

Create an initial Delaunay triangulation.
Queue all unacceptable simplices.
while the queue of simplices is nonempty **do**
    if it is safe to split the front simplex **then**
        Take an action based on the front simplex.
        Queue additional unacceptable simplices.
    end if
    Remove the front simplex from the queue.
    Dequeue any queued simplices which no longer exist.
end while

Action	Where should a vertex be inserted to "split" a simplex? Should other simplices be added to the queue?
Priority	In what order should the queue be processed?
Unacceptability	Which simplices are unacceptable?
Safety	Which simplices are safe to split?

3   Delaunay Refinement in 2D

Before describing the full 3D algorithms, analogous 2D Delaunay refinement algorithms are given. The resulting algorithms are similar to those typically used for

Delaunay refinement in the presence of acute input angles [17, 8], but avoid certain challenges which are difficult to extend to 3D.

We will describe the two steps in the refinement of an arbitrary PLC given in Algorithm 1: acute input angles are first protected, and then Delaunay refinement is performed. We assume that an appropriate estimate of the local feature size is available at each input vertex. Specifically, we require that for each q_0 which is the vertex of an acute input angle, we are given a distance d_{q_0} which satisfies $b \cdot \mathrm{lfs}(q_0) \leq d_{q_0} \leq \min(c_0 \cdot \mathrm{lfs}_0(q_0), c_1 \cdot \mathrm{lfs}(q_0))$ for some constants $b > 0$, $c_0 \in (0, .5)$ and $c_1 \in (0, 1)$.

3.1   Collar Protection Region

A collar protection region involves forming "collar" segments of equal length around each input vertex so that the Delaunay triangulation conforms to the input near this vertex. The subsequent Delaunay refinement algorithm will then prevent the insertion of any vertices which encroach this collar region.

$$\boxed{\text{(PROTECT) Formation of the Protection Region}}$$

For each q_0 which is the vertex of an acute input angle, each input segment containing q_0 is split at a distance d_{q_0} away from q_0. Figure 3.1 depicts an example of the points inserted during this step.

Each end segment containing the vertex of an acute input angle will be called a collar simplex and vertices inserted during this step are called collar vertices. First, we observe that the collar simplices are sufficiently far away from disjoint input features of C.

Lemma 1. *For any input point $q_0 \in \mathcal{P}$,*

$$\mathrm{dist}\left(B(q_0, d_{q_0}), B(q'_0, d_{q'_0})\right) \geq (1 - 2c_0)\,\mathrm{lfs}(q_0) \text{ for all } \mathcal{P} \ni q'_0 \neq q_0 \text{ and}$$

$$\mathrm{dist}\left(B(q_0, d_{q_0}), s\right) \geq (1 - c_1)\,\mathrm{lfs}(q_0) \text{ for all segments } s \not\ni q_0.$$

Let α be the smallest angle between adjacent segments in C and let $\bar{C}$ denote the refined PLC obtained after inserting all of the collar vertices. The next lemma quantifies the relationship between the local feature size of $\bar{C}$ and C.

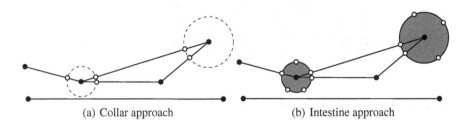

(a) Collar approach   (b) Intestine approach

Fig. 1. Two different protection approaches

Lemma 2. *There exists $K > 0$ depending only on b and c_0 such that for all x,*

$$\text{lfs}(x, \bar{C}) \leq \text{lfs}(x, C) \leq \frac{K}{\sin(\alpha)} \text{lfs}(x, \bar{C}).$$

With the protection region in place, Delaunay refinement can now be performed.

| (REFINE) Protected Delaunay Refinement |

This step is the Delaunay refinement algorithm described in Algorithm 3 (by specializing Algorithm 2). Each new end segment is "protected" during refinement: no vertices will be inserted in the diametral ball of these segments. To ensure this, circumcenters which encroach these end segments are rejected by the safety criteria. Lemma 1 ensures that no inserted midpoints encroach upon a collar simplex and thus the diametral disk of each collar simplex will be empty throughout the algorithm.

Algorithm 3. 2D Delaunay Refinement With Collar

Action	Insert the circumcenter of a simplex unless it causes a lower-dimensional simplex to be unacceptable. In this case, queue the lower-dimensional simplex.
Priority	Segments are given higher priority than triangles.
Unacceptability	A segment with a non-empty diametral disk is unacceptable.
	A triangle with radius-edge ratio larger than τ is unacceptable.
Safety	Collar simplices are not safe to split.

The termination of the algorithm and properties of the resulting mesh are described in Theorems 1 and 2. The first theorem ensures that the algorithm terminates and the resulting mesh is graded to the local feature size, and the second theorem asserts that the mesh conforms to the input PLC and specifies which triangles near collar simplices may have poor quality.

Theorem 1. *For any $\tau > \sqrt{2}$, there exists $K > 0$ depending only upon τ, b, and c_0 such that for each vertex q inserted by Algorithm 3,*

$$\text{lfs}(q, C) \leq \frac{K}{\sin(\alpha)} r_q.$$

Remark 1. The inequality $\text{lfs}(q, \bar{C}) \leq K r_q$ is shown using an argument identical to the standard analysis of Ruppert's algorithm. Then Lemma 2 yields the desired inequality.

Theorem 2. *Algorithm 3 produces a conforming Delaunay triangulation of C. The circumcenter of any remaining triangle with radius-edge ratio larger than τ lies in the diametral disk of a collar simplex.*

3.2 Intestine Protection Region

The intestine protection region yields the added result that no triangles in the resulting mesh have angles larger than $\pi - 2\kappa$, where $\kappa := \sin^{-1}\left(\frac{1}{2\tau}\right)$ is the minimum angle corresponding to the radius-edge threshold τ.

(PROTECT) Formation of the Protection Region

For each input vertex q_0 at an acute input angle, all input segments containing q_0 are split at a distance d_{q_0} away from q_0. Additionally, vertices are added such that all arcs of the circle centered at q_0 with radius d_{q_0} are no larger than $\frac{\pi}{2}$. This ensures that the diametral ball of each arc of the circle does not contain q_0 and requires at most three additional vertices per input vertex.

We will now consider a PSC $\hat{C}$ defined by the input PLC, vertices inserted on each segment at distance d_{q_0} from each input vertex q_0 (as in the collar protection region), and the boundary arcs of each disk $B(q, d_q)$ as depicted in Figure 3.1. The essential property of the PSC $\hat{C}$ is that all acute angles between features occur between segments of C and are contained in $\bigcup_{q_0} B(q_0, d_{q_0})$. Let α again denote the smallest angle between adjacent segments of C. The local feature sizes with respect to C and $\hat{C}$ are related, as described in the next lemma.

Lemma 3. *There exists $K > 0$ depending only on b, c_0, and c_1 such that for all x,*

$$\mathrm{lfs}(x, \hat{C}) \leq \mathrm{lfs}(x, C) \leq \frac{K}{\sin(\alpha)} \, \mathrm{lfs}(x, \hat{C}).$$

A suitably sized protection region has been formed and now the subsequent Delaunay refinement algorithm can be described and analyzed.

(REFINE) Protected Delaunay Refinement

Ruppert's algorithm can be performed *outside* of $\bigcup_{q_0} B(q_0, d_{q_0})$ and each of the boundary arcs of any disk $B(q_0, d_{q_0})$ is protected by the diametral disk of its endpoints. This is described completely in Algorithm 4. Refinement of general PSCs in 2D by algorithms similar to Ruppert's has been considered [1, 2, 12] and our analysis follows these developments.

Algorithm 4 terminates and produces a conforming graded mesh as described in the following two theorems.

Algorithm 4. 2D Delaunay Refinement With Intestine

Action	Insert the circumcenter of a simplex unless it causes a lower-dimensional simplex or arc to be unacceptable. In this case, queue the lower-dimensional object. Insert the midpoint of an arc.
Priority	Segments and arcs are given higher priority than triangles.
Unacceptability	A segment or arc with a non-empty diametral disk is unacceptable. A triangle with radius-edge ratio less than τ is unacceptable.
Safety	All simplices and arcs are safe to split.

Theorem 3. *For any* $\tau > \sqrt{2}$, *there exists* $K > 0$ *depending only upon* τ, b, c_0, *and* c_1 *such that for each vertex* q *inserted by Algorithm 4,*

$$\mathrm{lfs}(q, C) \leq \frac{K}{\sin(\alpha)} r_q.$$

Remark 2. Unlike Theorem 1, the proof of Theorem 3 is substantially more involved than the usual proof for Ruppert's algorithm. This is a result of the smooth input features of $\hat{C}$. Using the techniques of Theorem 1, Theorem 3 can be shown with the strong restriction that $\tau > 2$.

Theorem 4. *Algorithm 4 produces a conforming Delaunay triangulation of* C. *Any remaining triangle with radius-edge ratio larger than* τ *is inside* $B(q_0, d_{q_0})$ *for some input vertex* q_0. *The resulting triangulation contains no angles larger than* $\pi - 2\kappa$.

4 Delaunay Refinement in 3D

Producing a conforming Delaunay tetrahedralization of a 3D PLC requires a consistent mesh along segments between acutely adjacent features. To initially form this consistent mesh we require the feature size to be known along segments of the input mesh. Given a PLC $C = (\mathcal{P}, \mathcal{S}, \mathcal{F})$, we will assume that we have a refinement $C_1 = (\mathcal{P}_1, \mathcal{S}_1, \mathcal{F})$ such that

(H1) $(\mathcal{P}' \setminus \mathcal{P}) \setminus (\cup_{s \in \mathcal{S}} s) = \emptyset$,
(H2) for any $q_0 \in \mathcal{P}$, all $s_1 \in \mathcal{S}_1$ such that $q_0 \in s_1$ have equal length satisfying $|s_1| \leq c_0 \cdot \mathrm{lfs}_0(q_0)$, and
(H3) for all $s_1 \in \mathcal{S}_1$, $b \cdot \min(\mathrm{fs}_1(s_1), \mathrm{lfs}(s_1)) < |s_1| < c_1 \cdot \min(\mathrm{fs}_1(s_1), \mathrm{lfs}(s_1))$,

where $b > 0$, $c_0 \in (0, .5)$, and $c_1 \in (0, 1)$ are some constants.

4.1 Collar Protection Region

| (PROTECT) Formation of the Protection Region |

For each input face, the collar is formed by inserting vertices according to the following rules.

1. If s and s' are adjacent non-end segments which meet at vertex q, then a vertex p is inserted at distance $\frac{\max(|s|,|s'|)}{2}$ from q, in any direction into the face perpendicular to s.
2. If s is an end segment and s' is an adjacent non-end segment, both containing vertex q, then insert vertex p at the intersection of any line parallel to s in the face at distance $\frac{|s'|}{2}$ away from s and on the circle of radius $|s|$ around the input point on s.
3. For any input vertex q_0 on a segment s, insert collar vertices such that the sphere of radius $|s|$ around q_0 restricted to the face has no arcs of angle larger than $\frac{\pi}{2}$.

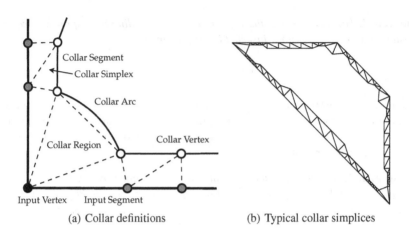

(a) Collar definitions   (b) Typical collar simplices

Fig. 2. Collar region

Below is a list of objects defined to describe the collar based on the vertices inserted during this step. These objects are depicted in Figure 2(a).

Collar Vertex	A vertex inserted during the (PROTECT) step or as a midpoint of a collar segment or arc during the (REFINE) step.
Collar Segment	A segment between collar vertices corresponding to adjacent vertices on an input segment.
Collar Arc	An arc between adjacent collar vertices corresponding to the same input vertex.
Collar Region	The region between input segments and collar segments and arcs.
Collar Simplex	A simplex in the Delaunay triangulation of the face which lies inside the collar region.

Following the insertion of the collar vertices, the resulting Delaunay tetrahedralization satisfies a number of properties given in the following lemma.

Lemma 4. *After inserting collar vertices, the following properties hold.*

(I) All adjacent collar segments and arcs meet at non-acute angles.

(II) The diametral disk of each collar segment contains no vertices in $\mathcal{P}'$.

(III) The circumball of any collar simplex contains no vertices in $\mathcal{P}'$.

(IV) The circumball of any collar simplex does not intersect any disjoint faces or segments.

Remark 3. The circumball of a simplex refers to the *smallest* open sphere such that all vertices of the simplex lie on the boundary of the sphere.

Since the circumball of each collar simplex is empty, the collar simplices conform to the input. Collar segments meet non-acutely and thus the complement of the collar region in each face is well-suited for Ruppert's algorithm. The final property is needed to guarantee that subsequent vertices inserted for conformity will not encroach upon disjoint collar simplices.

The collar divides each face into two regions: the collar region and the non-collar region. Let $\bar{C}$ be the PSC including each face divided into its collar and non-collar regions and all collar segments and arcs. Let α_1 be the smallest angle between an input segment and another adjacent input feature in the mesh and let α_2 be the smallest angle between adjacent input faces. The next lemma asserts that this augmented complex $\bar{C}$ preserves the initial local feature size, up to a factor depending on α_1 and α_2.

Lemma 5. *There exists a constant $K > 0$ depending only upon b, c_0, and c_1 such that*

$$\mathrm{lfs}(x, \bar{C}) \leq \mathrm{lfs}(x, C) \leq \frac{K}{\sin \alpha_1 \sin \alpha_2} \mathrm{lfs}(x, \bar{C}).$$

As usual, the protection procedure is followed by a Delaunay refinement algorithm.

> (REFINE) Protected Delaunay Refinement

The PSC $\bar{C}$ is now refined based on both quality and conformity criteria using a modified version of Ruppert's algorithm. Similarly to the non-acute case, any maximum radius-edge threshold $\tau > 2$ can be selected for determining poor quality tetrahedra. The Delaunay refinement algorithm is specified in Algorithm 5.

Algorithm 5. 3D Delaunay Refinement With Collar

Action	Insert the circumcenter of a simplex unless it causes a lower-dimensional simplex, collar segment, or collar arc to be unacceptable. In this case, queue the lower-dimensional object. Insert the midpoint of a collar arc.
Priority	Collar segments and arcs are given the highest priority. Other simplices are prioritized by dimension with lower-dimensional simplices processed first.
Unacceptability	A simplex, collar segment, or collar arc is unacceptable if it has a nonempty circumball. A tetrahedron is unacceptable if its radius-edge ratio is larger than τ.
Safety	It is not safe to split any collar simplex (this includes both triangles in input faces and subsegments of input segments).

The key difference between Algorithm 5 and the 3D version of Ruppert's algorithm is the safety criteria. This prevents the cascading encroachment associated with acutely adjacent segments and faces. Since collar arcs must be protected, analysis of the 3D refinement with the *collar* protection scheme is closely related to the 2D refinement with the *intestine* protection scheme.

During the algorithm, it is important to ensure that the properties of the collar in Lemma 4 continue to hold while allowing refinement of the non-collar region of each face to create a conforming mesh. In the 2D collar protection procedure, the collar simplices (i.e. the end segments) never change during Algorithm 3. In

3D however, the set of collar simplices does change. This occurs when the standard Delaunay refinement algorithm seeks to insert a vertex in a face that encroaches upon a collar segment or collar arc. Instead of adding this encroaching vertex, this collar segment or arc is split. This new vertex is a collar vertex and the collar segment or arc is replaced with two new collar segments or arcs. The collar *region* has not changed but the set of collar *simplices* has changed. Further, this new vertex may encroach upon the circumball of another collar simplex in an adjacent face. In this face, the collar segment associated with this encroached circumball is also split so that the collar simplices on adjacent faces again "line up." So conformity of the mesh is maintained by only splitting the collar segments and thus the algorithm never attempts to insert the circumcenter of an encroached collar simplex.

Several key properties hold throughout the algorithm whenever there are no collar segments or collar arcs on the queue.

Lemma 6. *If the queue of unacceptable simplices does not contain any collar segments or collar arcs, the following properties hold.*

(I) Adjacent collar segments and arcs meet at non-acute angles.

(II) The circumball of any collar element contains no vertices in $\mathcal{P}'$.

The first property is important to guarantee the termination of the algorithm, while the second property is important for ensuring the resulting tetrahedralization conforms to the input. These two facts are stated precisely in the next two theorems. Recall that α_1 is the smallest angle between an input segment and an adjacent feature while α_2 is the smallest angle between adjacent input faces.

Theorem 5. *For any $\tau > 2$, there exists $K > 0$ depending only upon τ, b, c_0, and c_1 such that for each vertex q inserted by Algorithm 5,*

$$\mathrm{lfs}(q, C) \le \frac{K}{\sin \alpha_1 \sin \alpha_2} r_q.$$

Remark 4. Since $\bar{C}$ includes smooth arcs, the proof of Theorem 5 involves many of the techniques used in Theorem 3.

Theorem 6. *Algorithm 5 produces a conforming Delaunay tetrahedralization of C. The circumcenter of any remaining tetrahedra with radius-edge ratio larger than τ lies in the circumball of a collar simplex.*

4.2   Intestine Protection Region

The intestine approach for protecting acute input angles mirrors that in 2D described in Section 3.2. Smooth features will be added to the input to isolate all input segments and vertices (or at least those contained in acutely adjacent features) from the region to be refined for tetrahedron quality.

$\boxed{\text{(PROTECT) Formation of the Protection Region}}$

The vertices and features which are added to the mesh in this step are a superset of those added during the (PROTECT) step of the collar approach (which created the PSC $\bar{C}$). In addition to features of $\bar{C}$, the following objects are included to form a new PSC $\hat{C}$.

- For each input vertex q_0 which belongs to some segment let d_{q_0} be the length of all segments containing q_0. Then $\hat{C}$ includes $\partial B(q_0, d_{q_0})$.
- For each collar segment s let c be the surface of revolution produced by revolving segment s about its associated input segment. The features c and ∂c are included in $\hat{C}$.

The region inside each sphere and cylindrical surface added to the mesh will be called the intestine region and the remaining volume is called the non-intestine region. This is depicted in Figure 3. This construction is designed to ensure the following fact.

Lemma 7. *The non-intestine region of the PSC $\hat{C}$ contains no acute angles between features.*

This lemma is necessary to ensure that the usual proof of termination and grading will apply to Delaunay refinement in the non-intestine region. Let α_1 and α_2 denote the smallest angles in the input as discussed previously.

Lemma 8. *There exists $K > 0$ depending only on b, c_0 and c_1 such that for all x,*

$$\text{lfs}(x, C) \le \text{lfs}(x, \hat{C}) \le \frac{K}{\sin \alpha_1 \sin \alpha_2} \text{lfs}(x, C).$$

Remark 5. Recall that $\bar{C}$ is the PSC containing the input and the collar construction. Lemma 8 is shown by first showing

$$\text{lfs}(x, \hat{C}) \le \text{lfs}(x, \bar{C}) \le K \text{lfs}(x, \hat{C}),$$

and then applying Lemma 5.

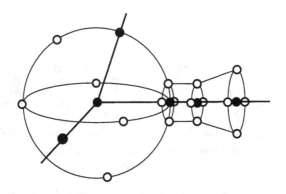

Fig. 3. Intestine Protection Region

(REFINE) Protected Delaunay Refinement

In a similar fashion to the Delaunay refinement algorithm of Cheng and Poon [5], the PSC $\hat{C}$ has been constructed without any acute angles in the non-intestine region so that Delaunay refinement can be performed. The analysis of this approach involves an understanding of the Delaunay refinement of smooth surfaces in 3D. The intermediate PSC $\bar{C}$ including the collar region is much simpler from this perspective as all 2D faces in the complex are affine. While the collar approach involved elements of the analysis for 2D PSCs, the analysis of the intestine approach more closely resembles the much less complete theory of the refinement of 3D PSCs [5, 3, 15, 7].

We now consider two different approaches to performing a quality refinement of the non-intestine region. The first is to perform the usual Delaunay refinement and split smooth surfaces by projecting the circumcenter of any Delaunay triangle in the face to the surface. This is described in Algorithm 6. This approach suffers from one minor drawback: the Delaunay tetrahedralization inside the cylindrical regions of the intestine may not conform to the input. To eliminate this issue, the second approach is to impose more structure on the refinement of these cylindrical regions. This algorithm is given in Algorithm 7. Figure 4 shows the difference between the refinement around required cylindrical surfaces of the two algorithms.

Algorithm 6. 3D Delaunay Refinement With Intestine - Unstructured

Action	Project the circumcenter of a simplex to its associated surface or curve and insert this vertex, unless it causes a lower-dimensional simplex to be unacceptable. In this case, queue the lower-dimensional object.
Priority	Simplices are prioritized by dimension, with lower-dimensional items processed first.
Unacceptability	A simplex in the non-intestine region is unacceptable if it has a nonempty circumball. A tetrahedron is unacceptable if its radius-edge ratio is larger than τ.
Safety	All simplices are safe to split.

 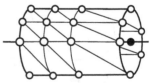

(a) Unstructured Approach of Algorithm 6

(b) Structured Approach of Algorithm 7

Fig. 4. Refinement of cylindrical surfaces around the intestine.

These algorithms terminate and produce meshes which are graded to the local feature size. This is summarized in the following theorem.

Theorem 7. *For any $\tau > 4$, there exists $K > 0$ depending only upon τ, b, c_0, and c_1 such that for each vertex q inserted by Algorithm 6 or Algorithm 7,*

$$\text{lfs}(q, C) \leq \frac{K}{\sin \alpha_1 \sin \alpha_2} r_q.$$

Remark 6. The restriction $\tau > 4$ is stronger than the restriction $\tau > 2$ seen in Theorem 5. The techniques of Theorem 3 have not yet been extended to the case of curved surfaces, and without these techniques, the stronger condition on τ is necessary. Extending this result to admit all $\tau > 2$ is a topic of ongoing research.

Algorithm 7 produces a conforming Delaunay tetrahedralization of the input. This is shown in the next theorem.

Theorem 8. *Algorithm 7 produces a conforming Delaunay tetrahedralization of C. All tetrahedra with radius-edge ratio larger than τ lie in the intestine region.*

The previous result does not hold for Algorithm 6, as the resulting mesh may not conform to the input. This may occur when a vertex on the boundary of the cylindrical region encroaches upon a triangle in a required face inside the intestine region.

However, a simple conforming (but not Delaunay) tetrahedralization of the intestine region does exist. The spheres around input vertices are tetrahedralized using the Delaunay tetrahedra. For the cylindrical sections, let p_1 and p_2 be the endpoints of the corresponding input segment. The tetrahedralization is produced with two types of tetrahedra.

- For any Delaunay triangle t on the boundary of the cylinder, include the tetrahedron with base t and vertex at p_1.

Algorithm 7. 3D Delaunay Refinement With Intestine - Structured

Action	Project the circumcenter of a simplex to its associated surface or curve and insert this vertex, unless it causes a lower-dimensional simplex to be unacceptable. In this case, queue the lower-dimensional object. EXCEPTION: when handling a triangle associated with a cylindrical region which did not yield to another simplex, divide this cylindrical region into two cylinders of equal length and include the new boundary circle in the PSC. Moreover, insert vertices on this circle in the same fashion as in the construction of the intestine region.
Priority	Simplices are prioritized by dimension, with lower-dimensional items processed first.
Unacceptability	A simplex in the non-intestine region is unacceptable if it has a nonempty circumball. A tetrahedron is unacceptable if its radius-edge ratio is larger than τ.
Safety	All simplices are safe to split.

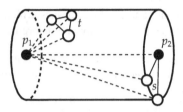

Fig. 5. Two types of tetrahedra are used to produce a conforming tetrahedralization of the intestine region following Algorithm 6

- For any arc s on the circle around p_2, include the tetrahedra with vertices p_1, p_2 and the endpoints of s.

These tetrahedra are depicted in Figure 5. This construction yields a mesh which conforms to the input. This is summarized in the following theorem.

Theorem 9. *Algorithm 6 produces a conforming Delaunay tetrahedralization of the non-intestine region of $\hat{C}$. The previous construction yields a conforming tetrahedralization of the intestine region of $\hat{C}$ which matches the Delaunay tetrahedralization on the boundary of the intestine region. All tetrahedra with radius-edge ratio larger than τ lie in the intestine region.*

5 Implementation Details and Examples

In 3D, we have implemented both collar and intestine based protection schemes. Our implementation relies on estimates of the local feature size given by a different Delaunay refinement algorithm [13, 14]. Algorithm 6 (rather than Algorithm 7) has been implemented and will be referred to as the intestine approach in the examples below. In the future, we hope to implement both algorithms and do a thorough comparison.

Figure 6 demonstrates both protection strategies on a very simple PLC: a single tetrahedra. Figure 7 shows the refinement of a single face of the pyramid during this refinement using the collar. The result looks very similar when using the intestine approach.

An essential method for reducing the number of vertices in the final mesh is to protect only input segments and vertices which are part of acute input angles. This yields a substantial improvement in the output mesh size. Figure 8 shows an input PLC, the resulting mesh when all segments are protected, and the resulting mesh when only acute input segments are protected. The resulting mesh with full protection contains 18079 vertices while the mesh with partial protection only contains 3216 vertices.

Finally, Figure 9 contains six examples produced by Algorithm 5. Data on the input and output sizes of the meshes produced for each of these examples is contained in Table 1. Each of the meshes produced only uses the partial collar described

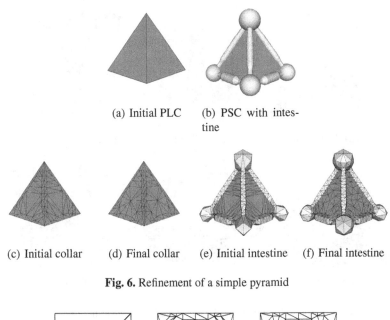

(a) Initial PLC (b) PSC with intestine

(c) Initial collar   (d) Final collar   (e) Initial intestine   (f) Final intestine

Fig. 6. Refinement of a simple pyramid

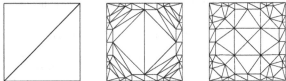

Fig. 7. Refinement of the base of the pyramid

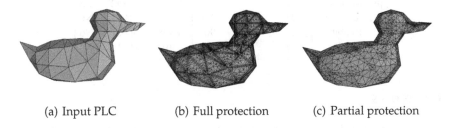

(a) Input PLC   (b) Full protection   (c) Partial protection

Fig. 8. Comparison of full and partial collar protection

above. While the refinement is performed in a bounding box, this bounding box was removed for the PLCs which enclose a volume. This is indicated in the "Box" column of Table 1.

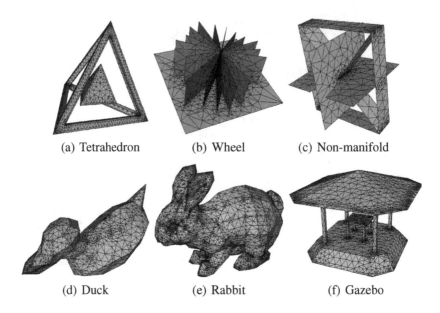

(a) Tetrahedron   (b) Wheel   (c) Non-manifold

(d) Duck   (e) Rabbit   (f) Gazebo

Fig. 9. Examples meshes produced by Algorithm 5

Table 1. Results of Algorithm 5 on six PLCs with acute angles.

Name	Input			Output		
	Vertices	Segments	Faces	Vertices	Tetrahedra	Box
Tetrahedron	24	54	28	3700	11476	No
Wheel	46	65	21	4397	27182	Yes
Non-manifold	22	35	10	2498	15142	Yes
Duck	93	273	182	3216	11001	No
Rabbit	453	1353	902	18968	69001	No
Gazebo	97	148	57	4868	15318	No

References

1. Boivin, C., Ollivier-Gooch, C.: Guaranteed-quality triangular mesh generation for domains with curved boundaries. International Journal for Numerical Methods in Engineering 55(10), 1185–1213 (2002)
2. Cardoze, D.E., Miller, G.L., Olah, M., Phillips, T.: A Bezier-based moving mesh framework for simulation with elastic membranes. In: Proceedings of the 13th International Meshing Roundtable, pp. 71–80 (2004)
3. Cheng, S.-W., Dey, T.K., Levine, J.A.: A practical Delaunay meshing algorithm for a large class of domains. In: Proceedings of the 16th International Meshing Roundtable, pp. 477–494 (2007)

4. Cheng, S.-W., Dey, T.K., Ramos, E.A.: Delaunay refinement for piecewise smooth complexes. In: Proceedings of the 18th Annual ACM-SIAM Symposium on Discrete Algorithms, pp. 1096–1105 (2007)
5. Cheng, S.-W., Poon, S.-H.: Three-dimensional Delaunay mesh generation. In: Proceedings of the 14th Annual ACM-SIAM Symposium on Discrete Algorithms, pp. 295–304 (2003)
6. Cohen-Steiner, D., de Verdière, E.C., Yvinec, M.: Conforming Delaunay triangulations in 3D. Computational Geometry: Theory and Applications 28(2-3), 217–233 (2004)
7. Hudson, B.: Safe Steiner points for delaunay refinement. In: Research Notes of the 17th International Meshing Roundtable (2008)
8. Miller, G.L., Pav, S.E., Walkington, N.J.: When and why Ruppert's algorithm works. In: Proceedings of the 12th International Meshing Roundtable, pp. 91–102 (2003)
9. Miller, G.L., Phillips, T., Sheehy, D.: Fast sizing calculations for meshing. In: 18th Fall Workshop on Computational Geometry (2008)
10. Murphy, M., Mount, D.M., Gable, C.W.: A point-placement strategy for conforming Delaunay tetrahedralization. International Journal of Computational Geometry and Applications 11(6), 669–682 (2001)
11. Pav, S.E., Walkington, N.J.: Robust three dimensional Delaunay refinement. In: Proceedings of the 13th International Meshing Roundtable, pp. 145–156 (2004)
12. Pav, S.E., Walkington, N.J.: Delaunay refinement by corner lopping. In: Proceedings of the 14th International Meshing Roundtable, pp. 165–181 (2005)
13. Rand, A., Walkington, N.: 3D Delaunay refinement of sharp domains without a local feature size oracle. In: Proceedings of the 17th International Meshing Roundtable, Pittsburgh, PA (2008)
14. Rand, A., Walkington, N.: Delaunay refinement algorithms for estimating local feature size (submitted, 2009)
15. Rineau, L., Yvinec, M.: Meshing 3D domains bounded by piecewise smooth surfaces. In: Proceedings of the 16th International Meshing Roundtable, pp. 443–460 (2007)
16. Ruppert, J.: A Delaunay refinement algorithm for quality 2-dimensional mesh generation. Journal of Algorithms 18(3), 548–585 (1995)
17. Shewchuk, J.R.: Mesh generation for domains with small angles. In: Proceedings of the 16th Annual Symposium on Computational Geometry, pp. 1–10 (2000)
18. Si, H.: On refinement of constrained Delaunay tetrahedralizations. In: Proceedings of the 15th International Meshing Roundtable, pp. 510–528 (2006)
19. Si, H., Gartner, K.: Meshing piecewise linear complexes by constrained Delaunay tetrahedralizations. In: Proceedings of the 14th International Meshing Roundtable, pp. 147–163 (2005)

An Analysis of Shewchuk's Delaunay Refinement Algorithm

Hang Si

Weierstrass Institute for Applied Analysis and Stochastics (WIAS), Berlin
si@wias-berlin.de

Abstract. Shewchuk's Delaunay refinement algorithm is a simple scheme to efficiently tetrahedralize a 3D domain. The original analysis provided guarantees on termination and output edge lengths. However, the guarantees are weak and the time and space complexity are not fully covered. In this paper, we present a new analysis of this algorithm. The new analysis reduces the original 90° requirement for the minimum input dihedral angle to $\arccos \frac{1}{3} \approx 70.53^o$. The bounds on output edge lengths and vertex degrees are improved. For a set of n input points with spread Δ (the ratio between the longest and shortest pairwise distance), we prove that the number of output points is $O(n \log \Delta)$. In most cases, this bound is equivalent to $O(n \log n)$. This theoretically shows that the output number of tetrahedra is small.

1 Introduction

Delaunay refinement is one of the classical techniques to generate Delaunay meshes for domains in $\mathbb{R}^d$ with well-shaped simplices and a small output size. It is first introduced by Chew [6] and Ruppert [18] for meshing 2D domains.

Shewchuk's Delaunay refinement algorithm [19] (abbreviated as Shewchuk's algorithm) is a 3D generalization of Ruppert's [18]. This algorithm has the features of being very simple and easy to implement. Practical implementations [20, 22] show that it is efficient and produces tetrahedral meshes with relatively small size compared with other size-optimal meshing algorithms [1, 14], see Fig. 1 for an example. Since its introduction, this algorithm has been generalized into a number of meshing algorithms: for handling small (acute) input angles [5, 4, 17], for meshing curved domains [15, 3], and for mesh adaptation [23]. However, there is no known improvement in its original analysis.

The original analysis of Shewchuk, which follows Ruppert's framework [18], proved several theoretical guarantees on its termination and output edge lengths. It is observed in practice that the behaviors of this algorithm greatly outperforms the proved estimates. For instances, the algorithm usually terminates on inputs with a dihedral angle as small as 60°, which is far from

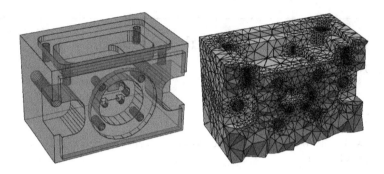

Fig. 1. A three-dimensional polyhedron (left) and a boundary conforming Delaunay tetrahedral mesh generated by Shewchuk's algorithm [19] (right).

the originally required 90°. The algorithm is able to produce a mesh whose tetrahedra having a radius-edge ratio less than $\sqrt{2}$ (or even smaller) instead of a value ≥ 2. The bounds on output edge lengths are obviously too large.

The time and space complexity of this algorithm remain largely unsolved. Ruppert [18] has shown that the number of output vertices of Delaunay refinement algorithms for a domain $\Omega \subset \mathrm{R}^d$ is $\Theta(\int_{\mathbf{x} \in \Omega} \frac{1}{\mathrm{lfs}(x)^d} \mathrm{d}\mathbf{x})$, where $\mathrm{lfs}(\mathbf{x})$ is the *local feature size* (explained in Section 3) at a point $\mathbf{x} \in \Omega$. However, this estimate does not relate to the input. Hudson et al [10] recently proposed a variant of Shewchuk's algorithm, the so-called Sparse Voronoi Refinement (SVR). For a set of n points in $\mathbb{R}^3$, SVR has runtime $O(n \log \Delta + m)$ and space usage $O(m)$, where m is the number of output vertices and Δ is the spread of the input [8]. Note that these bounds depend on the output mesh size m which is not predictable. Recently, Miller et al [12] and Hudson et al [11] showed that under some simple assumptions on the input point sets, m linearly depends on n.

In this paper, we present a new analysis of Shewchuk's algorithm. Our goal is to improve the original analysis and to gain more insight into the simple and elegant scheme of this algorithm. Practitioners of Shewhcuk's algorithm could be benefited from our analysis. For instances, one can avoid adding unnecessary protecting points in the problem of handling small input angles; it may guide the choice of the order of Steiner points to reduce the total number of output points; and it may help in designing new Delaunay refinement algorithms.

The rest of this paper is organized as follows. We briefly review Shewchuk's algorithm in Section 2. In Section 3, we show that an dihedral angle bound of $\arccos \frac{1}{3} \approx 70.53^o$ is sufficient to guarantee the termination. A useful tool in our analysis is a proper sequence of added points which will be defined in Section 4. Improved bounds for output edge lengths and vertex degrees are given in Section 5 and Section 6, respectively. Section 7 discusses the output mesh size. For a set of n points with spread Δ, we show that the number of output points is bounded by $O(n \log \Delta)$. We end our analysis by a discussion of open issues.

2   The Algorithm

This section presents Shewchuk's algorithm for the later analysis. Some preliminary definitions of the input and output objects are given first.

A *piecewise linear system* [24] (abbreviated as PLS) $\mathcal{X}$ is a collection of polytopes such that: *(i)* $P \in \mathcal{X}$ implies that all faces of P are in $\mathcal{X}$, and *(ii)* the intersections of any two polytopes of $\mathcal{X}$ are again polytopes of $\mathcal{X}$. The polytopes in a PLS are not necessarily convex. This definition of a PLS is generalized from Miller et al's [13]. The *dimension* of a PLS is the largest dimension of its polytopes. For an example, Fig. 1 left shows a 3D PLS which is a collection including a 3D polyhedron and all its faces. The *underlying space* $|\mathcal{X}|$ of $\mathcal{X}$ is the union of all polytopes of $\mathcal{X}$, i.e., $|\mathcal{X}| = \bigcup_{P \in \mathcal{X}} P$. Note that $|\mathcal{X}|$ is a topological subspace of $\mathbb{R}^3$. The collection $\mathcal{X}$ induces a topology on its underlying space $|\mathcal{X}|$.

A *tetrahedral mesh* of a 3D *PLS* $\mathcal{X}$ is a finite set $\mathcal{T}$ of simplices, e.g., vertices, edges, triangles, and tetrahedra, such that: *(i)* any two simplices of $\mathcal{T}$ are either disjoint or intersect at their common face, *(ii)* the union of $\mathcal{T}$ equals to $|\mathcal{X}|$, and *(iii)* each polytope $P \in \mathcal{X}$ is the union of a subset of $\mathcal{T}$. Fig. 1 right shows a tetrahedral mesh of a 3D PLS.

Let S be a finite set of points in $\mathbb{R}^d$. A simplex σ in S is *Delaunay* [7] if it has a circumscribed sphere Σ such that no other point of S lies inside Σ. Moreover, σ is *Gabriel* [9] if no other point of S lies inside the *diametrical sphere* of σ, i.e., the smallest circumscribed sphere of σ. A *boundary conforming Delaunay mesh* [24] of a 3D PLS $\mathcal{X}$ is a tetrahedral mesh of $\mathcal{X}$ such that *(i)* every simplex of $\mathcal{T}$ is Delaunay, and *(ii)* every simplex of $\mathcal{T}$ in a polytope $P \in \mathcal{X}$ and $\dim(P) < 3$ is Gabriel.

The *radius-edge ratio*, ρ, of a tetrahedron is the ratio between the radius of its circumscribed ball and the length of its shortest edge. The regular tetrahedron has the minimum value $\sqrt{6}/4 \approx 0.612$. Most of the badly shaped tetrahedra will have a large radius-edge ratio except *slivers*, which are nearly degenerate tetrahedra whose radius-edge ratio may be as small as $\sqrt{2}/2 \approx 0.707$. Hence, strictly speaking ρ is not a shape measure. Nevertheless, it is useful in the analysis of this algorithm.

The Algorithm: Let $\mathcal{X}$ be a 3D PLS. We call 1- and 2-polytopes of $\mathcal{X}$ *segments* and *facets*. Each segment and facet will be represented by a subcomplex of a mesh $\mathcal{T}$ of $\mathcal{X}$. We call 1- and 2-simplices of that subcomplex *subsegments* and *subfaces* to distinguish them from other simplices of $\mathcal{T}$. A subsegment (or a subface) σ is said to be *encroached* if it is not Gabriel in $\mathcal{T}$. The algorithm is given in Fig. 2.

DELAUNAYREFINEMENT $(\mathcal{X}, \rho_0)$
// $\mathcal{X}$ is a 3D PLS; ρ_0 is a radius-edge ratio bound.
1. initialize a DT $\mathcal{D}$ of the vertex set of $\mathcal{X}$;
2. **while** ($\exists$ encroached subsegment or subface)
3.   **or** ($\exists\ \tau \in \mathcal{D}$ such that $\rho(\tau) > \rho_0$), **do**
4.   create a new point $\mathbf{v}$ by rule i, $i \in \{1, 2, 3\}$;
5.   update $\mathcal{D}$ to be the DT of vert($\mathcal{D}$) $\cup \{\mathbf{v}\}$;
6. **endwhile**
7. $\mathcal{T} := \mathcal{D} \setminus \{\tau \mid \tau \in \mathcal{D} \text{ and } \tau \nsubseteq |\mathcal{X}|\}$
8. **return** $\mathcal{T}$;

Fig. 2. Shewchuk's Delaunay refinement algorithm [19]. It takes a 3D PLS $\mathcal{X}$ and a radius-edge ratio ρ_0 as inputs, generates a boundary conforming Delaunay mesh $\mathcal{T}$ of $\mathcal{X}$ such that no tetrahedra of $\mathcal{T}$ has radius edge ratio larger than ρ_0.

After the initialization, the algorithm runs in a loop (lines $2 - 6$). A new point $\mathbf{v}$ (in line 4) is found by one of the three *point generating rules*:

$R1$: If a subsegment is encroached, $\mathbf{v}$ is its midpoint. $\mathbf{v}$ is an $R1$-*vertex*.

$R2$: If a subface is encroached, $\mathbf{v}$ is the circumcenter of its diametric ball. $\mathbf{v}$ is an $R2$-*vertex*. However, if $\mathbf{v}$ encroaches upon some subsegments, then reject $\mathbf{v}$, and use $R1$ to return a $\mathbf{v}$.

$R3$: If a tetrahedron τ has radius-edge ratio $\rho(\tau) > \rho_0$, $\mathbf{v}$ is its circumcenter. $\mathbf{v}$ is an $R3$-*vertex*. However, if $\mathbf{v}$ encroaches upon some subsegments or subfaces, then reject $\mathbf{v}$, and use $R1$ or $R2$ to return a $\mathbf{v}$.

Among these rules, $R1$ has the highest priority, and $R3$ has the lowest priority. The priorities of the rules are important. They ensure that the new point $\mathbf{v}$ lies either inside the mesh domain or on its boundary.

3   Proof of Termination

The original analysis [19] requires that no facet angle (defined below) of the input PLS is less than $90°$. In this section, we show that this angle can be reduced.

Definitions: Let $\mathcal{X}$ be a 3D PLS. The *local feature size* [18] of a point $\mathbf{x} \in |\mathcal{X}|$ is a function lfs : $|\mathcal{X}| \to \mathbb{R}^+$, such that lfs($\mathbf{x}$) is the radius of the smallest ball centered at $\mathbf{x}$ that intersects at least two disjoint polytopes of $\mathcal{X}$. Note that lfs($\mathbf{x}$) only depends on $\mathcal{X}$, it does not change when new vertices are added in $|\mathcal{X}|$. A well-known property of lfs is that it is 1-Lipschitz, i.e., for any $\mathbf{x}, \mathbf{y} \in |\mathcal{X}|$, lfs($\mathbf{x}$) $\leq$ lfs($\mathbf{y}$) $+ \|\mathbf{x} - \mathbf{y}\|$, where $\|\mathbf{x} - \mathbf{y}\|$ denote the Euclidean distance between $\mathbf{x}$ and $\mathbf{y}$.

For each new point $\mathbf{v}$ we define a *parent* $p(\mathbf{v})$ which is a unique point responsible for the addition. If $\mathbf{v}$ is an $R1$- or $R2$-vertex, $p(\mathbf{v})$ is the encroaching point of $\mathbf{v}$. The point $p(\mathbf{v})$ may be a Delaunay vertex or a rejected circumcenter. If there are several encroaching points then $p(\mathbf{v})$ is the one closest to $\mathbf{v}$.

If two encroaching points are at the same distance to $\mathbf{v}$, choose the one which is either an input vertex or has been added earlier. If $\mathbf{v}$ is an $R3$-vertex, let τ be the tetrahedron $\mathbf{v}$ splits, then $p(\mathbf{v})$ is one of the endpoints of the shortest edge of τ which either is an input vertex or has been added earlier. If $\mathbf{v}$ is an input point, then $\mathbf{v}$ has no parent.

For each new point $\mathbf{v}$ we define the *insertion radius* $r(\mathbf{v}) = \|\mathbf{v} - p(\mathbf{v})\|$. For an input point $\mathbf{u} \in \mathcal{X}$ we define $r(\mathbf{u}) = \|\mathbf{u} - \mathbf{w}\|$, where $\mathbf{w} \in \mathcal{X}$ is the nearest input vertex of $\mathbf{u}$. Obviously, $r(\mathbf{u}) \geq \mathrm{lfs}(\mathbf{u})$.

An *input angle* of $\mathcal{X}$ is one of the three kinds: If two segments of $\mathcal{X}$ intersect, they formed an angle in $|\mathcal{X}|$, it is called a *segment-segment angle*; if a facet intersects a non-coplanar segment, any line inside the facet and the segment form an angle, the smallest such angle in $|\mathcal{X}|$ is called a *segment-facet angle*; if two facets intersect, they form a dihedral angle (i.e., the angle between their normals) in $|\mathcal{X}|$, it is called a *facet-facet angle*.

The following lemma is well-known. It is first proved in [19].

Lemma 1 ([19]). *Let $\mathbf{v}$ be an added vertex, $\mathbf{p} = p(\mathbf{v})$. Let θ_m denote the smallest input angle. Then:*

(r1) $r(\mathbf{v}) \geq \rho_0\, r(\mathbf{p})$, when $\mathbf{v}$ is an R3-vertex.
(r2) $r(\mathbf{v}) \geq \frac{1}{\sqrt{2}}\, r(\mathbf{p})$, when $\mathbf{v}$ is an R1- or R2-vertex, and $\theta_m \geq 90^\circ$.
(r3) $r(\mathbf{v}) \geq \frac{1}{2\cos\theta}\, r(\mathbf{p})$, when $\mathbf{v}$ is an R1- or R2-vertex, and $\theta_m \geq 45^\circ$, where θ_m is either a segment-segment or a segment-facet angle;

Below we will prove a lemma for the case when θ_m is an acute facet-facet angle. First we will need a geometrical fact which we prove it in the following lemma.

Lemma 2. *Let $\mathbf{abc}$ be a triangle with vertices $\mathbf{a}$, $\mathbf{b}$, and $\mathbf{c}$, and let $\theta = \angle \mathbf{abc}$ and $\theta < 90^\circ$, see Fig. 3 (a). If (i) $\frac{\|\mathbf{a}-\mathbf{c}\|}{\|\mathbf{a}-\mathbf{b}\|} \leq \sqrt{2}$, and (ii) $\theta \geq \arctan\sqrt{2} \approx 54.74^\circ$, then*

$$\frac{\|\mathbf{a}-\mathbf{c}\|}{\|\mathbf{b}-\mathbf{c}\|} \geq \frac{2}{3\sqrt{2}\cos\theta}. \tag{1}$$

Proof. We will prove this lemma in two steps. We first construct a case where the equality in (1) holds. We then show that this case indeed gives the smallest ratio among all possible triangles satisfying the two preconditions.

Place the edge $\mathbf{ab}$ on a horizontal line, and let $\mathbf{c}$ freely move above it. All possible locus of $\mathbf{c}$ form a region shown in Fig. 3 (a) (the shaded part). In the triangle $\mathbf{abp}$, $\frac{\|\mathbf{a}-\mathbf{p}\|}{\|\mathbf{a}-\mathbf{b}\|} = \sqrt{2}$, and $\theta = \arctan\sqrt{2}$. The point $\mathbf{q}$ locates on the line containing $\mathbf{ab}$ and the angle $\psi = \angle \mathbf{aqp} = 45^\circ$, hence $\frac{\|\mathbf{p}-\mathbf{b}\|}{\|\mathbf{p}-\mathbf{q}\|} = \frac{\sqrt{3}}{2}$. Then

$$\cos\theta = \frac{\|\mathbf{a}-\mathbf{b}\|}{\|\mathbf{p}-\mathbf{b}\|} = \frac{1/\sqrt{2}\,\|\mathbf{a}-\mathbf{p}\|}{\sqrt{3}/2\,\|\mathbf{p}-\mathbf{q}\|} = \frac{2}{\sqrt{6}}\frac{\|\mathbf{a}-\mathbf{q}\|}{\|\mathbf{p}-\mathbf{q}\|} = \frac{2}{\sqrt{6}}\cos\psi. \tag{2}$$

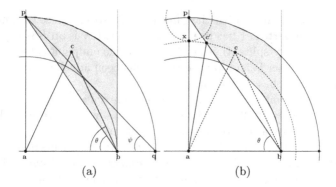

Fig. 3. Proof of Lemma 2.

In the triangle **aqp**, we have the equality

$$\frac{\|\mathbf{a} - \mathbf{p}\|}{\|\mathbf{p} - \mathbf{q}\|} = \frac{1}{2\cos\psi},$$

substitute $\cos\psi$ by eq. (2), and $\|\mathbf{p} - \mathbf{q}\|$ by $\frac{2}{\sqrt{3}}\|\mathbf{b} - \mathbf{p}\|$ in above, we get

$$\frac{\|\mathbf{a} - \mathbf{p}\|}{\|\mathbf{b} - \mathbf{p}\|} = \frac{1}{\sqrt{6}\cos\theta}\frac{2}{\sqrt{3}} = \frac{2}{3\sqrt{2}\cos\theta}.$$

Next we show that the above ratio is the smallest one for all possible choices of **c**. First of all, if **c** lies inside or on the circle centered at **a** with radius $\|\mathbf{a} - \mathbf{b}\|$, Pav [16] has proved that $\frac{\|\mathbf{a} - \mathbf{c}\|}{\|\mathbf{b} - \mathbf{c}\|} \geq \frac{1}{2\cos\theta}$. Hence our claim holds. In the following, we consider that **c** lies in the rest of the admissible region, see Fig. 3 (b).

For any point **c** in this region, we can find a point **c'** which is at the intersection of the line **bp** and the circle C centered at **a** with radius $\|\mathbf{a} - \mathbf{c}\|$, see Fig. 3 (b). It is easy to see that $\frac{\|\mathbf{a} - \mathbf{c}'\|}{\|\mathbf{b} - \mathbf{c}'\|} < \frac{\|\mathbf{a} - \mathbf{c}\|}{\|\mathbf{b} - \mathbf{c}\|}$. Now we show that $\frac{\|\mathbf{a} - \mathbf{c}'\|}{\|\mathbf{b} - \mathbf{c}'\|} > \frac{\|\mathbf{a} - \mathbf{p}\|}{\|\mathbf{b} - \mathbf{p}\|}$. Introduce an auxiliary point **x** at the intersection of the circle C and the edge **ap**, see Fig. 3 (b), clearly, $\|\mathbf{x} - \mathbf{p}\| < \|\mathbf{c}' - \mathbf{p}\|$. Then

$$\frac{\|\mathbf{a} - \mathbf{p}\|}{\|\mathbf{b} - \mathbf{p}\|} = \frac{\|\mathbf{a} - \mathbf{x}\| + \|\mathbf{x} - \mathbf{p}\|}{\|\mathbf{b} - \mathbf{c}'\| + \|\mathbf{c}' - \mathbf{p}\|} \leq \frac{\|\mathbf{a} - \mathbf{c}'\|}{\|\mathbf{b} - \mathbf{c}'\|}.$$

Since we have chosen **c** arbitrarily in the admissible region, so $\frac{2}{3\sqrt{3}\cos\theta}$ is the smallest ratio among all admissible choices of **c**.   ∎

The next lemma consider the case which is not given in Lemma 1.

Lemma 3. *Let* **v** *be an added R2-vertex on facet* F_1, $\mathbf{p} = p(\mathbf{v})$ *is an R2-vertex (not a rejected one) on facet* F_2. *Let* θ *be the facet-facet angle formed by* F_1 *and* F_2. *Then* *(r4)* $r(\mathbf{v}) \geq \frac{1}{3\cos\theta} r(\mathbf{p})$, *when* $\theta \geq \arctan\sqrt{2} \approx 54.74^\circ$.

Proof. Let $F_1 \ni \mathbf{v}$ and $F_2 \ni \mathbf{p}$ be the two facets which intersect at a common segment R and form a dihedral angle $\theta < 90^\circ$, see Fig. 4 (a). $\mathbf{xy}$ is a subsegment of R. Let $B_{\mathbf{xy}}$ be the diametric ball of $\mathbf{xy}$. Both $\mathbf{v}$ and $\mathbf{p}$ must lie outside $B_{\mathbf{xy}}$ (otherwise, $\mathbf{xy}$ is encroached and it must be split before we can add $\mathbf{p}$ or $\mathbf{v}$). Let $B_{\mathbf{v}}$ denote the ball centered at $\mathbf{v}$ with a radius r. $B_{\mathbf{v}}$ is the diametric ball of a subface on F_1 encroached by $\mathbf{p}$. Without loss of generality, we assume that $\mathbf{v}$ is closer to $\mathbf{x}$ than to $\mathbf{y}$. Hence $r(\mathbf{v}) = \|\mathbf{v} - \mathbf{p}\| \leq r \leq \|\mathbf{v} - \mathbf{x}\|$, and $r(\mathbf{p}) \leq \|\mathbf{p} - \mathbf{x}\|$.

Let $\mathbf{p}'$ be the projection of $\mathbf{p}$ onto R, see Fig. 4 (b). Note that $\|\mathbf{v} - \mathbf{p}'\| \geq \|\mathbf{v} - \mathbf{x}\|/\sqrt{2}$, so in the triangle $\mathbf{vp}'\mathbf{p}$, $\frac{\|\mathbf{v}-\mathbf{p}\|}{\|\mathbf{v}-\mathbf{p}'\|} \leq \frac{\|\mathbf{v}-\mathbf{x}\|}{\|\mathbf{v}-\mathbf{p}'\|} \leq \sqrt{2}$. Let θ' be the angle $\angle\mathbf{vp}'\mathbf{p}$. Note that $\theta' \geq \theta$ (since θ is the dihedral angle between the two facets). We now show that if $\theta \geq \arctan\sqrt{2}$, we can map the triangle $\mathbf{vp}'\mathbf{p}$ congruently to a triangle $\mathbf{abc}$ in Fig. 3 (a).

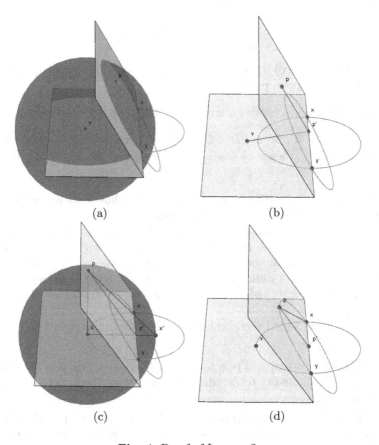

(a)   (b)

(c)   (d)

Fig. 4. Proof of Lemma 3.

Let $\mathbf{v} = \mathbf{a}$, $\mathbf{p}' = \mathbf{b}$, and $\theta' = \angle \mathbf{vp'p} = \angle \mathbf{abc}$. What remain is to map $\mathbf{p}$ to a point $\mathbf{c}$ in the shaded region in Fig. 3 (a). We start from a case which $\mathbf{v}$ lies just on the bisector of $B_{\mathbf{xy}}$ and touch $B_{\mathbf{xy}}$, $\mathbf{p}$ lies vertically on the top of $\mathbf{v}$ at a distance $\|\mathbf{v} - \mathbf{p}\| = \sqrt{2}\|\mathbf{v} - \mathbf{p}'\|$, see Fig. 4 (c). In this case, $\theta = \arctan\sqrt{2} \approx 54.74°$. Clearly $\mathbf{vp'p}$ is congruent to $\mathbf{abp}$ in Fig. 3 (a). Now if $\theta > \arctan\sqrt{2}$, since $\mathbf{p}$ must lie inside $B_{\mathbf{v}}$, and $\mathbf{v}$ can not be inside $B_{\mathbf{xy}}$, hence it must be $\frac{\|\mathbf{v}-\mathbf{p}\|}{\|\mathbf{v}-\mathbf{p}'\|} < \sqrt{2}$, there exists a point $\mathbf{c}$ in the shaded region in Fig. 3 (a) that can be mapped to $\mathbf{p}$, and the two triangles $\mathbf{vp'p}$ and $\mathbf{abc}$ are congruent.

So if $\theta \geq \arctan\sqrt{2}$, we can apply Lemma 2 on the triangle $\mathbf{vp'p}$ to get

$$\frac{\|\mathbf{v} - \mathbf{p}\|}{\|\mathbf{p}' - \mathbf{p}\|} \geq \frac{2}{3\sqrt{2}\cos\theta'} \geq \frac{2}{3\sqrt{2}\cos\theta}. \tag{3}$$

Note that $\|\mathbf{p} - \mathbf{p}'\| \geq \frac{1}{\sqrt{2}}\|\mathbf{p} - \mathbf{x}\|$. The case when the equality holds is shown in Fig. 4 (d), where $\mathbf{p}$ lies on the ball $B_{\mathbf{xy}}$. With the help of (3), we have

$$\frac{\|\mathbf{v} - \mathbf{p}\|}{\|\mathbf{x} - \mathbf{p}\|} \geq \frac{1}{\sqrt{2}}\frac{\|\mathbf{v} - \mathbf{p}\|}{\|\mathbf{p} - \mathbf{p}'\|} \geq \frac{1}{3\cos\theta}. \tag{4}$$

Remember that $r(\mathbf{v}) = \|\mathbf{v} - \mathbf{p}\|$, and $r(\mathbf{p}) \leq \|\mathbf{x} - \mathbf{p}\|$, our claim holds by substituting corresponding terms in (4). ∎

Lemma 3 allows us to prove an improved angle bound for the termination.

Theorem 1. *Let $\mathcal{X}$ be a 3D PLS with its smallest input angle $\theta_m \geq \arccos\frac{1}{3} \approx 70.53°$. Shewchuk's algorithm terminates on $\mathcal{X}$ for any value of $\rho_0 \geq 2$.*

Proof. Let $\text{lfs}_m = \min\{\text{lfs}(\mathbf{x}) \mid \mathbf{x} \in |\mathcal{X}|\}$. We will show that no output edge of this algorithm will have length shorter than lfs_m.

Suppose for the sake of contradiction that the algorithm introduces an edge shorter than lfs_m into the mesh. Let $\mathbf{xy}$ be the first such edge introduced. Clearly, $\mathbf{x}$ and $\mathbf{y}$ cannot both be input vertices, nor can they lie on non-incident boundaries. Assume $\mathbf{x}$ was added after $\mathbf{y}$. By assumption, no edge shorter than lfs_m exists before $\mathbf{x}$ was added, i.e., for any existing vertex $\mathbf{q}$, $r(\mathbf{q}) \geq \text{lfs}_m$.

Now let $\mathbf{x}$ be an added vertex, and $\mathbf{p} = p(\mathbf{x})$, we enumerate all the possible cases for $\mathbf{x}$ and $\mathbf{p}$:

- If $\mathbf{x}$ is an $R1$-vertex, and $\mathbf{p}$ is a rejected $R3$-vertex, i.e., $\mathbf{p}$ is the circumcenter of a bad quality tetrahedron. Let $\mathbf{g} = p(\mathbf{p})$, then $r(\mathbf{x}) \geq \frac{1}{\sqrt{2}}r(\mathbf{p}) \geq \frac{\rho_0}{\sqrt{2}}r(\mathbf{g}) \geq \text{lfs}_m$.
- If $\mathbf{x}$ is an $R1$-vertex, and $\mathbf{p}$ is a rejected $R2$-vertex, i.e., $\mathbf{p}$ is the circumcenter of a an encroached subface σ, and σ is encroached by a rejected $R3$-vertex. Let $\mathbf{g} = p(\mathbf{p})$, and $\mathbf{e} = p(\mathbf{g})$, then $r(\mathbf{x}) \geq \frac{1}{\sqrt{2}}r(\mathbf{p}) \geq \frac{1}{2}r(\mathbf{g}) \geq \frac{\rho_0}{2}r(\mathbf{e}) \geq \text{lfs}_m$.

- If $\mathbf{x}$ is an $R1$-vertex, and $\mathbf{p}$ is either an $R1$- or $R2$-vertex (not rejected), i.e., $\mathbf{p}$ lies on an incident segment or facet. Then $r(\mathbf{x}) \geq \frac{1}{2\cos\theta_m} r(\mathbf{p}) \geq \mathrm{lfs}_m$.
- If $\mathbf{x}$ is an $R2$-vertex, and $\mathbf{p}$ is a rejected $R3$-vertex. Let $\mathbf{g} = p(\mathbf{p})$, then $r(\mathbf{x}) \geq \frac{1}{\sqrt{2}}r(\mathbf{p}) \geq \frac{\rho_0}{\sqrt{2}}r(\mathbf{g}) \geq \mathrm{lfs}_m$.
- If $\mathbf{x}$ is an $R2$-vertex, and $\mathbf{p}$ is an $R1$-vertex (not rejected) lies on an incident segment. Then $r(\mathbf{x}) \geq \frac{1}{\sqrt{2}\cos\theta_m}r(\mathbf{p}) \geq \mathrm{lfs}_m$.
- If $\mathbf{x}$ is an $R2$-vertex, and $\mathbf{p}$ is an $R2$-vertex (not rejected) lies on an incident facet. Then $r(\mathbf{x}) \geq \frac{1}{3\cos\theta_m} r(\mathbf{p}) \geq \mathrm{lfs}_m$.
- If $\mathbf{x}$ is an $R3$-vertex, no matter what type of $\mathbf{p}$ has, $r(\mathbf{x}) \geq \rho_0 r(\mathbf{p}) \geq \mathrm{lfs}_m$.

Hence $r(\mathbf{x}) \geq \mathrm{lfs}_m$ in all cases, contradicting the assumption. It must be that no edge shorter than lfs_m is ever introduced, hence the algorithm will terminate.   ∎

4   Parent Sequences

Since every new vertex $\mathbf{v}$ has a unique parent point, we can form a sequence of points by tracing the parents. In this section, we derive relations of insertion radii of a new vertex $\mathbf{v}$ and points in its parent sequence.

Definitions: The *parent sequence* of a new vertex $\mathbf{v}$ is a sequence of points, $\{\mathbf{v}_i\}_{i=0}^{m+1}$, such that $\mathbf{v} = \mathbf{v}_0$, for $i = 1, 2, ..., m$, $\mathbf{v}_i$ is an existing $R3$-vertex (not a rejected one), $\mathbf{v}_{m+1}$ is either an $R1$-, or $R2$-vertex, or an input vertex (i.e., $\mathbf{v}_{m+1}$ is a boundary vertex), for $i = 1, 2, ..., m$, $\mathbf{v}_{i+1}$ is the parent of $\mathbf{v}_i$. Note that $\mathbf{v}_1$ may not be the parent of $\mathbf{v}_0$ when $\mathbf{v}_0$ is an $R1$- or $R2$-vertex. Denote $g(\mathbf{v}) = \mathbf{v}_{m+1}$ the *grandparent* of $\mathbf{v}$. A parent sequence of $\mathbf{v}$ has at least two vertices, namely $\mathbf{v}$ and $g(\mathbf{v})$. Note that m counts the number of $R3$-vertices in this sequence.

Any parent sequence whose last vertex is not an input vertex can be extended to a longer sequence. We define the *maximal sequence* of $\mathbf{v}$ as a sequence of parent sequences, $\{\{\mathbf{v}_i^j\}_{i=0}^{m_j+1}\}_{j=0}^{k-1}$, such that $k \geq 1$, for $j = 0, 1, ..., k-2$, $\{\mathbf{v}_i^j\}_{i=0}^{m_j+1}$ is a parent sequence of the vertex $\mathbf{v}_0^j$ (which has m_j $R3$-vertices), $\mathbf{v}_0^{j+1} = g(\mathbf{v}_0^j)$, and $\mathbf{v}_{m_{k-1}+1}^{k-1}$ (the last vertex) is an input vertex, it is called the *ancestor* of $\mathbf{v}$, denoted as $a(\mathbf{v})$. An expansion of the maximal sequence of $\mathbf{v}$ looks like follows

$$
k \begin{cases}
\mathbf{v} = \mathbf{v}^0, & \underbrace{\mathbf{v}_1^0, \mathbf{v}_2^0, \cdots,}_{m_0} & \mathbf{v}_{m_0+1}^0, \\
\mathbf{v}_{m_0+1}^0 = \mathbf{v}^1, & \underbrace{\mathbf{v}_1^1, \mathbf{v}_2^1, \cdots,}_{m_1} & \mathbf{v}_{m_1+1}^1, \\
\cdots, & \cdots, & \\
\mathbf{v}_{m_{k-2}+1}^{k-2} = \mathbf{v}^{k-1}, & \underbrace{\mathbf{v}_1^{k-1}, \mathbf{v}_2^{k-1}, \cdots,}_{m_{k-1}} & \mathbf{v}_{m_{k-1}+1}^{k-1} = a(\mathbf{v}).
\end{cases}
\tag{5}
$$

Given a geometric series with a common ratio $\frac{x}{\rho_0}$, let C_x^n denote the sum of its first n terms, i.e., $C_x^n = 1 + \frac{x}{\rho_0} + \left(\frac{x}{\rho_0}\right)^2 + \cdots + \left(\frac{x}{\rho_0}\right)^n = \frac{1-(x/\rho_0)^{n+1}}{1-(x/\rho_0)}$.
In particular, $C_x^0 = 1$ and $C_x^\infty = \frac{1}{1-(x/\rho_0)}$.

The following lemma derives the relations of the insertion radii of $\mathbf{v}$, and $g(\mathbf{v})$ for a new vertex $\mathbf{v}$.

Lemma 4. *Let $\{\mathbf{v}_i\}_{i=0}^{m+1}$ be the parent sequence of $\mathbf{v}$, and let $\mathbf{w} = g(\mathbf{v})$.*

(r5) If $\mathbf{v}$ is an R3-vertex, then $r(\mathbf{v}) \geq \rho_0^{m+1} r(\mathbf{w})$, and $r(\mathbf{v}) \geq \frac{1}{C_1^m}\|\mathbf{v} - \mathbf{w}\|$.

(r6) If $\mathbf{v}$ is an R2-vertex, then $r(\mathbf{v}) \geq \frac{1}{\sqrt{2}}\rho_0^{m+1} r(\mathbf{w})$, and $r(\mathbf{v}) \geq \frac{1}{1+\sqrt{2}C_1^m}\|\mathbf{v}-\mathbf{w}\|$.

(r7) If $\mathbf{v}$ is an R1-vertex, then $r(\mathbf{v}) \geq \frac{1}{2}\rho_0^{m+1} r(\mathbf{w})$, and $r(\mathbf{v}) \geq \frac{1}{1+\sqrt{2}+2C_1^m}\|\mathbf{v}-\mathbf{w}\|$.

Proof. If $\mathbf{v}$ is an R3-vertex, for $i = 0, ..., m + 1$, $r(\mathbf{v}_i) \geq \rho_0 r(\mathbf{v}_{i+1})$ (by *(r1)*). Then $r(\mathbf{v}_0) \geq \rho_0 r(\mathbf{v}_1) \geq \rho_0^2 r(\mathbf{v}_2) \geq \cdots \geq \rho_0^{m+1} r(\mathbf{v}_{m+1}) \implies r(\mathbf{v}) \geq \rho_0^{m+1} r(\mathbf{w})$, which is the first part of *(r5)*. The second part of *(r5)* can be proved by

$$\|\mathbf{v} - \mathbf{w}\| \leq \|\mathbf{v}_0 - \mathbf{v}_1\| + \|\mathbf{v}_1 - \mathbf{v}_2\| + \cdots + \|\mathbf{v}_m - \mathbf{v}_{m+1}\|$$
$$= r(\mathbf{v}_0) + r(\mathbf{v}_1) + r(\mathbf{v}_2) \cdots + r(\mathbf{v}_m)$$
$$\leq (1 + 1/\rho_0 + (1/\rho_0)^2 + \cdots + (1/\rho_0)^m) r(\mathbf{v}_0)$$
$$= C_1^m r(\mathbf{v}).$$

When $\mathbf{v}$ is an R1-vertex, $\mathbf{v}_1$ may not be the parent of $\mathbf{v}_0$. In the worst case, there will be two rejected vertices, $\mathbf{p}$ and $\mathbf{q}$, such that $\mathbf{p} = p(\mathbf{v}_0)$,

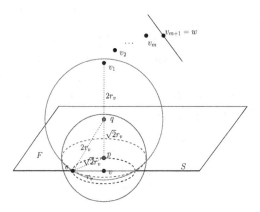

Fig. 5. Proof of Lemma 4. $\mathbf{v}$ is an R1-vertex on a segment S. $\mathbf{p}$ is a rejected R2-vertex on facet F (it encroaches upon the ball centered at $\mathbf{v}$), and $\mathbf{q}$ is a rejected R3-vertex (it encroaches upon the ball centered at $\mathbf{p}$). The rest vertices, $\mathbf{v}_1, ..., \mathbf{v}_{m+1}$, are in the parent sequence of $\mathbf{v}$.

$\mathbf{q} = p(\mathbf{p})$, and $\mathbf{v}_1 = p(\mathbf{q})$, see Fig. 5. Then by *(r1)*, $r(\mathbf{v}_0) \geq \frac{1}{\sqrt{2}} r(\mathbf{p}), r(\mathbf{p}) \geq \frac{1}{\sqrt{2}} r(\mathbf{q}), r(\mathbf{q}) \geq \rho_0 r(\mathbf{v}_1) \implies r(\mathbf{v}_0) \geq \frac{1}{2}\rho_0 r(\mathbf{v}_1)$. By *(r5)* we have $r(\mathbf{v}_1) \geq \rho_0^m r(\mathbf{v}_{m+1})$. Thus, $r(\mathbf{v}) \geq \frac{1}{2}\rho_0^{m+1} r(\mathbf{w})$. The gives the first half of *(r7)*. The second half of *(r7)* can be proved by

$$\|\mathbf{v} - \mathbf{w}\| \leq \|\mathbf{v}_0 - \mathbf{p}\| + \|\mathbf{p} - \mathbf{q}\| + \|\mathbf{q} - \mathbf{v}_1\| + \|\mathbf{v}_1 - \mathbf{v}_2\| + \cdots + \|\mathbf{v}_m - \mathbf{v}_{m+1}\|$$
$$= r(\mathbf{v}_0) + r(\mathbf{p}) + r(\mathbf{q}) + r(\mathbf{v}_1) + r(\mathbf{v}_2) \cdots + r(\mathbf{v}_m)$$
$$\leq (1 + \sqrt{2} + 2 + 2/\rho_0 + 2(1/\rho_0)^2 + \cdots + 2(1/\rho_0)^m) r(\mathbf{v}_0)$$
$$= (1 + \sqrt{2} + 2C_1^m) r(\mathbf{v}).$$

Finally, the proof of *(r6)* is similar to the proof of *(r7)* and is skipped. ∎

We consider the longest maximal sequence of an $R3$-vertex $\mathbf{v}$, see Fig. 6. It consists of a number of parent sequences on a facet and on a segment, and ends at an input vertex $\mathbf{a}$. We will derive the relation between $r(\mathbf{v})$ and $r(\mathbf{a})$. First we need two relations given in the following lemma.

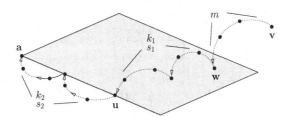

Fig. 6. A maximal sequence of an $R3$-vertex $\mathbf{v}$.

Lemma 5. *Let* $\{\{\mathbf{v}_i^j\}_{i=0}^{m_j+1}\}_{j=0}^{k-1}$ *be the maximal sequence of an $R3$-vertex $\mathbf{v}$ and $\mathbf{a} = a(\mathbf{v})$, see Fig. 6. Let $\mathbf{w}$ and $\mathbf{u}$ be the first $R2$-vertex and $R1$-vertex in this sequence, respectively. Then*

(r8) $r(\mathbf{w}) \geq \left(\frac{\rho_0}{\sqrt{2}}\right)^{k_1} \rho_0^{s_1} r(\mathbf{u})$, *and* $r(\mathbf{w}) \geq \dfrac{1}{(1+\sqrt{2}C_1^\infty)C_{\sqrt{2}}^{k_1-1}} \|\mathbf{w} - \mathbf{u}\|$.

(r9) $r(\mathbf{u}) \geq \left(\frac{\rho_0}{2}\right)^{k_2} \rho_0^{s_2} r(\mathbf{a})$, *and* $r(\mathbf{u}) \geq \dfrac{1}{(1+\sqrt{2}+2C_1^\infty)C_2^{k_2-1}} \|\mathbf{u} - \mathbf{a}\|$.

Where k_1, k_2, s_1, and s_2 are quantities determined in the maximal sequence. Table 1 gives the meaning of these parameters.

Proof. These two claims can be proved similarly. Only the proof of *(r9)* is given.

We can expand the sequence starting from $\mathbf{u}$ and ending at $\mathbf{a}$ (note that in this case it is just the maximal sequence of $\mathbf{u}$). The expansion is given in eq. (5) by substituting $\mathbf{v}$ by $\mathbf{u}$ and k by k_2. For each row in (5) we

Table 1. A summary of the parameters of Lemma 5

m	the number of $R3$-vertices between $\mathbf{v}$ and $\mathbf{w}$	$m \geq 0$
s_1	the number of $R3$-vertices between $\mathbf{w}$ and $\mathbf{u}$	$s_1 \geq 0$
s_2	the number of $R3$-vertices between $\mathbf{u}$ and $\mathbf{a}$	$s_2 \geq 0$
k_1	the number of parent sequences between $\mathbf{w}$ and $\mathbf{u}$	$k_1 \geq 1$
k_2	the number of parent sequences between $\mathbf{u}$ and $\mathbf{a}$	$k_2 \geq 1$

can apply *(r7)*: $r(\mathbf{u}^0) \geq \frac{\rho_0}{2}\rho_0{}^{m_0}r(\mathbf{u}^1), r(\mathbf{u}^1) \geq \frac{\rho_0}{2}\rho_0{}^{m_1}r(\mathbf{u}^2), ..., r(\mathbf{u}^{k_2-1}) \geq \frac{\rho_0}{2}\rho_0{}^{m_{k_2}-1}r(\mathbf{a}) \Longrightarrow r(\mathbf{u}) \geq \left(\frac{\rho_0}{2}\right)^{k_2}\rho_0^{s_2}r(\mathbf{a})$. This proves the first half of *(r9)*. Next

$$\|\mathbf{u} - \mathbf{a}\| \leq \|\mathbf{u}^0 - \mathbf{u}^1\| + \|\mathbf{u}^1 - \mathbf{u}^2\| + \cdots + \|\mathbf{u}^{k_2-1} - \mathbf{a}\| \qquad \text{(apply } (r7))$$

$$\leq (1 + \sqrt{2} + 2C_1^{m_0})r(\mathbf{u}^0) + (1 + \sqrt{2} + 2C_1^{m_1})r(\mathbf{u}^1) + \cdots +$$
$$(1 + \sqrt{2} + 2C_1^{m_{k_2-1}})r(\mathbf{u}^{k_2-1})$$

$$\leq (1 + \sqrt{2} + 2C_1^\infty)(r(\mathbf{u}^0) + r(\mathbf{u}^1) + \cdots + r(\mathbf{u}^{k_2-1})) \qquad \text{(apply } (r7))$$

$$\leq (1 + \sqrt{2} + 2C_1^\infty)$$
$$\left(1 + \frac{2}{\rho_0}\left(\frac{1}{\rho_0}\right)^{m_0} + \left(\frac{2}{\rho_0}\right)^2\left(\frac{1}{\rho_0}\right)^{m_0+m_1} + \cdots + \right.$$
$$\left. \left(\frac{2}{\rho_0}\right)^{k_2-1}\left(\frac{1}{\rho_0}\right)^{\sum_{i=0}^{k_2-2} m_i}\right) r(\mathbf{u})$$

$$\leq (1 + \sqrt{2} + 2C_1^\infty)C_2^{k_2-1}r(\mathbf{u}) \qquad \blacksquare$$

Comments on termination and mesh quality: The termination guarantee of this algorithm requires that $\rho_0 \geq 2$. This is explained by the data flow diagram in [19]. It is observed that the algorithm usually terminates at a much smaller ρ_0, see Fig. 7. This fact can be explained using the maximal sequence. The number m of $R3$-vertices in the parent sequences plays an important role which is not visible by the data flow diagram. From *(r9)* we see if there is at least one $R3$-vertex in the maximal sequence of $\mathbf{u}$, then $r(\mathbf{u}) \geq \frac{1}{2}\rho_0^2 r(\mathbf{a})$. This immediately relaxes the requirement of termination to be $\rho_0 \geq \sqrt{2}$. Moreover, if s_1 and s_2 are larger (there are more $R3$-vertices in the sequence), ρ_0 can be much smaller.

Note that D_3 is smaller if the number of $R3$-vertex in parent sequence (m) is large. This means vertices having a long parent sequence (which means they are far away from the boundary) will have longer edge lengths than those vertices close to boundary, see Fig. 7. This also explains why the algorithm will terminate fast.

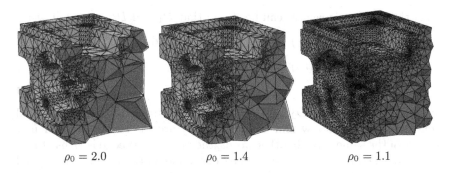

$$\rho_0 = 2.0 \qquad\qquad \rho_0 = 1.4 \qquad\qquad \rho_0 = 1.1$$

Fig. 7. The resulting meshes with different ρ_0. The input PLS has 994 points, 2988 segments, and 1995 triangles. From left to the right, the number of output mesh points are: 7884, 15044, and 42686.

5 Output Edge Lengths

Ruppert [18] first showed that any output edge of Delaunay refinement algorithms will has a bounded length which is no smaller than some constant divided by the local feature sizes of one of its endpoints. Shewchuk provided three constants, $\frac{(3+\sqrt{2})\rho_0}{\rho_0-2}$, $\frac{(1+\sqrt{2}\rho_0)+\sqrt{2}}{\rho_0-2}$, and $\frac{\rho_0+1+\sqrt{2}}{\rho_0-2}$, one for each type of new vertices. In this section, we derive new constants which bound output edge lengths using parent sequences. And we will compare them with Shewchuk's.

Theorem 2. *Let* $\mathbf{v}$ *be an R3-vertex. Consider the maximal sequence of* $\mathbf{v}$ *shown in Fig. 6. Then* $r(\mathbf{v}) \geq \frac{lfs(\mathbf{v})}{D_3}$, $r(\mathbf{w}) \geq \frac{lfs(\mathbf{w})}{D_2}$, *and* $r(\mathbf{u}) \geq \frac{lfs(\mathbf{u})}{D_1}$, *where* D_1, D_2, *and* D_3 *are three constants depending on* m, k_1, s_1, k_2, *and* s_2.

Proof. Denote $L_1(k_1) = (1 + \sqrt{2}C_1^\infty)C_{\sqrt{2}}^{k_1-1}$, $L_2(k_2) = (1 + \sqrt{2} + 2C_1^\infty)C_2^{k_2-1}$, and denote $\mu_x = \frac{x}{\rho_0}$. Then

$$
\begin{aligned}
lfs(\mathbf{v}) &\leq \|\mathbf{v} - \mathbf{a}\| + lfs(\mathbf{a}) \\
&\leq \|\mathbf{v} - \mathbf{w}\| + \|\mathbf{w} - \mathbf{u}\| + \|\mathbf{u} - \mathbf{a}\| + lfs(\mathbf{a}) \quad \text{(apply (r5), (r8), r(9))} \\
&\leq C_1^m r(\mathbf{v}) + L_1(k_1)r(\mathbf{w}) + L_2(k_2)r(\mathbf{u}) + r(\mathbf{a}) \\
&\leq C_1^m r(\mathbf{v}) + L_1(k_1)\mu_1^{m+1}r(\mathbf{v}) + L_2(k_2)\mu_{\sqrt{2}}^{k_1}\mu_1^{s_1}r(\mathbf{w}) + \mu_2^{k_2}\mu_1^{s_2}r(\mathbf{u}) \\
&\leq C_1^m r(\mathbf{v}) + L_1(k_1)\mu_1^{m+1}r(\mathbf{v}) + L_2(k_2)\mu_{\sqrt{2}}^{k_1}\mu_1^{s_1}\mu_1^{m+1}r(\mathbf{v}) \\
&\quad + \mu_2^{k_2}\mu_1^{s_2}\mu_{\sqrt{2}}^{k_1}\mu_1^{s_1}r(\mathbf{w}) \\
&\leq \left(C_1^m + L_1(k_1)\mu_1^{m+1} + L_2(k_2)\mu_{\sqrt{2}}^{k_1}\mu_1^{s_1}\mu_1^{m+1} + \mu_2^{k_2}\mu_1^{s_2}\mu_{\sqrt{2}}^{k_1}\mu_1^{s_1}\mu_1^{m+1} \right) r(\mathbf{v}).
\end{aligned}
$$

Hence D_3 is

$$D_3 = C_1^m + L_1(k_1)\mu_1^{m+1} + L_2(k_2)\mu_{\sqrt{2}}^{k_1}\mu_1^{s_1}\mu_1^{m+1} + \mu_2^{k_2}\mu_1^{s_2}\mu_{\sqrt{2}}^{k_1}\mu_1^{s_1}\mu_1^{m+1}. \quad (6)$$

Using the same approach, it can be shown that D_1 and D_2 are as follows:

$$D_2 = L_1(k_1) + L_2(k_2)\mu_{\sqrt{2}}^{k_1}\mu_1^{s_1} + \mu_2^{k_2}\mu_1^{s_2}\mu_{\sqrt{2}}^{k_1}\mu_1^{s_1}, \tag{7}$$

$$D_1 = L_2(k_2) + \mu_2^{k_2}\mu_1^{s_2}. \tag{8}$$

∎

Since the constants D_1, D_2, and D_3 are obtained from the longest maximal sequence of $\mathbf{v}$ (and for $\mathbf{w}$ and $\mathbf{u}$ as well), they are the largest ones for new vertices of the same type. In other words, they bound the edge lengths at new vertices of the same type. However, the exact values of these constants depend on their maximal sequences. Below we discuss the asymptotic behavior of these constants.

It can be shown that for $\rho_0 > 2$, D_1, D_2, and D_3 are upper bounded. Note that increasing s_1 and s_2 can only make D_3 smaller. The maximum of D_3 is attained when $m = s_1 = s_2 = 0$, $k_1 = 1$, and $K_2 \to \infty$, which is

$$D_{3,\max} = C_1^0 + (1 + \sqrt{2}C_1^\infty)C_{\sqrt{2}}^0\mu_1 + (1 + \sqrt{2} + 2C_1^\infty)C_2^\infty\mu_{\sqrt{2}}\mu_1$$
$$= 1 + (1 + \tfrac{\sqrt{2}\rho_0}{\rho_0-1})\tfrac{1}{\rho_0} + (1 + \sqrt{2} + \tfrac{2\rho_0}{\rho_0-1})\tfrac{\sqrt{2}}{(\rho_0-2)\rho_0}$$

The maximums of D_2 and D_1 are attained when $k_1 \to \infty$ and $k_2 \to \infty$, respectively,

$$D_{2,\max} = (1 + \sqrt{2}C_1^\infty)C_{\sqrt{2}}^\infty = (1 + \tfrac{\sqrt{2}}{\rho_0-1})\tfrac{\sqrt{2}}{\rho_0-\sqrt{2}}$$
$$D_{1,\max} = (1 + \sqrt{2} + 2C_1^\infty)C_2^\infty = (1 + \sqrt{2} + \tfrac{2\rho_0}{\rho_0-1})\tfrac{\rho_0}{\rho_0-2}$$

Compare to Shewchuk's constants. Suppose $\rho_0 = 2.5$. D_3 is in the range from 1.667 ($m = \infty$) to 5.33 ($m = s_1 = s_2 = 0$, $k_1 = 1$, $k_2 = 7$). Shewchuk's constant $(\tfrac{\rho_0+1+\sqrt{2}}{\rho_0-2})$ is about 9.83. D_2 is in the range from 3.35 ($k_1 = 1$) to 7.73 ($k_1 \geq 25$). While Shewchuk's constant $(\tfrac{1+\sqrt{2}\rho_0)+\sqrt{2}}{\rho_0-2})$ gives about 14.9. At last, D_1 is in the range from 5.75 ($k_2 = 1$) to 28.74 ($k_2 \geq 58$). Shewchuk's constant $(\tfrac{(3+\sqrt{2})\rho_0}{\rho_0-2})$ is about 22.07. Only $D_{1,\max}$ is larger than Shewchuk's constant. Note that it is obtained by ignoring all $R3$-vertices in the maximal sequence of $\mathbf{u}$ which is not realistic, i.e., the estimation for *(r9)* is much too large.

6   Vertex Degrees

Talmor [25] showed that each vertex of a tetrahedral mesh with bounded radius-edge ratio only belongs to at most some fixed number of edges, and the number only depends on the ratio. However, the constant given in [25] is miserably large[1]. In this section, we show a simple proof of this fact and bring the constant down to a reasonable size.

---

[1] In [25], Theorem 3.4.4, the proved constant is $(2C^2 + 1)^3$, where $C = C_2^{C_3}$, which is at least $(4\rho_0^2)^{C_3}$, and C_3 is the number of circular caps having a cone angle θ depending on ρ_0 that form a cover of a unit sphere.

The following lemma shows a fact about any tetrahedral mesh (which is not necessarily Delaunay) with a bounded radius-edge ratio ρ_0.

Lemma 6. *Let T be a tetrahedral mesh with a bounded radius-edge ratio ρ_0. Let* **ab** *be an edge in T. There is a cone which has* **a** *as its apex,* **ab** *as its axis, and with a cone angle $\eta_0 = \arcsin \frac{1}{2\rho_0}$, such that its interior contains no other edges of T.*

Proof. Consider the set of faces in T that contain edge **ab**. Let their apexes be $\mathbf{c_1}, \mathbf{c_2}, ..., \mathbf{c_n}$, see Fig. 8 left. Since no tetrahedra of the mesh has a radius-edge ratio larger than ρ_0, it holds on each face as well. This implies that: $\min\{\angle\mathbf{bac}_i \mid i = 1, ..., n\} \geq \arcsin\frac{1}{2\rho_0} = \eta_0$. Thus the cone must contain none of the edges $\mathbf{ac_1}, ..., \mathbf{ac_n}$ in its interior. ∎

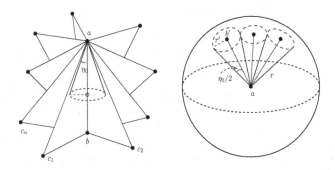

Fig. 8. Proof of Lemma 6 (left) and Theorem 3 (right).

Theorem 3. *Let T be a tetrahedral mesh with a bounded radius-edge ratio ρ_0. Then every vertex* **a** *of T belongs to at most δ_0 edges, where*

$$\delta_0 = \frac{2}{1 - \cos(\eta_0/2)} = \frac{2}{1 - \sqrt{1/2 - 1/2\sqrt{1 - 1/(4\rho_0^2)}}}. \tag{9}$$

Proof. By Lemma 6 every edge containing **a** has an empty cone whose angle is η_0. Particularly, the cones of edges at **a** with angle $\eta_0/2$ do not overlap each other. It turns out that the number of edges at **a** is never larger than the maximal number of non-overlapping cones whose angles are all at least $\eta_0/2$.

To bound the maximal number of such cones, we place a small sphere Σ centered at **a** with radius r. For each edge **ab** in the star of **a**, let $\mathbf{b'} \in \Sigma$ be the radial projection of **b**. Place a circular cap centered at each $\mathbf{b'}$ with a radius $r' = r\sqrt{2 - 2\cos(\eta_0/2)}$, see Fig. 8 right. Clearly, any two of such caps do not overlap. The area of each cap is $(1 - \cos(\eta_0/2))/2$ times the area of Σ, which implies there are at most $2/(1 - \cos(\eta_0/2)) = \delta_0$ caps. ∎

The constant δ_0 only depends on ρ_0. Here are some examples, for $\rho_0 = 4$, $\delta_0 \approx 1019$, for $\rho_0 = 2$, $\delta_0 \approx 251$, and for $\rho_0 = \sqrt{2}$, $\delta_0 \approx 123$.

7 Output Mesh Size

In this section, we analyze the output mesh size, i.e., the number of output vertices and tetrahedra of the Delaunay refinement algorithm.

We assume that the input is only a finite set S of points in $\mathbb{R}^3$. To distill the boundary effect, we assume that S is a *periodic point set* [2], i.e., $S \subseteq [0,1)^3$ is a finite set of points, and duplicated within each integer unit cube: $S' = S + \mathbb{Z}^3$, where $\mathbb{Z}^3$ is the three-dimensional integer grid. For refining such a point set, only the rule $R3$ is needed, i.e., to generate the circumcenters of bad quality tetrahedra. Let L denote the diameter of S, while s denotes the smallest pairwise distance among input features, the ratio $\Delta = L/s$ is referred as *spread* of S [8].

The following theorem shows that for a periodic point sets, the total number of output points depends on the input parameters.

Theorem 4. *Let S be a set of n periodic points in $\mathbb{R}^3$ with spread Δ. Let $\rho_0 > 1$. The output mesh of Shewchuk's algorithm has $O(n\lceil \log_{\rho_0} \Delta \rceil)$ vertices.*

Proof. Let V denote the set of output vertices. Let $|\cdot|$ denote the cardinality of a set. Hence $|S| = n$. We want to show that $|V| = O(n\lceil \log_{\rho_0} \Delta \rceil)$.

We first sort all new vertices produced by the algorithm into a collection of subsets of V by the following approach: Let $\mathbf{v}$ be a new vertex and $\{\mathbf{v}_i\}_{i=0}^{m+1}$ be its parent sequence. Then it is called a *rank-m* vertex, i.e., there are m $R3$-vertices between $\mathbf{v}$ and $g(\mathbf{v})$. Denote V_m be the set of all rank-m vertices. Let $\mathcal{V} = \{V_0, V_1, ..., V_{l-1}\}$ be the collection of all sets of rank-m vertices. Obviously, $V_i \cap V_j = \emptyset$ for any $V_i, V_j \in \mathcal{V}$. Hence we have

$$|V| = |S| + |V_0| + |V_1| + \cdots + |V_{l-1}|. \tag{10}$$

Next we show that the collection $\mathcal{V}$ has a finite size. Note that an edge connecting at a rank-m vertex $\mathbf{v}$ has a length at least $r(\mathbf{v})$. By *(r5)*, $r(\mathbf{v}) \geq \rho_0^{m+1} r(\mathbf{p})$, where $\mathbf{p} = g(\mathbf{v})$. Since here $\mathbf{p}$ is an input vertex, $r(\mathbf{p}) \geq \mathrm{lfs}(\mathbf{p}) \geq s$. Also note that $r(\mathbf{v}) \leq L$. Hence $L \geq r(\mathbf{v}) \geq \rho_0^{m+1} \geq s \implies \rho_0^{m+1} \leq L/s \implies m+1 \leq \log_{\rho_0}(L/s)$. The largest number of subsets in $\mathcal{V}$ is:

$$l = \lceil \log_{\rho_0} \Delta \rceil - 1. \tag{11}$$

Next we show a key fact, that is the set V_0 has the "capability" to have the largest cardinality among all sets in $\mathcal{V}$. Note it does not mean that the cardinality of V_0 must be the largest in $\mathcal{V}$. It means that no set in $\mathcal{V}$ will have a cardinality larger than the maximal possible cardinality of V_0.

Let $\mathcal{D}(S)$ be the Delaunay tetrahedralization for S, and let $\mathcal{V}(S)$ be the dual Voronoi diagram of $\mathcal{D}(S)$. $\mathcal{V}(S)$ forms a space partition of the convex hull, $\mathrm{conv}(S)$ of S. All vertices in $\mathcal{V}(S)$ are candidates of rank-0 vertices to be added by this algorithm. Let S_0 be the set formed by all vertices of S and only rank-0 vertices (no rank-1 vertices are added yet). $\mathcal{V}(S_0)$ is a new partition of $\mathrm{conv}(S)$. A candidate for rank-1 vertex is a Voronoi vertex in $\mathcal{V}(S_0)$ which is the circumcenter of a bad quality tetrahedron which must have a short edge formed by two rank-0 vertices. Obviously, the candidates for rank-1 vertices are locally restricted in $\mathcal{V}(S_0)$ since they depend on the locations of rank-0 vertices. This shows that the space which rank-0 vertices can be added are much larger than the space for adding rank-1 vertices. The same holds for vertices having higher ranks, since a new vertex is always added at the "locally sparest" location. The updated Voronoi diagram becomes sparser after new vertices are added. Fig. 9 illustrates this fact in 2D case. Hence V_0 has the capability to have the largest cardinality among other sets in $\mathcal{V}$.

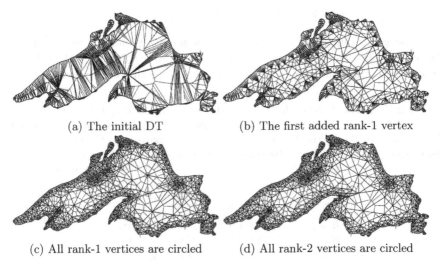

(a) The initial DT   (b) The first added rank-1 vertex

(c) All rank-1 vertices are circled   (d) All rank-2 vertices are circled

Fig. 9. Output mesh size analysis. From (a)-(d) are output meshes of a 2D Delaunay refinement algorithm (produced by `Triangle` [21]) on a planar straight line graph (Lake Superior). The initial Delaunay mesh and its dual Voronoi partition are shown in (a). All Voronoi vertices are candidates of rank-0 vertices. The circled vertex in (b) is the first rank-1 vertex added by this algorithm, it is the circumcenter of the shaded triangle. For this example, the algorithm has added 326 new vertices, in which 313 are rank-0 vertices, 11 rank-1 vertices and 2 rank-2 vertices. (c) and (d) show the locations of rank-1 and rank-2 vertices, respectively.

It remains to find out what is the largest possible number of rank-0 vertices. Let $\mathbf{v}$ be an input vertex. Let e be an edge connecting at $\mathbf{v}$, and e is the shortest edge of a bad quality tetrahedron t_0 in a Delaunay mesh $\mathcal{T}$. The worst radius-edge ratio of t_0, $\rho(t_0) \leq L/s = \Delta$. The Delaunay refinement algorithm will remove t_0 by adding its circumcenter $\mathbf{c}_0$. Let t_1 be the tetrahedron $t_1 \ni e$, created by the insertion of $\mathbf{c}_0$. It must be $\rho(t_1) \leq \frac{1}{2}\Delta$. If $\rho(t_1) > \rho_0$, then t_1 is again removed by adding its circumcenter $\mathbf{c}_1$. This will create a new tetrahedron $t_2 \ni e$, with $\rho(t_2) \leq \frac{1}{4}\Delta$. This process will repeat until a tetrahedron $t_k \ni e$ with $\rho(t_k) \leq (\frac{1}{2})^k \Delta \leq \rho_0$ is created. Hence a short edge e could produce at most $k \leq \lceil \log_2 \Delta \rceil$ new vertices.

Hence after adding at most $\lceil \log_2 \Delta \rceil$ rank-0 vertices, $\mathbf{v}$ gets an edge which does not belong to any bad quality tetrahedron. By Theorem 3, there are at most δ_0 edges connected at $\mathbf{v}$, thus, after inserting at most $\delta_0 \lceil \log_2 \Delta \rceil$ vertices, all edges connecting at $\mathbf{v}$ do not belong to any bad quality tetrahedron. Since there are n input vertices, thus, the total number of rank-0 vertices are:

$$|V_0| \le (\delta_0 \lceil \log_2 \Delta \rceil) n. \tag{12}$$

Note that $\delta_0 \lceil \log_2 \Delta \rceil$ is a constant which does not depend on n. The total size of V can be estimated by substituting (11) and (12) into (10),

$$|V| \le (l+1)(\delta_0 \lceil \log_2 \Delta \rceil) n = O(n \lceil \log_{\rho_0} \Delta \rceil), \qquad \blacksquare$$

The above theorem shows that for a periodic point set, the output size of the Delaunay refinement algorithm depends on n and a value $\log_{\rho_0} \Delta$. Table 2 shows some tests on different values of $\log_{\rho_0} \Delta$. In this experiment, we see

Table 2. Relations between $log_{\rho_0} \Delta$ and the output sizes. R_{10}, R_{100}, and R_{1k} are three PLSs. R_{10} is formed by two spheres, one inside the other, the outer sphere has radius $r = 10$ (see Fig. 10). R_{100} and R_{1k} are obtained from R_{10} by scaling the outer sphere to the radii $r = 100$ and $r = 1000$, respectively. n_{in}, n_{out}, and n_{add} are the number of input nodes, output nodes, and added nodes, respectively.

PLS	n_{in}	ρ_0	Δ	$\log_{\rho_0} \Delta$	n_{out}	n_{add}
R_{10}	772	2.0	130.10	7.022	1085	313
R_{100}			1301.08	10.350	1153	381
R_{1k}			13010.80	13.670	1208	436
R_{10}	772	1.4	130.10	14.050	1586	814
R_{100}			1301.08	20.700	1821	1049
R_{1k}			13010.80	27.345	2034	1262
R_{10}	772	1.2	130.10	26.700	2560	1788
R_{100}			1301.08	39.330	3380	2608
R_{1k}			13010.80	51.960	4081	3309

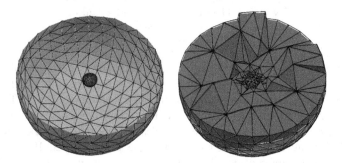

Fig. 10. Left: The PLS R_{10} (772 nodes, 1536 faces), formed by two spheres, one inside the other, the outer sphere has radius $r = 10$, the inner one has $r = 1$. Right: The output Delaunay mesh (1085 nodes, 5121 tetrahedra) at $\rho_0 = 2$.

that the value $\log_{\rho_0} \Delta$ approximately add a factor of $O(\log n)$ to the output number of points.

8 Conclusions

In this paper we reanalyzed Shewchuk's Delaunay refinement algorithm. Our new result on the termination condition $\arccos \frac{1}{3} \approx 70.53^o$ expands the acceptable class of inputs for this algorithm. The output mesh size bound $O(n \lceil \log_{\rho_0} \Delta \rceil)$ is a theoretical guarantee that this algorithm will produce small mesh size. Our new bounds on output edge lengths and vertex degrees better explain the practical behaviors of this algorithm.

However, our analysis is still far from complete. A number of interested issues remain to be investigated. A very interested question is: Can the 70.53^o bound be reduced? Our practical experience is that a 60° bound never hurts the termination. In the output mesh size proof, can we show that V_0 has the largest cardinality quantitatively among other sets in $\mathcal{V}$? Experiments showed that this algorithm is able to generate a true quality mesh if it also adds the circumcenters of slivers. In such case, what is the dihedral angle bound for all output tetrahedra?

References

1. Bern, M., Eppstein, D., Gilbert, J.R.: Provably good mesh generation. Journal of Computer and System Sciences 48(3), 384–409 (1994)
2. Cheng, S.-W., Dey, T., Edelsbrunner, H., Facello, M.A., Teng, S.-H.: Sliver exudation. J. Assoc. Comput. Mach. 47, 883–904 (2000)
3. Cheng, S.-W., Dey, T.K., Levine, J.A.: A practical Delaunay meshing algorithm for a large class of domains. In: Proc. 16th International Meshing Roundtable, pp. 477–494 (2007)
4. Cheng, S.-W., Dey, T.K., Ramos, E.A., Ray, T.: Quality meshing for polyhedra with small angles. International Journal on Computational Geometry and Applications 15, 421–461 (2005)
5. Cheng, S.-W., Poon, S.-H.: Three-dimensional Delaunay mesh generation. Discrete and Computational Geometry 36, 419–456 (2006)
6. Chew, P.L.: Guaranteed-quality triangular meshes. Technical Report TR 89-983, Dept. of Comp. Sci., Cornell University (1989)
7. Delaunay, B.N.: Sur la sphère vide. Izvestia Akademii Nauk SSSR, Otdelenie Matematicheskikh i Estestvennykh Nauk 7, 793–800 (1934)
8. Erickson, J.: Nice point sets can have nasty Delaunay triangulations. Discrete and Computational Geometry 30(1), 109–132 (2003)
9. Gabriel, K.R., Sokal, R.R.: A new statistical approach to geographic analysis. Systematic Zoology 18(3), 259–278 (1969)
10. Hudson, B., Miller, G.L., Phillips, T.: Sparse voronoi refinement. In: Proc. 15th International Meshing Roundtable, pp. 339–356 (2006)
11. Hudson, B., Miller, G.L., Phillips, T., Sheehy, D.: Size complexity of volume meshes vs. surface meshes. In: 20th Annual ACM-SIAM Symposium on Discrete Algorithms, pp. 1041–1047 (2009)

12. Miller, G.L., Phillips, T., Sheehy, D.: Linear-size meshes. In: 20th Canadian Conference on Computational Geometry (2008)
13. Miller, G.L., Talmor, D., Teng, S.-H., Walkington, N.J., Wang, H.: Control volume meshes using sphere packing: Generation, refinement and coarsening. In: Proc. 5th International Meshing Roundtable (1996)
14. Mitchell, S.A., Vavasis, S.A.: Quality mesh generation in higher dimensions. SIAM Journal on Computing 29, 1334–1370 (2000)
15. Oudot, S., Rineau, L., Yvinec, M.: Meshing volumes bounded by smooth surfaces. In: Proc. 14th International Meshing Roundtable, pp. 203–220 (2005)
16. Pav, S.E.: Delaunay Refinement Algorithm. PhD thesis, Department of Mathematical Sciences, Carnegie Mellon University, Pittsburgh, Pennsylvania (2003)
17. Pav, S.E., Walkington, N.J.: Robust three dimensional Delaunay refinement. In: Proc. 13th International Meshing Roundtable (2004)
18. Ruppert, J.: A Delaunay refinement algorithm for quality 2-dimensional mesh generation. Journal of Algorithms 18(3), 548–585 (1995)
19. Shewchuk, J.R.: Delaunay Refinement Mesh Generation. PhD thesis, Department of Computer Science, Carnegie Mellon University, Pittsburgh, Pennsylvania (1997)
20. Shewchuk, J.R.: Pyramid (2006) (Unpublished)
21. Shewchuk, J.R.: Triangle (2006), http://www.cs.cmu.edu/~quake/triangle.html
22. Si, H.: TetGen (2007), http://tetgen.berlios.de
23. Si, H.: Adaptive tetrahedral mesh generation by constrained delaunay refinement. International Journal for Numerical Methods in Engineering 75(7), 856–880 (2008)
24. Si, H.: Three dimensional boundary conforming Delaunay mesh generation. PhD thesis, Institute für Mathematik, Technische Universität Berlin (2008)
25. Talmor, D.: Well-spaced points for numerical methods. PhD thesis, Dept. of Computer Science, Carnegie Mellon University, Pittsburgh, Pennsylvania (1997)

Hexagonal Delaunay Triangulation

Gerd Sußner[1] and Günther Greiner[2]

[1] RTT AG, Munich, Germany
   Gerd.Sussner@rtt.ag
[2] Computer Graphics Group, University of Erlangen-Nuremberg, Germany
   guenther.greiner@cs.fau.de

Abstract. We present a novel and robust algorithm for triangulating point clouds in $\mathbb{R}^2$. It is based on a highly adaptive hexagonal subdivision scheme of the input domain. That hexagon mesh has a dual triangular mesh with the following properties:

- any angle of any triangle lies in the range between $43.9°$ and $90°$,
- the aspect ratio of triangles is bound to 1.20787,
- the triangulation has the Delaunay property,
- the minimum triangle size is bounded by the minimum distance between input points.

The iterative character of the hexagon subdivision allows incremental addition of further input points for selectively refining certain regions. Finally we extend the algorithm to handle planar straight-line graphs (PSLG). Meshes produced by this method are suitable for all kinds of algorithms where numerical stability is affected by triangles with skinny or obtuse angles.

Keywords: Unstructured Mesh Generation, Delaunay Triangulation, Guaranteed Angle Bounds, Hexagon Subdivision.

1   Introduction

Mesh generation is the subject of many articles due to its importance to computational geometry, computer graphics, numerical simulation and various other areas. Many problems are based on input data lying in a 2-dimensional domain. This data is often triangulated to allow computations on the mesh, e.g. for finite element methods. The result usually strongly depends on the quality of the triangular mesh, e.g. small or obtuse angles reduce the numerical stability for a high number of elements. The number of triangles also determines the runtime for solving the problem. Therefore it is desirable to have a well-shaped triangular mesh with as few triangles as possible.

There are several kinds of unstructured mesh generators. Some are based on grid-techniques, e.g. quad-trees, while others try to improve an existing triangulation iteratively. These and other approaches like advancing front techniques are surveyed by Owen [7].

In this paper we use a different approach. Instead of a quad-tree a hexagonal highly adaptive subdivision scheme is used for the input domain. When the domain is finally subdivided a dual mesh is extracted from the centers of the full hexagons. It turns out that this mesh has the Delaunay property, i.e. maximizing the minimum angle of any triangle, which is guaranteed to be 43.9° – **independent** of the input data. The method is also aware of different regions of interest, meaning that areas with less input points have larger triangles. The size of the triangles quickly decreases in regions of high interest. It is limited by the minimum distance among the input points which also ensures termination of the algorithm. Both properties lead to triangular meshes with a reasonable number of triangles.

After describing previous work, we give a short introduction into the hexagon subdivision scheme. Then refinement rules are defined to subdivide the input domain adaptively according to the location of the input points. From the hexagonal mesh a dual triangular mesh is extracted where any angle is between 30° and 120°. In order to achieve much tighter bounds the hexagons are classified according to their adjacent neighbors. Additional refinement rules and shifting the centers of the classified hexagons results in a triangular mesh with angles between 43.9° and 90°.

As an additional result the algorithm is extended to handle also PSLGs. This may be important to limit the domain only to valid input values. However at these boundaries the proven angle bounds are lost.

2   Previous Work

Baker et al. [1] introduced guaranteed shape properties for the resulting meshes so that all angles of the triangles lie within 13° and 90°, yielding an aspect ratio of at most 4.6. This is achieved by placing a square-grid over the polygons to include Steiner points. The size of the grid is determined by the smallest distance among the input points and edges. However the resulting meshes may be very large.

Bern et al. [2] use a quad-tree instead of a uniform grid. This method gives bounds to shape property and the number of triangles. Subdividing the domain by a quad-tree allows a local refinement with Steiner points where necessary while regions of low interest remain coarse. For point sets the angle bounds are 36° and 80° and for a planar straight-line graph (PSLG) the range is 18.4° and 153.2°. To achieve this, the key-idea is to move the corners of the quad-tree according to some patterns. The method of Neugebauer and Diekmann [6] uses rhombi instead of squares when subdividing the domain, yielding angle bounds of 30° and 90° for polygonal input.

A different approach is refining an already obtained Delaunay-triangulation of a point set by inserting Steiner points at certain positions iteratively which is called *Delaunay refinement*. The original approach of Chew [3] and Ruppert

[8] uses the circumcenter of badly shaped triangles while latest methods of Üngör [10] and Erten [4] use different locations . The guaranteed angle bounds for point sets are 30° and 120°. For PSLGs they are set to 20° and 160°. In practice often higher angles up to 34° are achieved, depending on the input data which was examined by Shewchuk [9]. The Off-Center approach even raises this limit up to 42°. However it is always possible to get input data, where the Delaunay refinement fails for high angles. Both methods are implemented in `Triangle` which is a robust software for creating Delaunay-refined meshes available at `http://www.cs.cmu.edu/~quake/triangle.html`.

3   Hexagonal Subdivision

Sußner et al. [5] use a bidirectional subdivision of a hexagonal mesh to adaptively refine the domain of huge height-fields for interactive purposes. The subdivision method follows simple and easy-to-implement rules and is therefore an efficient way to adaptively subdivide planar regions.

3.1   Subdividing a Hexagon

A hexagon is subdivided by scaling it to half size and filling the remaining space with semi-hexagons. The hexagons are organized in levels: Full hexagons are located in even levels while semi-hexagons only occur in odd levels. When subdividing a full hexagon, the scaled full hexagon is moved into the next even level while the semi-hexagons are put in the odd level in-between. After the operation each of the semi-hexagons is checked for a fitting adjacent semi-hexagon (see Fig. 1). If there is one both semi-hexagons are joined to form a new full hexagon one level above.

3.2   Reverse Operation

The above operation is reversible. In a first step adjacent full hexagons may be separated into two semi-hexagons. In order to break the right hexagons,

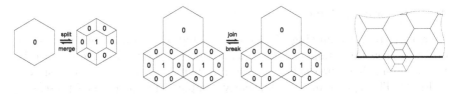

Fig. 1. The numbers denote the split-counter of each hexagon. On the left the split-/merge-operator is shown which increases/decreases the split-counter by one. The join-/break-operator in the middle creates a new full hexagon with a split-counter value of zero. As shown on the right side, a boundary semi-hexagon is treated like a virtual full hexagon.

each hexagon gets a subdivision counter which is set to zero when a new hexagon was created by joining two semi-hexagons. Every time a hexagon is subdivided the counter is increased by one and decreased if it was scaled up to a hexagon of lower level. Finally a hexagon may only be broken up into two semi-hexagons if the subdivision counter is zero.

3.3   Adaptive Refinement

For the sake of a smooth increase of level-of-detail, a simple balancing rule is applied. After each subdivision step the level difference of adjacent (semi-) hexagons must not be larger than one. If the difference is larger the hexagon of lower level is subdivided as well. If this is a semi-hexagon, the opposite neighbor, which must be a full hexagon, is subdivided. The balancing rule is illustrated in Fig. 2. Note that a single subdivision step may trigger the refinement of large parts of the hexagon mesh.

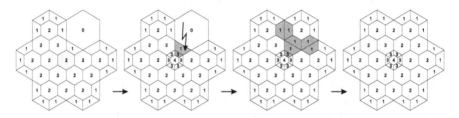

Fig. 2. The numbers represent the level counter of each hexagon. If the level difference is greater than one further splits and joins of semi-hexagons are forced.

4   Simple Refinement

In this section we present a simple version of the triangulation algorithm, with angle range of 30° to 120°. Beyond these angles triangles are usually considered as *bad*.

4.1   Refinement Rules

As a first step, a bounding hexagon is created around all input points. In order to keep the number of triangles as low as possible, a good choice is the minimum bounding sphere as the inner circle of the hexagon as shown in Fig. 3. This sphere may be efficiently computed by the method presented by Welzl [11]. In addition to the balancing rules of section 3.3 the following rules must be obeyed in order. A hexagon containing an input point is marked *occupied*. Given a set of hexagons $\mathcal{H}$, the rules are:

1. subdivide any hexagon $h_i \in \mathcal{H}$ occupied by several input points,
2. if $h_i \in \mathcal{H}$ is an occupied semi-hexagon, subdivide the full hexagon adjacent to the long edge of h_i,
3. subdivide an occupied hexagon $h_i \in \mathcal{H}$ if any adjacent neighbor is occupied as well,
4. if any of the adjacent neighbors is of lower level, subdivide the (full) hexagons adjacent to the long edge of the neighboring semi-hexagon,
5. subdivide an occupied hexagon $h_i \in \mathcal{H}$ if any adjacent neighbor is of higher level.

At each single subdivision step, i.e. splitting full hexagons and joining semi-hexagons, input points were re-distributed among the affected hexagons. In order to test the whole hexagon-mesh after each operation, a list of affected hexagons is maintained. Refinement stops if this list is empty. Termination is guaranteed since an input point is only associated to one hexagon. Fig. 3 shows examples of these rules. In the final hexagon mesh all occupied hexagons are full hexagons surrounded by a ring of full non-occupied hexagons.

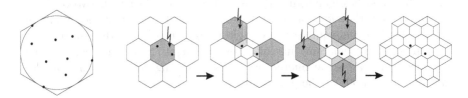

Fig. 3. Initial setup of input points on the left side, followed by a sequence of applied refinement rules: several input points, input point in semi-hexagon, adjacent lower level hexagons.

4.2 Extracting the Dual Mesh

For getting the dual triangulation from the hexagon mesh just add an edge from the center of each full hexagon to the centers of its adjacent hexagons. If the neighbor is a semi-hexagon connect the edge to the center of the opposite full hexagon of the neighbor (see Fig. 4). In a final step the center positions of the *occupied* hexagons are replaced by the location of corresponding assigned input points.

4.3 Proving Angles

The proof for a minimum angle of 30° is split into two parts. First the angles of the plain dual mesh, i.e. without relocated centers, are examined. In the second part we take a look at the situation of a relocated center.

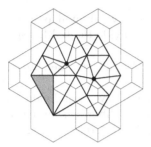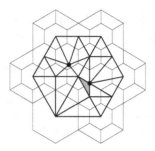

Fig. 4. The dual mesh consists of the edges among adjacent full hexagons. For semi-hexagons the opposite neighbor is taken. Note that the minimum angle of 30° is always in the triangulation due to the adaptive subdivision (grey triangle on the left). On the right side there is a triangle consisting of both angle bounds, 30° and 120°.

Angles of the Plain Dual Mesh

As shown in Fig. 4 the centers of three adjacent full hexagons form a equilateral triangle with three angles of 60°. If one of them is subdivided, the center of scaled hexagon still remains at the same position. If two of them are subdivided, they form a new full hexagon located right in the middle of them. Therefore the equilateral triangle is split into two with half angle at the center of the remaining un-subdivided hexagon. Since the center of the new hexagon lies on the middle of the two existing centers both triangles have an angle of 90° at the new center.

Angles at Relocated Centers

A hexagon containing a relocated center is surrounded by full hexagons as shown on the right side in Fig. 4. The input point may be located anywhere within the hexagon. In extremum the center is moved to a hexagon's vertex. In this case one triangle has two angles of 30°, but not less. This also means that the third angle must be 120°.

5   Extended Refinement

For simple refinement the dual mesh is extracted by connecting the centers among all full hexagons. But there is plenty of freedom to shift the centers' position to achieve tighter angle bounds. To reach this goal, two enhancements are necessary.

First the full hexagons are classified according to their adjacent (semi-)hexagons. Based on that classification the hexagon mesh is adaptively refined. In a second step, when the dual mesh is extracted, the centers of some classified hexagons are shifted to yield higher angle bounds of 43.9° and 90° respectively.

5.1   Classification of Hexagons

The classification of the full hexagons is based on the number of fins. A fin is a semi-hexagon which joins at its long edge to the full hexagon (see top row of Fig. 5). If there is any fin the hexagon $h_i \in \mathcal{H}$ is marked as follows:

MOVED_CENTER_1	- h_i has only one fin,
MOVED_CENTER_2	- h_i has only two consecutive fins,
MOVED_CENTER_3	- h_i has only three consecutive fins,
SUBDIVIDE	- else.

Full hexagons adjacent to the edge before the fin strip are called *left neighbor* h_{i_l}. Those adjacent to the edge after the fins are called *right neighbor* h_{i_r}. In case that h_i is of type MOVED_CENTER_3 the remaining adjacent full hexagon is called *top neighbor* h_{i_t}. Full hexagons $h_i \in \mathcal{H}$ without fins are marked as well (see bottom row of Fig. 5):

OCCUPIED_1	- input point lies within the half-sized hexagon of h_i,
OCCUPIED_2	- input point lies outside of half-sized hexagon of h_i,
FIXED_CENTER_IN_RING	- h_i is in 1-ring of a hexagon of type OCCUPIED_1 or in 2-ring of a hexagon of type OCCUPIED_2,
MOVED_CENTER_IN_RING	- h_i is in 1-ring of a hexagon of type OCCUPIED_2,
MOVED_CENTER_4	- h_i is top neighbor of a hexagon of type MOVED_CENTER_3,
FIXED_CENTER_REGULAR	- else.

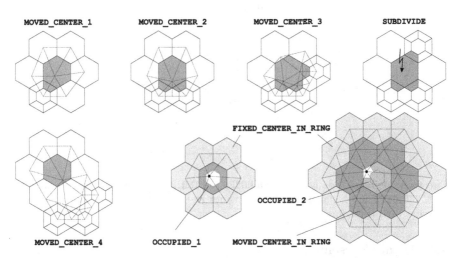

Fig. 5. Hexagons are classified to their number of fins. Hexagons containing an input point marked as OCCUPIED_[12] depending on the location of the input point within the hexagon (white areas). OCCUPIED_2-hexagons are surrounded by ones of type MOVED_CENTER_IN_RING. Input points are also surrounded by FIXED_CENTER_IN_RING-hexagons either in the 1-ring or 2-ring. The 1-ring of a hexagon h_i is defined by the full hexagons surrounding h_i. The 2-ring of h_i then consists of all full hexagons surrounding the hexagons of the 1-ring of h_i.

5.2   Extended Refinement Rules

According to the classified hexagons further refinement rules are defined:

- subdivide any hexagon $h_i \in \mathcal{H}$ of type SUBDIVIDE,
- subdivide any hexagon $h_i \in \mathcal{H}$ of type MOVED_CENTER_x lying in the 1-ring of a hexagon of type OCCUPIED_1,
- subdivide any hexagon $h_i \in \mathcal{H}$ of type MOVED_CENTER_x lying in the 1- or 2-ring of a hexagon of type OCCUPIED_2,
- subdivide any hexagon $h_i \in \mathcal{H}$ of type MOVED_CENTER_[123] if any full hexagon adjacent to its fins is not of type FIXED_CENTER_x,
- subdivide a hexagon h_i of type MOVED_CENTER_4 if it is the top neighbor of more than one non-consecutive hexagons of type MOVED_CENTER_3.
- Given a hexagon h_i of type MOVED_CENTER_3 with a left neighbor of the same type. If its right neighbor $h_j := h_{i_r}$ is of type MOVED_CENTER_2, subdivide right neighbor h_{j_r} of h_j. (Applied symmetrically with left neighbor!)

The last rule is necessary, since we have a special treatment for two or more consecutive hexagons of type MOVED_CENTER_3 when extracting the dual mesh. The consequences of this rather complicated rule are illustrated in Fig. 6. Note that six hexagons of type MOVED_CENTER_3 may form a ring - called *blossom*. This is exploited later in order to reduce the number of hexagons. Also note that each sequence of MOVED_CENTER_2/MOVED_CENTER_3- hexagons, except for a blossom, is enclosed by hexagons of type MOVED_CENTER_1. Fig. 7 shows an example of a valid subdivision.

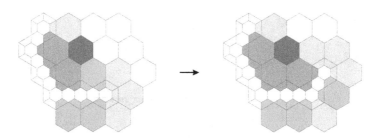

Fig. 6. Angle bounds do not hold if a MOVED_CENTER_2-hexagon is adjacent to a pair of MOVED_CENTER_3-hexagons. Subdividing its right/left neighbor creates another hexagon with three fins.

5.3   Local Coarsening

Applying above rules may lead to a subdivision with greater areas of hexagons of same size. In order to avoid an unnecessary large number of triangles, a local coarsening step is performed according to *as fine as necessary but as coarse as possible* (see Fig. 7). For this the neighborhood of each hexagon h_i with a split-counter of 0 is examined whether it is possible to execute all necessary break-/merge-operations to form a blossom. That is all of the

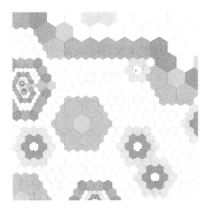 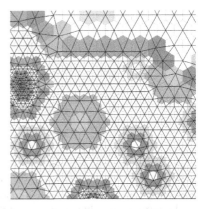

Fig. 7. This subdivision shows almost all possible combinations among the different hexagon types. Blossoms are created at a final local coarsening step in order to avoid large regions of small sized hexagons. Note that two input points may share hexagons of type FIXED_CENTER_IN_RING, e.g. in the lower right quarter. The corresponding dual mesh is drawn as an overlay on the right side.

hexagons in the 4-ring of h_i must be of type FIXED_CENTER_REGULAR. By this operation 37 hexagons are replaced by 7. In the dual mesh a blossom is represented by 60 triangles instead of 96 for the plain region.

5.4   Modifications to the Dual Mesh

In this section we show how to adapt the center positions for the dual mesh. For hexagons of type MOVED_CENTER_3 it is additionally necessary to insert moved centers of their fins.

Lower Angle Bound

First the lower angle bound is determined since this angle is used to compute some of the relocated centers. Have a look at the left side of Fig. 8 where a hexagon of type OCCUPIED_1 with radius r is shown. The input vertex is located at a vertex of the half-sized hexagon, causing extremal angles. The smallest angle in this configuration is α and is determined analytically:

$$\tan \alpha = \frac{y}{x} = \frac{\frac{5}{4}r}{\frac{3}{2}r\frac{\sqrt{3}}{2}} = \frac{5}{3\sqrt{3}} \quad \Rightarrow \alpha \approx 43.8979$$

Centers of MOVED_CENTER_IN_RING-hexagons

In case that the input point does not lie in the half-sized hexagon of h_i, the closest vertex v_{c_i} to the input point of h_i is determined. As illustrated at the right side in Fig. 8 the input point may be located anywhere in the grey

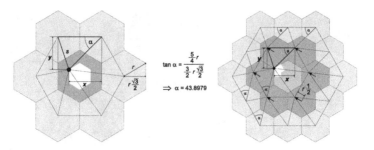

Fig. 8. On the left side the input point is located at a vertex of the half-sized hexagon (white area). The smallest angle α is determined by x and y. On the right side the input point is located outside of the half-sized hexagon h_i. The centers of the surrounding hexagons are shifted by vector of $\frac{1}{2}r$ from the center of h_i towards the nearest corner of the input point. For that corner the input point may lie anywhere in the white area. Note that in either configuration all triangles are acute.

region. In order to maintain angle bounds, the half-sized vector between v_{c_i} and the center C_{h_i} is added to each of the surrounding hexagons which are of type MOVED_CENTER_IN_RING.

Centers of MOVED_CENTER_1-hexagons

Given a hexagon h_i of type MOVED_CENTER_1 and a hexagon $h^{f_i^0}$ of type FIXED_CENTER_x adjacent to the only fin f_i^0 of h_i. The new center position $\widehat{C}_{h_i}$ of h_i lies on the axis between the original center position C_{h_i} and the center $C_{h^{f_i^0}}$ of $h^{f_i^0}$:

$$\widehat{C}_{h_i} := \frac{5}{6}C_{h_i} + \frac{1}{6}C_{h^{f_i^0}}.$$

The ratios are shown on the right side of Fig. 9 together with special configuration of MOVED_CENTER_3-hexagons.

Centers of MOVED_CENTER_2-hexagons

Given a hexagon h_j of type MOVED_CENTER_2 and three hexagons $h^{f_j^0}$, $h^{f_j^{01}}$ and $h^{f_j^1}$ of type FIXED_CENTER_x adjacent to the fins f_j^0 and f_j^1 of h_j. The new center position $\widehat{C}_{h_j}$ of h_j lies on the axis between the original center position C_{h_j} and the center $C_{h^{f_j^{01}}}$ of $h^{f_j^{01}}$. Looking at Fig. 9 $x := \frac{\sqrt{3}}{2}r$ is the edge between $C_h^{f_j^0}$ and $C_h^{f_j^{01}}$, i.e.

$$\tan\alpha = \frac{x}{y} \quad \Rightarrow y = \frac{\frac{\sqrt{3}}{2}r}{\frac{5}{3\sqrt{3}}} = \frac{9}{10}r.$$

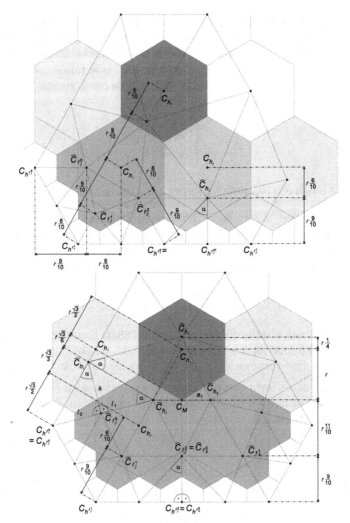

Fig. 9. In this picture the ratios for computing the new location of the centers are illustrated. On the left side the configuration with a single MOVED_CENTER_3-hexagon and on the right side a pair of them. Note that the ratio for a 1-fin-hexagon is derived from the right side.

The distance between the centers of h_j and $h^{f_j^{01}}$ is $\frac{3}{2}r$. Hence the hexagon's center is relocated to

$$\widehat{C}_{h_j} := \frac{3}{5}C_{h_j} + \frac{2}{5}C_{h^{f_j^{01}}}$$

Centers of single MOVED_CENTER_3-hexagons

Given a hexagon h_i of type MOVED_CENTER_3 and five hexagons $h^{f_i^0}$, $h^{f_i^{01}}$, $h^{f_i^1}$, $h^{f_i^{12}}$ and $h^{f_i^2}$ of type FIXED_CENTER_x adjacent to the fins f_i^0, f_i^1 and f_i^2 of

h_i. If it is a single MOVED_CENTER_3-hexagon, i.e. neither the left or nor the right neighbor is of same type, the original center is neglected. Instead the three centers of the fins are shifted and included in the triangulation. For computing the new positions of three centers the same ratios are used as for MOVED_CENTER_2-hexagons. However for the middle one the ratio is swapped.

$$\widehat{C}_{f_i^0} := \tfrac{3}{5}C_{h_i} + \tfrac{2}{5}C_{h f_i^0}$$
$$\widehat{C}_{f_i^1} := \tfrac{2}{5}C_{h_i} + \tfrac{3}{5}C_{h f_i^1}$$
$$\widehat{C}_{f_i^2} := \tfrac{3}{5}C_{h_i} + \tfrac{2}{5}C_{h f_i^2}$$

Centers of consecutive MOVED_CENTER_3-hexagons

As mentioned in Sec. 5.2 we have different topologies for two or more consecutive hexagons of type MOVED_CENTER_3. In addition to the previous case, the center of h_j is included in the triangulation. As shown in Fig. 9, it lies on the axis between the original center position C_{h_j} and the center of its top neighbor $C_{h_j^t}$, i.e.

$$\widehat{C}_{h_j} := \frac{2}{3}C_{h_j} + \frac{1}{3}C_{h_j^t}.$$

Suppose the left neighbor is of type MOVED_CENTER_1, then the right neighbor must be of type MOVED_CENTER_3 as shown on the right side of Fig. 9. The center $C_{f_j^1}$ lies on the line through $\widehat{C}_{h_j}$ and $C_{h_j^{f^0}}$. Since we want an upper angle bound of 90°, it must form two right triangles together with center of the left neighbor. Both triangles have the same height and one angle of α, then let l_0, l_1 and l be defined as:

$$l_0 := \frac{5}{3\sqrt{3}}h \qquad l_1 := \frac{3\sqrt{3}}{5}h \qquad l := \left\| \widehat{C}_{h_j} - C_{h f_j^0} \right\|.$$

Solving this for l_0 and l_1 leads to a ratio of 25 : 27, i.e. the new center is

$$\widehat{C}_{f_j^0} := \frac{25}{52}\widehat{C}_{h_j} + \frac{27}{52}C_{h f_j^0}.$$

Note that this configuration also determines the ratio for hexagons of type MOVED_CENTER_1. The center of the second fin is set to

$$\widehat{C}_{f_j^1} := \frac{3}{5}C_{h_j} + \frac{2}{5}C_{h f_j^1},$$

in contrast to the single configuration where the ratio was swapped. At the common edge of the two MOVED_CENTER_3-hexagons the corresponding neighbors coincide and are set to

$$\widehat{C}_{f_j^2} := \frac{9}{20}C_M + \frac{11}{20}C_{h f_j^2},$$

with C_M as the average of the two hexagons' new centers.

Centers of `MOVED_CENTER_4`-hexagons

The relocated position depends on the number n of adjacent hexagons of type `MOVED_CENTER_3`. If it is forming a blossom, the center is not relocated. In any other case the most left hexagon h_{k0} of the consecutive `MOVED_CENTER_3`-hexagons is determined:

$$\widehat{C}_{h_i} := \tfrac{3}{5}C_{h_i} + \tfrac{2}{5}C_{h_{k0}} \qquad\qquad (n = 1),$$
$$\widehat{C}_{h_i} := \tfrac{5}{4}C_{h_i} - \tfrac{1}{4}C_M, \quad C_M := \tfrac{1}{2}(C_{h_{k0}} + C_{h_{k1}}) \ (n = 2),$$
$$\widehat{C}_{h_i} := \tfrac{7}{6}C_{h_i} - \tfrac{1}{6}C_{h_{k1}} \qquad\qquad (n = 3),$$
$$\widehat{C}_{h_i} := \tfrac{4}{3}C_{h_i} - \tfrac{1}{3}C_M, \quad C_M := \tfrac{1}{2}(C_{h_{k1}} + C_{h_{k2}}) \ (n = 4).$$

6 Properties

6.1 Angle Bounds

The minimum angle bound was already explained in the previous section where it was used to relocate the hexagon centers so that each triangle has no angle below that value. The center relocation also took care of right angles, i.e. no triangle is created with an angle greater than 90°. Yet there is still missing an analytic proof of all possible combinations which is omitted due to space constraints. We rather show in Fig. 10 a sample of almost all possible configurations and and angles. It is left to the reader to verify them by computing the single angles of each triangle.

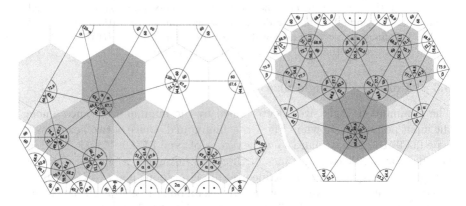

Fig. 10. This figure shows almost all possible hexagon-configurations, missing only `MOVED_CENTER_3`-strips of size 3 and 4 and 6 (blossoms). α is the minimum angle ($\tan\alpha = \frac{5}{3\sqrt{3}} \Rightarrow \alpha = 43.9$) and β is its counterpart in a right triangle ($\tan\beta = \frac{3\sqrt{3}}{5} \Rightarrow \beta = 46.1$).

6.2   Aspect Ratio

Both angle bounds also limit the aspect ratio of a triangle, i.e. the ratio between the diameter of incircle and radius of the circumcircle. Given a triangle T having both bounds as angles as shown in Fig. 11 and the shortest edge has length s. The radii of the circumcircle and the incircle of a right triangle are defined as

$$\varphi := \frac{c}{2} = \frac{1}{2}s\sqrt{1^2 + \left(\frac{3\sqrt{3}}{5}\right)^2} = \frac{1}{2}s\sqrt{\frac{52}{25}}.$$

$$\rho := \frac{a+b-c}{2} = \frac{s\left(1 + \frac{3\sqrt{3}}{5} - \sqrt{\frac{52}{25}}\right)}{2}$$

Thus the aspect ratio of T is $\frac{2\rho}{\varphi} \approx 1.20786$.

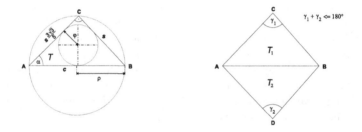

Fig. 11. On the left side the circumcircle reaches its maximum extent at the angle bounds of 43.9° and 90°. This also defines the maximum aspect ratio of the triangle T. The Delaunay property is achieved if for every inner edge (AB) the sum of γ_1 and γ_2 is not greater than 180° as shown on the right side.

6.3   Minimum Triangle Size

The minimum triangle size is determined by the minimum distance among the input points. As shown in Fig. 7 the minimum distance of two hexagons of type OCCUPIED_2 may share some hexagons of type FIXED_CENTER_IN_RING, i.e. they have at least a distance of three full hexagons. Since the point-in-hexagon-test may produce undefined results if an input point lies exactly on an edge between two hexagons or on a vertex among three adjacent hexagons, the vertex is assigned to the first hexagon to be tested. Therefore an additional distance of three full hexagons is added. The shortest edge in the triangular mesh is the one marked in Fig. 8 in the 1-ring of a hexagon h with radius $r_{\min}$ of type OCCUPIED_1 and has a length of

$$l_{\min} := r_{\min}\sqrt{\left(\frac{3}{2}\frac{\sqrt{3}}{2}\right)^2 + \left(\frac{1}{4}\right)^2} = \frac{r_{\min}}{2}\sqrt{7}.$$

The radius $r_{\min}$ is defined by the minimum distance $d_{\min}$ among the input points and a number of at most six full hexagons in between, i.e.

$$r_{\min} := \frac{d_{\min}}{6 \cdot 2\frac{\sqrt{3}}{2}} \quad \Rightarrow \quad l_{\min} := \frac{d_{\min}\sqrt{7}}{12\sqrt{3}}.$$

6.4 Delaunay Property

The upper angle bound of $90°$ automatically implies the Delaunay property. Note that a triangular mesh has the Delaunay property if no other vertex lies within the circumcircle of any triangle. This is equivalent that for each inner edge the sum of the apex angles of the adjacent triangles is not greater than $180°$ (see Fig. 11). Since there is no larger angle than $90°$ this is obviously satisfied.

7   Triangulating Planar Straight-Line Graphs

Domains are often bounded by a sequence of straight line segments, e.g. to restrict the computation to valid input values or reduce the number of computed elements. For this purpose we present a method to integrate a PSLG in the hexagon subdivision assuming that line segments do not intersect each other. However guaranteed angle bounds are lost at the line segments - but are guaranteed in the interior of the bound region.

7.1   Input Line Segment

Additionally to the input points one or several line segments are associated to the hexagons as shown in Fig. 12. During splitting-/join-operations line segments are re-assigned to the children, according to a hexagon-intersects-line test. Note that a line segment will never be split into several pieces while refining the hexagon mesh. When extracting the dual mesh, edges of the triangles are moved onto the line segment to ensure that all input lines are represented in the final triangular mesh.

7.2   Extended Refinement Rules

All hexagons intersecting a line segment must not be of type MOVED_CENTER_[1234]. These kinds of hexagons are neither allowed in the 1-ring of the intersected hexagons. Additionally a hexagon may be intersected at most by two line segments, but only if they are consecutive (see Fig. 12). Therefore the following rules for an intersected hexagon h_i are added to the refinement rules of Sec. 5.2:

- if any neighbor h_j of h_i is of higher level subdivide h_i,
- if any neighbor h_j of h_i is of lower level subdivide h_j,
- if h_i is intersected by two or more non-consecutive line segments subdivide h_i.

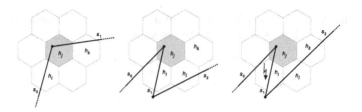

Fig. 12. Similar to input points line segments are assigned to hexagons. On the left side s_0 is assigned to h_i and h_j. s_1 is assigned to h_j and h_k. A hexagon is allowed to have at most two assigned input segments if they are consecutive. On the right side h_i gets all three line segments assigned and must be subdivided.

7.3   Line intersections

The basic idea is to relocate the centers of intersected hexagons. For this consider the three main axis of a hexagon which are orthogonal to the corresponding edges of the hexagon as shown in Fig. 13. Pick the axis forming the largest angle with respect to the line segment and compute the intersection of this axis and the line segment. If the intersection is still within the hexagon the new center location and the corresponding distance to the start point of the segment is kept in a list for that segment.

If the hexagon does not contain an input point but is intersected by two line segments, e.g. at acute input segments, a link to its neighbor intersection is stored in the record list as well. Finally for each line segment the associated list records are sorted with respect to the distance to the start point s_0 of the line segment. Afterwards for each line segment the affected hexagons are traversed according to the sorted intersection entries and their centers are shifted accordingly. For hexagons with two intersections a new point and new triangles have to be inserted into the triangulation as explained in the following subsection.

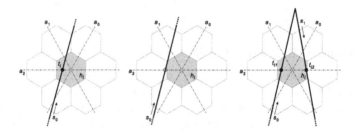

Fig. 13. The axis a_2 of h_i forms the largest angle with line segment s_0. On the left side the intersection lies within h_i and therefore an intersection record with t_i is created. In contrast to the situation in the middle. Here the intersection lies in the neighbor hexagon and the center of h_i will not be relocated. On the right side a second intersection record is created for s_1. For these kinds of hexagons additional triangles have to be added to the dual mesh.

7.4   Adding Triangles

Hexagons intersected only by one line segment are treated as non-intersected ones. However hexagons with two projected centers require additional triangles. The original single center is split into two, creating an inner region in between the two line segments. This *inner region* is filled with two new triangles as seen in Fig. 14. The triangles outside of the *inner region* are built in a regular way by connecting the two centers with the centers of the adjacent hexagons.

Special care is also needed for consecutive projected centers. This happens for example at input points lying outside of the half-sized hexagon in Fig. 14. In this case degenerated triangles would be created. The situation is solved by omitting the projected center of the concerned hexagon if the distance between the moved center and the projected center is greater than the radius r of the hexagon.

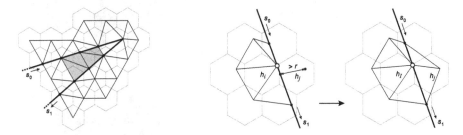

Fig. 14. The original center is split into two, opening a gap between the two line segments. This gap is filled by four new triangles. The edges outside the gap are connected as usual. On the other side in case of almost co-linear segments, degenerated triangles may occur. The projected center is not used if the distance between the moved center of h_j and the projected center is greater than the radius r of h_j.

8   Results

The method was implemented and tested on a regular PC with an Intel Core2 Duo processor at 2.6 GHz using only a single thread. The implementation is a *proof-of-concept*, i.e. it uses a conservative approach for the refinement loop with a runtime complexity of $O(n^2)$ which makes the current implementation unsuitable for larger data sets.

In this section we compare the results of our method with the current version of `Triangle` (v1.6). We have chosen 6 data sets. The first data set – a popular benchmark – consists of two single points with a close distance to each other, 0.02 units in this case. It is located within a hexagon with unit radius. The second data set consists of 100 random points lying on a straight line. In the third data set the 100 points are distributed randomly

Fig. 15. Four data sets triangulated with our method. In the first row two close points were triangulated. In this case the bounding hexagon was not adapted to the bounding sphere of input points. In the second row a random set of 100 points lying on a straight line is shown, while in the next row the 100 points were randomly distributed. The last row shows the inclusion of line segments. The density of the triangles increases with level-of-detail of the line segments. The right picture shows a close-up of the triangulation.

Table 1. `Triangle` was executed in single precision as well as in double precision while our method uses just single precision. Per trial-and-error the angle was increased by 0.25 degrees until the `Triangle` fails. When adding a PSLG the angle bounds of the hexagon method are lost. For the IMR-letters the minimum angle is 13.7° and the maximum angle is 133.4° near the line segments. In the interior regions the mininum and the maximum angle are 43.9° and 90°. Timings for `Triangle` are below 0.03 seconds for all data sets. The timings of our method are listed in parenthesis behind the angle numbers.

# triangles of	Triangle v1.6 (single)	Triangle v1.6 (double)	Hexagon Delaunay
2 close pts	122 at 36.75°	131 at 37.00°	684 at 43.9° (0.005 s)
100 pts on line	4890 at 35.00°	5289 at 35.00°	18667 at 43.9° (0.13 s)
100 pts on plane	1149 at 34.50°	1133 at 34.50°	12748 at 43.9° (0.21 s)
IMR-letters	2569 at 34.50°	1770 at 34.50°	30551 (1.03 s)
1k pts on plane	10423 at 34.5°	14414 at 34.75°	127729 at 43.9° (30.4 s)
2k pts on plane	19377 at 34.25	19244 at 34.25	268138 at 43.9° (123 s)

on the plane. The fourth data set consists of the sampled letters `IMR` where the level-of-detail increases from letter `I` to letter `R`. Finally we tested the algorithm with slightly larger data sets consisting of 1k and 2k resp. random points. The timings clearly show the quadratic runtime complexity. However it should be possible to implement a more sophisticated refinement loop to reach a runtime complexity of $O(n \log n)$ or slightly above.

The resulting triangulations of the first four data sets are shown in Fig. 15. In Table 1 we have listed the number of triangles for each data set according to the maximum reachable angle bounds. For `Triangle` we increase the angle by 0.25 degrees step by step until the program fails.

9 Conclusion and Future Work

We have presented a new approach of generating meshes of high quality by triangulating point clouds. All angles of the triangles lie within the range of 43.9° and 90°. We have also shown that the triangular meshes have the Delaunay property.

The triangulation is locally adaptive, i.e. regions of interest have more triangles than regions with less input points. By balancing the underlying hexagon-subdivision, the triangles grow quickly from high to low detailed regions. Furthermore the minimum distance among the input points determines the minimum size of the triangles, which may be pre-computed in advance.

In order to limit the domain to certain regions we provided a simple method to include a planar straight-line graph in the triangulation. Although angle bounds may fail directly at these line segments, they are still valid in the interior regions, i.e. one hexagon away from the lines.

Future work will be concentrated on finding a refinement loop implementation with a better runtime complexity than $O(n^2)$ and an integration of a PSLG with guaranteed angle bounds, preferably in the range of 30° and 90°.

An open problem is the extension to 3D space. It is not clear which 3D-polytope will substitute the 2D-hexagon. A promising candidate seems to be the octahedron.

References

1. Baker, B.S., Grosse, E., Rafferty, C.S.: Nonobtuse triangulation of polygons. Discrete Comput. Geom. 3(2), 147–168 (1988)
2. Bern, M., Eppstein, D., Gilbert, J.: Provably good mesh generation. J. Comput. Syst. Sci. 48(3), 384–409 (1994)
3. Chew, L.: Guaranteed-quality triangular meshes. Tech. Rep. TR 89-983. Cornell University (1989)
4. Erten, H., Üngör, A.: Triangulations with locally optimal steiner points. In: SGP 2007: Proceedings of the fifth Eurographics symposium on Geometry processing, Aire-la-Ville, Switzerland, pp. 143–152. Eurographics Association (2007)
5. Sußner, G., Dachsbacher, C., Greiner, G.: Hexagonal LOD for Interactive Terrain Rendering. In: Vision Modeling and Visualization, pp. 437–444 (2005)
6. Neugebauer, F., Diekmann, R.: Improved mesh generation: Not simple but good. In: 5th Int. Meshing roundtable, pp. 257–270. Sandia National Laboratories (1996)
7. Owen, S.J.: A survey of unstructured mesh generation technology. In: IMR, pp. 239–267 (1998)
8. Ruppert, J.: A new and simple algorithm for quality 2-dimensional mesh generation. In: SODA 1993: Proceedings of the fourth annual ACM-SIAM Symposium on Discrete algorithms, Philadelphia, PA, USA, pp. 83–92. Society for Industrial and Applied Mathematics (1993)
9. Shewchuk, J.R.: Triangle: Engineering a 2d quality mesh generator and delaunay triangulator. In: FCRC 1996/WACG 1996: Selected papers from the Workshop on Applied Computational Geormetry, Towards Geometric Engineering, London, UK, pp. 203–222. Springer, Heidelberg (1996)
10. Üngör, A.: Off-centers: A new type of steiner points for computing size-optimal quality-guaranteed delaunay triangulations. In: Farach-Colton, M. (ed.) LATIN 2004. LNCS, vol. 2976, pp. 152–161. Springer, Heidelberg (2004)
11. Welzl, E.: Smallest enclosing disks (balls and ellipsoids). In: Maurer, H.A. (ed.) New Results and New Trends in Computer Science. LNCS, vol. 555. Springer, Heidelberg (1991)

Tetrahedral Mesh Improvement Using Multi-face Retriangulation

Marek Krzysztof Misztal[1], Jakob Andreas Bærentzen[1], François Anton[1], and Kenny Erleben[2]

[1] Informatics and Mathematical Modelling, Technical University of Denmark
{mkm,jab,fa}@imm.dtu.dk
[2] Department of Computer Science, University of Copenhagen
kenny@diku.dk

Abstract. In this paper we propose a simple technique for tetrahedral mesh improvement without inserting Steiner vertices, concentrating mainly on boundary conforming meshes. The algorithm makes local changes to the mesh to remove tetrahedra which are poor according to some quality criterion. While the algorithm is completely general with regard to quality criterion, we target improvement of the dihedral angle. The central idea in our algorithm is the introduction of a new local operation called multi-face retriangulation (MFRT) which supplements other known local operations. Like in many previous papers on tetrahedral mesh improvement, our algorithm makes local changes to the mesh to reduce an energy measure which reflects the quality criterion. The addition of our new local operation allows us to advance the mesh to a lower energy state in cases where no other local change would lead to a reduction. We also make use of the edge collapse operation in order to reduce the size of the mesh while improving its quality. With these operations, we demonstrate that it is possible to obtain a significantly greater improvement to the worst dihedral angles than using the operations from the previous works, while keeping the mesh complexity as low as possible.

1 Introduction and Motivation

For many types of physical simulation, the tetrahedral mesh representation is the natural choice. For instance, finite element computations in 3D usually employ tetrahedral meshes which are far better at adapting to boundaries and changes in scale than e.g. regular voxel grids.

For 2D triangulations, Delaunay triangulation is often a natural choice since it leads to a mesh which is optimal in the sense that the minimal angles are maximized which is a reasonable quality criterion in 2D. In 3D however, it is less clear what quality criterion we should strive for and a 3D Delaunay tetrahedralization may contain very flat sliver tetrahedra with extreme dihedral angles, and extreme dihedral angles are often precisely what we wish to avoid since they may lead to problems, such as great interpolation errors or ill-conditioned stiffness matrices in some finite element computations (although in the anisotropic case they might be desirable) or problems with interpolation accuracy [21].

Consequently, in this paper and in other recent work [12], the goal is to optimize a tetrahedralization obtained through either Delaunay or other methods in order to improve some criterion – particularly dihedral angles. However, little is known about globally optimal meshes in the sense that the smallest dihedral angle is maximal or that the largest dihedral angle is minimal. Consequently, one strives instead for a set of simple, local transformations which improve the mesh by removing poorly shaped tetrahedra. The best one can hope for in this case is a good local minimum, and whether one attains such a minimum is highly dependent on one's vocabulary of local transformations. It is this vocabulary which we extend by the addition of two local transformations which are highly beneficial to the mesh quality yet have not previously been used in tetrahedral mesh optimization.

The most powerful way of improving triangle or tetrahedral meshes is through the insertion of more vertices (as shown in [12]). Indeed this is sometimes the only way to improve quality. Unfortunately, one pays the price of adding (sometimes significantly) more tetrahedra, and finding the optimal place to put a vertex can be hard. Besides, many applications (such as dynamic meshes) require their own Steiner vertex insertion routines. For these reasons, we opine that it is very worthwhile to explore to what extent our mesh improvement vocabulary can be augmented without adding vertices.

Our main contribution is the *multi-face retriangulation* operation. Assume a set of tetrahedra which we can divide into *upper* and *lower* tetrahedra. Any upper tetrahedron shares a face with precisely one lower tetrahedron (and vice versa) and the upper tetrahedra all share a vertex (the upper vertex) as do the lower tetrahedra (the lower vertex). We can say that the set of tetrahedra is *sandwiched* between the upper and the lower vertex (as illustrated in Figure 1). The union of the triangular faces shared between upper and lower tetrahedra can be seen as a triangulation of a polygon. Our proposed operation simply retriangulates this polygon to obtain better sets of upper and lower tetrahedra. Multi-face retriangulation can also be seen as a composition of the known *multi-face removal* and *edge removal* operations (as shown in Figure 1) [9, 12]. However, multi-face retriangulation is more powerful than the concatenation of these two operations: in the case of some configurations multi-face removal followed by edge removal would never be selected because very poor or inverted tetrahedra would result from the multi-face removal operation (as illustrated in Figure 2). Additionally, multi-face retriangulation works on boundaries whereas concatenation of multi-face removal and edge removal does not.

The other contribution is the use of the well known *edge collapse* operation. Curiously, to the best of our knowledge, this operation has not been incorporated into any tetramesh improvement algorithm previously. It significantly reduces the complexity of the mesh and it also might improve the worst quality within the set of affected tetrahedra.

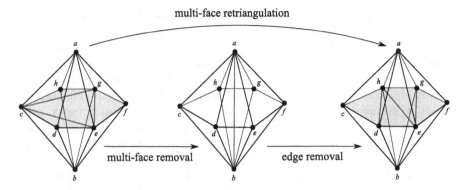

Fig. 1. Multi-face removal, edge removal and their superposition – multi-face retriangulation.

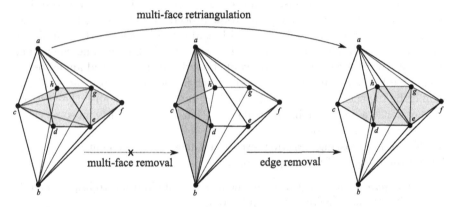

Fig. 2. Configuration in which multi-face removal would not be performed. Vertices a, b, c and d are nearly coplanar. Performing multi-face removal in such a configuration would lead to creating a very poorly shaped tetrahedron $abcd$ (highlighted in orange) of extremely low quality. Also, by perturbing vertex a or b we can easily create a situation in which tetrahedron $abcd$ would be inverted. In both situations multi-face removal would not be performed by a greedy algorithm – hence the arrow is crossed out. The strength of multi-face retriangulation is in tunneling through these kinds of hills in the energy landscape.

2 Related Work

Clearly, whether mesh improvement is needed depends on how the mesh was generated. Broadly speaking, there are three ways of producing tetrahedral meshes from a boundary representaion of an object. First, if the boundary is a piecewise linear complex (in particular – triangulated manifold), we could use constrained Delaunay tetrahedralization to produce a conforming mesh [19, 22]. Alternatively, we could use an advancing fronts method which would build the tetrahedralization out from the boundary. As mentioned, the former approach will often have problems with sliver tetrahedra even after Delaunay refinement, unless the boundary satisfies a set of strict conditions [3,19] limiting the practical applications of this approach, and the latter tends to produce some bad tetrahedra around areas where the front collides on itself [16]. These problems are compounded if the boundary mesh has poorly shaped triangles. An alternative approach is the centroidal Voronoi tessellation based Delaunay tetrahedralization [6] which can, however, still leave some poorly shaped tetrahedra. We conclude that the mesh optimization is likely to be useful as a step following both Delaunay based methods and also advancing fronts based methods.

A third and alternative strategy is to force the boundary to conform to an isosurface of an implicit function rather than a mesh. The spatial domain is first divided into tetrahedra, and a subset which approximates the shape well is selected. In a subsequent compression step, the boundary vertices of this subset are forced to lie precisely on the isosurface [15]. However, we note that the compression step is an optimization procedure because, generally, not only the boundary vertices are moved but also the interior vertices in order to improve the quality of the mesh. In recent work, Labelle and Shewchuk were able to demonstrate good provable bounds on the dihedral angles using such a method [13]. However, methods which fit meshes to isosurfaces [15,13] cannot be expected to capture sharp edges and corners because the vertices are not constrained to lie in particular positions. Consequently, in some cases they simply do not apply.

Most of the existing work for tetrahedral mesh improvement uses the following three types of mesh operations:

1. *Mesh smoothing* – relocation of the mesh points in order to improve mesh quality without changing mesh topology.
2. *Topological operations* – reconnection of the vertices in the mesh (without displacing them).
3. *Vertex insertion* – adding extra vertices into the mesh (through eg. splitting of the edges, faces or tetrahedra) and reconnecting affected regions of the mesh.

2.1   Mesh Smoothing

One of the best known smoothing methods, Laplacian smoothing, in which a vertex is moved to the centroid of the vertices to which it is connected, is a popular and quite effective choice for triangular meshes. In tetrahedral meshes, however, it often produces poor tetrahedra [7]. Optimal (with regard to linear interpolation error) Delaunay vertex placement has been investigated by Chen and Xu [2]. More general mesh smoothing algorithms are based on numerical optimization. One of the most popular local algorithms for mesh smoothing was suggested by Freitag et al. [8]. This method relocates one vertex at a time. Given one vertex, its new position is found, so that the minimum quality of all the tetrahedra adjacent to this vertex is maximized (this requires non-smooth optimization). This procedure is performed for each vertex in the mesh and can be iterated until a stable configuration is attained. It can also be performed on the boundary of the mesh, given extra constraints for the position of the vertex. Another optimization based approach, using generalized linear programming, was presented by Amenta et al. [1], but this one is not as general as Freitag's and is not well suited for dihedral angles optimization. Mesh smoothing can also be performed by continuous optimization in the space of coordinates of all vertices of the mesh (as in [10]), but Freitag's method has advantages over this approach – it is easier to use with non-smooth quality measures, and its characterised by stable behavior even if the initial quality of the mesh is very low.

2.2 Topological Operations

Reconnection of the mesh can be pictured as picking a set of adjacent tetrahedra and replacing them with another set of tetrahedra, of higher minimum quality, filling in the same volume. This can be performed in a more or less arbitrary manner (as small polyhedron re-tetrahedralization in [14]), or can be organized into a set of *topological operations*, such as:

- *2-3 flip* and its inverse, *3-2 flip*, as shown in Figure 3.
- *4-4 flip* and its version for boundary configuration, *2-2 flip*, illustrated in in Figure 3 – ambiguous, requires specifying which edge pair of vertices is going to be connected after the operation.
- *Edge removal* is illustrated in Figure 1 – generalizes 3-2 flip, 4-4 flip and 2-2 flip; ambiguous, requires specifying the final triangulation of the link of the removed edge, which can be performed by using triangulation templates, as in [9] or by using Klincsek's algorithm [11] in order to maximize the minimal quality of the created set of tetrahedra, as in [12]. Edge removal can be performed for boundary edges.
- *Multi-face removal* of de Cougny and Shephard [5] is the inverse to the edge removal, as shown Figure 1 – generalizes 2-3 flip and 4-4 flip; requires dynamic programming in order to select the subset of faces sandwiched between two vertices, which gives the best improvement.

The original paper of Freitag and Ollivier-Gooch [9] uses the first three operations. It can easily be noticed, that multi-face removal can actually be decomposed into a sequence of a single 2-3 flip followed by a certain number of 3-2 flips. However, this can not always be performed in the *hill-climbing* approach (which is usually the choice for the tetrahedral mesh-improvement algorithms) if one of the operations in the sequence decreases quality locally. Klingner and Shewchuk [12] use all the operations from the list above.

2.3 Vertex Insertion

Klingner and Shewchuk showed in [12] that mesh improvement is far more effective with the inclusion of transformations that introduce Steiner vertices.

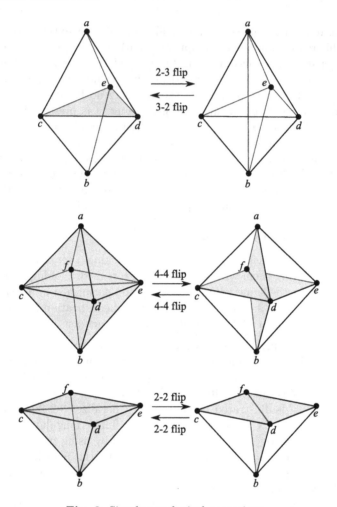

Fig. 3. Simple topological operations.

Proper placement of Steiner vertices is a hard problem. Klingner and Shewchuk describe a sophisticated and rather complex algorithm for vertex insertion which mimics Delaunay vertex insertion and, together with optimization based smoothing and topological operations, allows them to improve the meshes so that all dihedral angles are between 31° and 149°, or, using a different objective function, between 23° and 136°.

3   Tetrahedral Mesh Quality Improvement

Our mesh improvement algorithm is based on the algorithm proposed by Klingner and Shewchuk [12] (which in turn extends one by Freitag and Ollivier-Gooch [9]) which uses vertex smoothing by Freitag et al. [8], edge removal, multi-face removal and vertex insertion (most of the operations they use can be performed on the boundary of the mesh). In turn, our algorithm uses the following set of operations:

- Vertex-smoothing as in Freitag et. al [8],
- Topological operations:
  - edge removal,
  - multi-face removal,
  - multi-face retriangulation.
- Edge collapse.

Vertex smoothing and edge removal can be performed for the boundary vertices and edges, if the boundary is sufficiently flat around them. Additionally, vertex smoothing can be performed along *straight* ridges on the boundary of the mesh, if the surface patches separated by the ridge are sufficiently flat. Multi-face removal and edge removal are implemented essentially the same way as in [20].

3.1   Multi-face Retriangulation

Multi-face retriangulation can be seen as a composition of multi-face removal and edge removal, however, it can be also performed on the boundary of the mesh. It includes the 4-4 and 2-2 flips. Multi-face retriangulation does not change the number of tetrahedra in the mesh. So far as we know, it has never appeared in the literature.

The reasons in favor of using MFRT alongside multi-face removal and edge removal are:

- In some cases, the configuration produced by multi-face removal is of lower quality, as illustrated in Figure 2. Thus a greedy approach would not select that configuration even if the subsequent edge removal led to a state of lower energy than the initial configuration.
- In some cases the configuration produced by multi-face removal includes inverted tetrahedra, and no approach would select that (also shown in Figure 2). However, MFRT cannot produce inverted tetrahedra, as the best triangulation of the multi-face cannot be worse than the initial one and we assume we run our algorithm on valid tetrahedral meshes.
- MFRT can be applied to boundary configurations. To see this, let us only consider a set of lower tetrahedra in Figure 1. In such a configuration, multi-face consists of boundary faces and it cannot be removed using multi-face removal, but it can easily be retriangulated. However, if the multi-face is not sufficiently flat, which is the case when the angles between the normals to the faces are greater than $0°$, MFRT can change the geometry of the boundary of the mesh, which is usually not desirable.
- MFRT does not change the number of tetrahedra. This property is a direct consequence of a well known fact that every triangulation of a polygon has the same number of triangles.

In our implementation, the input is a single face f we wish to remove. We find the apices a and b of the two tetrahedra adjoining f. Among the set of all faces sandwiched between a and b we find the connected component that contains f. For the multi-face defined like that, we find the optimal triangulation of its bounding polygon using Klincsek's algorithm. The routine is similar for a boundary face f, although in this case we have to make sure that the retriangulated multi-face is sufficiently flat (otherwise geometry of the boundary might change).

3.2 Edge Collapse

Edge collapse (also known as *edge contraction* or *half-edge contraction*) is a well known mesh operation which has been used as a primary tool for simplifying 2D and 3D meshes in numerous works, such as [4,17]. It identifies one of the vertices of an edge e with the other, removes e and all faces and tetrahedra which contain it. This can, however, lead to invalid configurations (violating the simplicial complex criterion) or alter the surface geometry of the mesh, unless certain conditions are fulfilled, described in detail by Natarajan and Edelsbrunner in [17]. If edge collapse is not performed for the boundary edges, which is the case in our implementation, those conditions simplify to the following:

$$\mathrm{Lk}(e) = \mathrm{Lk}(a) \cap \mathrm{Lk}(b),$$

where a and b are the vertices of the edge e, and $\mathrm{Lk}(\sigma)$ denotes the link of a simplex σ which, in tetrahedral meshes, can be defined as a set of those simplices (vertices, edges and faces) in the mesh, that do not intersect with σ, but are contained by the one of the tetrahedra containing σ. In our implementation this is performed if the minimum quality of the set of tetrahedra affected by the operation increases, or if it does not decrease below a certain quality threshold q_{min}, which is a global parameter of our algorithm.

3.3 Quality Measures

Both the smoothing algorithm and the topological operations which we are using are indifferent to the tetrahedron quality measure. In order to be able to compare our results to those provided in [12] and [9], we are using:

- The *minimum sine measure* – the minimum sine of a tetrahedron's six dihedral angles, penalizes both small and large dihedral angles.
- The *minimum biased sine measure*, which is like the minimum sine measure, but if a dihedral angle is obtuse, its sine is multiplied by 0.7 (before choosing the minimum). This quality measure penalizes large angles more aggressively than the small angles.

Many quality measures have been proposed for tetrahedral meshes reviewed by [21], [10]. Our two choices are well behaved and very intuitive, although non-smooth.

4   Implementation

Our mesh improvement schedule follows that of Klingner and Shewchuk [12] (pseudo code is shown in Algorithm 1). Same as in their work, we use a short list of *quality indicators* in order to measure progress in lowest quality tetrahedra improvement. Those are: the quality of the worst tetrahedron in the entire mesh and seven *thresholded means* of the qualities of all the tetrahedra in the mesh. A mean $\bar{q}_\theta$ with threshold θ is computed the following way:

$$\bar{q}_\theta = \frac{1}{\#\{\text{tetrahedra in } M\}} \sum_{t \in M} \min(q(t), \theta),$$

where M is the mesh and q is the tetrahedron quality measure we use. For our quality measures we use thresholded means with thresholds $\sin(1°)$, $\sin(5°)$, $\sin(10°)$, $\sin(15°)$, $\sin(25°)$, $\sin(35°)$ and $\sin(45°)$. A quality indicator designed like that is a good measure of how narrow the distribution of the tetrahedron qualities is and allows us to detect the quality improvement even if the minimum quality does not change. The minimum quality alone is much less efficient as a mesh quality indicator – it leads to premature termination of the mesh improvement algorithm and significantly worse final results. We consider the improvement in the mesh quality *sufficient* if either the quality of the worst tetrahedron improves, or if one of the thresholded means increases by at least 0.0001.

We begin mesh improvement with a vertex smoothing pass, followed by a topological pass. In the topological pass, pseudo code of which is shown in Algorithm 2, we first obtain the list of all the tetrahedra in the mesh and then try to remove every tetrahedron t on the list by first trying to remove its edges using the edge remove operation and then, if we have not succeeded, by trying to remove its faces using multi-face retriangulation followed by multi-face removal. Such an ordering of the operations is justified by the fact that first performing multi-face retriangulation still leaves room for extra improvement through multi-face removal, while it does not work the other way round. The optimal multi-face for multi-face removal is chose using dynamic programming, accordingly to an algorithm described in [20]. Any of those operations are performed only if they locally improve the quality. If that happens, we proceed to the next tetrahedron on the list. Every topological operation that we use can destroy more tetrahedra than the one for which it was called, so before attempting to remove any tetrahedron we have to make sure that it still exists in the mesh.

After two initial passes we begin the main loop, in which we first smooth all the vertices until there is no more improvement detected by our mesh quality indicators. Then we start the topological pass again. If it improves the quality of the mesh sufficiently, we start the loop again, otherwise we start the thinning pass (pseudo code is shown in Algorithm 3). In the thinning pass we attempt to collapse every edge which is not a boundary one, does not connect two boundary vertices and fulfills the edge collapse feasibility

---

Algorithm 1. IMPROVE(M)

{M is a mesh}

1: Smooth each vertex of M.
2: TOPOLOGICALPASS(M)
3: $failed \Leftarrow 0$
4: **while** $failed < 3$ **do**
5:      Smooth each vertex of M.
6:      **if** M improved sufficiently **then**
7:          $failed \Leftarrow 0$
8:      **else**
9:          TOPOLOGICALPASS(M)
10:         **if** M improved sufficiently **then**
11:             $failed \Leftarrow 0$
12:         **else**
13:             THINNINGPASS(M)
14:             **if** M improved sufficiently **then**
15:                 $failed \Leftarrow 0$
16:             **else**
17:                 $failed \Leftarrow failed + 1$
18:             **end if**
19:         **end if**
20:     **end if**
21: **end while**

Algorithm 2. TOPOLOGICALPASS(M)

1: Create the list of all tetrahedra in M.
2: **for** each tetrahedron t on the list, that still exists in M **do**
3:      **for** each edge e of t (if t still exists) **do**
4:          Attempt to remove edge e.
5:      **end for**
6:      **for** each face f of t (if t still exists) **do**
7:          Attempt to remove face f by first using multi-face retriangulation
             followed by multi-face removal.
8:      **end for**
9: **end for**

---

condition. We perform the collapse only if it improves the quality locally or if the quality of the affected tetrahedra after the operation is not smaller than a threshold value $q_0 = 0.5$. If the thinning pass improved the quality of the mesh sufficiently, we start the loop again, otherwise we record that the sequence of smoothing, topological and thinning passes did not manage to improve the quality of the mesh. If that happens three times in a row, the algorithm stops.

Algorithm 3. THINNINGPASS(M)

1: **for** each edge $e \in M$ that is not on the boundary **do**
2:     **if** e still exists **then**
3:         Find the vertices a and b of e.
4:         **if** b is not a boundary vertex **then**
5:             Attempt to collapse edge e: $a \leftarrow b$.
6:             **if** success **then**
7:                 Smooth a.
8:             **end if**
9:         **end if**
10:         **if** a is not a boundary vertex **and** e still exists **then**
11:             Attempt to collapse edge e: $b \leftarrow a$.
12:             **if** success **then**
13:                 Smooth b.
14:             **end if**
15:         **end if**
16:     **end if**
17: **end for**

---

5 Tests and Results

We tested our schedule on the following meshes:

- BOID, TEAPOT and DEER are Delaunay tetrahedralizations generated by TetGen [22] with extremely bad dihedral angles due to the lack of interior vertices.
- RAND1 – used by Freitag and Ollivier-Gooch and also by Klingner and Shewchuk to evaluate their mesh improvement algorithms.
- P and TFIRE – used by Klingner and Shewchuk to evaluate their mesh improvement algorithm.
- GLASS – medium size mesh generated using TetGen [22] with few interior vertices and low quality boundary triangles.

Unfortunatelly, Klingner and Shewchuk published the results of mesh improvement without vertex insertion only for a very few meshes, so the possibility of comparing our results to theirs was limited.

The results of mesh improvement for those meshes are presented in the Tables 1, 2 and 3. For the BOID mesh we tried to maximize the minimum biased sine quality measure for this mesh. The boundary of the mesh is nowhere flat so smoothing and topological operations are not allowed on the boundary. There are no interior vertices, so in fact smoothing and thinning cannot take place at all, as they would alter the surface geometry. Not much improvement can be achieved without vertex insertion in this case, but still we can see that the topological pass with MFRT is significantly more effective at fighting the worst dihedral angles than the topological pass without MFRT. The situation and the results are similar in the case of the TEAPOT mesh.

Table 1. Mesh quality improvement results for meshes: BOID, TEAPOT and DEER. Minimum biased sine measure was used for the first two and minimum sine quality measure was used for the last one. Pictures in the first row show the initial surface geometry of our meshes. Surface geometry remains the same after the mesh improvement, although the tesselation might change in flat regions. Histograms show, from the top to the bottom, the distribution of all dihedral angles in the original mesh, mesh improved without MFRT and mesh improved using MFRT.

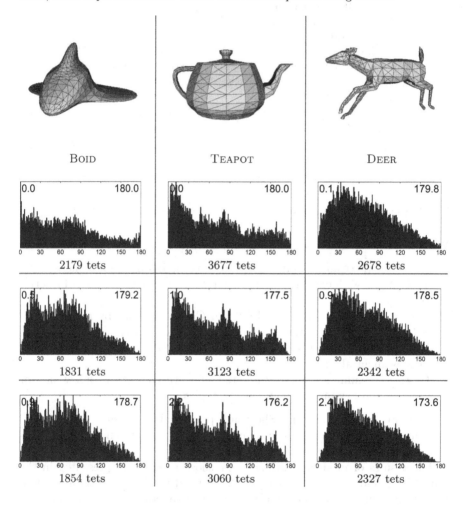

We also obtain a significant extra improvement (6.4°) by the use of MFRT for the DEER mesh, while in this case we tried to maximize the minimum sine quality measure.

For RAND1 the use of MFRT allows us to narrow the dihedral angles range by as much as 8° for both sine and biased sine quality measures. Additionally, edge collapse allows us to decrease the complexity of the meshes by almost 35% and to narrow the dihedral angles range by almost 3° for sine quality

Table 2. Mesh quality improvement results for meshes: P (using minimum biased sine quality measure) and RAND1 (left column – using minimum biased sine quality measure, right column - using minimum sine quality measure). Histograms show, from the top to the bottom, the dihedral angle distribution in the original mesh, mesh improved without MFRT, mesh improved using MFRT, meshed improved using MFRT and thinning. Red bars indicate particularily abundant dihedral angles and were scaled down to increase the readability of the histograms.

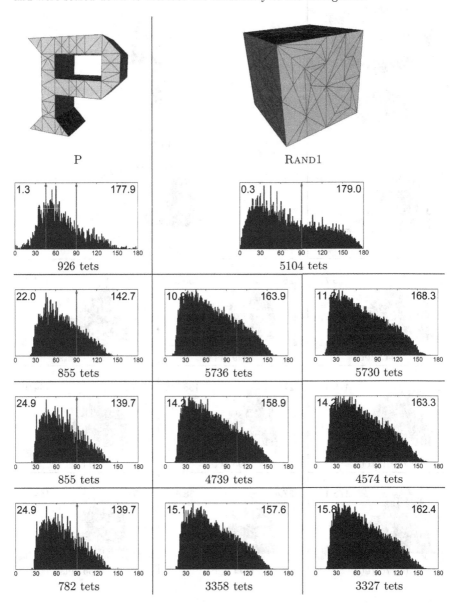

Table 3. Mesh quality improvement results for meshes: GLASS (using minimum biased sine quality measure) and TFIRE (left column – using minimum biased sine quality measure, right column - using minimum sine quality measure). Histograms show, from the top to the bottom, the dihedral angle distribution in the original mesh, mesh improved without MFRT, mesh improved using MFRT, meshed improved using MFRT and thinning.

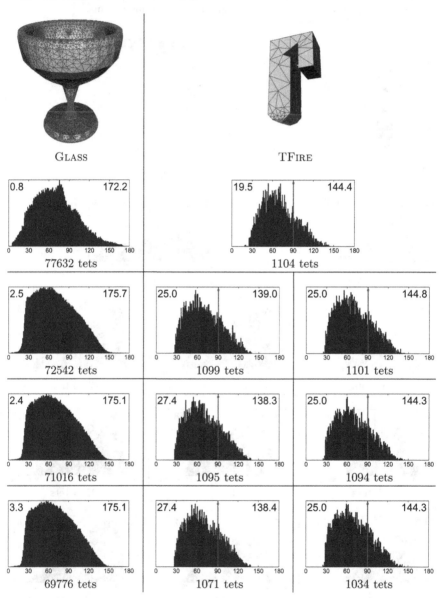

measure – ultimately we obtain 15.8°–162.4°, and by 2° for biased sine quality measure – ultimately we obtain 15.1°–157.6°. For comparison, the best results Freitag and Ollivier-Gooch [9] obtained for the same mesh was 15.0°–166.7° for minmax cosine quality measure (and 12.5°-167.3° for sine quality measure). Mesh P also benefits significantly from the use of MFRT – it narrows the dihedral range by 6°, but in this case the thinning pass does not improve the extreme quality values.

The TFIRE mesh also benefit from adding the MFRT and the edge collapse operation, although not as significantly as the previous ones. Still, our result 24.9°–139.7° is better than 21.3°–147.1° obtained by Klingner and Shewchuk [12].

In case of the GLASS mesh, we can notice that our mesh improvement algorithm actually expands the dihedral angles range. This is due to the lack of extremely obtuse angles in the original mesh, and due to the fact, that the mesh operations we use choose to "sacrifice" good quality tetrahedra in order to locally improve the worst tetrahedron. However, we can notice that the we still benefit from inclusion of MFRT and thinning in the mesh improvement algorithm.

6   Discussion and Future Work

Our results show that using the multi-face retriangulation operation alongside smoothing and topological operations from the previous works can lead to better improvement of the dihedral angles and should be included in the standard repertoire of the topological operations for tetrahedral meshes. For the meshes we tested, we obtained a narrowing of the range of dihedral angles by up to 8° without inserting a single Steiner vertex. Additionally, edge collapse can also improve the worst dihedral angles and decrease the complexity of the mesh by up to 35%, esspecially when the initial quality of the mesh is very poor.

However, during our experiments we have noticed that the mesh improvement algorithm is still prone to get stuck in the local minima, even if we use multi-face retriangulation – in a few cases, running the algorithm with some mesh operations "switched off" (for instance operations on the boundary of the mesh) leads to better results than running the algorithm with the full repertoire of mesh operations. This is, of course, a consequence of using a greedy, hill-climbing approach. This could possibly be improved by applying a randomized approach.

It is also important to notice that our algorithm is designed for valid input meshes. If the initial mesh has inverted tetrahedra the algorithm might fail to remove them. Also, the tetrahedron quality measures we used are not particularily well suited for meshes with inverted tetrahedra, since they lose continuity as the tetrahedron gets inverted.

In the future we are going to further investigate the possibilities of mesh improvement without Steiner vertex insertion, also with other quality measures, such as the *volume-length* measure [18] V/l_{rms}^3, where V is the signed volume of a tetrahedron and l_{rms} is the root-mean-squared edge length.

Acknowledgments

We thank Frederik Gottlieb for the implementation of the core part of the data structure that we are using, Bryan Klingner, Jonathan Richard Shewchuk and Mads Fogtmann Hansen for meshes, geometric models and discussion. We would also like to thank Vedrana Andersen for helping us give this paper its final shape and anonymous reviewers for useful suggestions.

References

1. Amenta, N., Bern, M., Eppstein, D.: Optimal point placement for mesh smoothing. In: Proceedings of the eighth annual ACM-SIAM symposium on Discrete algorithms, pp. 528–537. Society for Industrial and Applied Mathematics, Philadelphia (1997)
2. Chen, L., Xu, J.: Optimal delaunay triangulation. J. Comp. Math. 22, 299–308 (2004)
3. Chew, L.P.: Guaranteed-quality delaunay meshing in 3d (short version). In: Proceedings of the thirteenth annual symposium on Computational geometry, pp. 391–393. ACM, New York (1997)
4. Cutler, B., Dorsey, J., McMillan, L.: Simplification and improvement of tetrahedral models for simulation. In: SGP 2004: Proceedings of the 2004 Eurographics/ACM SIGGRAPH symposium on Geometry processing, pp. 93–102. ACM, New York (2004)
5. de Cougny, H.L., Shephard, M.S.: Refinement, derefinement and optimization of tetrahedral geometric triangulations in three dimensions (1995) (Unpublished manuscript)
6. Du, Q., Wang, D.: Tetrahedral mesh generation and optimization based on centroidal voronoi tessellations. Int. J. Numer. Meth. Eng 56, 1355–1373 (2002)
7. Freitag, L.A.: On combining laplacian and optimization-based mesh smoothing techniques. In: Trends in Unstructured Mesh Generation, pp. 37–43 (1997)
8. Freitag, L.A., Jones, M., Plassmann, P.: An efficient parallel algorithm for mesh smoothing. In: Proceedings of the Fourth International Meshing Roundtable (1995)
9. Freitag, L.A., Ollivier-Gooch, C.: Tetrahedral mesh improvement using swapping and smoothing. International Journal for Numerical Methods in Engineering 40, 3979–4002 (1997)
10. Hansen, M.F., Bærentzen, J.A., Larsen, R.: Generating quality tetrahedral meshes from binary volumes. In: Proceedings of VISAPP 2009 (2009)
11. Klincsek, G.T.: Minimal triangulations of polygonal domains. Annals of Discrete Mathematics 9, 121–123 (1980)
12. Klingner, B.M., Shewchuk, J.R.: Agressive tetrahedral mesh improvement. In: Proceedings of the 16th International Meshing Roundtable, October 2007, pp. 3–23 (2007)

13. Labelle, F., Shewchuk, J.R.: Isosurface stuffing: Fast tetrahedral meshes with good dihedral angles. ACM Transactions on Graphics 26(3), 57 (2007)
14. Liu, J., Sun, S.: Small polyhedron reconnection: A new way to eliminate poorly-shaped tetrahedra. In: Proceedings of the 15th International Meshing Roundtable, pp. 241–257 (2006)
15. Molino, N., Bridson, R., Teran, J., Fedkiw, R.: A crystalline, red green strategy for meshing highly deformable objects with tetrahedra. In: Proc. International Meshing Roundtable (2003)
16. Möller, P., Hansbo, P.: On advancing front mesh generation in three dimensions. International Journal for Numerical Methods in Engineering 38(21), 3551–3569 (1995)
17. Natarajan, V., Edelsbrunner, H.: Simplication of three-dimensional density maps. IEEE Transactions on Visualization and Computer Graphics 10, 587–597 (2004)
18. Parthasarathy, V.N., Graichen, C.M., Hathaway, A.F.: A comparison of tetrahedron quality measures. Finite Elements in Analysis and Design 15(3), 255–261 (1994)
19. Shewchuk, J.R.: Tetrahedral mesh generation by Delaunay refinement. In: Proceedings of the fourteenth annual symposium on Computational geometry, pp. 86–95. ACM, New York (1998)
20. Shewchuk, J.R.: Two discrete optimization algorithms for the topological improvement of tetrahedral meshes (Unpublished manuscript) (2002)
21. Shewchuk, J.R.: What is a good linear finite element? interpolation, conditioning, anisotropy, and quality measures (2002) (Unpublished manuscript)
22. Si, H.: Tetgen, a quality tetrahedral mesh generator and three-dimensional delaunay triangulator, v1.3 user's manual. Technical report, WIAS (2004)

Automatic All Quadrilateral Mesh Adaption through Refinement and Coarsening

Bret D. Anderson[1], Steven E. Benzley[1], and Steven J. Owen[2,*]

[1] Brigham Young University, Provo, Utah, USA
   bret.d.anderson@gmail.com, seb@et.byu.edu
[2] Sandia National Laboratories, Albuquerque, New Mexico, USA
   sjowen@sandia.gov

Abstract. This work presents a new approach to conformal all-quadrilateral mesh adaptation. Most current quadrilateral adaptivity techniques rely on mesh refinement or a complete remesh of the domain. In contrast, we introduce a new method that incorporates both conformal refinement and coarsening strategies on an existing mesh of any density or configuration. Given a sizing function, this method modifies the mesh by combining template-based quadrilateral refinement methods with recent developments in localized quadrilateral coarsening and quality improvement into an automated mesh adaptation routine. Implementation details and examples are included.

Keywords: quadrilateral, adaptivity, refinement, coarsening.

1 Introduction

The ability to automatically adapt a finite element mesh based on a sizing function is an important component of an automatic modeling and simulation process. Although it is not a new concept, its principal application has been to triangle and tetrahedral-based methods. Quadrilateral meshes are often preferred by analysts for improved accuracy over triangle-based methods. In spite of this, adaptive quadrilateral techniques are not as prevalent in the literature.

A truly general mesh adaptation scheme must have the ability to both enhance (refine) and simplify (coarsen) a mesh to provide sufficient accuracy and efficiency in the analysis. While there are numerous methods currently used, relatively few provide for both refinement and coarsening. Additionally, no current algorithm has the ability to adapt an all-quadrilateral mesh with refinement and a coarsening technique not constrained to de-refining.

This work presents a unique all-quadrilateral mesh adaptation algorithm that modifies a given mesh by adding and removing elements and employs a

---

* Sandia is a multiprogram laboratory operated by Sandia Corporation, a Lockheed Martin Company, for the United States Department of Energys National Nuclear Security Administration under contract DE-AC04-94AL85000.

coarsening process that is not limited to undoing previous steps of refinement. This algorithm combines template-based quadrilateral refinement techniques [1] with recent developments in coarsening [2] and quadrilateral improvement [3] to adapt an existing mesh. Additionally, to provide an algorithm that will meet conformity and element type requirements of finite element solvers, this method guarantees a fully conformal, all-quadrilateral mesh.

2   Background

To meet the ever increasing computational demands of complex finite element models, mesh adaptation has become a valuable area of study. There are three basic classes of adaptation, commonly referred to as $r-$, $h-$, and $p-$adaptation. $r-$adaptation, sometimes known as smoothing, refers to methods that alter element geometry by repositioning the nodes, but do not change the topology of the mesh. $h-$adaptation, refers to methods which change both the geometry of the elements and the topology of the mesh by adding and/or removing elements. $p-$adaptation, involves methods that do not alter the geometry or topology of individual elements; instead, these techniques change the degree of the elements in the mesh. While recognizing the importance of both $r-$ and $p-$adaptive methods we focus this work on the $h-$adaptive technique, concentrating specifically on the conformal refinement and coarsening of all-quadrilateral meshes.

2.1   Current Methods

Initial adaptive mesh generation is perhaps the easiest way to build an adapted finite element mesh because it simply employs the given sizing function in the original creation of the mesh. This is a widely used method and is available in many mesh generation schemes, including paving [4]. The major drawback of initial adaptive mesh generation is that it requires significant foresight into the probable results of the analysis which are used to determine element sizes and an appropriate distribution of element density across the mesh. Because of this required foresight, initial mesh generation techniques that incorporate sizing are particularly useful when based on geometric characteristics of the model [5].

Closely related to initial adaptive mesh generation is adaptive mesh regeneration; a mesh adaptation scheme in which the mesh is analyzed, a sizing function is determined, and the entire mesh or the region of the mesh requiring modification is removed and reconstructed according to the new sizing function. A significant amount of research has been done in this area and numerous algorithms have been presented employing adaptive mesh regeneration [6] [7] [8]. Although regenerating the mesh does not have the same drawbacks with respect to required foresight as does initial adaptive mesh generation, deleting and re-creating the mesh can be inefficient when only a small region requires adaption. In addition, the initial mesh generation may specify detailed user controls requiring specific local sizing and quality. Regenerating a complete mesh in these circumstances when only a small portion is required to be adapted may not be a feasible option.

2.2   Concurrent Refinement and Coarsening

While there are many effective methods that have been proposed in the literature [9] - [13] to adapt a mesh, none address the ability to provide effectively coupled coarsening and refinement for quadrilateral meshes. Utilization of both coarsening and refinement in mesh adaptation greatly increases the ability to modify a mesh to provide an appropriate element density without the need to know the coarsest state of the model.

Hierarchical adaptation methods [14] have been developed that are able to adapt all-quadrilateral meshes, while maintaining conformity, by using quadtree refinement with transition templates. Coarsening in hierarchical adaptation is accomplished by simply removing quadtrees from parent elements; however, a major limitation of this method is that there is no way for the mesh to be coarsened further than the initial base mesh. By taking advantage of new coarsening techniques the algorithm presented in this work provides adaptation that includes coarsening not limited to undoing previous refinement.

3   Automated Mesh Adaptation

3.1   Sizing Functions

The first step in creating an adaptive mesh is to provide an appropriate sizing function across the mesh domain. Sizing functions are typically based on error estimates derived from the solution of a finite element analysis, geometric characteristics of the model, or other user defined constraints. A solution based sizing function might specify an increased element density in regions of high stress or strain gradients. Geometry based sizing functions, such as a skeleton sizing function [5] [15], consider feature size as well as surface or boundary curvature and specify an appropriate element density throughout the mesh.

In addition to specifying the desired size of elements throughout the mesh, sizing functions must also take into account mesh gradation, the rate at which the element sizes change across the mesh [16] [17]. Gradation control is an important part of ensuring high shape quality of elements in a conformal mesh by not allowing a large change in size between adjacent elements. Although it is an important area of study, the development of sizing functions is not part of this research and it is assumed that an appropriate sizing function is provided as input to each adaptive meshing problem.

3.2   Tools and Requirements

The adaptation technique presented in this work employs a combination of quadrilateral refinement, coarsening, and quality improvement operations to adapt a given mesh. Since some finite element solvers require a conformal mesh and others do not allow hybrid meshes with more than one element type, only adaptation operations that preserve a conformal all-quadrilateral mesh are used. Additionally, these operations must be able to be applied locally to allow for concurrent coarsening and refinement.

Since the primary goal of adaptation is to ensure accurate results, refinement is usually required. This adaptation algorithm employs a refinement method that subdivides faces in the refinement region using a four element quadtree, referred to as 2-refinement, with templates inserted into the transition zone to maintain a conformal all-quadrilateral mesh [18] (see figure 1(b)). While nine element quadtrees or 3-refinement (see figure 1(c)) are sometimes used to refine quadrilateral faces, 2-refinement was chosen because it offers more control over the number of elements added to the mesh.

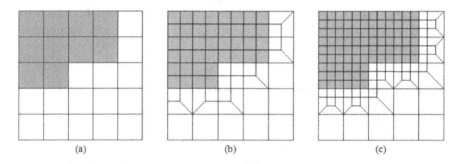

Fig. 1. Quadrilateral refinement with transition templates. (a) Original mesh and shaded refinement region. (b) Mesh refined with 2-refinement. (c) Mesh refined with 3-refinement.

This adaptation algorithm uses the Automated Quadrilateral Coarsening by Ring Collapse (AQCRC) algorithm recently developed by Dewey [2]. This coarsening method provides completely localized coarsening by selecting and removing rings of adjacent faces from within a specified coarsening region as shown in figure 2. One consequence of the removal of these coarsening rings is the creation of poor quality faces and quadrilateral improvement (cleanup) is a necessary step in this process. Although the AQCRC algorithm is a very effective local coarsening technique, it assumes that the coarsening rings are closed rings and does not have any provisions for coarsening of the mesh boundaries. Because of this limitation we employ the removal of dual chords [19] to coarsen the boundaries as illustrated in figure 3.

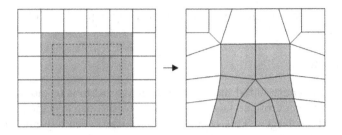

Fig. 2. AQCRC coarsening.

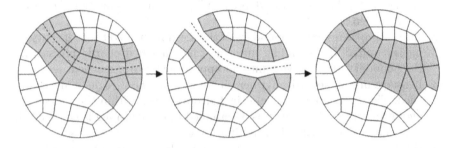

Fig. 3. Quadrilateral chord removal.

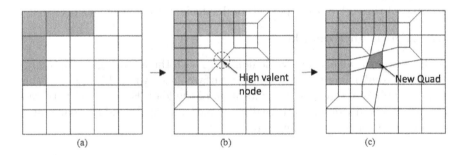

Fig. 4. One example of quality improvement required by refinement. (a) Shaded refinement region. (b) Seven-valence node formed from 2-refinement. (c) Mesh after quality improvement operation (face open on high-valence node).

In addition to the use of clean-up within the AQCRC algorithm, poor quality faces may form as a result of the refinement of irregular regions, making quadrilateral improvement a necessary step in this adaptation procedure. As a result, we also introduce new cleanup procedures [3], extending procedures introduced in [20], for improving the quality of an adapted mesh. For example, figure 4 shows a case where a concave refinement region forms a node with a valence of seven. This high-valence node is then removed with a face open procedure, resulting in a more structured mesh. The removal of high valence nodes from a mesh is one of many capabilities of this quality improvement method.

3.3 Algorithm

The goal of this automated mesh adaptation algorithm is to modify an existing mesh so that all of the faces are as close to the size specified by a sizing function but no larger. This ensures that solution accuracy and resolution are not sacrificed for decreased computation time. The algorithm flowchart is shown in figure 5 and described by the following steps:

1. A quadrilateral mesh to be adapted and an accompanying sizing function are provided as input.
2. Each curve defining the boundary of the mesh is checked to see if coarsening is required. If a bounding curve must be coarsened, chords intersecting the boundary are removed until an appropriate size is reached.
3. If coarsening is needed anywhere in the mesh, those regions are coarsened. The clean-up algorithm is included as a step within the coarsening algorithm. If at any point, it is determined that coarsening is not needed, this step is skipped in all future iterations.
4. If refinement is needed anywhere in the mesh, those regions are refined.
5. Following the refinement of elements in the mesh, the entire mesh surface is cleaned-up.
6. Steps 3 through 5 are repeated until sufficient refinement has occurred.

3.4 Algorithm Example

To illustrate this algorithm, we demonstrate with the following example simulating a circular line load on a planar surface as shown in Figure 6. In this case, the area of interest is at the location of the load. For simplicity, the algorithm can be divided into three distinct parts; input, boundary coarsening, and iterative coarsening/refinement.

Input

This algorithm requires an already meshed surface and an appropriate sizing function to be provided as input. In this case the surface is a flat 10 x 10 plate, meshed with a perfectly structured 10 x 10 quadrilateral mesh also shown in figure 6.

The size of each quadrilateral face, h_a is the average length of its four edges:

$$h_a = \frac{1}{4} \sum_{i=1}^{4} l_i \tag{1}$$

where l_i is the length of the ith edge of the quadrilateral face. Therefore, the size of each face in this initial mesh is 1.0. The sizing function for this example specifies a very high element density with an element size of 0.1 at the location

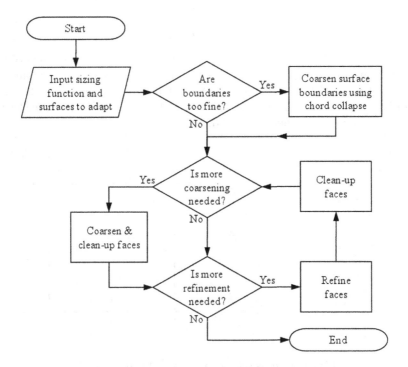

Fig. 5. Algorithm flowchart

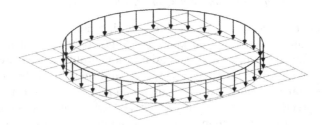

Fig. 6. Circular load on plane surface.

of the applied load. This specified element density gets progressively lower, varying linearly, as we move further away from the load, eventually reaching a recommended element size of nearly 5 at the center of the plate.

Boundary Coarsening

Since the AQCRC algorithm developed by Dewey does not allow boundary coarsening, it is achieved with simple chord removals in areas of the boundary that require larger element sizes. The edge length ratio f_l, defined in Equation 2, is the ratio of actual edge size l_a to desired edge size l_d and is calculated for all boundary edges and used to select chords for removal.

$$f_l = \frac{l_a}{l_d} \tag{2}$$

In this example, boundary coarsening is necessary near the corners of the mesh. Four chords, shown in the left panel of Figure 7 are chosen for removal. The right panel of Figure 7 shows the mesh following the removal of the four chords. To maintain a more isotropic mesh structure, each bounding curve that was coarsened is then smoothed.

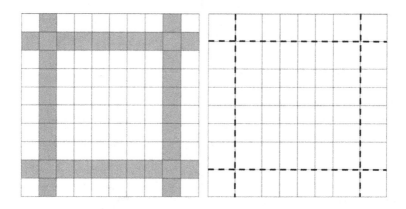

Fig. 7. Chords selected for removal and removed.

Iterative Coarsening, Refining, and Quality Improvement

The remainder of the algorithm modifies the interior of the mesh by itera-tively coarsening and refining elements until the goal has been reached. To provide a balanced approach to the adaptation problem, each iteration of this algorithm alternates between coarsening and refinement. Since coarsen-ing tends to make the mesh less structured and refinement tends to make the mesh more structured, the algorithm always begins with coarsening which is followed by refinement. Refining after coarsening also helps to achieve the goal of ensuring that the elements in the mesh are smaller than specified by the sizing function.

Each iteration begins by calculating the size ratio of each face f_s as:

$$f_s = \frac{h_a}{h_d} \tag{3}$$

where h_a is the actual face size as defined in Equation 1 and h_d is the desired face size as specified by the sizing function. A size ratio greater than 1.0 indicates that the face is too large and should be refined; a size ratio less than 1.0 indicates that the face is too small and can be coarsened.

To better control the amount of coarsening that takes place and to ensure that the coarsening operation does not overshadow refinement requirements, a dynamic threshold t_c, given as:

$$t_c = 1.0 - 0.2n_c \tag{4}$$

where $n_c = 0, 1, 2, ...$ is an integer defined by the number of iterations.

Any face with a size ratio less than the coarsening threshold is considered too small and should be coarsened. For example, in the first iteration, any face with a size ratio less than 1.0 is considered for coarsening. In later iterations, the coarsening threshold is relaxed until the sixth iteration when it becomes zero and disappears. Also, if at any point in the adaptation process, fewer than 10% of the faces in the mesh require coarsening, the coarsening is considered complete and is skipped in all future iterations. This requirement helps to ensure that the same mesh regions are not being repeatedly coarsened and refined. Figure 8(a) shows the faces selected for coarsening in the example. This shaded region is coarsened and the resulting mesh topology can be seen in Figure 8(b).

Since the coarsening algorithm requires a contiguous region of quadrilateral faces to create coarsening rings, a lone face requiring coarsening surrounded by faces that do not require coarsening will automatically be neglected by the AQCRC algorithm, making sure coarsening does not extend outside of the desired region.

After the first iteration of coarsening, the size ratio is re-calculated for all of the faces in the mesh in preparation for refinement. Any face with a size ratio greater than 1.0 is considered too large and requires refinement. This limit of 1.0 is not relaxed at any time throughout the algorithm; however, a dynamic refinement threshold t_r, defined in Equation 5, is used to separate the elements requiring refinement into two categories.

$$t_r = 1.25 - 0.05n_r \tag{5}$$

where $n_r = 0, 1, 2, ...$ is an integer defined by the number of iterations.

Any element with a size ratio greater than 1.0 and greater than the refinement threshold is considered a high-refine face, while any face with a size ratio greater than 1.0 but less than the refinement threshold is considered a low-refine face. In the refinement step, only the high-refine faces are refined, unless there are none, in which case the low-refine faces are refined. The purpose of the separation between faces is that the low-refine faces are often very close to the high-refine faces and fall within their transition zones which are refined by means of transition template insertion. Similar to the coarsening threshold, the refinement threshold gradually shrinks the allowable range of low-refine face until the refinement threshold equals 1.0 and disappears. At this point, all faces with a size ratio greater than 1.0 are high-refine elements.

If at any time, less than 3% of the faces are considered low-refine faces and less than 0.5% of the faces are considered high-refine faces, refinement is deemed sufficient. If these criteria are met, however, future iterations of refinement are not precluded as future iterations of coarsening are.

In Figure 8(b) the high-refine faces are shaded dark gray and the low-refine faces are shaded light gray. As expected, the low-refine faces are in close

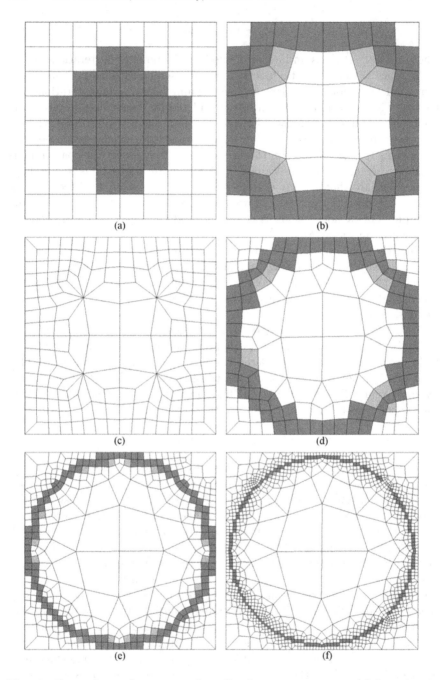

Fig. 8. Coarsening, refinement, and quality improvement steps. (a) Iteration 1, coarsening region. (b) Iteration 2, refinement regions. (c) Iteration 2, before quality improvement. (d) Iteration 3, refinement regions. (e) Iteration 4, refinement regions. (f) Iteration 5, refinement regions.

proximity to the high-refine faces and, in fact, fall in the transition zone. The mesh topology resulting from the refinement of the high-refine faces is shown in Figure 8(c). Note the creation of four 9-valence nodes as a result of the refinement of the irregular high-refine region. Because of situations like this, the quadrilateral improvement algorithm is applied after each instance of refinement. In the clean-up procedure, nodes with a valence greater than 6 are considered unacceptable and mesh topology is changed to remove the high valence. The new mesh topology following the clean-up algorithm is shown in Figure 8(d) where the unacceptable 9-valence nodes have been reduced to acceptable 5-valence nodes.

The algorithm then iteratively coarsens and refines the mesh until sufficient coarsening and refinement have both taken place. At this point in the example, after the first refinement step, it was determined that fewer than 10% of the faces had a size ratio that warranted coarsening; therefore, coarsening is now considered complete for all future iterations. Following this completion of coarsening, only refinement steps occur. Figures 8(d-f) show the successive refinement iterations for the remainder of this example. The final adapted mesh is shown in Figure 9.

Table 1 provides the distribution of size ratios of the faces in the final mesh. Note that nearly all of the faces have a size ratio less than 1.0 and are therefore smaller than desired. Since the goal of this adaptation is to provide elements close to the desired size, but not larger, this is a good result. It is not surprising, however, that some of the elements are too large since this

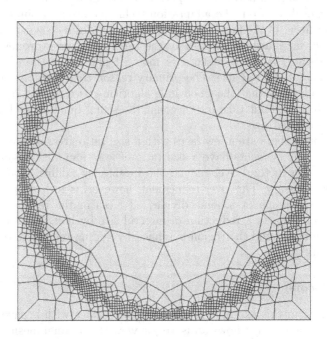

Fig. 9. Adapted mesh of circular load on plane surface.

Table 1. Results of circular load adaptation

Size Ratio f_s	Number of Faces	Percent of Total	Size Ratio f_s	Number of Faces	Percent of Total
<0.5	1485	61.0%	1.0 - 1.1	36	1.5%
0.5 - 0.8	708	29.1%	1.1 - 1.2	14	0.6%
0.8 - 0.9	114	4.7%	1.2 - 1.5	0	0.0%
0.9 - 1.0	77	3.2%	>1.5	1	0.0%
	2384	97.9%		51	2.1%

method uses an isotropic smoothing scheme which ignores the desired size specified by the sizing function.

4   Examples

We show two additional examples to illustrate the results of this new quadrilateral adaptation scheme. Both of the examples show the initial mesh and a contour plot of the sizing function, as well as the mesh after adaptation and a table with data showing the results of the adaptation. The final example, a plate with a hole, also provides results from a finite element analysis of the plate under a tensile load. In each example the original mesh was created with the paving algorithm in the mesh generation software package, CUBIT [21].

In these examples, and other experiments not included here [22], the size ratio data resulting from the adaptation technique are very similar. In each case, 2% or fewer of the faces are larger than their desired size and nearly all of the larger faces are within 10% of the target. This is a promising result considering the goal is to make sure most, if not all, of the elements are smaller than the desired size. The primary reason that there are some faces that are too large is that the smoothing algorithm used as part of the cleanup operations does not take into account the sizing function and may work against the desired size.

Even though there are a few faces with a size ratio greater than 1.0, more than 80% of the elements have a size ratio of less than 0.8, suggesting that this adaptation method over-refined the meshes by adding more elements than were necessary. This over-refinement, however, is to be expected since the quadtree refinement scheme divides all faces in the refinement region into four, resulting in a reduction of interval size by half. Additionally, the transition zone around the region is refined by adding templates to ensure a conformal mesh.

4.1   Nosecone

The nosecone in this example is a non-planar surface with a paved quadrilateral mesh. Figure 10 shows an isotropic view the original mesh and a side view of the sizing function. The original element size in this example was 1.0

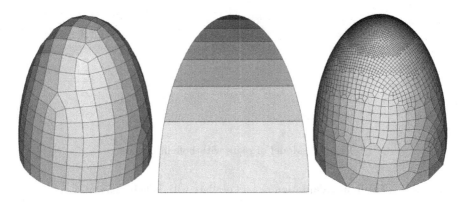

Fig. 10. Original mesh of nosecone (left) with contour plot of sizing function (middle) and final addapted mesh (right).

and the desired size ranged from 0.1 to 3. The sizing function used for this adaptation specified a high element density at the tip of the nosecone where the curvature is high and low element density away from the tip at the base. This example also illustrates how a sizing function might be used to adapt a mesh based on geometric characteristics of the model. In addition to the refinement near the tip of the object, note the difference in element size along the curve at the base of the nosecone.

Table 2 shows the distribution of size ratios through the mesh. In this example, fewer than 3% of the faces are too large and fewer than 7% of the faces are less than half of the desired size. This example provided very good results for not over-refining the mesh.

Table 2. Adaptation results of nosecone example.

Size Ratio f_s	Number of Faces	Percent of Total	Size Ratio f_s	Number of Faces	Percent of Total
<0.5	132	6.5%	1.0 - 1.1	50	2.4%
0.5 - 0.8	1494	73.0%	1.1 - 1.2	1	0.0%
0.8 - 0.9	233	11.4%	1.2 - 1.5	0	0.0%
0.9 - 1.0	136	6.6%	>1.5	1	0.0%
	1995	97.5%		51	2.5%

4.2 Plate with Hole in Tension

This example models a plate with a hole under a tensile loading, as shown in Figure 11. Due to symmetry of both the geometry and loads, this problem can be reduced to an analysis of a quarter of the plate, denoted by the shaded region. The three locations, A, B, and C, have been marked on the diagram

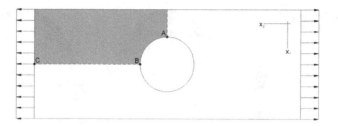

Fig. 11. Model of plate with hole in tension.

Table 3. Adaptation results of plate with hole in tension.

Size Ratio f_s	Number of Faces	Percent of Total	Size Ratio f_s	Number of Faces	Percent of Total
<0.5	93	7.9%	1.0 - 1.1	7	0.6%
0.5 - 0.8	924	79.0%	1.1 - 1.2	0	0.0%
0.8 - 0.9	120	10.3%	1.2 - 1.5	0	0.0%
0.9 - 1.0	26	2.2%	>1.5	0	0.0%
	1163	99.4%		7	0.6%

where displacement results (see table 4) have been recorded after an analysis using the finite element program, ADINA [23].

The initial mesh of this example is shown in Figure 12 with an average element size of about 0.6. The sizing function used to adapt this mesh was based off of the stress error estimates determined from the analysis of the original mesh also shown in Figure 12. In this figure, the darker colors represent more error. The sizing function determined from this analysis specified an element size of 0.04 in the areas of highest error to an element size of 1.5 in areas with small error. The mesh resulting from the adaptation procedure is shown in Figure 13 and size ratio results of the adaptation are provided in Table 3.

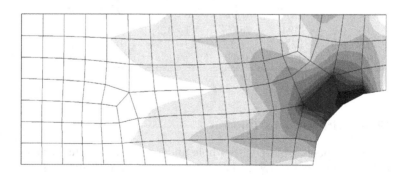

Fig. 12. Band plot of stress error used to define sizing function for plate with hole in tension.

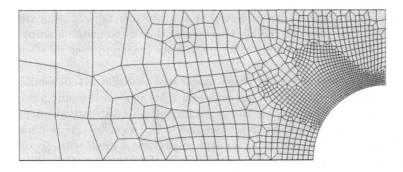

Fig. 13. Adapted mesh of plate with hole in tension.

Table 4. Results of finite element analysis.

	Mesh Info		Effective Stress		Displacements		
	Nodes	Faces	Max Value	Max Error	Δx_1 at A	Δx_2 at B	Δx_2 at C
Coarse	121	97	94.3	32.08	0.00143	0.00279	0.00467
Medium	1770	1677	106.5	19.23	0.00154	0.00287	0.00474
Fine	6973	6788	107.9	12.33	0.00155	0.00288	0.00475
Adapted	1232	1170	108.3	7.53	0.00154	0.00287	0.00474

In addition to the initial coarse mesh and the adapted mesh, the quarter-plate was also meshed with two other meshes, each much more fine than the coarse base mesh. These meshes were used to help show convergence to a solution as well as to compare error between analyses of each of the meshes. Results shown in table 4 indicate that stress error is much more uniform across the domain in the adapted mesh than in any of the other analyzed meshes.

The maximum estimated stress error is significantly reduced by the adapted mesh. These results are significant especially when considering that the adapted mesh has only 70% of the nodes in the medium mesh and fewer than 20% of the nodes in the fine mesh. Not only does the adapted mesh provide virtually equal displacement values and superior stress values, it does so with fewer elements while reducing the estimated error. Although this is a very small problem and the time savings were negligible, the savings of computational effort on a large problem can be significant.

The results of this analysis not only show the effectiveness of this adaptation algorithm in providing an efficient solution to a computational mechanics problem, but also the importance of mesh adaptation generally in finite element problems.

5   Conclusion

The ability to adapt a finite element mesh is critical to providing an efficient analysis to many finite element problems. Although there are currently effective quadrilateral adaptation techniques available, none of them are truly general in that they can modify the element density to match a sizing function by adding and removing elements without having to re-mesh all or part of the domain.

Recent developments in localized, automated quadrilateral coarsening have made it possible to combine quadrilateral refinement, coarsening, and quality improvement techniques into a conformal, all-quadrilateral adaptation method. Given a sizing function, this new adaptation technique iteratively coarsens and refines the mesh domain to provide a mesh with an element density reasonably close to that specified by the sizing function. As shown in examples, this method is an effective way to streamline the computational analysis of a finite element mesh by providing high element density in areas of the mesh that require high accuracy or geometric resolution while removing elements in less important areas of the mesh to decrease element density and save computation time.

5.1   Further Research

The adaptation technique described in this work effectively adds and removes elements resulting in an adapted mesh that improves accuracy or resolution where needed while improving the efficiency of the analysis by removing elements away from the area of interest. Although the results shown in this work are promising, there are still improvements that can be made and more research that can be done.

One way to improve this algorithm is to provide adaptive smoothing with refinement and coarsening. The smoothing technique currently employed in this algorithm attempts to improve the quality of the mesh by re-distributing the nodes, but does not take into account the desired element size specified by the given sizing function. In fact, the smoothing algorithm may work against the size function by attempting to create a uniformly sized mesh while the sizing function has specified a mesh with varying element density. Coupling this $h-$adaptation method with an $r-$adaptation technique that considers the element size specified by the sizing function [24] would be a major improvement to this algorithm.

In some finite element applications, particularly computational fluid dynamics problems, anisotropic elements with a high aspect ratio are desired at mesh boundaries. This adaptation technique does not account for isotropy and adds or removes elements based solely on their size. By more selectively choosing where to add elements, or even applying chord dicing capabilities, this adaptation method could be modified to allow for anisotropy.

One purpose of this research was to provide a springboard into the development of an automated all-hexahedral mesh adaptation algorithm. Recent developments have been made in conformal hexahedral refinement [1] [25] that provide localized refinement and are robust on both structured and unstructured hexahedral meshes. Additional developments have been made in automated hexahedral coarsening as well. Woodbury [26] recently introduced a new method that provides localized conformal coarsening to an all-hexahedral mesh. Woodbury's method isolates the coarsening region through pillowing and then uses chord-collapse operations to redirect hexahedral sheets so they are located entirely within the desired coarsening region. Additionally, this method does account for boundary and surface coarsening and therefore does not have the same limitations as the AQCRC algorithm used for quadrilateral adaptation in this work. One potential difficulty in the development of an automated hexahedral adaptation scheme, however, is providing quality improvement operations to ensure a high quality mesh. The improvement operators used in quadrilateral mesh improvement do not extend directly into 3-dimensions and topological restrictions in hexahedra make local topology changes very difficult.

References

1. Parrish, M., Borden, M., Staten, M.L., Benzley, S.E.: A Selective Approach to Conformal Refinement of Unstructured Hexahedral Finite Element Meshes. In: Proceedings of 16th International Meshing Roundtable, pp. 251–268 (2007)
2. Dewey, M.W.: Automated Quadrilateral Coarsening by Ring Collapse, Master's Thesis, Brigham Young University, Provo (2008)
3. Anderson, B.D., Shepherd, J.F., Daniels, J., Benzley, S.E.: Quadrilateral Mesh Improvement, Research Note. In: 17th International Meshing Roundtable (2008)
4. Blacker, T.D., Stephenson, M.B.: Paving: A New Approach to Automated Quadrilateral Mesh Generation. International Journal for Numerical Methods in Engineering 32(4), 811–847 (1991)
5. Quadros, W.R., Vyas, V., Brewer, M., Owen, S.J., Shimada, K.: A Computational Framework for Generating Sizing Function in Assembly Meshing. In: Proceedings of 14th International Meshing Roundtable, pp. 55–72 (2005)
6. Borouchaki, H., Frey, P.J.: Optimization Tools for Adaptive Surface Meshing, AMD Trends in Unstructured Mesh Generation. ASME 220, 81–88 (1997)
7. Borouchaki, H., Frey, P.J.: Adaptive Triangular Quadrilateral Mesh Generation. International Journal for Numerical Methods in Engineering 41, 915–934 (1998)
8. Feng, Y.T., Peric, D.: Coarse Mesh Evolution Strategies in the Galerkin Multigrid Method with Adaptive Remeshing for Geometrically Non-linear Problems. International Journal for Numerical Methods in Engineering 49, 547–571 (2000)
9. Botkin, M.E., Wang, H.P.: An Adaptive Mesh Refinement of Quadrilateral Finite Element Meshes Based upon an a posteriori Error Estimation of Quantities of Interest: Modal Response. Engineering with Computers 20, 38–44 (2004)
10. Baehmann, P.L., Shephard, M.S.: Adaptive Multiple-Level h-Refinement in Automated Finite Element Analyses. Engineering with Computers 5, 235–247 (1989)
11. Branets, L., Carey, G.F.: Smoothing and Adaptive Redistribution for Grids with Irregular Valence and Hanging Nodes. In: Proceedings of 13th International Meshing Roundtable, pp. 333–344 (2004)
12. Zhang, Y., Bajaj, C.: Adaptive and Quality Quadrilateral/Hexahedral Meshing From Volumetric Data. In: Proceedings of 13th International Meshing Roundtable, pp. 365–376 (2004)

13. Jiao, X., Colombi, A., Ni, X., Hart, J.C.: Anisotropic Mesh Adaptation for Evolving Triangulated Surfaces. In: Proceedings of 15th International Meshing Roundtable, pp. 173–190 (2006)
14. Sandhu, J.S., Menandro, F.C.M., Liebowitz, H.: Hierarchical Mesh Adaptation of 2D Quadrilateral Elements. Engineering Fracture Mechanics 50, 727–735 (1995)
15. Albuquerque, N.M.: Cubit 11.1 user documentation, Sandia National Laboratories (2009), http://cubit.sandia.gov/documentation.html
16. Borouchaki, H., Hecht, F., Frey, P.J.: Mesh Gradation Control. International Journal for Numerical Methods in Engineering 13, 1143–1165 (1998)
17. Persson, P.-O.: Mesh Size Functions for Implicit Geometries and PDE-based Gradient Limiting. Engineering with Computers 22(2), 95–109 (2006)
18. Schneiders, R.: Refining Quadrilateral and Hexahedral Element Meshes. Numerical Grid Generation in Computational Field Simulations 1, 679–688 (1996)
19. Staten, M.L., Benzley, S.E., Scott, M.: A Methodology for Quadrilateral Finite Element Mesh Coarsening. Engineering with Computers 24, 241–251 (2008)
20. Kinney, P.: Cleanup: Improving Quadrilateral Finite Element Meshes. In: Proceedings of 6th International Meshing Roundtable, pp. 449–461 (1997)
21. CUBIT 11.1 Geometry and Mesh Generation Toolkit, Sandia National Laboratories (2009), http://cubit.sandia.gov
22. Anderson, B.D.: Automated All Quadrilateral Mesh Adaptation through Refinement and Coarsening, Master's Thesis, Brigham Young University, Provo (2009)
23. ADINA-AUI 8.5.2, ADINA R & D Inc. (2008), http://www.adina.com
24. Hansen, G., Zardecki, A.: Unstructured Surface Mesh Adaptation Using the Laplace-Beltrami Target Metric Approach. Journal of Computational Physics 225, 165–182 (2007)
25. Harris, N.: Conformal Refinement of All-Hexahedral Finite Element Meshes, Master's Thesis, Brigham Young University, Provo (2004)
26. Woodbury, A.: Localized Coarsening of Conforming All-Hexahedral Meshes, Master's Thesis, Brigham Young University, Provo (2008)

Optimal 3D Highly Anisotropic Mesh Adaptation Based on the Continuous Mesh Framework

Adrien Loseille[1,2] and Frédéric Alauzet[2]

[1] CFD Center, Dept. of Computational and Data Sciences, College of Science, MS 6A2, George Mason University, Fairfax, VA 22030-4444, USA
`aloseill@gmu.edu`
[2] INRIA Paris-Rocquencourt, Projet Gamma, Domaine de Voluceau, BP 105, 78153 Le Chesnay cedex, France
`Frederic.alauzet@inria.fr`

Abstract. This paper addresses classical issues that arise when applying anisotropic mesh adaptation to real-life 3D problems as the loss of anisotropy or the necessity to truncate the minimal size when discontinuities are present in the solution. These problematics are due to the complex interaction between the components involved in the adaptive loop: the flow solver, the error estimate and the mesh generator. A solution based on a new continuous mesh framework is proposed to overcome these issues. We show that using this strategy allows an optimal level of anisotropy to be reached and thus enjoy the full benefit of unstructured anisotropic mesh adaptation: optimal distribution of the degrees of freedom, improvement of the ratio accuracy with respect to cpu time, ...

Keywords: Anisotropy; multi-scale mesh adaptation; metric-based mesh adaptation; continuous mesh; convergence order; Euler equations.

Introduction

Nowadays, there is no more need to recall the benefits of metric-based mesh adaptation when dealing with anisotropic physical phenomena. A lot of 3D successful examples on real-life problems have already proved its efficiency [3, 6, 11, 18, 19]. However, one question remains: are the adaptive computations really anisotropic or optimal? Apart from its simplicity, this question raises, as we will see, many other capital issues: assessment of the numerical solution, convergence of the computation at the theoretical order, automatic detection and capturing of all the scales of the solution, ... Consequently, answering positively to these questions is not straightforward as we face both theoretical and practical difficulties. Indeed, claiming that a mesh is optimal requires at least the definition of a proper cost function along with the possibility of differentiating it. A classical cost function is the interpolation error. However, problems occur when attempting to differentiate it with respect to a discrete mesh in order to derive the optimal one. Indeed, the differentiation is not defined on the space of discrete meshes. Despite this weakness, if we now assume that a specification of the optimal can be exhibited, it is of main importance to derive a numerical algorithm to generate it practically. Being able to guarantee the convergence of the algorithm to the optimal continuous solution is a supplementary difficulty.

We first recall the formulation of the mesh adaptation problem for minimizing the interpolation error. The problematics arising when the function becomes a numerical solution are then illustrated on a simple example.

An initial ill-posed problem. In its more general form, the problem of mesh adaptation consists in finding the mesh $\mathcal{H}$ of Ω that minimizes a given error for a given function u. For the sake of simplicity, we consider here the linear interpolation error $u - \Pi_h u$ controlled in $\mathbf{L}^p$ norm. Note that considering other norms works as well [13]. The problem is thus stated in an *a priori* way:

$$\text{Find } \mathcal{H}_{opt} \text{ having } N \text{ nodes such that } E(\mathcal{H}_{opt}) = \min_{\mathcal{H}} \|u - \Pi_h u\|_{\mathbf{L}^p(\Omega)} . \quad (1)$$

As it, Problem (1) is a global combinatorial problem which turns out to be intractable practically. Indeed, both topology and vertices location need to be optimized. Consequently, simpler problems are considered to approximate the solution. A common simplification is to perform a **local analysis** of the error instead of considering the global problem. A first set of methods consists in deriving the optimal element shape [2]. A second set consists in deriving a local bound of the interpolation error. This bound is then transformed into a metric-based estimate [8, 9, 13]. Direct minimization of the error can be also considered by using the interpolation error as a cost function directly in the mesh generator [14]. All these strategies have in common the resolution of a local problem as they act in the vicinity of an element. Consequently, such error minimizations are equivalent to a steepest descent algorithm that converges only to a **local minimum** with poor convergence properties. This drawback arises because of considering directly the **minimization on a discrete mesh**.

Loss of anisotropy when turning to numerical solutions. When the solution u becomes a numerical solution u_h provided by a solver, additional problematics arise in the resolution of Problem (1). The choice of the error estimate used to derive the metric becomes of main importance. To illustrate this point, we consider a standard metric-based error estimate as in [4], *i.e.*, the control of the interpolation error in $\mathbf{L}^\infty$ norm, for the accurate capturing of shock-waves inside a scramjet. This is done by considering a recovered Hessian of one variable of the flow field (mach, pressure, . . .). The final result is shown in Figure 1. If the final adapted mesh seems perfectly anisotropic (left), a closer view around a shock reveals a complete loss of the anisotropy (right). A second problem is that such a strategy does not capture the small-scale features of the flow. Several modifications of this Hessian-based estimate

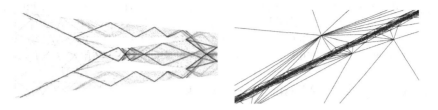

Fig. 1. Scramjet adaptive computation based on a control of the interpolation error in $\mathbf{L}^\infty$ norm

have been considered to overcome this issue. For instance, the following local normalizations:

$$\frac{|H|}{|u|}, \quad \frac{|H|}{|u| + h\,\|\nabla u\|_2}, \quad \frac{|H|}{\gamma|u| + (1-\gamma)h\|\nabla u\|_2}, \tag{2}$$

were introduced in [4, 15, 9], respectively, as an attempt to capture all the scales of the solution. However, the control of the interpolation error remains in $\mathbf{L}^\infty$ norm. If more scales of the solution are captured, the **loss of anisotropy** remains and the request for a **minimal size** prescription is still **necessary**.

Continuous mesh framework and multi-scale mesh adaptation. To overcome the previous issues, a complete duality between discrete entities and continuous ones is introduced using classical concepts of Differential Geometry as Riemannian metric space. In the proposed continuous framework, notions of continuous mesh, continuous element and continuous interpolation operator naturally appear. This discrete-continuous duality is demonstrated from equivalence formula. In this framework, Problem (1) can be recast as a continuous optimization problem. Contrary to discrete-based study, the continuous formulation succeeds in solving **globally** the optimal interpolation error problem by using powerful mathematical tools as the calculus of variations. When dealing with numerical simulations, the use of $\mathbf{L}^p$ norm interpolation error control enables us to **capture all the scales** of the numerical solution. Numerical experiments show that solution scales that have an amplitude 1 000 times lower than the largest one are still captured and refined. From a practical point of view, prescribing a **minimal size is no more required**. This results in the generation of highly anisotropic meshes. Moreover, the analysis for regular functions predicts a second order convergence for the mesh adaptation algorithm. We show that this order is preserved on numerical solutions even when they are issued from flows with shocks with a modern high-order shock capturing solver. Verifying numerically this second order of convergence is a first assessment of computations.

Outline. The main results associated with the continuous mesh framework are reviewed in Section 1. Section 2 deals with the discrete and continuous equivalence for the local interpolation error. Then, the optimal continuous mesh minimizing the global interpolation error in $\mathbf{L}^p$ norm is derived in Section 3. Practical implementation of the continuous mesh framework and theory assessment on complex numerical simulations are done in Section 4.

1   Continuous Mesh Framework

All the notions are introduced in 3D even if all the concepts extend to nD as well. Most of the time the complete proofs are skipped for conciseness purposes. However, they are all available in [16, 17]. In the following, a metric tensor $\mathcal{M}$ is a 3×3 positive symmetric matrix. When the metric field varies over the domain $\Omega \subset \mathbb{R}^3$, a Riemannian metric space $\mathbf{M} = (\mathcal{M}(\mathbf{x}))_{\mathbf{x} \in \Omega}$ of Ω is defined. Rewriting locally metric tensor $\mathcal{M}$ gives a new insight of the possibility of metric-based mesh adaptation. In particular, a duality between discrete and continuous views appear clearly. We exemplify in this section the set of meshes that are represented by Riemannian metric space $\mathbf{M}$. The study is first done locally for an element and then generalized to the whole computational domain Ω. These considerations are based on the concept of unit mesh [10], recalled hereafter.

Local duality. An element K (a tetrahedron in 3D) is **unit** with respect to a constant metric tensor $\mathcal{M}$ if the length of all its edges is unit in metric $\mathcal{M}$. If K is given by its list of edges $(\mathbf{e_i})_{i=1..6}$, then :

$$\forall i = 1, ..., 6, \ \ \ell_\mathcal{M}(\mathbf{e_i}) = 1 \text{ with } \ell_\mathcal{M}(\mathbf{e_i}) = \sqrt{{}^t\mathbf{e_i} \mathcal{M} \, \mathbf{e_i}}.$$

If K is composed only of unit length edges then its volume $|K|_\mathcal{M}$ in $\mathcal{M}$ is constant equal to:

$$|K|_\mathcal{M} = \frac{\sqrt{2}}{12} \text{ and } |K| = \frac{\sqrt{2}}{12}(\det(\mathcal{M}))^{-\frac{1}{2}},$$

where $|K|$ is its Euclidean volume. The function *unit with respect to* defines classes of equivalences of discrete elements.

Proposition 1 (Equivalence classes). *Let $\mathcal{M}$ be a constant metric tensor, there exists a non-empty infinite set of unit elements with respect to $\mathcal{M}$. Conversely, given an element $K = (\mathbf{e_i})_{i=1..6}$ such that $|K| \neq 0$, then there is a unique metric tensor $\mathcal{M}$ for which element K is unit with respect to $\mathcal{M}$.*

The previous proposition induces deeper relationships between $\mathcal{M}$ and the set of unit discrete elements. These relations write as geometric invariants [16].

Proposition 2 (Geometric invariant). *Let $\mathcal{M}$ be a constant metric tensor and K be a unit element with respect to $\mathcal{M}$. We denote by $(\mathbf{e}_i)_i$ its edges list. Then, the following invariant holds for all symmetric matrix H:*

$$\sum_{i=1}^{6} {}^t\mathbf{e}_i \, H \mathbf{e}_i = 2 \operatorname{trace}(\mathcal{M}^{-\frac{1}{2}} H \mathcal{M}^{-\frac{1}{2}}). \tag{3}$$

Global duality. When dealing with a Riemannian metric space $\mathbf{M} = (\mathcal{M}(\mathbf{x}))_{\mathbf{x} \in \Omega}$, the main complexity is to take into account the variations of the function $\mathbf{x} \mapsto \mathcal{M}(\mathbf{x})$. To simplify the analysis, $\mathbf{M}$ is first rewritten in order to distinguish local properties from global ones.

Proposition 3. *A Riemannian metric space* $\mathbf{M} = (\mathcal{M}(\mathbf{x}))_{\mathbf{x} \in \Omega}$ *locally writes:*

$$\forall \mathbf{x} \in \Omega, \quad \mathcal{M}(\mathbf{x}) = d^{\frac{2}{3}}(\mathbf{x}) \, \mathcal{R}(\mathbf{x}) \begin{pmatrix} r_1^{-\frac{2}{3}}(\mathbf{x}) & & \\ & r_2^{-\frac{2}{3}}(\mathbf{x}) & \\ & & r_3^{-\frac{2}{3}}(\mathbf{x}) \end{pmatrix} {}^t\mathcal{R}(\mathbf{x}),$$

where

- *the density d is equal to:* $d = (h_1 h_2 h_3)^{-1} = (\lambda_1 \lambda_2 \lambda_3)^{\frac{1}{2}}$, *with λ_i the eigenvalues of $\mathcal{M}$ and $h_i = \lambda_i^{-\frac{1}{2}}$*
- *the anisotropic quotients r_i are equal to:* $r_i = h_i^3 (h_1 h_2 h_3)^{-1}$
- *$\mathcal{R}$ is the eigenvectors matrix of $\mathcal{M}$.*

The anisotropy is given by the anisotropic quotients, the level of accuracy is given by the density and the orientation by the orthonormal matrix $\mathcal{R}$. Global properties of $\mathbf{M}$ can be deduced by integrating these local quantities on Ω. When integrating d over Ω, the complexity of $\mathbf{M}$ is defined:

$$\mathcal{C}(\mathbf{M}) = \int_\Omega d(\mathbf{x}) \, d\mathbf{x} = \int_\Omega \sqrt{\det(\mathcal{M}(\mathbf{x}))} \, d\mathbf{x}.$$

This quantity can be viewed as the continuous counterpart of the number of vertices of a mesh. In the context of error estimation, this notion enables the study of the order of convergence with respect to a sequence of Riemannian metric spaces having an increasing complexity. Consequently, the complexity $\mathcal{C}(\mathbf{M})$ is also the continuous counterpart of the classical parameter h used for uniform meshes while studying convergence order.

The set of discrete meshes represented by $\mathbf{M}$ is more complex to describe than the class of unit elements. The problem arises from the impossibility to tessellate $\mathbb{R}^3$ uniquely with the regular tetrahedron, see discussions in [17]. Consequently, the notion of unit element does not extend as well to a mesh. In order to ensure existence, the notion of quasi-unit element is devised. This definition takes into account the variations of the continuous mesh:

Definition 1 (Quasi-unit element). *A tetrahedron K defined by its list of edges* $(\mathbf{e_i})_{i=1\dots6}$ *is said quasi-unit for $\mathcal{M}$ if*

$$\forall i \in [1, 6], \quad \ell_\mathcal{M}(\mathbf{e_i}) \in \left[\frac{1}{\sqrt{2}}, \sqrt{2}\right] \quad and \quad Q_\mathcal{M}(K) \in [\alpha, 1] \quad with \quad \alpha > 0, \quad (4)$$

where

$$Q_{\mathcal{M}}(K) = \frac{36}{3^{\frac{1}{3}}} \frac{|K|_{\mathcal{M}}^{\frac{2}{3}}}{\sum_{i=1}^{6} \ell_{\mathcal{M}}^2(e_i)} \in [0,1], \quad with \ |K|_{\mathcal{M}} = \int_K \sqrt{\det(\mathcal{M}(\mathbf{x}))} \, d\mathbf{x},$$

$$and \ \ell_{\mathcal{M}}(e_i) = \int_0^1 \sqrt{{}^t\mathbf{ab} \, \mathcal{M}(\mathbf{a} + t\,\mathbf{ab}) \, \mathbf{ab}} \ dt, \quad with \ \ \mathbf{e}_i = \mathbf{ab}.$$

$$(5)$$

The quality function $Q_{\mathcal{M}}$ ensures that the volume and the shape of the elements are controlled while generating elements with quasi-unit edges lengths. The integral in the computation of $\ell_{\mathcal{M}}$ given by (5) is necessary to take into account the variations of $\mathbf{M}$ along each edge $\mathbf{e}_i$. A discrete mesh is **unit** with respect to $\mathbf{M}$ when it is only composed of quasi-unit elements.

Propositions 1 and 3 highlights a duality between discrete entities and continuous ones. It results that, in the proposed continuous framework, a metric tensor $\mathcal{M}$ is assimilated to a continuous element and a **continuous mesh** of a domain Ω is defined by a collection of continuous elements $\mathbf{M} = (\mathcal{M}(\mathbf{x}))_{\mathbf{x} \in \Omega}$, *i.e.*, a Riemannian metric space. In particular, for an element, this duality is justified by strict analogy between discrete and continuous notions: orientation vs. matrix $\mathcal{R}$, stretching vs. r_i and size vs. d. For a mesh, we point out the duality between the number of vertices and $\mathcal{C}(\mathbf{M})$. Proposition 2 also illustrates a duality between geometric quantities. This duality will be even reinforced in the next section by studying the interpolation error.

In what follows, the continuous terminology is employed to emphasize the exhibited duality.

2   Interpolation Error: Discrete-Continuous Duality

As our intent is to propose a fully discrete-continuous duality, it is not enough to derive only the optimal mesh arising from an interpolation error bound as in classical studies on interpolation error [4, 9, 13]. Instead, we want to evaluate the interpolation error for any functions on any continuous meshes without imposing some optimality conditions as alignment, equi-distribution, ... We show in this section that the interpolation error can be computed analytically for a given function on a given continuous mesh. We start with an estimate for quadratic functions. The general case is deduced from this study.

An error estimate for quadratic functions. In this section, we consider a quadratic function u defined on a domain $\Omega \subset \mathbb{R}^3$. The function is given by its matrix representation:

$$\forall \mathbf{x} \in \Omega, \quad u(\mathbf{x}) = \frac{1}{2} {}^t\mathbf{x} \, H \, \mathbf{x},$$

where H is a symmetric matrix representing the Hessian of u. For every symmetric matrix H, $|H|$ denotes the positive symmetric matrix deduced from

H by taking the absolute values of its eigenvalues. The function u is linearly interpolated on a tetrahedron K. We denote by $\Pi_h u$ the linear interpolate of u on K. We can now state the following result:

Proposition 4. *For every quadratic function u, its linear interpolation error in $\mathbf{L}^1$ norm on a tetrahedron K verifies:*

$$\|u - \Pi_h u\|_{\mathbf{L}^1(K)} \leq \frac{|K|}{40} \sum_{i=1}^{6} {}^t \mathbf{e}_i |H| \mathbf{e}_i,$$

where $(\mathbf{e}_i)_{i=1,6}$ is the set of edges of K.
  The previous inequality becomes an equality when u is elliptic or parabolic.

If K is now assumed to be unit with respect to $\mathcal{M}$, the following theorem holds:

Theorem 1. *For all unit elements K with respect to $\mathcal{M}$, the interpolation error of u in $\mathbf{L}^1$ norm does not depend on the element shape and is only a function of the Hessian H of u and of $\mathcal{M}$. In 3D, for all unit tetrahedra K in $\mathcal{M}$, the following equality holds:*

$$\|u - \Pi_h u\|_{\mathbf{L}^1(K)} = \frac{\sqrt{2}}{240} \det(\mathcal{M}^{-\frac{1}{2}}) \operatorname{trace}(\mathcal{M}^{-\frac{1}{2}} H \mathcal{M}^{-\frac{1}{2}}). \tag{6}$$

It is important to note that Relation (6) links an infinite set of elements (on the left-hand side) to a unique entity: $\mathcal{M}$ (on the right-hand side). Moreover, it shows that whatever the choice of unit-element made by the mesh generator, the resulting interpolation error is always the same as it is only function of metric $\mathcal{M}$. Consequently, this theorem demonstrates the possibility to evaluate the interpolation error for continuous element $\mathcal{M}$ associated with discrete element K. When u is no more quadratic and when the interpolation error is computed on a continuous mesh $\mathbf{M}$, the following continuous discrete local equivalence is proved:

Theorem 2 (Discrete-continuous duality). *Let u be a twice continuously differentiable fonction of a domain Ω and $(\mathcal{M}(\mathbf{x}))_{\mathbf{x} \in \Omega}$ be a continuous mesh of Ω. Then, there exists a unique function $\pi_\mathcal{M}$ such that:*

$$\forall \mathbf{a} \in \Omega, \quad |u - \pi_\mathcal{M} u|(\mathbf{a}) = 2 \frac{\|u_Q - \Pi_h u_Q\|_{\mathbf{L}^1(K)}}{|K|},$$

for every K unit element with respect to $\mathcal{M}(\mathbf{a})$ and where u_Q is the quadratic model of u at $\mathbf{a}$.

This theorem underlines another discrete-continuous duality by pointing out a continuous counterpart of the interpolation error. For this reason, we propose the following formalism: $\pi_\mathcal{M}$ is called continuous linear interpolate and $|u - \pi_\mathcal{M} u|$ represents the continuous dual of the interpolation error. The continuous linear interpolate is defined by:

$$\pi_{\mathcal{M}} u(\mathbf{a}) = u(\mathbf{a}) + \nabla u(\mathbf{a}) + \frac{1}{c_n} \mathrm{trace}(\mathcal{M}^{-\frac{1}{2}}(\mathbf{a})\, H(\mathbf{a})\, \mathcal{M}^{-\frac{1}{2}}(\mathbf{a})),$$

where c_n is $1/8$ in 2D and $1/10$ in 3D [17]. This result allows us to compute interpolation errors analytically. The following examples give a comparison between continuous and discrete interpolation errors evaluation.

Continuous examples. The set of 2D continuous meshes $\mathbf{M}(\alpha) = (\mathcal{M}_\alpha(\mathbf{x}))_{\mathbf{x}\in\Omega}$ is defined on the square domain $\Omega = [0,1] \times [0,1]$ by:

$$\mathcal{M}_\alpha(x,y) = \alpha \begin{pmatrix} h_1^{-2}(x,y) & 0 \\ 0 & h_2^{-2}(x,y) \end{pmatrix},$$

where $h_1(x,y) = 0.1(x+1) + 0.05(x-1)$ and $h_2(x,y) = 0.2$. The parameter α is used to control the level of accuracy of the mesh. The continuous mesh becomes coarser when α decreases but anisotropic quotients and orientations remain constant. This trend is given by the computation of the complexity $\mathcal{C}(\mathbf{M}(\alpha))$:

$$\mathcal{C}(\mathbf{M}(\alpha)) = N(\alpha) = \iint_\Omega \frac{1}{h_1 h_2}(x,y)\, dxdy = \frac{200}{3} \ln(2)\alpha.$$

The continuous interpolation error on $\mathbf{M}(\alpha)$ is computed for two analytical functions: $u_1(x,y) = 6x^2 + 2xy + 4y^2$ and $u_2(x,y) = e^{(2x^2+y)}$. As regards the function u_1, the point-wise continuous interpolation error on $\mathbf{M}(\alpha)$ is

$$(u_1 - \pi_{\mathcal{M}} u_1)(x,y) = \frac{27\,x^2 + 18\,x + 35}{800\,\alpha}.$$

The previous expression is then integrated over Ω:

$$\iint_\Omega |u_1 - \pi_{\mathcal{M}} u_1|(x,y)\, dxdy = \frac{53}{800\,\alpha} = \frac{53}{21} \frac{\ln(2)}{N(\alpha)}.$$

For the function u_2, the point-wise continuous interpolation error on $\mathbf{M}(\alpha)$ is:

$$(u_2 - \pi_{\mathcal{M}} u_2)(x,y) = \frac{e^{4x^2+y}}{8\,\alpha} \left((0.15x + 0.05)^2\, (4 + 16x^2) + 0.05 \right).$$

By a direct integration over Ω, it comes:

$$\iint_\Omega |u_2 - \pi_{\mathcal{M}} u_2|(x,y)\, dxdy \approx \frac{0.2050950191}{\alpha} \approx \frac{13.673\,\ln(2)}{N(\alpha)}.$$

These analytical evaluations of the continuous interpolation error are compared to the evaluation of the discrete interpolation error on a set of a unit meshes with respect to $\mathbf{M}(\alpha)$ for several values of α. These evaluations are plotted in Figure 2 where a perfect correlation is observed. The black-plain lines represent the extremal bound of the interpolation error due to the relaxed notion of quasi-unit elements, cf. Definition 1, while considering uniform

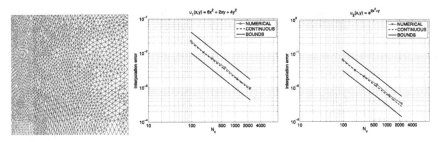

Fig. 2. Left, a unit mesh with respect to $\mathbf{M}(\alpha)$ for $\alpha = 32$. Comparison between continuous interpolation error $\|u - \pi_{\mathcal{M}}u\|_{\mathbf{L}^1(\Omega)}$ and discrete interpolation error $\|u - \Pi_h u\|_{\mathbf{L}^1(\Omega)}$ evaluations for the functions u_1 (middle) and u_2 (right) on $\mathbf{M}(\alpha)$.

meshes in $\mathbf{M}(\alpha)$ with edges length equal to $\sqrt{2}/2$ (lower line) and 2 (upper line).

We now consider the set of 3D continuous meshes $\mathbf{M}(\alpha) = (\mathcal{M}_\alpha(\mathbf{x}))_{\mathbf{x}\in\Omega}$ defined on the domain $\Omega = [0,1] \times [0,1] \times [0,1]$ which are given by:

$$\mathcal{M}_\alpha(x,y,z) = \alpha \begin{pmatrix} h_1^{-2}(x,y,z) & 0 & 0 \\ 0 & h_2^{-2}(x,y,z) & 0 \\ 0 & 0 & h_3^{-2}(x,y,z) \end{pmatrix},$$

where $h_1(x,y,z) = 0.1(x+1) + 0.05(x-1)$,
$h_2(x,y,z) = 0.2$ and $h_3(x,y,z) = 0.2(z+2)$.

We consider the interpolation error of the function $u_3(x,y,z) = e^{2x+y+z}$. The continuous linear interpolation error is (see [17] for details):

$$\iiint_\Omega |u_3 - \pi_{\mathcal{M}} u_3|(x,y,z)\,\mathrm{d}x\mathrm{d}y\mathrm{d}z \approx \frac{0.73}{\alpha} \approx \frac{126.215}{N(\alpha)^{\frac{2}{3}}}.$$

Comparisons between continuous and discrete interpolation errors evaluations for the set of continuous meshes are depicted in Figure 3. As previously, the matching between both evaluations is excellent.

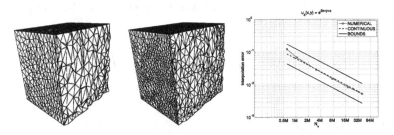

Fig. 3. 3D unit meshes with respect to $\mathbf{M}(\alpha)$ for $\alpha = \{16, 32\}$ (left and middle). Right, comparison between continuous interpolation error $\|u_3 - \pi_{\mathcal{M}} u_3\|_{\mathbf{L}^1(\Omega)}$ and discrete interpolation error $\|u_3 - \Pi_h u_3\|_{\mathbf{L}^1(\Omega)}$.

These examples justifies the continuous terminology as the continuous interpolation computation is equivalent to the discrete one. Then, we use this equivalence to derive a global optimal minimizing the continuous interpolation error.

3   Optimal Control of the Interpolation Error in L^p Norm

Using the definition of the linear continuous interpolate $\pi_{\mathcal{M}}$ of Section 2, the following 3D point-wise interpolation error for u on $\mathbf{M} = (\mathcal{M}(\mathbf{x}))_{\mathbf{x}\in\Omega}$ is deduced:

$$e_{\mathcal{M}}(\mathbf{x}) = (u - \pi_{\mathcal{M}}u)(\mathbf{x}) = \frac{1}{10} \sum_{i=1}^{3} h_i^2(\mathbf{x})|^t\mathbf{v}_i(\mathbf{x})\, H(\mathbf{x})\, \mathbf{v}_i(\mathbf{x})|,$$

where H is the Hessian of u, $(\mathbf{v}_i)_{i=1,3}$ the local eigen-directions of $\mathbf{M}$ and $(h_i)_{i=1,3}$ the local sizes of $\mathbf{M}$ along these directions. It is then possible to set the well-posed global optimization problem of finding the optimal continuous mesh minimizing the continuous interpolation error in $\mathbf{L}^p$ norm:

$$\text{Find } \mathbf{M_{L^p}} = \min_{\mathbf{M}} E_{\mathbf{L}^p}(\mathbf{M}) = \left(\int_{\Omega} e_{\mathcal{M}}^p \right)^{\frac{1}{p}} = \left(\int_{\Omega} (u - \pi_{\mathcal{M}}u)^p \right)^{\frac{1}{p}}, \quad (7)$$

under the constraint

$$\mathcal{C}(\mathbf{M}) = \int_{\Omega} d = N.$$

The constraint on the complexity is added to avoid the trivial solution where all h_i are zero which provides a null error. Contrary to discrete analysis, this problem can be solved globally by using the calculus of variations as it is well-defined on the space of continuous meshes. In [16], it is proved that Problem (7) admits a unique solution. In addition, the following properties hold:

Theorem 3. *Let u be a twice continuously differentiable function defined on $\Omega \subset \mathbb{R}^3$, the optimal continuous mesh $\mathbf{M_{L^p}}(u)$ minimizing Problem (7) reads locally:*

$$\mathcal{M}_{\mathbf{L^p}} = D_{\mathbf{L^p}} \det(|H|)^{\frac{-1}{2p+3}} |H|, \ \text{with } D_{\mathbf{L^p}} = N^{\frac{2}{3}} \left(\int_{\Omega} \det(|H|)^{\frac{p}{2p+3}} \right)^{-\frac{2}{3}}.$$

$$(8)$$

It verifies the following properties:

- $\mathbf{M_{L^p}}(u)$ *is unique*
- $\mathbf{M_{L^p}}(u)$ *is locally aligned with the eigenvectors basis of H and has the same anisotropic quotients as H*

- $\mathbf{M_{L^p}}(u)$ *provides an optimal explicit bound of the interpolation error in* $\mathbf{L}^p$ *norm:*

$$\|u - \pi_{\mathcal{M}_{L^p}} u\|_{\mathbf{L}^p(\Omega)} = 3\, N^{-\frac{2}{3}} \left(\int_\Omega \det\left(|H|\right)^{\frac{p}{2p+3}} \right)^{\frac{2p+3}{3p}}. \tag{9}$$

- *For a sequence of continuous meshes having an increasing complexity with the same orientation and anisotropic quotients* $(\mathbf{M}_{L^p}^N(u))_{N=1...\infty}$, *the asymptotic order of convergence verifies:*

$$\|u - \pi_{\mathcal{M}_{L^p}^N} u\|_{\mathbf{L}^p(\Omega)} \le \frac{Cst}{N^{2/3}}. \tag{10}$$

Relation (10) points out a global second order of mesh convergence.

Note that Bound (9) has been also derived in [5]. However, in our case, all the constants of (9) are explicitly given. In addition, a second order of convergence is predicted. Last but not least, the final difference is that we are able practically to generate a discrete mesh approximating the continuous optimal solution by using any metric-based adaptive mesh generators as soon as the generated meshes verify (5). In addition, this approach is fully compatible with steepest descent methods discussed in the introduction. Indeed, the unit mesh with respect to the global optimal continuous mesh can be used as an initialization, then a discrete steepest descent method can be used to converge toward an optimal discrete mesh.

Examples. We first give an example that illustrates why the $\mathbf{L}^\infty$ norm is not well-suited for flow solutions involving different scales. The considered function is:

$$\forall (x, y) \in [0, 1]^2, \quad f(x, y) = 0.1 \sin(50x) + \text{atan}\left(\frac{0.1}{\sin(5y) - 2x}\right).$$

It is composed of a main shock induced by the atan function with variations of small amplitudes given by the sine, see Figure 4 (left). Two optimal adapted meshes have been generated: one controlling the $\mathbf{L}^1$ norm and the other controlling the $\mathbf{L}^\infty$ norm of the interpolation error. Both meshes are composed of 100 000 vertices. All the scales are refined with the $\mathbf{L}^1$ norm , see Figure 4 (middle), whereas only the main shock is refined with the $\mathbf{L}^\infty$ norm, see Figure 4 (right).

The second example illustrates the convergence of the adaptive scheme for a 1D discontinuous function, the step function f_H, with and without the introduction of an artificial dissipation. Indeed, modern shock capturing schemes that are not compressive generally introduce such dissipation [1]. Figure 5 represents on a uniform mesh the diffused step function f_h on two elements (left) and its linear interpolation Πf_H (middle), *i.e.*, its discrete representation without any dissipation. The right picture shows the evolution of the minimal size prescription at each iteration of the mesh adaptation loop for two different error thresholds ($\varepsilon = 0.1$ and $\varepsilon = 0.12$) for both functions. We observed that the

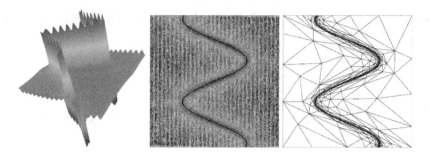

Fig. 4. From left to right, iso-values of function f, optimal meshes controlling the interpolation error in $\mathbf{L}^1$ norm and in $\mathbf{L}^\infty$ norm

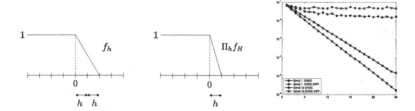

Fig. 5. Linear interpolate Πf_H of discontinuous function f_H and numerical diffused shock f_h. Whatever the level of error Err desired, the minimal size converges when adapting to f_h and diverges when adapting to f_H.

minimal size converge progressively towards zero for the step function without any dissipation f_H whereas it converges towards a fixed value for the diffused one f_h. Consequently, one may expect that the size in the normal direction to a numerical shock will not tend to zero during the refinement process if a dissipation is introduced. This feature of multi-scale mesh adaptation is verified in Section 4 for a modern shock-capturing scheme.

4   3D Numerical Validations

We first review the main modifications that arises when using the previous concept with numerical solutions. This concerns the recovery of derivatives of piecewise linear by element solutions, the adaptive loop and the computation of anisotropic ratios and quotients. Then, several 3D flow simulations involving highly anisotropic meshes are presented. For all the examples, a control of the interpolation error in $\mathbf{L}^2$ norm of the local Mach number is used and **no** minimal size is prescribed.

High-order approximation and hessian recovery. In our case, the flow solver provides a continuous piecewise linear by element representation of the solution u_h. Consequently, our analysis cannot be applied directly to the numerical solution. The idea is to build a higher order solution approximation u^* of the exact solution u from u_h which is twice continuously differentiable and to consider u^* in the optimal metric expression (8). Practically, only the Hessian of u^* is recovered [1]. We also mention that the Hessian recovery procedure from discrete data may results in a noisy recovered Hessian. Consequently, using a proper anisotropic mesh gradation is strongly advised to smooth the field of metric tensors.

A non linear loop. Anisotropic mesh adaptation is a non-linear problem, therefore, an iterative procedure is required to solve this problem. For stationary simulations, an adaptive computation is carried out *via* a mesh adaptation loop inside which an algorithmic convergence of the mesh-solution couple is sought. At each stage, a numerical solution is computed on the current mesh with the flow solver and is analyzed with a metric-based error estimate providing the optimal metric using (8). Next, an adapted mesh, *i.e.*, a unit mesh, is generated with respect to this metric. The mesh generator used is described in [7]. Finally, the solution is linearly interpolated on the new mesh. This procedure is repeated until convergence of the couple mesh-solution.

Measuring the anisotropy. We define some anisotropic measures computation. Anisotropic quotients have been introduced in Section 1 for a continuous element. Deriving this quantity for an element is straightforward. It relies on the fact that there always exists a unique metric tensor for which this element is unit, see Proposition 1. If $\mathcal{M}_K$ denotes the metric tensor associated with element K, solving the following linear system provides $\mathcal{M}_K$:

$$(S) \begin{cases} \ell^2_{\mathcal{M}_K}(\mathbf{e}_1) = 1 \\ \vdots \\ \ell^2_{\mathcal{M}_K}(\mathbf{e}_6) = 1, \end{cases}$$

where $(\mathbf{e}_i)_{i=1,6}$ is the edges list of K and $\ell^2_{\mathcal{M}_K}(\mathbf{e}_i) = {}^t\mathbf{e}_i \, \mathcal{M}_K \, \mathbf{e}_i$. Note that (S) admits a unique solution as soon as the volume of K is not null. Once $\mathcal{M}_K$ is computed, the anisotropic ratio and the anisotropic quotient are simply given by

$$\text{ratio} = \sqrt{\frac{\min_i \lambda_i}{\max_i \lambda_i}} = \frac{\max_i h_i}{\min_i h_i}, \quad \text{and} \quad \text{quo} = \frac{\max_i h_i^3}{h_1 h_2 h_3},$$

where $(\lambda_i)_{i=1,3}$ are the eigenvalues of $\mathcal{M}_K$ and $(h_i)_{i=1,3}$ are the corresponding sizes. The anisotropic ratio stands for the maximum elongation of a tetrahedron by comparing two eigen-directions. The anisotropic quotient represents the overall anisotropic ratio of a tetrahedron taking into account all the possible directions. It corresponds to the overall gain in three dimensions of **an anisotropic adapted mesh** as compared to **an isotropic adapted mesh**. The gain is of course even larger when compared to a uniform mesh. In the sequel, these measures are used to quantify the obtained level of anisotropy.

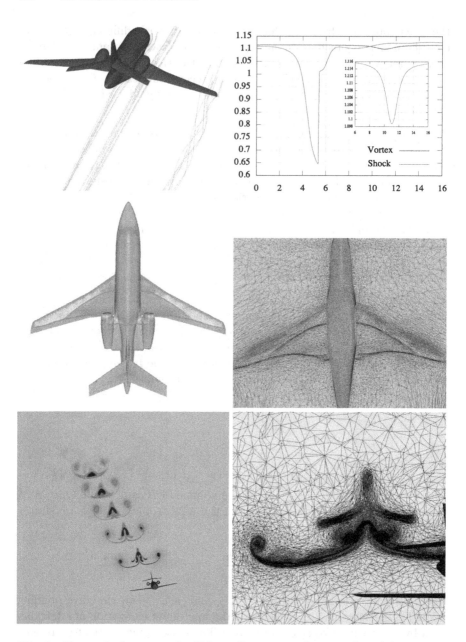

Fig. 6. Transonic flow around a Falcon. From top to bottom, from left to right, Falcon geometry along with speed streamlines, difference of amplitudes between the wings shocks and the tip vortices, Mach iso-values on the aircraft, cut in the volume mesh that shows how the wings shocks are captured, Mach iso-values in planes located at 100, 200, 300, 400 and 500 meters behind the Falcon and a cut in the final volume mesh behind the Falcon showing how vortices are captured in the mesh.

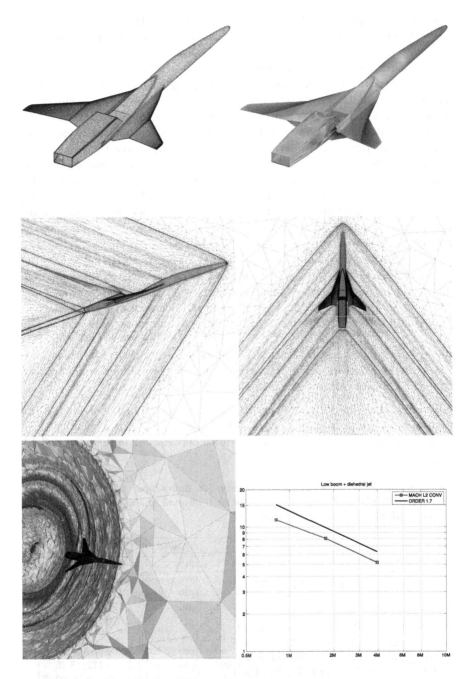

Fig. 7. Supersonic flow around a lowboom jet. From top to bottom, from left to right, aircraft geometry, Mach iso-values on the geometry, final anisotropic mesh in the symmetry plane and below the aircraft, Mach iso-values in a plane 50m behind the aircraft and order of convergence of the Mach number.

Transonic flow around a Falcon. The Falcon jet geometry is depicted in Figure 6 (top left). The aircraft is flying at Mach number 0.8 with an angle of attack of 3 degrees. The computational domain is a cylinder of radius 250m and of length 700m. As the aircraft is flying at a transonic speed, the flow is composed of both shocks and smooth vortices. These phenomena have different magnitudes and mathematical properties. Across a shock, most variables become discontinuous whereas a vortex corresponds to a smooth variation of the variables while having a very small amplitude. These features are exemplified in Figure 6 (top right). An extraction of the pressure across the wing extrados where a shock occurs (green curve) is superposed to the pressure variation in the wake across a vortex located 400m behind the aircraft (red curve). The amplitude of the vortex is less than 2% of the amplitude of the shock. Moreover, the smoothness property of the vortex is a supplementary difficulty as its derivatives involved in our estimate are also smooth. Consequently, vortices are difficult to detect and it is hard not to diffuse them. Detecting and preserving these vortices is still a challenge in the field of CFD. We show that the multi-scale approach detects these two phenomena, *i.e.*, the shocks on the wing, Figure 6 (middle) along with the vortices behind the aircraft, Figures 6 (bottom). The final anisotropic mesh depicted in Figure 6 is composed of 2 025 231 vertices and 11 860 697 tetrahedra providing a mean anisotropic ratio of 177 and a mean anisotropic quotient of 1 639. This example illustrates two main features of multi-scale anisotropic mesh adaptation: the accurate capturing of the shocks on the wings and the drastic reduction of the solver diffusion that allows us to still capture vortices 500m behind the Falcon.

Supersonic flow around a lowboom jet. The aircraft geometry is depicted in Figure 7 (top left). Its length is $L = 42$ meters and it has a wing span of 20 meters. The surface mesh accuracy varies between 0.2 millimeters and 12

Table 1. Supersonic flow around a lowboom jet: anisotropic ratios and quotients histograms for the final adapted mesh. For each interval, the number of tetrahedra is given with the corresponding percentage.

Anisotropic ratio				Anisotropic quotient		
$1 \leq$ ratio ≤ 2	38 740	0.07 %		$1 \leq$ quo ≤ 2	10 042	0.02 %
$2 \leq$ ratio ≤ 3	175 929	0.33 %		$2 \leq$ quo ≤ 3	50 171	0.09 %
$3 \leq$ ratio ≤ 4	274 955	0.51 %		$3 \leq$ quo ≤ 4	81 027	0.15 %
$4 \leq$ ratio ≤ 5	328 501	0.61 %		$4 \leq$ quo ≤ 5	100 385	0.19 %
$5 \leq$ ratio ≤ 10	1 554 625	2.89 %		$5 \leq$ quo ≤ 10	526 474	0.98 %
$10 \leq$ ratio ≤ 50	6 620 533	12.29 %		$10 \leq$ quo ≤ 50	1 989 374	3.69 %
$50 \leq$ ratio $\leq 10^2$	5 983 308	11.10 %		$50 \leq$ quo $\leq 10^2$	1 204 384	2.24 %
$10^2 \leq$ ratio $\leq 10^3$	34 830 344	64.64 %		$10^2 \leq$ quo $\leq 10^3$	7 408 172	13.75 %
$10^3 \leq$ ratio $\leq 10^4$	4 077 796	7.57 %		$10^3 \leq$ quo $\leq 10^4$	14 595 766	27.09 %
$10^4 \leq$ ratio $\leq 10^5$	131	0.00 %		$10^4 \leq$ quo $\leq 10^5$	20 999 034	38.97 %
$10^5 \leq$ ratio $\leq$	1	0.00 %		$10^5 \leq$ quo $\leq 10^6$	6 790 336	12.60 %
				$10^6 \leq$ quo	129 698	0.24 %

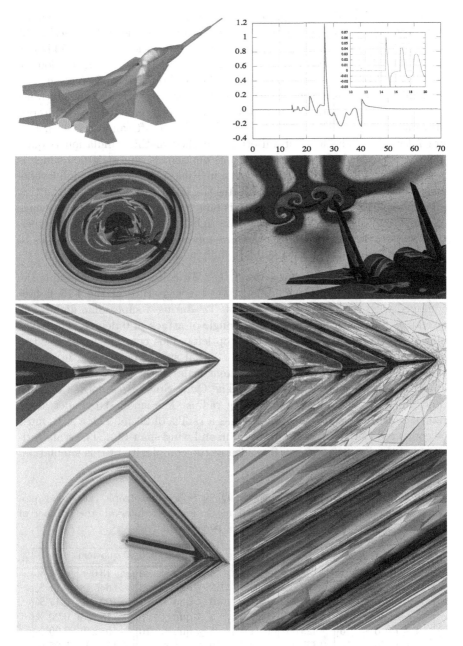

Fig. 8. F15 fighter equipped with the Quiet Spike. From top to bottom, from left to right, F15 geometry and Mach iso-values, pressure distribution 2m below the aircraft, Mach iso-values behind the aircraft, vortices details behind the F15, Mach iso-values near the Spike, anisotropic mesh near the spike, Mach iso-values in a 100m wide box and accuracy of the mesh 100m below the aircraft.

centimeters. This already represents a size variation of five orders of magnitude with respect to the aircraft size, and this only for the surface mesh. The computational domain is a cylinder of 2.25 kilometers length and 1.5 kilometers diameter. This represents a scale factor of 10^7 if the size of the domain is compared to the maximal accuracy of the low boom jet surface mesh. The final anisotropic mesh is composed of 9 083 53 vertices and 53 884 863 tetrahedra. This accuracy allows us to capture all the shocks emitted by the aircraft up to a distance of 16 times the length of the aircraft (almost 750m), Figure 7 (middle). The level of anisotropy reached in this simulation is quite impressive. Indeed, the mean anisotropic quotient is almost 50 000 which is very high. Detailed histograms are given in Table 1. Besides this high level of anisotropy, this simulation demonstrates the good correlation with the theory (10) as a 1.7 order of convergence is numerically verified on the sequence of adaptive meshes, see Figure 7 (bottom right). As in the previous examples, the multi-scale strategy reduces the flow solver dissipation that allows us to capture all solution scales while maintaining a high level of anisotropy.

F15 fighter equipped with the Quiet Spike. We consider in this example the accurate prediction of the mid-field pressure signature of the F15 fighter equipped with the Quiet Spike concept [12] during a supersonic flight. The aircraft is flying at Mach 1.8 with an angle of attack of 0 degree. This complex geometry is shown in Figure 8 (top left). This concept was devised to soften the sonic boom by splitting the initial strong bow shock in several shocks of smaller amplitudes. The different scales of the pressure distribution are depicted in Figure 8 (top right). The Quiet Spike is composed of three cones linked by cylinders of increasing radius. The smallest cylinder has a radius of 5cm while the greatest one has a radius of 20cm. These sizes must be compared to the aircraft length 19.3m and wing-span 13m. The scale variations of the geometry give a first idea of the complexity of this simulation.

Table 2. F15 fighter equipped with the Quiet Spike: anisotropic ratios and quotients histograms for the final adapted mesh. For each interval, the number of tetrahedra is given with the corresponding percentage.

Anisotropic ratio			Anisotropic quotient		
$1 \leq$ ratio $\leq$ 2	515 23	0.09 %	$1 \leq$ quo $\leq$ 2	13 096	0.02 %
$2 \leq$ ratio $\leq$ 3	245 783	0.41 %	$2 \leq$ quo $\leq$ 3	645 48	0.11 %
$3 \leq$ ratio $\leq$ 4	373 769	0.62 %	$3 \leq$ quo $\leq$ 4	102 693	0.17 %
$4 \leq$ ratio $\leq$ 5	429 601	0.71 %	$4 \leq$ quo $\leq$ 5	118 128	0.20 %
$5 \leq$ ratio $\leq$ 10	2 365 083	3.92 %	$5 \leq$ quo $\leq$ 10	598 219	0.99 %
$10 \leq$ ratio $\leq$ 50	17 823 972	29.57 %	$10 \leq$ quo $\leq$ 50	3 176 207	5.27 %
$50 \leq$ ratio $\leq 10^2$	15 172 581	25.17 %	$50 \leq$ quo $\leq 10^2$	2 676 990	4.44 %
$10^2 \leq$ ratio $\leq 10^3$	23 780 955	39.45 %	$10^2 \leq$ quo $\leq 10^3$	17 153 343	28.46 %
$10^3 \leq$ ratio $\leq$	37 332	0.06 %	$10^3 \leq$ quo $\leq 10^4$	26 329 992	43.68 %
			$10^4 \leq$ quo $\leq 10^5$	9 808 547	16.27 %
			$10^6 \leq$ quo	238 843	0.40 %

In the literature, this simulation is currently envisaged in a 2-stage process by coupling a structured solver with an unstructured one [12] which provides an accurate pressure field far below the aircraft with a limit at 70m. Here, the final adapted meshes is composed of 10 050 445 vertices and 60 280 606 tetrahedra featuring a mean anisotropic ratio of 110 and a mean anisotropic quotient of 6 400. This mesh comes up with an accurate signature 120m below the aircraft while using only unstructured meshes. Detailed histograms for anisotropic quotients and ratios are reported in Table 2.

Conclusion

A multi-scale mesh adaptation strategy has been introduced in this paper. It involves theoretical developments demonstrating that a field of metric tensors completely models discrete meshes and that the notion of interpolation error is well-defined in this continuous framework. Contrary to discrete classical approaches, the interpolation error can be computed analytically without any *a priori* hypothesis on the mesh. The optimal mesh minimizing the $\mathbf{L}^p$ norm of the interpolation error is then derived as a global optimum by a calculus of variations. The algorithm to derive a discrete optimal mesh is based on the definition of unit-mesh. Consequently, this method can be used with any metric-based mesh generators.

From a practical point a view, this approach automatically obtains adapted meshes with high level of anisotropy for realistic simulations. Optimal local Hessian normalization is set automatically and depends only on the choice of the norm. Prescribing a minimal size is no more necessary. In addition, numerical results show that all the scales of the solution are captured and refined when using an $\mathbf{L}^p$ norm error control: shocks, shear layers, ... There is no need to fix some parameters as in previous Hessian normalizations. During a simulation, verifying that a second order of convergence is reached as predicted by the theory gives a first assessment of the computations. Finally, the convergence to the most accurate solution is done in a natural way by increasing the complexity N which is, along with the $\mathbf{L}^p$ norm, the only parameter to set prior to a simulation.

References

1. Alauzet, F., Loseille, A.: High order sonic boom modeling by adaptive methods. RR-6845, INRIA (2009)
2. Babuska, I., Aziz, A.K.: On the angle condition in the finite element method. SIAM J. Numer. Anal. 13, 214–226 (1976)
3. Bottasso, C.L.: Anisotropic mesh adaption by metric-driven optimization. Int. J. Numer. Meth. Engng. 60, 597–639 (2004)
4. Castro-Díaz, M.J., Hecht, F., Mohammadi, B., Pironneau, O.: Anisotropic unstructured mesh adaptation for flow simulations. Int. J. Numer. Meth. Fluids 25, 475–491 (1997)

5. Chen, L., Sun, P., Xu, J.: Optimal anisotropic simplicial meshes for minimizing interpolation errors in L^p-norm. Math. Comp. 76(257), 179–204 (2007)
6. Coupez, T.: Génération de maillages et adaptation de maillage par optimisation locale. Revue Européenne des Éléments Finis 9, 403–423 (2000)
7. Dobrzynski, C., Frey, P.J.: Anisotropic delaunay mesh adaptation for unsteady simulations. In: Proc. of 17th Int. Meshing Rountable, pp. 177–194. Springer, Heidelberg (2008)
8. Formaggia, L., Perotto, S.: New anisotropic a prioiri error estimate. Numer. Math. 89, 641–667 (2001)
9. Frey, P.J., Alauzet, F.: Anisotropic mesh adaptation for CFD computations. Comput. Meth. Appl. Mech. Engrg. 194(48-49), 5068–5082 (2005)
10. Frey, P.J., George, P.-L.: Mesh generation. Application to finite elements, 2nd edn. ISTE Ltd and John Wiley & Sons (2008)
11. George, P.-L.: Gamhic3d, a fully automatic adaptive mesh generation method in three dimensions. Technical Note, INRIA (2001)
12. Howe, D.C., Waithe, K.A., Haering, E.A.: Quiet spike near field flight test pressure measurement with computational fluid dynamics comparisons. AIAA Paper, 2008-128 (2008)
13. Huang, W.: Metric tensors for anisotropic mesh generation. J. Comp. Phys. 204, 633–665 (2005)
14. Lagüe, J.-F., Hecht, F.: Optimal mesh for P_1 interpolation in H^1 semi-norm. In: Proc. of 15th Meshing Rountable, pp. 259–270. Springer, Heidelberg (2006)
15. Löhner, R.: Three-dimensional fluid-structure interaction using a finite element solver and adaptive remeshing. Computing Systems in Engineering 1(2-4), 257–272 (1990)
16. Loseille, A.: Adaptation de maillage 3D anisotrope multi-échelles et ciblé à une fonctionnelle. Application à la prédiction haute-fidélité du bang sonique. PhD thesis, Université Pierre et Marie Curie, Paris VI, Paris, France (2008)
17. Loseille, A., Alauzet, F.: Continuous mesh model and well-posed continuous interpolation error estimation. RR-6846, INRIA (2009)
18. Pain, C.C., Umpleby, A.P., de Oliveira, C.R.E., Goddard, A.J.H.: Tetrahedral mesh optimisation and adaptivity for steady-state and transient finite element calculations. Comput. Meth. Appl. Mech. Engrg. 190, 3771–3796 (2001)
19. Tam, A., Ait-Ali-Yahia, D., Robichaud, M.P., Moore, M., Kozel, V., Habashi, W.G.: Anisotropic mesh adaptation for 3D flows on structured and unstructured grids. Comput. Meth. Appl. Mech. Engrg. 189, 1205–1230 (2000)

Anisotropic Mesh Adaptation for Solution of Finite Element Problems Using Hierarchical Edge-Based Error Estimates

Abdellatif Agouzal[1], Konstantin Lipnikov[2], and Yuri Vassilevski[3]

[1] Universite de Lyon 1, Laboratoire d'Analyse Numerique
agouzal@univ-lyon1.fr
[2] Los Alamos National Laboratory, Theoretical Division
lipnikov@lanl.gov
[3] Institute of Numerical Mathematics
yuri.vassilevski@gmail.com

Abstract. We present a new technology for generating meshes minimizing the interpolation and discretization errors or their gradients. The key element of this methodology is construction of a space metric from edge-based error estimates. For a mesh with N_h triangles, the error is proportional to N_h^{-1} and the gradient of error is proportional to $N_h^{-1/2}$ which are the optimal asymptotics. The methodology is verified with numerical experiments.

1 Introduction

Unstructured simplicial meshes are ideally suited for adaptive finite element calculations. The simplexes can be aligned with solution features and cover the computational domain in an optimal way to equidistribute the error. This results in a smaller computational mesh and potentially faster calculations.

Generation of optimal adaptive meshes requires error estimates or error indicators that carry directional information about the solution. In this article, we use error estimates that are associated with mesh edges. We consider edge-based error estimates for the interpolation error and hierarchical error estimates for the discretization error [12]. In both cases, we employ the methodology developed in [1, 3, 4] for the interpolation error. This methodology results in a metric that captures correctly isotropic and anisotropic solution features. Here, we continue analysis of this metric, in particular, its smoothness and anisotropic alignment.

Other methods for generating a space metric are often based on the Hessian of the discrete solution [18, 15, 14, 13]. For such a metric, optimal error estimates for the interpolation error have been proved in [2, 8, 14, 17, 18]. The Hessian-based metric has been successfully applied to adaptive solution of PDEs [7, 13, 15]. However, its theoretical analysis requires to make an additional assumption that the discrete Hessian approximates the continuous one in the maximum norm. Despite the fact that this assumption is frequently violated in many Hessian recovery methods, the generated adaptive meshes still result in optimal error reduction.

The technology, which we proposed in [1, 3, 4] does not require the aforementioned approximation assumption. It can be also applied to adaptive solution of various finite element problems including problems with discontinuous finite element solutions. The price to pay is that error estimates or error indicators have to be prescribed to mesh edges. The finite element literature provides a number of ways to obtain these error estimates. For instance, in CFD literature, edge-based error estimates have appeared occasionally since mid of 1990's but have not received adequate theoretical treatment [5, 6]. In this article, we use the hierarchical error estimates from [12] and provide a numerical analysis of our methodology for the adaptive solution of finite element problems. The cornerstone of this methodology is construction of a space metric from edge-based error estimates.

We define a tensor metric $\mathfrak{M}$ such that the volume and the perimeter of a simplex measured in this metric control the norm of error or its gradient. The equidistribution principle, which can be traced back to D'Azevedo [11], suggests to balance $\mathfrak{M}$-volumes and $\mathfrak{M}$-perimeters. This leads to meshes that are quasi-uniform in the piecewise constant metric $\mathfrak{M}$. The piecewise constant metric may produce instabilities in an adaptive process, especially when the length of a mesh edge, measured in all metrics associated with simplexes sharing the edge, varies strongly. We show numerically that this variation is relatively small for our piecewise constant metrics. This allows us to convert the piecewise constant metric into a continuous one for additional robustness of the adaptive process.

A piecewise constant tensor metric that controls the error is not unique. Any such metric results in asymptotically optimal reduction of the error [1]. In this paper, we show how to build a metric that preserves solution anisotropy. The resulting quasi-optimal mesh equidistributes both the error over simplexes and the maximum norm of the error over edges of each simplex.

The paper outline is as follows. In Section 2, we derive metrics optimal for the interpolation errors. In Section 3, we present the algorithm for generating adaptive meshes. In Section 4, we apply the methodology for adaptive solution of finite element problems.

2   Interpolation Error Analysis

2.1   Edge-Based Error Estimates and a Tensor Metric

Let $\Omega \subset \Re^d$ be a bounded polyhedral domain and Ω_h be a conformal simplicial mesh with N_h simplexes. Let $\mathfrak{M}$ be a piecewise constant tensor metric on Ω_h. The volume of simplex Δ and the length of edge $\mathbf{e}$ in this metric are denoted by $|\Delta|_{\mathfrak{M}}$ and $|\mathbf{e}|_{\mathfrak{M}}$, respectively [2]. The total length of all edges of simplex Δ in denoted by $|\partial\Delta|_{\mathfrak{M}}$. We shall refer to $|\partial\Delta|_{\mathfrak{M}}$ as the perimeter of Δ in the metric $\mathfrak{M}$.

Let $\mathcal{I}_1 u$ be the piecewise linear Lagrange interpolant of u, and $\mathcal{I}_{1,\Delta} u$ be its restriction to Δ. Similarly, let $\mathcal{I}_2 u$ be the piecewise quadratic Lagrange interpolant of u, and $\mathcal{I}_{2,\Delta} u$ be its restriction to Δ. Our goal is to generate a mesh that minimizes (approximately) the L^p-norm, $p \in (0, \infty]$, of the interpolation error

$$e = u - \mathcal{I}_1 u$$

or its gradient ∇e. A sequence of meshes with increasing number of simplexes must provide the optimal reduction of this error. For instance, the Tichomirov result [16] implies that the optimal reduction of the L^∞-norm of error is proportional to N_h^{-1}.

Let us consider a particular d-simplex Δ with vertices $\mathbf{v}_i$, $i = 1, \ldots, d+1$, edge vectors $\mathbf{e}_k = \mathbf{v}_i - \mathbf{v}_j$, $1 \le i < j \le d+1$, and mid-edge points $\mathbf{c}_k$, $k = 1, \ldots, n_d$, where $n_d = d(d+1)/2$. Let λ_i, $i = 1, \ldots, d+1$, be the linear functions on Δ such that $\lambda_i(\mathbf{v}_j) = \delta_{ij}$ where δ_{ij} is the Kronecker symbol. For every edge $\mathbf{e}_k$, we define the quadratic bubble function $b_k = \lambda_i \lambda_j$.

Let u be a continuous function and $u_2 = \mathcal{I}_{2,\Delta} u$ be its quadratic approximation on Δ. We have

$$e_2 = u_2 - \mathcal{I}_{1,\Delta} u_2 = 4 \sum_{k=1}^{n_d} (u_2(\mathbf{c}_k) - \mathcal{I}_{1,\Delta} u_2(\mathbf{c}_k)) \, b_k \equiv \sum_{k=1}^{n_d} \gamma_k \, b_k. \qquad (1)$$

The L^2-norm of the error e_2 is given by

$$\|e_2\|_{L^2(\Delta)}^2 = |\Delta| \, (\mathbb{B}\boldsymbol{\gamma}, \boldsymbol{\gamma}), \qquad (2)$$

where $\boldsymbol{\gamma}$ is a vector with n_d components γ_k and $\mathbb{B}$ is the $n_d \times n_d$ symmetric positive definite scaled Gramm matrix with positive entries $\mathbb{B}_{k,l} = |\Delta|^{-1} \int_\Delta b_k b_l \, dV$. Note that (2) is only a number; therefore, it does not provide any directional information. To recover this information, we split the error into n_d pieces associated with edges of Δ:

$$\|e_2\|_{L^2(\Delta)} = |\Delta|^{1/2} \sum_{k=1}^{n_d} \alpha_k \qquad \text{and} \qquad \sum_{k=1}^{n_d} \alpha_k = (\mathbb{B}\boldsymbol{\gamma}, \boldsymbol{\gamma})^{1/2}. \qquad (3)$$

Not careful selection of α_k may result in loss of directional information. In the sequel, we motivate the following choice of α_k:

$$\alpha_k = |\gamma_k| \, (\mathbb{B}\boldsymbol{\gamma}, \boldsymbol{\gamma})^{1/2} \left(\sum_{k=1}^{n_d} |\gamma_k| \right)^{-1}. \qquad (4)$$

We repeat the above derivations for the gradient of the error. The L^2-norm of ∇e_2 is given by

$$\|\nabla e_2\|_{L^2(\Delta)}^2 = \| \sum_{k=1}^{n_d} \gamma_k \nabla b_k \|_{L^2(\Delta)}^2 = |\Delta| (\widetilde{\mathbb{B}} \boldsymbol{\gamma}, \boldsymbol{\gamma}),$$

where $\widetilde{\mathbb{B}}$ is the $n_d \times n_d$ symmetric positive definite matrix with entries $\widetilde{\mathbb{B}}_{k,l} = |\Delta|^{-1} \int_\Delta \nabla b_k \cdot \nabla b_l \, dV$. Again, we split this error (a number) into n_d edge-based error estimates $\tilde{\alpha}_k \geq 0$ such that

$$\|\nabla e_2\|^2_{L^2(\Delta)} = |\Delta| \sum_{k=1}^{n_d} \tilde{\alpha}_k \quad \text{and} \quad \sum_{k=1}^{n_d} \tilde{\alpha}_k = (\widetilde{\mathbb{B}}\,\boldsymbol{\gamma}, \boldsymbol{\gamma}). \tag{5}$$

Again, a proper choice of α_k is required to preserve directional information in a final metric. In the sequel, we motivate the following choice of $\tilde{\alpha}_k$:

$$\tilde{\alpha}_k = |\gamma_k| \, (\widetilde{\mathbb{B}}\,\boldsymbol{\gamma},\,\boldsymbol{\gamma}) \left(\sum_{k=1}^{n_d} |\gamma_k| \right)^{-1}. \tag{6}$$

In both cases, the edges-based errors α_k and $\tilde{\alpha}_k$ are proportional to $|\gamma_k|$ which is the same for all simplexes sharing the edge $\mathbf{e}_k$. This observation is a key to understanding 'smoothness' of the metric whose element-by-element construction is based on the following result [1, 4].

Lemma 1. *Let* α_k, $k = 1, \ldots, n_d$, *be values prescribed to edges of a d-simplex Δ such that*

$$\alpha_k \geq 0 \quad \text{and} \quad \sum_{k=1}^{n_d} \alpha_k > 0.$$

Then, there exists a constant tensor metric $\mathfrak{M}_\Delta$ such that

$$\left(\frac{d!}{(d+1)(d+2)} \right)^{1/d} |\Delta|^{2/d}_{\mathfrak{M}_\Delta} \leq \sum_{k=1}^{n_d} \alpha_k \leq |\partial\Delta|^2_{\mathfrak{M}_\Delta}. \tag{7}$$

The proof [1, 4] of Lemma 1 provides the constructive way to define the metric $\mathfrak{M}_\Delta$. Due to its importance, we present a shortened proof here.

<u>Proof.</u> Let us define the quadratic function

$$v_2 = -\frac{1}{2} \sum_{k=1}^{n_d} \alpha_k b_k.$$

The trace of v_2 on $\mathbf{e}_k$ is a quadratic function w_2 vanishing at endpoints $\mathbf{v}_i$, $\mathbf{v}_j$ of $\mathbf{e}_k$ with an extremum at $\mathbf{c}_k$. Therefore, $w_2'(\mathbf{c}_k) = 0$ and $\nabla v_2(\mathbf{c}_k) \cdot \mathbf{e}_k = 0$. Let $\mathbb{H}$ be the Hessian of v_2. Applying the multi-point Taylor formula [9, 10] for v_2 at endpoints $\mathbf{v}_i$ and $\mathbf{v}_j$ of $\mathbf{e}_k$, we get

$$0 = v_2(\mathbf{v}_i) = v_2(\mathbf{c}_k) - \frac{1}{2}\nabla v_2(\mathbf{c}_k) \cdot \mathbf{e}_k + \frac{1}{8}(\mathbb{H}\,\mathbf{e}_k, \mathbf{e}_k), \tag{8}$$

$$0 = v_2(\mathbf{v}_j) = v_2(\mathbf{c}_k) + \frac{1}{2}\nabla v_2(\mathbf{c}_k) \cdot \mathbf{e}_k + \frac{1}{8}(\mathbb{H}\,\mathbf{e}_k, \mathbf{e}_k).$$

Thus,
$$(\mathbb{H}\,\mathbf{e}_k, \mathbf{e}_k) = \alpha_k.$$

The Hessian $\mathbb{H}$ may be indefinite and hence cannot be used to define the metric $\mathfrak{M}_\Delta$. In order to make it positive semidefinite, we take its spectral module:
$$|\mathbb{H}| = W^T |\Lambda|\, W,$$

where $\mathbb{H} = W^T \Lambda W$ is the spectral decomposition of the symmetric matrix $\mathbb{H}$.

If $\det \mathbb{H} \neq 0$, we set $\mathfrak{M}_\Delta = |\mathbb{H}|$. The upper bound follows from

$$|\partial\Delta|^2_{|\mathbb{H}|} = \left(\sum_{k=1}^{n_d} (|\mathbb{H}|\mathbf{e}_k, \mathbf{e}_k)^{1/2}\right)^2 \geq \sum_{k=1}^{n_d} (|\mathbb{H}|\mathbf{e}_k, \mathbf{e}_k) \geq \sum_{k=1}^{n_d} |(\mathbb{H}\,\mathbf{e}_k, \mathbf{e}_k)| = \sum_{k=1}^{n_d} \alpha_k.$$

To estimate the lower bound, we use formula for the Cayley-Menger determinant generalized to the case $\mathbb{H} \neq \mathbb{I}$ (for its proof we refer to [1]):

$$\det(\mathbb{H})\, |\Delta|^2 = \frac{(-1)^{d-1}}{2^d (d!)^2} \det(K(\mathbb{H})), \tag{9}$$

where

$$K(\mathbb{H}) = \begin{pmatrix} (\mathbb{H}\mathbf{v}_{11}, \mathbf{v}_{11}) & \cdots & (\mathbb{H}\mathbf{v}_{1d_1}, \mathbf{v}_{1d_1}) & 1 \\ \vdots & \ddots & \vdots & \vdots \\ (\mathbb{H}\mathbf{v}_{d_1 1}, \mathbf{v}_{d_1 1}) & \cdots & (\mathbb{H}\mathbf{v}_{d_1 d_1}, \mathbf{v}_{d_1 d_1}) & 1 \\ 1 & \cdots & 1 & 0 \end{pmatrix} \tag{10}$$

and $\mathbf{v}_{ij} \equiv \mathbf{v}_i - \mathbf{v}_j$. Therefore,

$$|\Delta|^2_{|\mathbb{H}|} = \det(|\mathbb{H}|)\, |\Delta|^2 = \frac{(-1)^{d-1}}{2^d (d!)^2} \det(K(|\mathbb{H}|))$$
$$\leq \frac{1}{2^d (d!)^2} \sup_{\boldsymbol{\alpha} \in \Re^{n_d}} \frac{|\det K(\mathbb{H})|}{\max\limits_{1 \leq k \leq n_d} \alpha_k^d} \left(\sum_{k=1}^{n_d} \alpha_k\right)^d. \tag{11}$$

For a square matrix $K(\mathbb{H})$ with elements $k_{i,j}$, it holds

$$|\det(K(\mathbb{H}))| \leq \left| \sum_{\boldsymbol{\sigma}} \prod_{i=1}^{d+2} k_{i,\sigma_i}\right| \leq (d+2)! \max_{\boldsymbol{\sigma}} \left| \prod_{i=1}^{d+2} k_{i,\sigma_i}\right|,$$

where the summation is performed over all possible permutations $\boldsymbol{\sigma}$ of matrix rows and columns. Since $k_{i,j} = (\mathbb{H}\,\mathbf{e}_k, \mathbf{e}_k) = \alpha_k$, $1 \leq i < j \leq d_1$, from (10) we derive that $\det(K(\mathbb{H}))$ is a homogeneous polynomial of degree d of α_k and

$$\sup_{\boldsymbol{\alpha} \in \Re^{n_d}} \frac{|\det K(\mathbb{H})|}{\max\limits_{1 \leq k \leq n_d} \alpha_k^d} \leq (d+2)! \sup_{\boldsymbol{\alpha} \in \Re^{n_d}} \frac{\max\limits_{1 \leq k \leq n_d} \alpha_k^d}{\max\limits_{1 \leq k \leq n_d} \alpha_k^d} \leq (d+2)!.$$

Therefore, we conclude from (11) that

$$|\Delta|_{|\mathbb{H}|}^2 \leq \frac{1}{2^d} \frac{(d+1)(d+2)}{d!} \left(\sum_{k=1}^{n_d} \alpha_k \right)^d$$

which implies the lower bound in (7).

If $\det(\mathbb{H}) = 0$, the Hessian $\mathbb{H}$ cannot be used to generate a metric. In this case, we *modify* α_k to get a new quadratic function v_2 with a non-degenerate Hessian such that (7) is still satisfied. For the sake of simplicity, we restrict ourselves to the case $0 \leq \alpha_1 \leq \alpha_2 \leq \cdots \leq \alpha_{n_d}$ and $\alpha_{n_d} \neq 0$. The modified edge data are

$$\tilde{\alpha}_k = \alpha_k, \quad k = 1, \ldots, n_d - 1, \qquad \tilde{\alpha}_{n_d} = (1 + \delta)\alpha_{n_d},$$

where $\delta \in]0,1]$.

Let $\tilde{v}_2(\delta) = -\frac{1}{2} \sum_{k=1}^{n_d} \tilde{\alpha}_k b_k$ be the modified quadratic function and $\widetilde{\mathbb{H}}(\delta)$ be its Hessian. Formulas (9) and (10) imply that $p(\delta) \equiv \det(\widetilde{\mathbb{H}}(\delta))$ is a polynomial of degree two. Since $p(0) = \det(\mathbb{H}) = 0$, there exists $\delta_0 \in]0,1]$ such that $\det(\widetilde{\mathbb{H}}(\delta_0)) \neq 0$. We set $\mathfrak{M}_\Delta = |\widetilde{\mathbb{H}}(\delta_0)|$ and check that

$$\sum_{k=1}^{n_d} \alpha_k \leq \sum_{k=1}^{n_d} \tilde{\alpha}_k \leq \sum_{k=1}^{n_d} (|\widetilde{\mathbb{H}}(\delta_0)|\mathbf{e}_k, \mathbf{e}_k) \leq \left(\sum_{k=1}^{n_d} (|\widetilde{\mathbb{H}}(\delta_0)|\mathbf{e}_k, \mathbf{e}_k)^{1/2} \right)^2 = |\partial\Delta|_{\mathfrak{M}_\Delta}^2$$

and

$$\sum_{k=1}^{n_d} \alpha_k \geq \frac{1}{2} \sum_{k=1}^{n_d} \tilde{\alpha}_k \geq \left(\frac{(d+1)(d+2)}{d!} \right)^{-\frac{1}{d}} |\Delta|_{\mathfrak{M}_\Delta}^{\frac{2}{d}}.$$

This proves the assertion of the lemma. □

Using Lemma 1 and norm definition (3), we build the auxiliary metric $\mathfrak{M}_\Delta$ for error e_2. Similarly, using Lemma 1 and norm definition (5), we build the auxiliary metric $\widetilde{\mathfrak{M}}_\Delta$ for ∇e_2. These metrics do not provide a geometric representation of the error since the error estimates involve also the volume of simplex in the Cartesian metric. This mismatch is fixed in the following section.

2.2   Metrics for the L^p-norm of Error and Its Gradient

In this section, we consider L^p-norms, $p \geq 1$, of the errors as well as L^p-quasi-norms, $0 < p < 1$. As described in [18, 1, 4], as well as in [14], the metrics controlling various L^p-norms differ by a scaling factor. Let

$$\mathfrak{M}_{\Delta,p} = (\det(\mathfrak{M}_\Delta))^{-1/(d+2p)} \mathfrak{M}_\Delta \quad \text{and} \quad \widetilde{\mathfrak{M}}_{\Delta,p} = (\det(\widetilde{\mathfrak{M}}_\Delta))^{-1/(d+p)} \widetilde{\mathfrak{M}}_\Delta.$$

The following estimates are proved in [18, 1, 4].

Lemma 2. *Let $\mathfrak{M}_{\Delta,p}$ and $\widetilde{\mathfrak{M}}_{\Delta,p}$ be the constant tensor metrics defined above. Then,*

$$c_p|\Delta|_{\mathfrak{M}_{\Delta,p}}^{2/d+1/p} \leq \|e_2\|_{L^p(\Delta)} \leq C_p|\Delta|_{\mathfrak{M}_{\Delta,p}}^{1/p}|\partial\Delta|_{\mathfrak{M}_{\Delta,p}}^2 \tag{12}$$

and

$$\tilde{c}_p|\Delta|_{\widetilde{\mathfrak{M}}_{\Delta,p}}^{1/d+1/p} \leq \|\nabla e_2\|_{L^p(\Delta)} \leq \tilde{C}_p|\Delta|_{\widetilde{\mathfrak{M}}_{\Delta,p}}^{1/p}|\partial\Delta|_{\widetilde{\mathfrak{M}}_{\Delta,p}}, \tag{13}$$

where constants c_p, C_p, $\tilde{c}_p$, $\tilde{C}_p$ depend only on d and p.

For brevity, we confine ourselves to the case $p = \infty$. In this case, the constants in Lemma 2 depend only on d. Moreover, the metrics generated by Lemma 1 are optimal, i.e. $\mathfrak{M}_{\Delta,\infty} = \mathfrak{M}_\Delta$ and $\widetilde{\mathfrak{M}}_{\Delta,\infty} = \widetilde{\mathfrak{M}}_\Delta$.

2.3 Extension to General Functions

For a given continuous function u, we use the computable error e_2 to estimate the true error e:

$$e = u - \mathcal{I}_{1,\Delta}u.$$

Let $\mathcal{F}$ be the space of symmetric $d \times d$ matrices and $|\mathbb{H}|$ be the spectral module of $\mathbb{H} \in \mathcal{F}$. We introduce the following notations:

$$\||e_k\||_{|\mathbb{H}|}^2 = \max_{\mathbf{x}\in\Delta}(|\mathbb{H}(\mathbf{x})|\,e_k, e_k) \qquad \text{and} \qquad \||\partial\Delta\||_{|\mathbb{H}|}^2 = \sum_{k=1}^{n_d}\||e_k\||_{|\mathbb{H}|}^2.$$

The following result is proved in [1, 4].

Lemma 3. *Let $u \in C^2(\bar{\Delta})$. Then,*

$$\frac{d+1}{2d}\|e_2\|_{L^\infty(\Delta)} \leq \|e\|_{L^\infty(\Delta)} \leq \|e_2\|_{L^\infty(\Delta)} + \frac{1}{4}\inf_{\mathbb{F}\in\mathcal{F}}\||\partial\Delta\||_{|\mathbb{H}-\mathbb{F}|}^2.$$

and

$$\|\nabla e_2\|_{L^\infty(\Delta)} - \mathrm{osc}(\mathbb{H}, \Delta) \leq \|\nabla e\|_{L^\infty(\Delta)} \leq \|\nabla e_2\|_{L^\infty(\Delta)} + \mathrm{osc}(\mathbb{H}, \Delta), \tag{14}$$

where the oscillation term is

$$\mathrm{osc}(\mathbb{H}, \Delta) = C_{osc}\frac{|\partial\Delta|^{d-1}}{|\Delta|}\inf_{\mathbb{F}\in\mathcal{F}}\||\partial\Delta\||_{|\mathbb{H}-\mathbb{F}|}^2$$

and C_{osc} depends only on d.

The oscillation terms are conventional in the contemporary error analysis. Their value depend on the simplex and particular features of the function. For smooth solutions and shape-regular simplexes, the oscillation terms are much smaller than the error value.

2.4 On Selection of α_k

Let $\mathbb{H}(u_2)$ be definite. The derivation of metric $\mathfrak{M}_\Delta$ suggests a simple motivation for the choices (4) and (6). Since the bubble function b_k is non-zero only on one edge, we get

$$|(\mathfrak{M}_\Delta \mathbf{e}_k, \mathbf{e}_k)| = -\frac{1}{2}\alpha_k(\mathbb{H}(b_k)\mathbf{e}_k, \mathbf{e}_k).$$

The last term is the second derivative in direction $\mathbf{e}_k$. The inner product is a constant independent of the edge length and shape of the simplex. The definition of γ_k in (1) implies that the maximum norm of error on edge $\mathbf{e}_k$ is simply $|\gamma_k|/4$. Using this, we get

$$|(\mathfrak{M}_\Delta \mathbf{e}_k, \mathbf{e}_k)| = 4\frac{\alpha_k}{|\gamma_k|}\|e_2\|_{L^\infty(\mathbf{e}_k)}.$$

Let us consider a mesh that is uniform in metric $\mathfrak{M}_\Delta$. For such a mesh, we immediately get the following equalities:

$$\frac{\alpha_1}{|\gamma_1|}\|e_2\|_{L^\infty(\mathbf{e}_1)} = \frac{\alpha_2}{|\gamma_2|}\|e_2\|_{L^\infty(\mathbf{e}_2)} = \cdots = \frac{\alpha_{n_d}}{|\gamma_{n_d}|}\|e_2\|_{L^\infty(\mathbf{e}_{n_d})}.$$

Thus, the optimal mesh equidistributes $\|e_2\|_{L^\infty(\mathbf{e}_k)}$ over all edges of each simplex Δ. It explains our choice for (4). Similar arguments are used to motivate the choice (6).

The choice (4) has another advantage. In spite of the local metric construction, we have approximate equality of edge length measured in different metrics $\mathfrak{M}_\Delta$ coming from simplexes Δ sharing the edge. In other words, the recovered cell-based metrics are globally consistent. This results is verified with numerical experiments in Section 4.1.

2.5 Error Estimates as Functions of N_h

The error equidistribution principle suggests to build meshes that are quasi-uniform in metric $\mathfrak{M}$, for the interpolation error, or in metric $\widetilde{\mathfrak{M}}$, for the gradient of the interpolation error. Let Ω_h and $\widetilde{\Omega}_h$ be simplicial meshes with N_h cells that balance the volume and perimeter of cells:

$$N_h^{-1}|\Omega|_{\mathfrak{M}_p} \simeq |\Delta|_{\mathfrak{M}_{\Delta,p}} \simeq |\partial\Delta|_{\mathfrak{M}_{\Delta,p}}^d \qquad \forall \Delta \in \Omega_h$$

and

$$N_h^{-1}|\Omega|_{\widetilde{\mathfrak{M}}_p} \simeq |\Delta|_{\widetilde{\mathfrak{M}}_{\Delta,p}} \simeq |\partial\Delta|_{\widetilde{\mathfrak{M}}_{\Delta,p}}^d \qquad \forall \Delta \in \widetilde{\Omega}_h,$$

where $a \simeq b$ means that $ca \le b \le Ca$ with constants depending only on d and p. On such meshes, the following error estimates are held:

$$\|e\|_{L^p(\Omega)} = \left(\sum_{\Delta \in \Omega_h} \|e\|_{L^p(\Delta)}^p\right)^{\frac{1}{p}} \lesssim \left(\sum_{\Delta \in \Omega_h} |\Delta|_{\mathfrak{M}_{\Delta,p}}^{1+\frac{2p}{d}}\right)^{\frac{1}{p}} \lesssim |\Omega|_{\mathfrak{M}_p}^{\frac{1}{p}+\frac{2}{d}} N_h^{-\frac{2}{d}}$$

and

$$\|\nabla e\|_{L^p(\Omega)} = \left(\sum_{\Delta \in \widetilde{\Omega}_h} \|\nabla e\|_{L^p(\Delta)}^p \right)^{\frac{1}{p}} \lesssim \left(\sum_{\Delta \in \widetilde{\Omega}_h} |\Delta|_{\widetilde{\mathfrak{M}}_{\Delta,p}}^{1+\frac{p}{d}} \right)^{\frac{1}{p}} \lesssim |\Omega|_{\widetilde{\mathfrak{M}}_p}^{\frac{1}{p}+\frac{1}{d}} N_h^{-\frac{1}{d}}.$$

Thus, the $\mathfrak{M}_p$ (resp., $\widetilde{\mathfrak{M}}_p$)-quasi-uniform meshes provide asymptotically optimal rate for reduction of the interpolation error (resp., the gradient of the error).

3 Mesh Adaptation Algorithm

To build a continuous metric from a piecewise constant metric, we employ the method of shifts. For every node $\mathbf{a}_i$ in Ω_h, we define the superelement σ_i as the union of all d-simplices sharing $\mathbf{a}_i$. Then, to every node $\mathbf{a}_i$, we assign the metric with the largest determinant among all metrics associated with the superelement σ_i.

We use Algorithm 1 to build an adaptive mesh minimizing the L^p-norm of error or its gradient.

Algorithm 1. Adaptive mesh generation

1: Generate an initial mesh Ω_h, compute a piecewise constant metric $\mathfrak{M}_p$, and apply the method of shifts to get a continuous metric still denoted by $\mathfrak{M}_p$.
2: **loop**
3:     Generate a $\mathfrak{M}_p$-quasi-uniform mesh Ω_h.
4:     Recompute the metric $\mathfrak{M}_p$.
5:     If Ω_h is $\mathfrak{M}_p$-quasi-uniform, then exit the loop.
6: **end loop**

To generate a $\mathfrak{M}$-quasi-uniform mesh, we use a sequence of local mesh modifications [2, 7, 17] that gradually increase the measure of mesh quasi-uniformity. The local modifications of mesh topology include edge swapping, node relocation, insertion and deletion. These operations are implemented in package Ani2D (sourceforge.net/projects/ani2d).

4 Numerical Results

4.1 Interpolation Problems

In this section, we demonstrate with numerical experiments that the recovered piecewise-constant metric is sufficiently 'smooth' and reflects anisotropic features of the interpolated function. Let $\mathcal{E}^0$ be the set of interior mesh edges.

In the two dimensional case, we define the measure of metric discontinuity as follows:

$$V(\mathfrak{M}) = \frac{1}{N(\mathcal{E}^0)} \sum_{e \in \mathcal{E}^0} V_e(\mathfrak{M}), \qquad V_e(\mathfrak{M}) = \left| \frac{\|e\|_{\mathfrak{M}_\Delta} - \|e\|_{\mathfrak{M}_{\Delta'}}}{\|e\|_{\mathfrak{M}_\Delta} + \|e\|_{\mathfrak{M}_{\Delta'}}} \right|,$$

where $N(\mathcal{E}^0)$ is the number of interior edges and $\mathfrak{M}_\Delta$, $\mathfrak{M}_{\Delta'}$ are two triangles with the common edge e. Note that $V(\mathfrak{M})$ is zero for a continuous metric. For a sequence of refined shape-regular meshes and corresponding piecewise constant metrics approximating a continuous metric, $V(\mathfrak{M})$ is converging to zero.

In the first experiment, we calculate $V(\mathfrak{M})$ on a sequence of quasi-optimal meshes built with Algorithm 1. In the unit square $\Omega = [0, 1]^2$, we consider the analytical function proposed in [11]:

$$u(x, y) = \frac{(x - 0.5)^2 - (\sqrt{10}y + 0.2)^2}{((x - 0.5)^2 + (\sqrt{10}y + 0.2)^2)^2}.$$

The function has an anisotropic singularity at point $(0.5, -0.2/\sqrt{10})$ located outside the computational domain but close to its boundary. Table 1 shows that $V(\mathfrak{M})$ is roughly 0.1, i.e. the length of edge $\mathbf{e}$ varies roughly 20% when measured in metrics $\mathfrak{M}_\Delta$ and $\mathfrak{M}_{\Delta'}$ associated with this edge. Lack of convergence of $V(\mathfrak{M})$ to 0 as $N_h \to \infty$ may be related to the fact that the mesh is only quasi-uniform in metric $\mathfrak{M}$.

The L^∞-norm of the interpolation error is proportional to N_h^{-1}, while the L^∞-norm of its gradient is proportional to $N_h^{-0.5}$. Note that the meshes minimizing the interpolation error and its gradient are different (see Fig. 1). The figure indicates sharper features of the gradient of the error, which is the expected result.

In the second experiment, we consider the Texas-shape domain inscribed in $[-\frac{3}{2}; \frac{3}{2}]$ and shown in Fig 2. We consider the analytical function

$$u(x, y) = (x^2 y + y^3)/16^3 + \tanh(2(\sin(6y) - 3x)(\sin(6x) - 3y)) \qquad (15)$$

that has a spider-like distinguished feature highlighted by the mesh anisotropy. The results of numerical experiments collected in Table 2 confirm conclusions

Table 1. Experiment 1: convergence of the interpolation error and its gradient

N_h	Interpolation error		Gradient of interpolation error	
	$\|e\|_{L^\infty(\Omega)}$	$V(\mathfrak{M})$	$\|\nabla e\|_{L^\infty(\Omega)}$	$V(\mathfrak{M})$
1000	8.29e-2	0.122	5.41e+1	0.119
4000	2.36e-2	0.114	2.70e+1	0.097
16000	6.59e-3	0.115	1.42e+1	0.096
64000	1.83e-3	0.113	7.71e+0	0.099
rate	0.92		0.48	

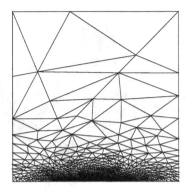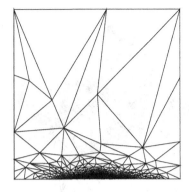

Fig. 1. Experiment 1: The adaptive meshes with roughly 2000 triangles minimizing the maximum norm of the interpolation error (left) and its gradient (right).

Table 2. Experiment 2: convergence of the interpolation error and its gradient.

N_h	Interpolation error		Gradient of interpolation error	
	$\|e\|_{L^\infty(\Omega)}$	$V(\mathfrak{M})$	$\|\nabla e\|_{L^\infty(\Omega)}$	$V(\mathfrak{M})$
1000	1.03e-1	0.197	1.75e-0	0.220
4000	2.09e-2	0.122	7.72e-1	0.146
16000	5.38e-3	0.098	3.76e-1	0.104
64000	1.39e-3	0.090	1.93e-1	0.090
rate	1.03		0.53	

that we made in the previous experiment. We observe the first-order convergence rate for the maximum norm of the error and the half-order convergence rate for the gradient of this error. The measure of metric discontinuity $V(\mathfrak{M})$ is slowly decreasing; however, its convergence to zero is questionable. Fig. 2 shows that the meshes minimizing the interpolation error and its gradient are different, which is the expected result.

Actual numerical values of $V(\mathfrak{M})$ cause slight but yet unpleasant instabilities in the adaptive process. We found numerically that the adaptation is more robust for a continuous tensor metric that provides faster convergence and results in a smoother mesh. That is why we use the method of shifts to generate of a continuous metric.

4.2 Applications to PDEs

In this section, we apply the developed methodology to adaptive solution of finite element problems. We consider problems with isotropic and anisotropic solutions.

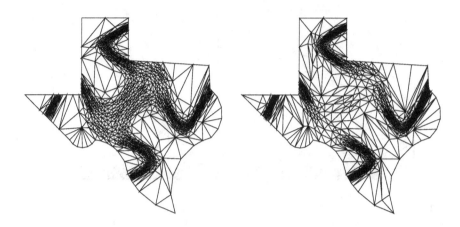

Fig. 2. Experiment 2: The adaptive meshes with roughly 2000 triangles minimizing the maximum norm of the interpolation error (left) and its gradient (right)

Hierarchical error estimates

We describe briefly a discretization error estimator based on enrichment of the linear finite element space with a space of piecewise quadratic finite element functions (bubbles) associated with edges of Ω_h [12]. The extended finite element problem results in a system of algebraic equations:

$$\begin{bmatrix} A_{LL} & A_{LQ} \\ A_{QL} & A_{QQ} \end{bmatrix} \begin{bmatrix} u_L \\ u_Q \end{bmatrix} = \begin{bmatrix} f_L \\ f_Q \end{bmatrix},$$

where subscripts L and Q stand for linear and quadratic terms.

Let u_L^* be an approximate solution of the original P_1 finite element problem $A_{LL} u_L^* = f_L$. We define the deviation $d_L = u_L - u_L^*$ and the discretization error $d_Q = u_Q$. They satisfy

$$\begin{bmatrix} A_{LL} & A_{LQ} \\ A_{QL} & A_{QQ} \end{bmatrix} \begin{bmatrix} d_L \\ d_Q \end{bmatrix} = \begin{bmatrix} r_L \\ r_Q \end{bmatrix}, \tag{16}$$

with

$$r_L = f_L - A_{LL} u_L^* \quad \text{and} \quad r_Q = f_Q - A_{QL} u_L^*.$$

The exact solution of (16) is too expensive. In order to estimate the discretization error d_Q, equation (16) is replaced with a simpler equation

$$\begin{bmatrix} A_{LL} & 0 \\ 0 & A_{QQ} \end{bmatrix} \begin{bmatrix} \tilde{d}_L \\ \tilde{d}_Q \end{bmatrix} = \begin{bmatrix} r_L \\ r_Q \end{bmatrix}. \tag{17}$$

Using a local finite element analysis, one can show that the diagonal matrix in (17) is spectrally equivalent to the matrix in (16). Therefore, the energy norm of the discretization error d_Q can be estimated using the energy norm of $\tilde{d}_Q$. The matrix A_{QQ} is well-conditioned for shape-regular meshes; therefore, the vector $\tilde{d}_Q$ can be efficiently calculated with a simple conjugate gradient method.

The entry of vector $\tilde{d}_Q$ associated with an edge $\mathbf{e}_k$ of a simplex Δ plays the role of interpolation error γ_k in formula (1). Thus, we can use the above methodology to generate quasi-optimal meshes.

Problem with a point singularity

Let Ω be a unit disk with a radial cut. We consider the classical crack problem with the exact solution

$$u(r, \theta) = r^{1/4} \sin(\theta/4),$$

where (r, θ) are polar coordinates, $r > 0$ and $\theta \in [0, 2\pi)$. The crack line S is defined by points $(r, 0)$. We consider the following boundary value problem:

$$\begin{aligned}
\Delta u &= 0 && \text{in} \quad \Omega \setminus S, \\
u &= \sin\frac{\theta}{4} && \text{on} \quad \partial\Omega \setminus S, \\
u &= 0 && \text{on} \quad S^+, \qquad \frac{\partial u}{\partial n} = 0 \quad \text{on} \quad S^-,
\end{aligned} \tag{18}$$

where S^+ and S^- denote the crack line when it is approached from regions $\theta \to +0$ and $\theta \to 2\pi$, respectively.

Table 3 demonstrates the half-order convergence of the gradient of the discretization error. A similar convergence is observed for the gradient of a finite element function $\tilde{d}_h$ corresponding to vector $\tilde{d}_Q$, which confirms the theory of hierarchical error estimates on shape-regular meshes. The difference in error values indicates that the constant of spectral equivalence of energy norms of d_Q and $\tilde{d}_Q$ is approximately 6.

This theory does not guarantee a similar connection between L^2-norms of these errors, which is also clear from the second and third columns in Table 3. For this norm, we have to use the finite element function d_h corresponding to d_Q; however, calculation of this function is rather expensive.

Problem with anisotropic singularities

Let Ω be the unit square $\Omega = (0, 1)^2$. We consider the following boundary value problem:

$$\begin{aligned}
-\operatorname{div}(\mathbb{K}\operatorname{grad} u) &= 1 && \text{in} \quad \Omega, \\
u &= 0 && \text{on} \quad \partial\Omega,
\end{aligned}$$

where

$$\mathbb{K}(x, y) = \mathbb{R}_\theta^T \begin{bmatrix} 1 & 0 \\ 0 & 10^3 \end{bmatrix} \mathbb{R}_\theta, \qquad \theta = 250\,(x + y),$$

and $\mathbb{R}_\theta$ is the rotation matrix by angle θ. The analytical solution is unknown and the discretization error cannot be computed. However, some features

Table 3. Experiment 3: convergence of the discretization error and its gradient

N_h	Discretization error		Gradient of discretization error	
	$\|\tilde{d}_h\|_{L^2(\Omega)}$	$\|e\|_{L^2(\Omega)}$	$\|\nabla \tilde{d}_h\|_{L^2(\Omega)}$	$\|\nabla e\|_{L^2(\Omega)}$
1000	1.10e-3	7.44e-3	1.94e-2	1.17e-1
4000	2.84e-4	3.10e-3	1.01e-2	6.08e-2
16000	7.04e-5	1.30e-3	5.35e-3	2.96e-2
64000	1.76e-5	9.48e-4	2.62e-3	1.59e-2
rate	1.00	0.51	0.48	0.48

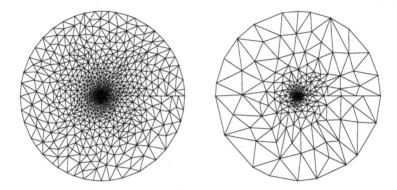

Fig. 3. Experiment 3: The adaptive meshes with roughly 2000 triangles minimizing the L^2-norm of the hierarchical error estimator (left) and its gradient (right)

Table 4. Experiment 4: convergence of the hierarchical error estimators $\tilde{d}_h$, d_h and their gradients

N_h	Discretization error		Gradient of discretization error	
	$\|\tilde{d}_h\|_{L^2(\Omega)}$	$\|d_h\|_{L^2(\Omega)}$	$\|\nabla \tilde{d}_h\|_{L^2(\Omega)}$	$\|\nabla d_h\|_{L^2(\Omega)}$
1000	4.98e-6	5.55e-3	1.88e-4	6.52e-4
4000	1.48e-6	3.02e-3	1.16e-4	3.52e-4
16000	3.88e-7	1.82e-3	5.94e-5	1.67e-4
64000	9.87e-8	1.03e-3	2.94e-5	9.07e-5
rate	0.95	0.40	0.45	0.48

of the solution can be extracted from the mesh structure shown in Fig. 4. Table 4 shows that the gradient of hierarchical edge-based a posteriori error estimator, $\|\nabla \tilde{d}_h\|_{L^2(\Omega)}$, correlates with $\|\nabla d_h\|_{L^2(\Omega)}$ even on anisotropic meshes. Similarly to the previous experiment, the L^2-norms of these estimators exhibit different behavior.

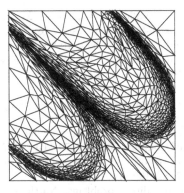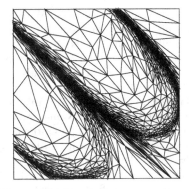

Fig. 4. Experiment 4: The adaptive meshes with roughly 2000 triangles minimizing the L^2-norm of the hierarchical error estimator $\tilde{d}_h$ (left) and its gradient (right)

5 Conclusion

We presented a new technology for generating meshes minimizing the interpolation and discretization errors or their gradients. The cornerstone of this methodology is construction of a space metric from edge-based error estimates. For the interpolation error, these estimates were computed explicitly. For the discretizations error, we used the hierarchical error estimators based on enrichment of the linear finite element space with quadratic bubble functions associated with mesh edges. We proved and verified with numerical experiments, that for a mesh with N_h triangles, the error is proportional to N_h^{-1} and the gradient of this error is proportional to $N_h^{-1/2}$, which are optimal asymptotics.

Acknowledgments. Research of the third author has been supported partially by the RFBR project 08-01-00159-a.

References

1. Agouzal, A., Vassilevski, Y.: Minimization of gradient errors of piecewise linear interpolation on simplicial meshes. Submitted to SIAM J. Numer. Anal. (2008)
2. Agouzal, A., Lipnikov, K., Vassilevski, Y.: Adaptive generation of quasi-optimal tetrahedral meshes. East-West J. Numer. Math. 7, 223–244 (1999)
3. Agouzal, A., Lipnikov, K., Vassilevski, Y.: Generation of quasi-optimal meshes based on a posteriori error estimates. In: Brewer, M., Marcum, D. (eds.) Proceedings of 16th International Meshing Roundtable, pp. 139–148. Springer, Heidelberg (2007)
4. Agouzal, A., Lipnikov, K., Vassilevski, Y.: Hessian-free metric-based mesh adaptation via geometry of interpolation error. Comp. Math. Math. Phys. 49(11) (to appear, 2009)

5. Ait-Ali-Yahia, D., Habashi, W.G., Tam, A., Vallet, M.-G., Fortin, M.: Directionally adaptive methodology using an edge-based error estimate on quadrilateral grids. Inter. J. Numer. Meth. Fluids 23(7), 673–690 (1996)
6. Bono, G., Awruch, A.M.: An adaptive mesh strategy for high compressible flows based on nodal re-allocation. J. Brazilian Soc. Mech. Sci. Engr. 30(3), 189–196 (2008)
7. Buscaglia, G.C., Dari, E.A.: Anisotropic mesh optimization and its application in adaptivity. Inter. J. Numer. Meth. Engrg. 40, 4119–4136 (1997)
8. Chen, L., Sun, P., Xu, J.: Optimal anisotropic meshes for minimizing interpolation errors in L^p-norm. Mathematics of Computation 76, 179–204 (2007)
9. Ciarlet, P., Wagschal, C.: Multipoint Taylor formulas and applications to the finite element method. Numer. Math. 17, 84–100 (1971)
10. Ciarlet, P.: The finite element method for elliptic problems. North-Holland, Amsterdam (1978)
11. D'Azevedo, E.: Optimal triangular mesh generation by coordinate transformation. SIAM J. Sci. Stat. Comput. 12, 755–786 (1991)
12. Deuflhard, P., Leinen, P., Yserentant, H.: Concepts of an adaptive hierarchical finite element code. IMPACT 1, 3–35 (1989)
13. Frey, P.J., Alauzet, F.: Anisotropic mesh adaptation for CFD computations. Comput. Meth. Appl. Mech. Eng. 194, 5068–5082 (2005)
14. Huang, W.: Metric tensors for anisotropic mesh generation. J. Comp. Physics. 204, 633–665 (2005)
15. Lipnikov, K., Vassilevski, Y.: Parallel adaptive solution of 3D boundary value problems by Hessian recovery. Comput. Meth. Appl. Mech. Eng. 192, 1495–1513 (2003)
16. Tichomirov, V.M.: The widths of sets in functional spaces and theory of optimal approximations. Uspehi Matem. Nauk 15(3), 81–120 (1960)
17. Vassilevski, Y., Lipnikov, K.: Adaptive algorithm for generation of quasioptimal meshes. Comp. Math. Math. Phys. 39, 1532–1551 (1999)
18. Vassilevski, Y., Agouzal, A.: An unified asymptotic analysis of interpolation errors for optimal meshes. Doklady Mathematics 72, 879–882 (2005)
19. Zhu, J.Z., Zienkiewicz, O.C.: Superconvergence recovery technique and a posteriori error estimators. Inter. J. Numer. Meth. Engrg. 30, 1321–1339 (1990)

On 3D Anisotropic Local Remeshing for Surface, Volume and Boundary Layers

Adrien Loseille[1] and Rainald Löhner[1]

CFD Center, Dept. of Computational and Data Sciences, College of Science, MS 6A2, George Mason University, Fairfax, VA 22030-4444, USA
{aloseill,rlohner}@gmu.edu

Abstract. A simple strategy for generating anisotropic meshes is introduced. The approach belongs to the class of metric-based mesh adaptation procedures where a field of metric tensors governs the adaptation. This development is motivated by the need of generating anisotropic meshes for complex geometries and complex flows. The procedure may be used advantageously for cases where global remeshing techniques become either unfeasible or unreliable. Each of the local operations used is checked in a variety of ways by taking into account both the volume and the surface mesh. This strategy is illustrated with surface mesh adaptation and with the generation of meshes suited for boundary layers analysis.

Two simple mesh operators are used to recursively modify the mesh: edge collapse and point insertion on edge. It is shown that using these operators jointly with a quality function allows to quickly produce an quality anisotropic mesh. Each adaptation entity, *ie* surface, volume or boundary layers, relies on a specific metric tensor field. The metric-based surface estimate is used to control the deviation to the surface and to adapt the surface mesh. The volume estimate aims at controlling the interpolation error of a specific field of the flow. The boundary layers metric-based estimate is deduced from a level-set distance function.

Keywords: Anisotropic Mesh Adaptation; Surface Remeshing; Metric-Based Error Estimate; Boundary Layers Mesh Generation; Level-Set Function.

1 Introduction

Generating a valid tetrahedral mesh for a given domain Ω of $\mathbb{R}^3$ of any arbitrary complexity is still a tedious task. The difficulty increases mainly with the complexity of the boundary $\partial\Omega$ of Ω. In addition, a second factor that impacts substantially the complexity is the way $\partial\Omega$ is meshed. Isotropically-meshed surfaces with a smooth element-size variation are generally easier to mesh than anisotropically-meshed surfaces with strong size variations. This is particularly true when one considers the set of methods that have demonstrated a good efficiency and reliability to produce a volume mesh from given complex surface mesh: advancing front method [28, 34], constraint global Delaunay [4, 18, 19] or a combination of both [33]. These methods have now attained a sufficient level of maturity to handle very complex geometries as long as the surface mesh of $\partial\Omega_h$ is **isotropic**. When dealing with anisotropic surface meshes, the frontal methods generally do not succeed to close the front, while the Delaunay-based generally will fail during the boundary recovery phase. Consequently, being able to certify that during an adaptive procedure any of this methods will succeed is not at all guaranteed. Consequently, it is of great interest to develop techniques that modify a mesh locally [9, 13, 36] while keeping a valid 3D mesh. In this paper, the emphasis is put on anisotropic meshes that correspond to the case where previous methods are the most susceptible to fail.

We concentrate on mesh-relative classical tasks required for a complete adaptive CFD (Computational Fluid Dynamics) run: the surface remeshing, the volume remeshing and the boundary layers mesh generation. Each task on its own has blossomed into a large field of research in the mesh-generation community. However, these tasks are often studied independently and without the constraints of keeping a valid 3D mesh. The approach followed here is to deal conjointly with these three tasks starting from an initial mesh and using the metric-based framework. The paper is organized as follow. In Section 1, the classical metric-based framework is recalled. Then, the main implementation choices for the adaptive mesh generator are discussed (Section 2). Section 3 deals with the derivation of metric-based estimates for controlling the interpolation error of a solution field, the deviation to a surface and for monitoring the generation of a boundary layers grid. Mesh modification operators are introduced in Section 4, along with numerical examples.

2   Metric-Based Anisotropic Local Remeshing

We briefly recall the main concepts of metric-based mesh adaptation. This is done in a generic way without having a specific problem at hand. Thereafter, the use of the local mesh generator is considered in the context of the full adaptive process.

2.1   Metric Tensors in Mesh Adaptation

Metric-based mesh adaptation is an elegant concept introduced in the pioneering works [10, 20]. It (theoretically) allows to transform any unstructured uniform mesh generator into an anisotropic one. This is done by computing the distance in a Riemannian space instead of the classical Euclidean metric space. The adaptive mesh generator aims at creating a unit-mesh in this space. In the following, we recall the continuous definition inherited from differential geometry considerations [7, 12] with their implementation in the mesh generator [17]. Each continuous definition is then accompanied with its discrete counterpart.

A metric tensor field of Ω is a Riemannian metric space denoted by $(\mathcal{M}(\mathbf{x}))_{\mathbf{x}\in\Omega}$, where $\mathcal{M}(\mathbf{x})$ is a 3×3 symmetric positive definite matrix. Taking this field at each vertex $\mathbf{x}_i$ of a mesh $\mathcal{H}$ of Ω defines the discrete field

$\mathcal{M}_i = \mathcal{M}(\mathbf{x}_i)$. If N denotes the number of vertices of $\mathcal{H}$, the linear discrete metric field is denoted by $(\mathcal{M}_i)_{i=1...N}$. As $\mathcal{M}(\mathbf{x})$ and $\mathcal{M}_i$ are symmetric definite positive, they can be diagonalized in an orthonormal frame, such that

$$\mathcal{M}(\mathbf{x}) = {}^t\mathcal{R}(\mathbf{x})\Lambda(\mathbf{x})\mathcal{R}(\mathbf{x}) \text{ and } \mathcal{M}_i = {}^t\mathcal{R}_i\Lambda_i\mathcal{R}_i,$$

where $\Lambda(\mathbf{x})$ and Λ_i are diagonal matrices composed of strictly positive eigenvalues $\lambda(\mathbf{x})$ and λ_i and $\mathcal{R}$ and $\mathcal{R}_i$ orthonormal matrices verifying ${}^t\mathcal{R}_i = (\mathcal{R}_i)^{-1}$. Setting $h_i = \lambda_i^{-1}$ allows to define the sizes prescribed by $\mathcal{M}_i$ along the principal directions given by $\mathcal{R}_i$. Note that the set of points verifying the implicit equation ${}^t\mathbf{x}\,\mathcal{M}_i\,\mathbf{x} = 1$ defines a unique ellipsoid. This ellipsoid is called the unit-ball of $\mathcal{M}_i$ and is used to represent geometrically $\mathcal{M}_i$ as in Figure 1.

Fig. 1. Some unit-elements with respect to a 3D metric represented by its unit-ball.

The two fundamental operations in a mesh generator are the computation of length and volume. The distance of an edge $\mathbf{e} = [\mathbf{x}_i, \mathbf{x}_j]$ and the volume of an element K are continuously evaluated in $(\mathcal{M}(\mathbf{x}))_{\mathbf{x}\in\Omega}$ by:

$$\ell_\mathcal{M}(\mathbf{e}) = \int_0^1 \sqrt{{}^t\mathbf{e}\,\mathcal{M}(\mathbf{x}_i + t\,\mathbf{e})\,\mathbf{e}}\,\mathrm{d}t \text{ and } |K|_\mathcal{M} = \int_K \sqrt{\det(\mathcal{M}(\mathbf{x}))}\,\mathrm{d}\mathbf{x}$$

From a discrete point view, the metric field needs to be interpolated [17] to compute approximate length and volume. For the volume, we consider a linear interpolation of $(\mathcal{M}_i)_{1...N}$ and the following edge length approximation is used:

$$|K|_\mathcal{M} \approx \sqrt{\det\left(\frac{1}{4}\sum_{i=1}^4 \mathcal{M}_i\right)}|K| \text{ and } \ell_\mathcal{M}(\mathbf{e}) \approx \sqrt{{}^t\mathbf{e}\,\mathcal{M}_i\,\mathbf{e}}\,\frac{r-1}{r\ln(r)}, \quad (1)$$

where $|K|$ is the Euclidean volume of K and r stands for the ratio $\sqrt{{}^t\mathbf{e}\,\mathcal{M}_i\,\mathbf{e}}/\sqrt{{}^t\mathbf{e}\,\mathcal{M}_j\,\mathbf{e}}$. The approximated length arises from considering a geometric approximation of the size variation along end-points of $\mathbf{e}$: $\forall t \in [0,1]\,h(t) = h_i^{1-t}\,h_j^t$.

The task of the adaptive mesh generator is then to generate a unit-mesh with respect to $(\mathcal{M}(\mathbf{x}))_{\mathbf{x}\in\Omega}$. A mesh is said unit when it is only composed of

unit-volume elements and unit-length edges. Practically, these two requirements are combined in a quality function computed in the metric field. A
mesh $\mathcal{H}$ is said to be unit with respect to $(\mathcal{M}(\mathbf{x}))_{\mathbf{x} \in \Omega}$ when each tetrahedron
$K \in \mathcal{H}$ defined by its list of edges $(\mathbf{e}_i)_{i=1\ldots6}$ verifies:

$$\forall i \in [1,6], \quad \ell_{\mathcal{M}}(\mathbf{e}_i) \in \left[\frac{1}{\sqrt{2}}, \sqrt{2} \right] \quad \text{and} \quad Q_{\mathcal{M}}(K) \in [\alpha, 1] \text{ with } \alpha > 0, \quad (2)$$

with:

$$Q_{\mathcal{M}}(K) = \frac{36}{3^{\frac{1}{3}}} \frac{|K|_{\mathcal{M}}^{\frac{2}{3}}}{\sum_{i=1}^{6} \ell_{\mathcal{M}}^2(\mathbf{e}_i)} \in [0,1]. \quad (3)$$

A classical and admissible value of α is 0.8. This value arises from some
discussions on the possible tessellation of $\mathbb{R}^3$ with unit-elements [30]. Some
unit-elements with respect to a 3D metric are depicted in Figure 1.

2.2   Adaptive CFD Simulations

The complete adaptive algorithm for steady simulations is composed of the
following steps.

1. Compute the flow field (i.e. converge the flow solution on the current
   mesh);
2. Compute the metric estimates: surface, volume, boundary layers, etc.
3. Generate a unit mesh with respect to these metric fields;
4. Re-project the surface mesh onto the true geometry using the CAD data;
5. Interpolate the flow solution on the new adapted mesh;
6. Goto 1.

We briefly describe steps 1., 4. and 5. while steps 2. and 3. are discussed in
details in Sections 3 and 4. Note that all operations are done on a volume
mesh. Furthermore, any of these operations delivers a valid volume mesh,
even for the surface mesh adaptation or the projection of the continuous
surface using the CAD.

Flow solver. The flow solver employed is FEFLO. FEFLO was conceived as
a general-purpose CFD code based on the following general principles:

1 Use of unstructured grids (automatic grid generation and mesh refinement);
2 Linear finite element discretization of space (one element-type code for
   simplicity and speed);
3 Edge-based data structures for speed;
4 Separate flow modules for compressible and incompressible flows;
5 ALE formulation for moving grids;
6 Embedded surface or immersed body options for complex, dirty geometries;
7 Overlapping grids or gliding regions for rotating bodies;

8   Bottom-up coding from the subroutine level to assure an open-ended, expandable architecture to consider new turbulence models, Riemann solvers and limiters;

9   Optimal data structures for different architectures.

The code has had a long history of relevant applications [5, 6], and has been ported to both shared memory [24] and distributed memory [38] machines.

Surface representation. The surface representation is given by either analytical surfaces (Coon's patches, planes, ...) or discrete triangulations (STL, tri-files, ...). Each patch is bound by a set of unique lines that are shared between patches. After introducing of new points on the surface, the surface is interrogated and the point is placed on the correct surface position. This may yield elements with negative volumes, particularly if the mesh is coarse. Therefore, an iterative algorithm is employed to project smoothly the point to the true surface while keeping only positive-volume elements.

Solution interpolation. The solution interpolation step uses a simple linear interpolation scheme.

Comments on the data structures. During step 3., a discrete geometry is used, so that (costly) requests to CAD are only done during step 4.. Normals, tangents, ridges and corners are stored on each boundary point. Then elements surrounding elements and triangles surrounding triangles are used. With these data structures, the ball of elements for a vertex, the shell of an edge, and the topological neighbors of a vertex are recovered on-the-fly very quickly. We refer to [17] for the practical algorithms.

3   Metric-Based Estimates

As stated in Section 2, using a metric-based adaptive mesh generator provides an elegant way to keep the mesh generator independent of the problem at hand. In the sequel, we illustrate and review the derivation of several metric fields for the following tasks: adapting the mesh to a solution field by controlling the interpolation error, adapting the mesh to control the deviation to a surface, and, finally, adapting a mesh to create a boundary layers mesh as required in RANS (Reynolds-Average Navier-Stokes) simulations. We first introduce two techniques intensively used the sequel: the anisotropic mesh gradation and the Log-Euclidean framework.

3.1   Techniques for Enhancing Robustness and Performance

The metric field provided has a direct, albeit complex, impact on the quality of the resulting mesh. A smooth and well-graded metric field makes the generation of the anisotropic mesh generation easier and generally improves the final quality. We consider two techniques that tend to give a substantial positive impact on the quality of the resulting mesh: The **anisotropic**

mesh gradation tends to smooth the metric field, while the **Log-Eucidean** interpolation allows to properly define metric tensors interpolation, thereby preserving the anisotropy even after a numerous interpolations.

Anisotropic mesh gradation. The mesh gradation is a process that smoothes the initial metric field that is generally noisy as it is derived from discrete data. Gradation strategies for anisotropic meshes are available in [1, 8, 22]. From a continuous point of view, the mesh gradation process consists in verifying the uniform continuity of the metric field:

$$\forall(\mathbf{x}, \mathbf{y}) \in \Omega^2 \; \|\mathcal{M}(\mathbf{y}) - \mathcal{M}(\mathbf{x})\| \leq C\|\mathbf{x} - \mathbf{y}\|_2,$$

where C is a constant and $\|.\|$ a matrix norm. This requirement is far more complex that imposing only the continuity of $(\mathcal{M}(\mathbf{x}))_{\mathbf{x}\in\Omega}$. From a practical point of view, this done that by ensuring that for all couples $(\mathbf{x}_i, \mathcal{M}_i)$ defined on $\mathcal{H}$ verify:

$$\forall(\mathbf{x}_i, \mathbf{y}_j) \in \mathcal{H}^2 \; \mathcal{N}(\|\mathbf{x}_i - \mathbf{y}_j\|_2) \, \mathcal{M}_i \cap \mathcal{M}_j = \mathcal{M}_j \text{and} \mathcal{N}(\|\mathbf{x}_i - \mathbf{y}_j\|_2) \, \mathcal{M}_j \cap \mathcal{M}_i = \mathcal{M}_i,$$

where $\mathcal{N}(.)$ is a matrix function defining a growth factor and $\cap$ is the classical metric intersection based on simultaneous reduction [17]. This standard algorithm has $O(N^2)$ complexity. Consequently, less CPU-intensive correction strategies need to be devised; we refer to [1] for some suggestions. Note that bounding the number of corrections to a fixed value is usually sufficient to correct the metric field near strongly anisotropic areas as the shocks. Two options are used in this paper giving either an isotropic growth or an anisotropic growth acting:

$$\mathcal{N}(d_{ij})\,\mathcal{M}_i = \begin{pmatrix} \eta_1(d_{ij})\,\lambda_1 & & \\ & \eta_2(d_{ij})\,\lambda_2 & \\ & & \eta_3(d_{ij})\,\lambda_3 \end{pmatrix}$$

with

$$(i) \; \eta_k(d_{ij}) = (1 + \sqrt{{}^t\mathbf{e}_{ij}\,\mathcal{M}_i\,\mathbf{e}_{ij}} \, \log(\beta))^{-2} \; \text{ or } \; (a) \; \eta_k(d_{ij}) = (1 + \lambda_k\,d_{ij}\,\log(\beta))^{-2}, \tag{4}$$

where $d_{ij} = \|\mathbf{x}_j - \mathbf{x}_i\|_2$, $\mathbf{e}_{ij} = \mathbf{x}_j - \mathbf{x}_i$ and β the gradation parameter > 1. The isotropic growth is given by law (i) while the anisotropic by law (a). Note that (i) is identical for all directions, contrary to anisotropic law (a) that depends on each eigenvalue along its principal direction. In the sequel, we use the gradation to smooth the transition between the various metric fields: surface and volume, surface and boundary layers, etc.

Log-Euclidean framework. After each point insertion or during the computation of edge-lengths, a metric field must be interpolated. Interpolation schemes based on the simultaneous reduction [17] lack several desirable theoretical properties. For instance, the unicity is not guaranteed. A framework introduced in [3] proposes to work in the logarithm space as if one were in the Euclidean one. Consequently, a sequence of n metric tensors can be interpolated in any order while providing a unique metric. Given a sequence of points $(\mathbf{x}_i)_{i=1\ldots k}$ and their respective metrics $\mathcal{M}_i$, then the interpolated metric in $\mathbf{x}$ verifying

$$\mathbf{x} = \sum_{i=1}^{k} \alpha_i\, \mathbf{x}_i, \text{ with } \sum_{i=1}^{k} \alpha_i = 1,$$

is

$$\mathcal{M}(\mathbf{x}) = \exp\left(\sum_{i=1}^{k} \alpha_i \ln(\mathcal{M}_i)\right). \tag{5}$$

On the space of metric tensors, logarithm and exponential operators are acting on metric's eigenvalues directly:

$$\ln(\mathcal{M}_i) = {}^t\mathcal{R}_i \ln(\Lambda_i)\, \mathcal{R}_i \quad \text{and} \quad \exp(\mathcal{M}_i) = {}^t\mathcal{R}_i \exp(\Lambda_i)\, \mathcal{R}_i.$$

Numerical experiments confirm that using this framework during interpolation allow to preserve the anisotropy. Note that the evaluation of length given by (1) corresponds to the Log-Eucldiean interpolation between the two metrics of the edge extremities.

3.2 $\mathbf{L}^p$ Norm Interpolation Error

Controlling the linear interpolation error of a given flow field allows to derive a simple anisotropic metric-based estimate [10] by considering an error bound involving a recovered Hessian [37] of the numerical solution. Note that this approach has already demonstrated its efficiency on numerous 3D real-life problems [2, 9, 14, 31, 35, 39, 41]. In this paper, instead of classical error equidistribution issued from an $\mathbf{L}^\infty$ norm, we prefer to control the $\mathbf{L}^p$ norm of the interpolation error. Such control allows to recover the order of convergence of the scheme on flows with shocks and to capture of the scales of the numerical solution [31].

Given a numerical solution u_h (density, pressure, mach numbers, ...), the point-wise metric tensor is given by:

$$\mathcal{M}_{\mathbf{L}^p}(u_h) = \det(|H_R(u_h)|)^{\frac{-1}{2p+3}} |H_R(u_h)|, \tag{6}$$

where $H_R(.)$ stands for an operator that from u_h recovers some approximated second derivatives of u_h. Then $|H_R(u_h)|$ is deduced from $H_R(u_h)$ by taking the absolute value of the eigen-values of $H_R(u_h)$. Most common operators are deduced from a double L^2 projection or by the use of the Green formula. A numerical review of H_R operators is given in [42]. When applied to a given smooth continuous function u, it has been proven [11, 29] that for any unit-mesh $\mathcal{H}$ of Ω_h with respect to $\mathcal{M}_{\mathbf{L}^p}$ will verify the following bound:

$$\|u - \Pi_h u\|_{\mathbf{L}^p(\Omega_h)} \leq C\,N^{-\frac{2}{3}} \left(\int_\Omega \det(|H(u)|)^{\frac{p}{2p+3}} \right)^{\frac{2p+3}{p}}, \qquad (7)$$

where here $H(u)$ is the true Hessian of u, $\Pi_h u$ the linear interpolate of u on $\mathcal{H}$ and C a constant that only depends on the quality (computed in $\mathcal{M}_{\mathbf{L}^p}$) of $\mathcal{H}$. Note that (7) gives a practical way to control the level of error ε that is desired. Estimating the right-hand-side of (7) with $H_R(u_h)$ instead of $H(u)$ gives a first ε_0 error level so that to get an ε level of error, it is sufficient to scale (6):

$$\mathcal{M}_{\mathbf{L}^p}(u_h, \varepsilon) = \left(\frac{\varepsilon_0}{\varepsilon} \right) \det(|H_R(u_h)|)^{\frac{-1}{2p+3}} |H_R(u_h)|,$$

In the sequel, the interpolation error is controlled in $\mathbf{L}^2$ norm exclusively, while the H_R operator is based on the Green formula.

3.3 Geometric Estimate for Surfaces

Controlling the deviation to a surface has been studied in previous works. We may cite [23] for isotropic remeshing and [16, 15] for anisotropic remeshing. Apart from their efficiency, these methods were initially thought to work only on surface meshes, i.e without keeping a valid volume mesh. This additional requirement just implies another constraint that consists in verifying that each modification engenders a valid mesh (with a positive volume element check). Following these previous works, we introduce a metric-based error estimate such that the length in these surface metric measures the distance to the surface. Moreover, we assume that the initial mesh has an in-homogeneous error level control to the surface deviation. This assumption is particularly true in many engineering application where designers know a priori areas of interest. For instance, in aerodynamics the wings are generally meshed finer than the fuselage. The proposed error estimate is thought to preserve this initial in-homogeneity in the error distribution during the adaptation process.

We recall that the surface remeshing is done by considering only discrete data in order to avoid requests to CAD (done in another phase). Prior to surface remeshing, normals and tangents are assigned to each boundary point. We denote by $\mathbf{n}_i$ the normal of the vertex $\mathbf{x}_i$. As in [16], a quadratic surface model is computed locally around a surface point $\mathbf{x}_i$. Starting from the topological neighbors of $\mathbf{x}_i$, the coordinates of each point are mapped onto the local orthonormal Frenet frame $(\mathbf{u}_i, \mathbf{v}_i, \mathbf{n}_i)$ centered in $\mathbf{x}_i$. Vectors $(\mathbf{u}_i, \mathbf{v}_i)$ lie in the orthogonal plane to $\mathbf{n}_i$. We denote by $(u_j, v_j, \sigma_j) = ({}^t\mathbf{x}_j.\mathbf{u}_i, {}^t\mathbf{x}_j.\mathbf{v}_i, {}^t\mathbf{x}_j.\mathbf{n}_i)$ the new coordinates of vertex $\mathbf{x}_j$. $\mathbf{x}_i$ is set as the new origin so that $(u_i, v_i, \sigma_i) = (0, 0, 0)$. The surface model consists in computing by a least squares approximation a quadratic surface:

$$\sigma(u, v) = au^2 + bv^2 + cuv, \quad \text{where } (a, b, c) \in \mathbb{R}^3. \qquad (8)$$

The least squares problem gives the solution minimizing

$$\min_{(a,b,c)} \sum_j |\sigma_j - \sigma(u_j, v_j)|^2,$$

where j is the set of neighbors of $\mathbf{x}_i$. Note that 3 neighbors points are necessary to recover the surface model. With our insertion strategy (see Section 4), the degree of the surface point is 4. Even if this number seems sufficient, some information are added in order to be more robust. The normals (that are not recovered from discrete data except from discrete attached surface type) are then added. To this end, mid-edge points are recovered from the following quadratic formula:

$$\mathbf{x} = (1-t)^2(1+2t)\mathbf{x}_1 + t(1-t)^2\mathbf{r}_1 + t^2(3-2t)\mathbf{x}_2 - t^2(1-t)\mathbf{r}_2, \text{ with}$$

$$\mathbf{r}_i = \|\mathbf{e}\|_2 \frac{\mathbf{n}_i \times (\mathbf{e} \times \mathbf{n}_i)}{\|\mathbf{n}_i \times (\mathbf{e} \times \mathbf{n}_i)\|_2} \text{ and } t \in [0,1], \tag{9}$$

where $\mathbf{e}$ is an edge issued from $\mathbf{x}_i$ and $\mathbf{x}_j$ a neighbor of $\mathbf{x}_i$. Finally, if the degree of $\mathbf{x}_i$ is d the size of the linear system to solve becomes $2d$. The linear system involving the d neighbors and d mid-points is:

$$A X = B \iff \begin{pmatrix} u_1^2 & v_1^2 & u_1 v_1 \\ \vdots & \vdots & \vdots \\ u_d^2 & v_d^2 & u_d v_d \\ u_{\frac{1}{2}}^2 & v_{\frac{1}{2}}^2 & u_{\frac{1}{2}} v_{\frac{1}{2}} \\ \vdots & \vdots & \vdots \\ u_{\frac{d}{2}}^2 & v_{\frac{d}{2}}^2 & u_{\frac{d}{2}} v_{\frac{d}{2}} \end{pmatrix} \begin{pmatrix} a \\ b \\ c \end{pmatrix} = \begin{pmatrix} \sigma_1 \\ \vdots \\ \sigma_d \\ \sigma_{\frac{1}{2}} \\ \vdots \\ \sigma_{\frac{d}{2}} \end{pmatrix},$$

where $(u_{\frac{i}{2}}, v_{\frac{i}{2}}, \sigma_{\frac{i}{2}})$ are mid-points local coordinates recovered using (9). The least square formulation consists in solving ${}^t A\, A = {}^t A\, B$. From this point, one may applied the surface metric given in [16]. We propose here a simplified version. We can first remark that the orthogonal distance from the plane $\mathbf{n}_i^{\perp}$ onto the surface is given by $\sigma(u,v)$ by definition. The trace of $\sigma(u,v)$ on $\mathbf{n}_i^{\perp}$ is a function that gives directly the distance to the surface. The 2D surface metric $\mathcal{M}_S^{2D}$ such that the length $\ell_{\mathcal{M}^{2D}}((u,v))$ is constant equal to ε is easy to find starting from the diagonalization of the quadratic function (8). Geometrically, it consists in finding the maximal area metric included in the level-set ε of the distance map. We assume that $\mathcal{M}_S^{2D}$ admits the following decomposition:

$$\mathcal{M}_S^{2D} = (\bar{\mathbf{u}}_S, \bar{\mathbf{v}}_S) \begin{pmatrix} \lambda_{1,S} & 0 \\ 0 & \lambda_{2,S} \end{pmatrix} {}^t(\bar{\mathbf{u}}_S, \bar{\mathbf{v}}_S), \text{ with } (\bar{\mathbf{u}}_S, \bar{\mathbf{v}}_S) \in \mathbb{R}^{2 \times 2}.$$

If we want to achieve the same error as the initial mesh, we compute $\varepsilon = \min_j |\sigma(u_j, v_j)|$ among the neighbors of $\mathbf{x}_i$. The 2D metric achieving an ε error becomes:

$$\mathcal{M}_S^{2D}(\varepsilon) = \frac{1}{\varepsilon}\mathcal{M}_S^{2D}.$$

The final 3D surface metric in $\mathbf{x}_i$ is:

$$\mathcal{M}_S(\varepsilon) = (\mathbf{u}_S, \mathbf{v}_S, \mathbf{n}_i)\begin{pmatrix} \dfrac{\lambda_{1,S}}{\varepsilon} & 0 & \\ 0 & \dfrac{\lambda_{2,S}}{\varepsilon} & 0 \\ 0 & 0 & h_{max}^{-2} \end{pmatrix}{}^t(\mathbf{u}_S, \mathbf{v}_S, \mathbf{n}_i), \begin{cases} \mathbf{u}_S = \bar{\mathbf{u}}_S(1)\mathbf{u}_i + \bar{\mathbf{u}}_S(2)\mathbf{v}_i, \\ \mathbf{v}_S = \bar{\mathbf{v}}_S(1)\mathbf{u}_i + \bar{\mathbf{v}}_S(2)\mathbf{v}_i. \end{cases}$$

(10)

The parameter h_{max} is initially chosen very large (e.g. 1/10 of the domain size). This normal size is corrected during various steps. A first anisotropic gradation using $(4)(a)$ is applied on surface edges only. The surface metric is then intersected with any computation metrics as given by (6). These two steps set automatically a proper element size in the normal direction.

3.4   Boundary Layers Metric

Boundary layers mesh generation has been devised to capture accurately the speed profile around a body during a viscous simulation. The width of the boundary layer depends on the local reynolds number [26]. So far, the generation of the boundary layer grids has been carried out by an extrusion of the initial surface along the normals to the surface [25] or by local modification of the mesh [32]. Note that using the normals as sole information requires several enrichments to obtain a smooth layers transition on complex surfaces [27]. In this paper, boundary layer mesh generation is based on a continuous field: the distance to the body. A classical adaptive strategy is then devised to recover by local modification the boundary layers. This distance to the body allows automatically to deactivate the boundary layers mesh generation on geometry details that are smaller than the boundary layer size. Using the gradient of the distance map allows to approximate the normals to the initial body surface, whatever the initial position in space. This strategy can be used on an existing volume mesh that could be adapted. Note that typical studies in areodynamics consists in running a fisrt computation without the boudary layers mesh (Euler mesh). The viscous simulation is done in a second step. Consequently, it may be of interest to be able to generate, for complex geometries, a boundary layer while keeping intact the previous adaptation issued from non viscous simulations.

We now introduce the required steps to compute the boundary layers metric ensuing from a body:

- Compute the distance map Φ to the body,
- Recover the surface mesh metric with a mean size in the normal direction,
- Compute the boundary layers metric $\mathcal{M}_{bl}$.

Step 1. is done using classical algorithms of level-set methods [26, 40]. This step can be done quickly and has generally a complexity of $O(N \ln(N))$ where

N is the number of points in the current mesh. (Furthermore, note that from a practical point of view, this function is evaluated only in the vicinity of the body). From this scalar field, its linear gradient is recovered using a $\mathbf{L}^2$ projection. Note that we have $\|\nabla\Phi(\mathbf{x})\|_2 = 1, \forall \mathbf{x} \in \Omega$. The gradient is used to emulate normals to the body. For no extra cost, the body's face for which the minimum distance is reached is also stored for each point of the volume.

The surface metric recovery of step 2 takes advantage of the Log-Euclidean framework. Starting from the ball of surface elements $(K)_{P \in K}$ of body point P, the unique surface metric tensor $\mathcal{M}_K$ (for which K is unit) is computed by solving the following 3×3 linear system:

$$(S) \begin{cases} \ell^2_{\mathcal{M}^{2D}_K}(\mathbf{e_1}) = 1 \\ \ell^2_{\mathcal{M}^{2D}_K}(\mathbf{e_2}) = 1 \\ \ell^2_{\mathcal{M}^{2D}_K}(\mathbf{e_3}) = 1 . \end{cases}$$

where $(\mathbf{e}_i)_{i=1,3}$ are elements edges expressed in the local surface plane coordinates. (S) has a unique solution as long as the aera of K is not null. 2D metrics $(\mathcal{M}^{2D}_K)_{P \in K}$ metrics are transcribed into 3D metrics $(\mathcal{M}_K)_{P \in K}$ by prescribing a mean size in the normal direction to the face. The logarithm of each metric is computed so that a classical Euclidean mean weighted by the elements' area is done. Finally, the body point metric $\mathcal{M}_P$ is mapped back using the exponential operator:

$$\mathcal{M}_P = \exp\left(\frac{\sum_{P \in K} |K| \ln(\mathcal{M}_K)}{\sum_{P \in K} |K|}\right).$$

Step 3 gives the final boundary layers metric. We describe it for a continuous exponential law of the form $h_0 \exp(\alpha\phi(.))$, where h_0 is the initial boundary layer size and α the growing factor. Note that its application for any discrete law is straightforward to implement. For a volume point $\mathbf{x}_i$, the boundary layers metric depends on the body point P_i for which the minimum distance is reached. The following operations conclude this step:

3.1 Compute the local Frenet frame $(\mathbf{u}_i, \mathbf{v}_i, \nabla\Phi(\mathbf{x}_i))$ associated with $\nabla\Phi(\mathbf{x}_i)$

3.2 Set the size in the normal direction to $h_{\mathbf{n}_i} = h_0 \exp(\alpha\,\Phi(\mathbf{x}_i))$, the sizes in the orthogonal plane to:

$$h_{\mathbf{u}_i} = (^t\mathbf{u}_i\,\mathcal{M}_{P_i}\,\mathbf{u}_i)^{-2} ,$$

$$h_{\mathbf{v}_i} = (^t\mathbf{v}_i\,\mathcal{M}_{P_i}\,\mathbf{v}_i)^{-2} ,$$

3.3 The final metric is given by:

$$\mathcal{M}_{bl}(\mathbf{x}_i) = {}^t(\mathbf{u}_i, \mathbf{v}_i, \nabla\Phi(\mathbf{x}_i)) \begin{pmatrix} h_{\mathbf{u}_i}^{-2} & & \\ & h_{\mathbf{v}_i}^{-2} & \\ & & h_{\mathbf{n}_i}^{-2} \end{pmatrix} (\mathbf{u}_i, \mathbf{v}_i, \nabla\Phi(\mathbf{x}_i)). \quad (11)$$

4   Quality-Driven Local Mesh Operators

This section describes the local operators used to adapt the mesh. For each operator, numerical results using the metric estimates derived in Section 2 are shown.

4.1   Insertion and Collapse

To generate a unit-mesh in a given metric field $(\mathcal{M}_i)_{i=1...N}$, two operations are recursively used: edge collapse and point insertion on edge.

The starting point for the insertion of a new point on an edge $\mathbf{e}$ is the shell of $\mathbf{e}$ composed of all elements sharing this edge. Each element of the shell is then divided into two new elements. The new point is accepted if each new tetrahedron has a positive volume. When a point is inserted on an boundary edge, either a linear approximation of the surface is used or a quadratic recovery using the edge point normals (9).

The edge collapse starts from the ball of the vertex to be deleted. Again, for the deletion of points inside the volume, the only possible rejection is the creation of a negative volume element. A special care is also required to avoid the creation of an element that already exists, see Figure 2 (left and middle). The rejections are more complicated in the case of a surface point. We first avoid each collapse susceptible to modify the topology of the object. This is simply done by assigning an order on each surface point types: corner, ridge (line), inside surface. The collapse can also be rejected if the normal deviation between old and new normals becomes too large. Currently, if $\mathbf{n}$ denotes the normals to an old face, we allow the collapse if each new normal $\mathbf{n}_i$ verifies ${}^t\mathbf{n}\,\mathbf{n}_i > \cos(\pi/4)$. Note that the control to the surface deviation is given by the surface metric and so it does not need to be handled directly in the collapse operation.

With these operations, the core of the adaptive algorithm consists in scanning each edge of the current mesh and, depending on its length, creating a new point on the edge or collapsing the edge. An edge is declared too small or too large according to the bounds given in (2). Without any more considerations, such adaptive mesh generator is known not to be efficient and to require a lot of CPU consuming optimizations as point smoothing and edge swapping. This inefficiency is simply due to the locality of these operations. Comparing to an anisotropic Delaunay kernel [13], when an edge needs to be refined, the metric lengths along the orthogonal directions are controlled by the creation of the cavity. Consequently, in one shot, the area of refinement must be large. With the present approach, the size is controlled along one direction only (along the edge being scanned). Consequently, one can reach intractable configurations where the same initial edge is refined successively to get the desired size whereas the sizes in the other directions get worse. A typical configuration is depicted in Figure 2 (right).

A simple way to overcome this major drawback is to use the quality function (3) together with the unit-length check. This supplementary check can

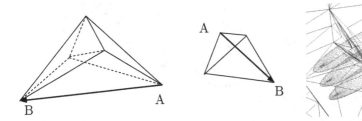

Fig. 2. Left and middle, volume and surface collapse of edge AB leading to the creation of an element that already exists . Right, Example where an edge is recursively refined to get a unit-length without checking the length requirement in the edge's orthogonal direction; the configuration may lead to edges acting as a barrier for future refinement.

be done at no cost since a lot of information can be re-used: the volume is already computed, as well as the length of the edges. By simply computing the quality function, we give to these operators the missing information on the orthogonal directions of the current scanned edge. For an optimal performance, two parameters are added in the rejection cases: a relative quality tolerance $q_r \geq 1$ and a global quality tolerance q_a. Indeed, it seems particularly interesting not to try to implement a full descent direction by imposing the quality to increase on each operation. We prefer to allow the quality to decrease in order to get out of possible local minima. Consequently, a new configuration of elements is accepted if:

$$q_r\, Q_{\mathcal{M}}^{ini} \leq Q_{\mathcal{M}}^{new} \quad \text{and} \quad Q_{\mathcal{M}}^{new} < q_a,$$

where $Q_{\mathcal{M}}^{ini}$ is the worse element quality of the initial configuration and $Q_{\mathcal{M}}^{new}$ is the worse quality of the new configuration. This approach is similar to the simulated annealing global optimization technique [21]. Note that the current version does not fully implement the classical metropolis algorithm where the rejection is based on a random probability. To ensure the convergence of the algorithm, the relative tolerance q_r is decreased down to 1 after each pass of insertions and collapses. At the end of the process, the absolute tolerance q_a is set up to the current worse quality among all elements. We now give some illustrative examples using the quality-driven insertion and collapse.

Figures 3, 4, 5 and 6 give anisotropic meshes obtained by applying only these two operators during the refinement process. The first example is a supersonic flow inside an inlet. It only involves the control of the interpolation error (6) on the Mach variable. The surface adapted mesh and the Mach number iso-values are depicted in Figure 3. The final mesh is composed of 70 000 tetrahedra. 5 iterations were performed to reach this accuracy. Despite the small number of elements, most of the features of the flow are well captured: strong amplitudes shocks are refined so as contact discontinuities emitted from the inlet spike. The second example is a supersonic flow inside

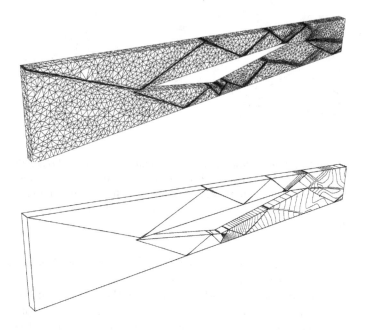

Fig. 3. Supersonic flow inside a scramjet. Only quality-driven insertion and collapse were used. Top, surface anisotropic mesh and bottom, mach iso-values. Both strong amplitude shocks and small amplitude shear layers are captured.

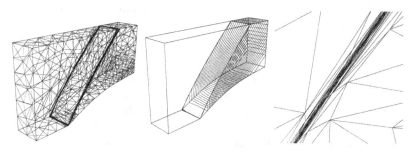

Fig. 4. Supersonic flow simulation inside a curved wedge. Only quality-driven insertion and collapse were used. From left to right, adapted surface mesh, density solution field and closer view of the surface mesh around the shock.

a curved wedge geometry. It involves the surface metric-based estimate (10) along with the control of the interpolation error (6) of the density variable. The results are depicted in Figure 4. The final mesh is composed of 8 000 tetrahedra with a resolution in the shock of 0.001m, see Figure 4 (right). This example illustrates how metric-based mesh adaptation gives an optimal distribution of the degree of freedom even though a very small number of elements is used.

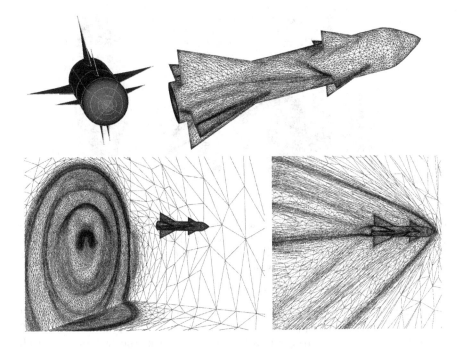

Fig. 5. Supersonic flow around a model missile. Only quality-driven insertion and collapse with anisotropic mesh gradation were used. Top, from left to right, CAD of the missile composed of 8 winglets having a quasi-null thickness and adapted surface mesh. Bottom, cut in the volume mesh behind the missile (left) and cut along the symmetry plane in the volume mesh (right).

In the following examples, an anisotropic volume gradation (4)(a) is performed on the volume metric field prior to the refinement. We now consider the accurate prediction of the flow field around the supersonic missile model where the CAD is depicted in Figure 5 (top left). The surface and the volume mesh are adapted to the Mach number in $\mathbf{L}^2$ norm. The deviation to the surface is controlled by using surface metric (10) with $\epsilon = 0.001$. A specific anisotropic re-meshing of the leading edges is also added. The cruise speed of the missile is Mach 2. The final mesh is composed of almost 600 000 tetrahedra. Surface mesh adaptation and volume mesh adaptation are perfectly combined. In particular, the complexity of the flow on the missile geometry appears clearly in Figure 5 (top right and bottom). Similarly, the complexity of the flow field in the volume is also well captured, see Figure 5 (bottom). Note that the supersonic missile model offers a large panel of challenges both for mesh adaptation and flow computation. Indeed, the very small thickness of the wings is one of the typical difficulties when attempting to mesh it in an anisotropic way using global methods. The last example is a transsonic flow around Onera M6 wing. The CAD is depicted in Figure 6 (left). The wing

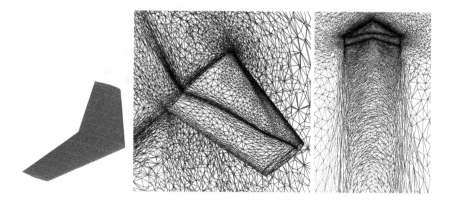

Fig. 6. Transsonic flow around Onera M6 wing. Only quality-driven insertion and collapse with anisotropic metric gradation were used. From left to right, CAD of the geometry, adapted surface mesh with a cut along the symmetry plane, and volume adapted mesh in the wake of the wing.

is flying at Mach 0.8 with an angle of attack of 1.5 degrees. At this speed regime, a strong shock appears on the wing profile as depicted in Figure 6 (middle). Despite the strong amplitude of the shock, the wake is also well captured when using multi-scale metric (6), see Figure 6 (right). The final adapted mesh is composed of 170 000 vertices and 950 000 tetrahedra.

4.2   Using the Boundary Layers Metric

We illustrate in this section the practical use of the boundary layers metric derived in Section 3.4.

Using (11) directly to generate a unit mesh is usually intractable due to the small required sizes (around 10^{-6}m). As pointed out by the previous examples, a classical iterative mesh adaptation procedure can reach quite easily a precision of 10^{-3}m near a shock after several adaptation steps. An example of a unit-mesh with respect to (11) for a minimal size of 0.005m is depicted in Figure 7 (left). In comparison with structured boundary layers grids, this result does not seem optimal in term of number of nodes and edges alignment. A first improvement to this direct approach is to build the boundary layers mesh layer by layer. The gradient of the distance map to the body is used as a relevant information for points location and edges alignment. In this respect, instead of generating a unit mesh in (11) directly, the quality computed in (11) is used to recover locally quasi-structured elements using swaps of edge and point smoothing, see Figure 11 (middle). More precisely, the algorithm for the current layer starts from the upper surface mesh of the previous layer and is composed of the following steps:

- Insert point at the current layer size along the closest edge to the gradient of the distance map Φ using the insertion of Section 4;

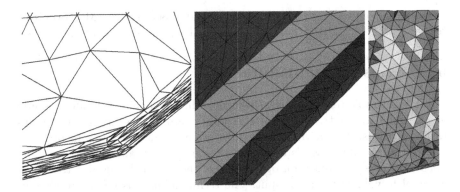

Fig. 7. From left to right, boundary layers mesh generation with a unit mesh approach using (11), structured layer generated and recovered using quality computed in (11), and structured layer with the uniform mesh from which it was generated.

- Align edges selected for refinement with the gradient of the distance map. This step is equivalent to the classical point smoothing [17];
- Optimize the mesh by swapping edges while controlling the quality computed using (11) in order to recover a structured layer.

A first layer created from a uniform mesh is depicted in Figure 7 (middle). The regular structure of the layer is fully recovered. The interaction between the structured layer and the uniform mesh is depicted in Figure 7 (right). This algorithm allows the creation of a boundary layers mesh from an adapted mesh while preserving the anisotropy. An example is depicted in Figure 8 where 10 boundary layers are added to an initial anisotropic mesh. The minimal size is 10^{-5}m. Almost 25 000 tetrahedra are added to the

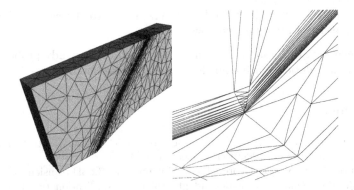

Fig. 8. Anisotropic mesh where 10 boundary layers were generated with a minimal size of 10^{-5}m (left), footprint of the boundary layers mesh near the shock (right).

initial mesh composed of 55 000 tetrahedra and 10 800 vertices. Moreover, this recovery process is faster than the traditional unit-mesh generation process.

5 Conclusions and Future Work

Local anisotropic remeshing has been introduced in this preliminary study as a reliable, alternative solution to global remeshing. To this end, the adaptation of the surface and the volume are done simultaneously in order to ensure that a valid mesh is available for computation after each remeshing phase. All mesh modification operators are thought as being able to handle a complete volume mesh as input. The technique used is based on edge insertion and collapse. Apart from their inherent simple formulation, they turn out to be efficient once they are monitored by a quality function. In terms of complexity, this approach seems much more simple that the generalization of the Delaunay kernel for anisotropy. It appears to provide a good compromise between simplicity and efficiency. For generality, the mesh generator uses the classical metric-based framework. It allows to take as an input various metric fields issued from differents tasks. Several metric fields for controlling the adaptation of the surface mesh or creating boundary layers have been derived. Currently work is directed at:

- The full interaction between the boundary layers metric and the interpolation error metric;
- Improving the robustness of surface remeshing for complex geometries,
- Improvements in boundary layers mesh generation, ie, taking into account the curvature of the distance map in order to simulate multi-normals behavior;
- Better vector-based edge alignment in order to reduce the number of nodes in the boundary layers, thereby tending to the number of nodes given by a truly structured grid;
- The application of the boundary layers metric to the case of shocks or any physical features of the flow.

More generally, this approach is currently tested on unsteady-problems and on RANS simulations with embedded and immersed bodies.

References

1. Alauzet, F.: Size gradation control of anisotropic meshes. Finite Elements in Analysis and Design (2009) (Published online)
2. Alauzet, F., Frey, P.J., George, P.-L., Mohammadi, B.: 3D transient fixed point mesh adaptation for time-dependent problems: Application to CFD simulations. J. Comp. Phys. 222, 592–623 (2007)
3. Arsigny, V., Fillard, P., Pennec, X., Ayache, N.: Log-Euclidean metrics for fast and simple calculus on diffusion tensors. Magnetic Resonance in Medicine 56(2), 411–421 (2006)

4. Baker, T.: Three-dimensional mesh generation by triangulation of arbitrary point sets. AIAA Paper, 87-1124 (1987)
5. Baum, J.D., Löhner, R.: Numerical simulation of pilot/seat ejection from an F-16. AIAA Paper, 93-0783 (1993)
6. Baum, J.D., Luo, H., Löhner, R.: Numerical simulation of blast in the world trade center. AIAA Paper, 95-0085 (1995)
7. Berger, M.: A panoramic view of Riemannian geometry. Springer, Berlin (2003)
8. Borouchaki, H., Hecht, F., Frey, P.J.: Mesh gradation control. Int. J. Numer. Meth. Engrg. 43(6), 1143–1165 (1998)
9. Bottasso, C.L.: Anisotropic mesh adaption by metric-driven optimization. Int. J. Numer. Meth. Engng. 60, 597–639 (2004)
10. Castro-Díaz, M.J., Hecht, F., Mohammadi, B., Pironneau, O.: Anisotropic unstructured mesh adaptation for flow simulations. Int. J. Numer. Meth. Fluids 25, 475–491 (1997)
11. Chen, L., Sun, P., Xu, J.: Optimal anisotropic simplicial meshes for minimizing interpolation errors in L^p-norm. Math. Comp. 76(257), 179–204 (2007)
12. do Carmo, M.: Differential geometry of curves and surfaces. Prentice-Hall, Englewood Cliffs (1976)
13. Dobrzynski, C., Frey, P.J.: Anisotropic delaunay mesh adaptation for unsteady simulations. In: Proc. of 17th Int. Meshing Rountable, pp. 177–194. Springer, Heidelberg (2008)
14. Dompierre, J., Vallet, M.G., Fortin, M., Bourgault, Y., Habashi, W.G.: Anisotropic mesh adaptation: towards a solver and user independent cfd. AIAA Paper, 97-0861 (1997)
15. Frey, P.J., Borouchaki, H.: Surface meshing using a geometric error estimate. Int. J. Numer. Methods Engng. 58(2), 227–245 (2003)
16. Frey, P.J.: About surface remeshing. In: Proc. of 15th Meshing Rountable 15, pp. 123–136. Springer, Heidelberg (2000)
17. Frey, P.J., George, P.-L.: Mesh generation. Application to finite elements, 2nd edn. ISTE Ltd and John Wiley & Sons (2008)
18. George, P.L., Borouchaki, H.: Delaunay triangulation and meshing: application to finite elements. Hermès Science, Paris (1998)
19. George, P.L., Hecht, F., Saltel, E.: Fully automatic mesh generator for 3d domains of any shape. Impact of Comuting in Science and Engineering 2, 187–218 (1990)
20. Hecht, F., Mohammadi, B.: Mesh adaptation by metric control for multi-scale phenomena and turbulence. AIAA Paper, 97-0859 (1997)
21. Kirkpatrick, S., Gelatt, C.D., Vecchi, M.P.: Optimization by simulated annealing. Science 220(4598), 671–680 (1983)
22. Li, X., Remacle, J.-F.: Anisotropic mesh gradation control. In: Proc. of 13th Meshing Rountable, Williamsburg, VA, USA (2004)
23. Löhner, R.: Regridding surface triangulations. J. Comp. Phys. 126, 1–10 (1996)
24. Löhner, R.: Renumbering strategies for unstructured-grid solvers operating on shared-memory, cache-based parallel machines. Comput. Meth. Appl. Mech. Engrg. 163, 95–109 (1998)
25. Löhner, R.: Generation of unstructured grids suitable for RANS calculations. AIAA Paper, 99-0662 (1999)
26. Löhner, R.: Applied CFD techniques. Wiley, New-York (2001)
27. Löhner, R.: Generation of viscous grids with ridges and corners. AIAA Paper, 07-3832 (2007)

28. Löhner, R., Parikh, P.: Three-dimensionnal grid generation by the advancing-front method. Int. J. Numer. Meth. Fluids 8(8), 1135–1149 (1988)
29. Loseille, A.: Adaptation de maillage 3D anisotrope multi-échelles et ciblé à une fonctionnelle. Application à la prédiction haute-fidélité du bang sonique. PhD thesis, Université Pierre et Marie Curie, Paris VI, Paris, France (2008)
30. Loseille, A., Alauzet, F.: Continuous mesh model and well-posed continuous interpolation error estimation. RR-6846, INRIA (2009)
31. Loseille, A., Alauzet, F., Dervieux, A., Frey, P.J.: Achievement of second order mesh convergence for discontinuous flows with adapted unstructured mesh adaptation. AIAA Paper, 07-4186 (2007)
32. Marcum, D.L.: Adaptive unstructured grid generation for viscous flow applications. AIAA Journal 34(8), 2440–2443 (1996)
33. Marcum, D.L.: Efficient generation of high-quality unstructured surface and volume grids. Engrg. Comput. 17, 211–233 (2001)
34. Mavriplis, D.J.: An advancing front delaunay triangulation algorithm designed for robustness. J. Comp. Phys. 117, 90–101 (1995)
35. Pain, C.C., Umpleby, A.P., de Oliveira, C.R.E., Goddard, A.J.H.: Tetrahedral mesh optimisation and adaptivity for steady-state and transient finite element calculations. Comput. Meth. Appl. Mech. Engrg. 190, 3771–3796 (2001)
36. Park, M.A.: Adjoint-based, three-dimensional error prediction and grid adaptation. AIAA Paper 42(9), 1854–1862 (2006)
37. Peraire, J., Vahdati, M., Morgan, K., Zienkiewicz, O.C.: Adaptive remeshing for compressible flow computations. J. Comp. Phys. 72, 449–466 (1987)
38. Ramamurti, R., Löhner, R.: Simulation of flow past complex geometries using a parallel implicit incompressible flow solver. AIAA Paper, CP-933 (1993)
39. Schall, E., Leservoisier, D., Dervieux, A., Koobus, B.: Mesh adaptation as a tool for certified computational aerodynamics. Int. J. Numer. Meth. Fluids 45, 179–196 (2004)
40. Sethian, S.: Level-set methods and fast marching methods. Cambridge University Press, Cambridge (1999)
41. Tam, A., Ait-Ali-Yahia, D., Robichaud, M.P., Moore, M., Kozel, V., Habashi, W.G.: Anisotropic mesh adaptation for 3D flows on structured and unstructured grids. Comput. Meth. Appl. Mech. Engrg. 189, 1205–1230 (2000)
42. Vallet, M.-G., Manole, C.-M., Dompierre, J., Dufour, S., Guibault, F.: Numerical comparison of some hessian recovery techniques. Int. J. Numer. Meth. Engrg. 72, 987–1007 (2007)

A Comparison of Gradient- and Hessian-Based Optimization Methods for Tetrahedral Mesh Quality Improvement[*]

Shankar Prasad Sastry[1] and Suzanne M. Shontz[1]

Department of Computer Science and Engineering,
The Pennsylvania State University
University Park, PA 16802
{sps210,shontz}@cse.psu.edu

1 Introduction

Discretization methods, such as the finite element method, are commonly used in the solution of partial differential equations (PDEs). The accuracy of the computed solution to the PDE depends on the degree of the approximation scheme, the number of elements in the mesh [1], and the quality of the mesh [2, 3]. More specifically, it is known that as the element dihedral angles become too large, the discretization error in the finite element solution increases [4]. In addition, the stability and convergence of the finite element method is affected by poor quality elements. It is known that as the angles become too small, the condition number of the element matrix increases [5].

Recent research has shown the importance of performing mesh quality improvement before solving PDEs in order to: (1) improve the condition number of the linear systems being solved [6], (2) reduce the time to solution [7], and (3) increase the solution accuracy. Therefore, mesh quality improvement methods are often used as a post-processing step in automatic mesh generation. In this paper, we focus on mesh smoothing methods which relocate mesh vertices, while preserving mesh topology, in order to improve mesh quality.

Despite the large number of papers on mesh smoothing methods (e.g., [8, 9, 10, 11, 12, 13, 14]), little is known about the relative merits of using one solver over another in order to smooth a particular unstructured, finite element mesh. For example, it is not known in advance which solver will converge to an optimal mesh faster or which solver will yield a mesh with better quality in a given amount of time. It is also not known which solver will most aptly handle mesh perturbations or graded meshes with elements of heterogeneous volumes. The answers may likely depend on the context. For

---

[*] This work was funded in part by NSF grant CNS 0720749, a Grace Woodward grant from The Pennsylvania State University, and an Institute for CyberScience grant from The Pennsylvania State University.

example, one solver may find an approximate solution faster than the others, whereas another solver may improve the quality of meshes with heterogeneous elements more quickly than its competitors.

To answer the above questions, we use Mesquite [15], a mesh quality improvement toolkit, to perform a numerical study comparing the performance of several local mesh quality improvement methods to improve the global objective function representing the overall mesh quality as measured with various shape quality metrics. We investigate the performance of the following gradient-based methods: steepest descent [16] and Fletcher-Reeves conjugate gradient [16], and the following Hessian-based methods: quasi-Newton [16], trust-region [16], and feasible Newton [17]. Mesh quality metrics used in this study include the aspect ratio [18], inverse mean ratio [19, 20], and vertex condition number metrics [21]. The optimization solvers are compared on the basis of efficiency and ability to smooth several realistic unstructured tetrahedral finite element meshes to both accurate and inaccurate levels of mesh quality. We used Mesquite in its native state with the default parameters. Only Mesquite was employed for this study so that differences in solver implementations, data structures, and other factors would not influence the results.

In this paper, we report the results of an initial exploration of the factors stated above to determine the circumstances when the various solvers may be preferred over the others. In an effort to make the number of experiments manageable, we limit the number of free parameters. Hence, we consider a fixed mesh type and objective function. In particular, we use unstructured tetrahedral meshes and an objective function which sums the squared qualities of individual tetrahedral elements. The free parameters we investigate are the problem size, initial mesh configuration, heterogeneity in element volume, quality metric, and desired degree of accuracy in the improved mesh.

The main results of this study are as follows: (1) the behavior of the optimization solvers is influenced by the degree of accuracy desired in the solution and the size of the mesh; (2) most of the time, the gradient-based solvers exhibited superior performance compared to that of the Hessian-based solvers; (3) the rank-ordering of the optimization solvers depends on the amount of random perturbation applied; (4) the rank-ordering of the optimization solvers is the same for the affine perturbation meshes; (5) the rank-ordering of the majority of the solvers is the same for graded meshes; however, the rank of conjugate gradient is a function of time; (6) graded meshes are sensitive to changes in the mesh quality metric.

2   Problem Statement

2.1   Element and Mesh Quality

Let V and E denote the vertices and elements, respectively, of an unstructured mesh, and let $|V|$ and $|E|$ denote the numbers of vertices and elements, respectively. Define V_B and V_I to be the set of boundary and interior mesh vertices. Let $x_v \in \mathbb{R}^n$ denote the coordinates for vertex $v \in V$. For the purposes of this paper, $n = 3$. Denote the collection of all vertex coordinates by $x \in \mathbb{R}^{n \times |V|}$. Let e be an element in E. Finally, let $x_e \in \mathbb{R}^{n \times |e|}$ the matrix of vertex coordinates for e.

We associate with the mesh a continuous function $q : \mathbb{R}^{n \times |e|} \to \mathbb{R}$ to measure the mesh quality as measured by one or more geometric properties of elements as a function of their vertex positions. In particular, let $q(x_e)$ measure the quality of element e. We assume a *smaller* value of $q(x_e)$ indicates a better quality element. A specific choice of q is an element quality metric. There are various metrics to measure shape, size, and orientation of elements [22].

The overall quality of the mesh is a function of the individual element qualities. The mesh quality depends on both the choice of the element quality metric q and the function used to combine them.

2.2 Aspect Ratio Quality Metric

An important parameter in this study is the choice of mesh quality metric. In general, we expect that the results could vary significantly depending on the choice of mesh quality metric. Thus, we consider three mesh quality metrics in this study, starting with the aspect ratio.

Various formulas have been used to compute the aspect ratio. The aspect ratio definition we employ is the one implemented in Mesquite. In particular, it is the average edge length divided by the normalized volume. Thus for tetrahedra, the aspect ratio is defined as follows:

$$\left(\frac{l_1^2 + l_2^2 + \cdots + l_6^2}{6} \right) / \left(\text{vol} \times \frac{12}{\sqrt{2}} \right),$$

where $l_i, i = 1, 2, \ldots, 6$ represent the six edge lengths, and vol represents its volume.

2.3 Inverse Mean Ratio Quality Metric

In order to derive the inverse mean ratio mesh quality metric, we let $a, b, c,$ and d denote the four vertices of a tetrahedron labeled according to the right-hand rule. Next, define the matrix A by fixing the vertex a and denoting by $e_1, e_2,$ and e_3 the three edge vectors emanating from a towards the remaining three vertices. Then, $A = [e_1; \; e_2; \; e_3] = [b - a; \; c - a; \; d - a]$. Next, define W to be the incidence matrix for the ideal element which is an equilateral tetrahedron in the isotropic case. In this case,

$$W = \begin{pmatrix} 1 & \frac{1}{2} & \frac{1}{2} \\ 0 & \frac{\sqrt{3}}{2} & \frac{\sqrt{3}}{6} \\ 0 & 0 & \frac{\sqrt{2}}{\sqrt{3}} \end{pmatrix}.$$

Next, let $T = AW^{-1}$ transform the ideal element to the physical element. Finally, the inverse mean ratio of a tetrahedral element is as follows:

$$\frac{\|T\|_F^2}{3|\det(T)|^{\frac{2}{3}}}.$$

2.4 Vertex Condition Number Quality Metric

In order to specify the vertex condition number quality metric, we first define some notation. Let x be any vertex of an element. Let x_k denote the kth neighboring vertex, for $k = 1, 2, \ldots, n$. Define k edge vectors $e_k = x_k - x$. Then the Jacobian of the element is given by the matrix $A = [e_1 \; e_2 \; \cdots \; e_n]$. Using A, we can define its vertex condition number as follows:

$$\|A\|_F \|A^{-1}\|_F,$$

where $\| \cdot \|_F$ denotes the Frobenius matrix norm.

All three mesh quality metrics range from 1 (for an equilateral tetrahedron) to ∞ (for a degenerate element). Invalid elements can be detected by the inverse mean ratio mesh quality metric when a complex value results.

2.5 Quality Improvement Problem

To improve the overall quality of the mesh, we assemble the local element qualities as follows: $Q = \sum_e q(x_e)^2$, where Q denotes the overall mesh quality, and $q(x_e)$ is the quality of element e. We compute an $x^* \in \mathbb{R}^{n \times |V|}$ such that x^* is a locally optimal solution to

$$\min_x Q(x) \tag{1}$$

subject to the constraint that $x_{v_B} = \overline{x_{v_B}}$, where $\overline{x_{v_B}}$ are the initial boundary vertex coordinates. In addition, we require that the initial mesh and subsequent meshes to be noninverted. This translates to the constraint $det(A^{(i)}) > 0$ for every element. In order to satisfy the two constraints, Mesquite fixes the boundary vertices and explicity checks for mesh inversion at each iteration.

3 Improvement Algorithms

In this paper, we consider the performance of five numerical optimization methods, namely, the steepest descent, conjugate gradient, quasi-Newton, trust-region, and feasible Newton methods, as implemented in Mesquite. The steepest descent and conjugate gradient solvers are gradient-based, whereas the remaining three are Hessian-based, i.e., they employ both the gradient and Hessian in the step computation. We describe each method below.

3.1 Steepest Descent Method

The steepest descent method [16] is a line search technique which takes a step along the direction $p_k = -\nabla f(x_k)$ at each iteration. In Mesquite the steplength, α_k, is chosen to satisfy the Armijo condition [23], i.e.,

$$f(x_k + \alpha_k p_k) \leq f(x_k) + c_1 \alpha_k \nabla f(x_k)^T p_k$$

for some constant $c_1 \in (0,1)$, which ensures that the step yields sufficient decrease in the objective function.

3.2 Conjugate Gradient Method

The conjugate gradient method [16] is a line search technique which takes a step in a direction which is a linear combination of the negative gradient at the current iteration and the previous direction, i.e.,.

$$p_k = -\nabla f(x_k) + \beta_k p_{k-1},$$

where $p_0 = -\nabla f(x_0)$. Conjugate gradient methods vary in their computation of β_k. The Fletcher-Reeves conjugate gradient method implemented in Mesquite computes

$$\beta_k^{FR} = \frac{\nabla f(x_k)^T \nabla f(x_k)}{\nabla f(x_{k-1})^T \nabla f(x_{k-1})}.$$

Care is taken in the line search employed by Mesquite to compute a steplength yielding both a feasible step (i.e., one which does not result in a tangled mesh) and an approximate minimum of the objective function along the line of interest.

3.3 Quasi-Newton Method

Quasi-Newton methods [16] are line search (or trust-region) algorithms which replace the exact Hessian in Newton's method with an approximate Hessian in the computation of the Newton step. Thus, quasi-Newton methods solve $B_k p_k = -\nabla f(x_k)$, for some $B_k \approx \nabla^2 f(x_k)$ at each iteration in an attempt to find a stationary point, i.e., a point where $\nabla f(x) = 0$. The quasi-Newton implementation in Mesquite [15] is a line search that approximates the Hessian using the gradient and true values of the diagonal blocks of the Hessian.

3.4 Trust-Region Method

Trust-region methods [16] are generalizations of line search algorithms in that they allow the optimization algorithm to take steps in any direction provided that the steps are no longer than a maximum steplength. Steps are computed by minimizing a quadratic model of the function over the trust region. The trust region is expanded or contracted at each iteration depending upon how reflective the model is of the objective function at the given iteration.

3.5 Feasible Newton Method

The feasible Newton method [17] is a specialized method for mesh quality improvement. In particular, it uses an inexact Newton method [24, 16] with an Armijo line search [23] to determine the direction in which to move the vertex coordinates. At each iteration, the algorithm solves the Newton equations via a conjugate gradient method with a block Jacobi preconditioner [24]. The solver also obtains good locality of reference.

4 Numerical Experiments

In this section, we report results from four numerical experiments designed to determine when each of the five solvers are preferred according to their time to convergence for local mesh smoothing. All solvers are implemented in Mesquite 2.0, the Mesh Quality Improvement Toolkit [15], and were run with their default parameter values. We solve the optimization problem (1) on a series of tetrahedral meshes generated with the CUBIT [25] and Tetgen [26] mesh generation packages. We consider the following geometries: distduct, foam, gear, hook [27] and cube. Sample meshes are shown in Figure 1. In the first three experiments, we study the effects of three different problem parameters on the time taken to reach x^*, a locally optimal solution. The problem parameters of interest are: problem size, initial mesh configuration, and grading of mesh elements. For each of the three parameters studied, we create a set of test meshes in which we isolate the parameter of interest and allow it to vary; these experiments were inspired by [28, 29]. Particular attention was paid to ensure that the remaining parameters were held as constant as possible. Due to space limitations, we have omitted most of the tables of initial mesh quality statistics which demonstrate this. In the fourth experiment, we investigate the effect that mesh quality metric has on solver performance.

Because the objective functions used for our experiments are non-convex, the optimization techniques may converge to different local minima. To ensure that this did not effect our study, we verified for each experiment whether or not the solvers converge to the same optimal mesh by comparing vertex coordinates of the optimal meshes.

In the following subsections, we describe the problem characteristics of the test meshes in terms of the numbers of vertices and elements, initial mesh quality (according to the mesh quality metric of interest), and parameter values of interest (such as magnitude of perturbation). We then specify performance results for the five optimization solvers. In all cases, the solution is considered optimal when it has converged to six significant digits. The machine employed for this study is equipped with an Intel P4 processor (2.67 GHz). The 32-bit machine has 1GB of RAM, a 512KB L2 cache, and runs Linux.

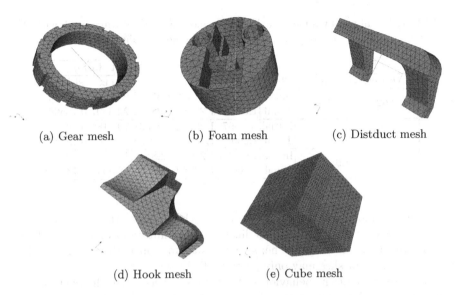

(a) Gear mesh   (b) Foam mesh   (c) Distduct mesh

(d) Hook mesh   (e) Cube mesh

Fig. 1. Sample meshes on the gear, foam, distduct, hook, and cube geometries. Geometries (a)-(d) were provided to us by Dr. Patrick Knupp of Sandia National Laboratories [27].

4.1   Increasing Problem Size

To test the effect that increasing the problem size has on optimization solver performance, we used CUBIT to generate a series of tetrahedral meshes with an increasing number of vertices while maintaining uniform mesh quality and element size. A series of meshes were generated for the distduct, foam, gear, and hook geometries shown in Figures 1(a) through 1(d); for each series of meshes, the number of elements is increased from approximately 5000 to 500,000 elements.

In the creation of the test meshes, care was taken to ensure that, for each mesh geometry, we achieve our goal of maintaining roughly uniform element size and mesh quality distributions. Table 1 shows the initial and final aspect ratio quality before and after conjugate gradient method was applied on three of the meshes. Such changes in mesh quality were typical of the results seen in this experiment.

For each mesh geometry, when the aspect ratio mesh quality metric was employed, the time to convergence required increased linearly with an increase in problem size. Figure 2 illustrates this trend for the use of the various solvers on the distduct geometry. Solver behavior was identical on the remaining geometries; in particular, the solvers also converged to the same optimal meshes. Thus, additional figures have been omitted. This is expected as the number of iterations to convergence is more or less a constant, and the time per iteration increases linearly with the number of elements used for

Table 1. Initial and final mesh quality after smoothing the distduct mesh with the conjugate gradient method using the aspect ratio mesh quality metric

Distduct Mesh			Mesh Quality (Aspect Ratio)				
# Vertices	# Elements	Phase	min	avg	rms	max	std dev
1,262	5,150	Initial	1.00557	1.33342	1.35118	2.71287	0.218363
		Final	1.00077	1.27587	1.28932	2.83607	0.185684
19,602	99,895	Initial	1.0007	1.28014	1.29531	10.3188	0.197718
		Final	1.00065	1.21742	1.22755	4.8624	0.157424
92,316	498,151	Initial	1.00009	1.27055	1.28513	18.5592	0.193054
		Final	1.00004	1.18949	1.1977	18.5592	0.139968

local mesh smoothing. There are instances where a deviation from linearity is seen in larger meshes. These are likely due to limitations on the size of the mesh which can fit in the cache; small meshes may fit entirely in the cache, whereas larger meshes may only partially fit in the cache.

We now examine the behavior of the various solvers on the distduct meshes with the use of the aspect ratio quality metric. For engineering applications, a highly accurate solution is not often needed or even desired. Thus, we consider partially-converged as well as fully-converged solutions. In each case, we consider smoothing with 85%, 90%, and 100%-converged solutions; the results are shown in Figure 2. The legend for the remaining plots in the paper is as follows: 'circle' (steepest descent), 'triangle' (conjugate gradient), 'diamond' (quasi-Newton), 'square' (trust-region), and 'star' (feasible Newton).

In all the cases, i.e., for the 85%−, 90%-, and 100%−converged solutions, the five optimization solvers converged towards the same optimal mesh. For the 85%−converged solutions, feasible Newton is the fastest method to reach an optimal solution (see Figure 2(a)); few iterations were required since the initial CUBIT-generated meshes were of fairly good quality. Feasible Newton was possibly the quickest method since it takes fewer iterations than the other methods; however, each iteration takes a greater amount of time than the other solvers. The ranking of all solvers in order of fastest to slowest on the larger meshes is: feasible Newton < steepest descent < conjugate gradient < trust-region < quasi-Newton. For the smaller meshes, the rank-ordering is: conjugate gradient < feasible Newton < steepest descent < trust-region < quasi-Newton. In general, the gradient-based solvers (i.e, steepest descent and conjugate gradient) performed better than the Hessian-based solvers (trust-region and quasi-Newton). However, feasible Newton, which is a Hessian-based solver, performed very competitively. This is likely due to the fact that local mesh smoothing was performed with a highly-tuned solver. In addition, the rank ordering of the solvers depends on the mesh size as noted above.

In the majority of the 90%-converged solution cases (see Figure 2(b)), the conjugate gradient algorithm reached convergence faster than the other methods. This was followed by the steepest descent, feasible Newton, trust-region,

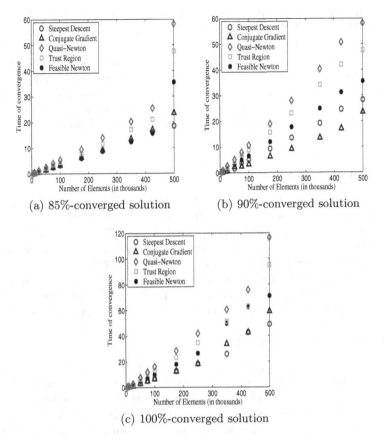

(a) 85%-converged solution   (b) 90%-converged solution

(c) 100%-converged solution

Fig. 2. Mesh smoothing to various convergence levels: (a) 85%-converged solution; (b) 90%-converged solution; (c) 100%-converged solution. Results are for the distduct meshes with the aspect ratio quality metric.

and quasi-Newton methods, respectively. This ordering is different than that which was obtained for the 85% case. Because local mesh smoothing was performed, only one vertex in the mesh is moved at a time. The steepest descent and conjugate gradient methods use only the gradient of the objective function to move a vertex to its optimal location. The other methods also use the Hessian of the objective function to move the vertex. The calculation of the Hessian adds computational expense, making the Hessian-based methods comparatively slower. However, Hessians may effect local mesh smoothing results less than global mesh smoothing results where the Hessian matrices are much larger. The conjugate gradient method is superior to steepest descent since it uses gradient history to determine the optimal vertex position.

In the majority of the 100%-converged solution case (see Figure 2(c)), the conjugate gradient algorithm was the fastest to reach convegence for smaller meshes; however, the steepest descent method proved to be faster for larger meshes. This is probably due to the increase in memory which is required for larger meshes. Eventually the increased requirements on the performance of the cache may slow down the conjugate gradient algorithm relative to the steepest descent algorithm since it must store and access an additional vector.

In conclusion, the behavior of the optimization solvers is influenced by the degree of accuracy desired in the solution and the size of the mesh. Most of the time, the gradient-based optimization solvers exhibited superior performance to that of the Hessian-based solvers.

4.2   Initial Mesh Configuration

In order to investigate the effect that the initial mesh configuration (as measured by distance from optimal mesh) had on the performance of the five solvers, a series of perturbed meshes, based on the 500,000 element distduct, foam, gear, and hook meshes from the previous experiment, were designed. In particular, the meshes were smoothed initially using the aspect ratio mesh quality metric. Then, random or systematic perturbations were applied to the interior vertices of the optimal mesh. For all experiments, the perturbations were applied to all interior vertices and to a randomly chosen subsets of vertices of size 5%, 10%, 25%, and 100% of the interior vertices. The formulas for the perturbations are as follows:

Random: $x_v = x_v + \alpha_v r$, where r is a vector of random numbers generated using the rand function, and α_v is a multiplicative factor controlling the amount of perturbation. For our experiments, we chose a random value for α_v; the resulting meshes were checked to verify that they were of poor quality.

Translational: $x_v = x_v + \alpha s$, where s is a direction vector giving the coordinates to be shifted, and α is a multiplicative factor controlling the degree of perturbation. In this case, we consider the shift with $s = [1\ 0\ 0]^T$ and α values ranging from 0.016 to 1.52 were used to maximize the amount of perturbation a particular mesh could withstand before the elements became inverted. Thus, the specific value of α chosen for a mesh depended upon the size of the elements.

Random Perturbations

The results obtained here differ somewhat from the results obtained from the scalability experiment above. They are similar in that the gradient-based methods performed better than the Hessian-based methods. This can be attributed to the greater computational expense of computing the Hessian matrices for a smaller payoff in terms of a decrease in the objective function. The main difference here is that, in almost all cases, the steepest descent algorithm performs better than the conjugate gradient algorithm.

For this experiment, the meshes to be smoothed were perturbed from the fully optimized CUBIT-generated meshes. Thus, the initial meshes are of poorer quality. Starting with poor quality meshes, i.e., far away from an optimal mesh, had a very significant impact on the performance of the solvers. There are cases when the conjugate gradient method does better than the steepest descent method when the quality of the input mesh is reasonably good. In this case, all solvers converged to the same optimal mesh.

However, when we start with a poor quality initial mesh, a coarse-scale improvement in the the mesh is needed. Once the mesh has been sufficiently smoothed, fine-scale improvements can be obtained through the use of superior solvers. In most cases, because the perturbation was large, coarse-scale smoothing was needed. As a result, the performance of steepest descent was the best (also due to the lower complexity of the algorithm). When the perturbations were small, fine-scale smoothing requirements imply that superior methods will converge faster. This was indeed seen when the perturbations were small. The conjugate gradient method's performance was better than that of steepest descent in such cases. However, the Hessian-based methods were slower because of their inherent computational complexity. Figure 3(a) shows typical objective function versus time plots for our experiments.

The behavior of the trust region method was distinctly different than that of the other algorithms. For small perturbations from the optimal mesh, the behavior of the trust-region method almost coincided with that of the other methods in the quality versus time plots. Figure 3(b) below illustrates an example of such behavior.

However, when the perturbations were large, the trust-region method was much slower than the other methods in terms of time to convergence. This is due to the constraint of the spherical trust-region bounding the maximum acceptable steplength at each iteration. This conservative approach slows the time to convergence of the trust-region method. It was also observed that, for large perturbations, the steepest descent method does not converge to the same optimal mesh as the other methods. In particular, it converges to an optimal mesh with a higher objective function value. The plot shown in Figure 3(c) is a good example of the dismal performance of the trust-region and steepest descent methods in the large perturbation case.

In conclusion, the rank-ordering of the optimization solvers depends upon the amount of random perturbations applied to the initial meshes in the context of mesh smoothing using the aspect ratio quality metric. In particular, all five methods performed competitively for the small perturbation case; however, the steepest descent and conjugate gradient methods performed the best. In the case of medium-sized perturbations, the steepest descent method performed the best, and the trust-region method performed very slowly. The other three methods exhibited average performance. Finally, for the case of large perturbations, the trust-region method is very slow to converge, and the steepest descent method may converge to a mesh of poorer quality.

Affine Perturbations

In order to determine the effect that affine perturbations had on the performance of the optimization solvers, the affine (translation) perturbation

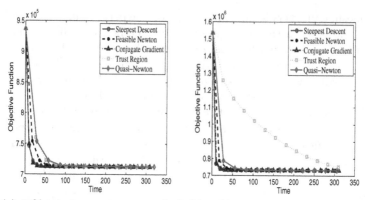

(a) 10% interior vertices perturbed (b) 10% interior vertices perturbed

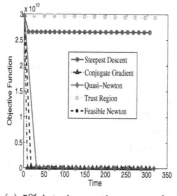

(c) 5% interior vertices perturbed

Fig. 3. Typical results from the random perturbation experiment. Results were obtained by smoothing the 500,000 element meshes using the aspect ratio quality metric. (a) The result is for the gear mesh with 10% of its vertices perturbed; because the perturbation was small, the behavior of the trust-region method was almost coincident with that of the other solvers. (b) The result is for the distduct mesh with 10% of its vertices perturbed; here the trust-region method is competitive when the initial mesh is of reasonable quality due to the medium-size perturbation. (c) The result is for the distduct mesh with 5% of its vertices perturbed. Because the perturbations were large, the steepest descent and trust-region methods performed very poorly.

shown above was applied to all interior mesh vertices once the appropriate initial 500,000 element distduct, foam, gear, and hook meshes were smoothed according to the aspect ratio mesh quality metric.

The qualities of the interior elements of the perturbed meshes were still fairly good since the transformation applied was affine; however the

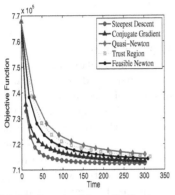
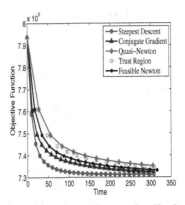

(a) Mesh smoothing of affinely perturbed distduct mesh

(b) Mesh smoothing of affinely perturbed gear mesh

Fig. 4. Typical results for the affine perturbation experiment using the aspect ratio for local mesh smoothing. The results are for smoothing the distduct and hook meshes with 500,000 elements after all interior vertices were affinely perturbed.

qualities of the boundary elements was much worse. It is expected that the convergence plots for all of the solvers will start with rapid decrease in the objective function and will end with a small decrease in the objective function. This is because the initial meshes were created by applying as large an affine perturbation as possible before mesh inversion occurred, thus generating meshes rather far away from the optimal ones. This behavior is typical and is observed in the plots shown in Figure 4. The time taken per nonlinear iteration varies with the computational complexity of the algorithm. However, the objective function values (for the various solvers) remain rather similar over the first few iterations until, eventually, more vertex movement occurs, and the objective function values become less predictable. However, all solvers did converge to the same optimal mesh.

The steepest descent method, being the least computationally expensive method, spends less time per iteration and converges to an optimal mesh fairly quickly. The ranking of the optimization solvers for the affine perturbation meshes is as follows: steepest descent < conjugate gradient < feasible Newton < trust-region < quasi-Newton. This rank-ordering demonstrates that methods for which every iteration is faster converge before methods for which each iteration is slower.

In conclusion, the optimization solvers exhibited a distinct rank-ordering for the affine perturbation meshes in the context of local mesh smoothing using the aspect ratio quality metric. In particular, the rank-ordering was as follows: steepest descent < conjugate gradient < feasible Newton < trust-region < quasi-Newton.

4.3 Graded Meshes

Our second test set was generated using Tetgen in order to test the effect that grading of mesh elements has on the performance of the five optimization solvers, as graded meshes have a larger distribution of element mesh qualities. For this experiment, three sets of structured tetrahedral meshes were generated which contain the same numbers of vertices and elements but whose elements have different volumes. The meshes were constructed on a cube domain having a side length of 20 units. In the first set of meshes, the vertices were evenly distributed in two of the three axes, but, for the other axis, half of the vertices were placed in first 10%, 20%, 30%, or 40% of the volume. Two additional sets of test meshes were created with the density of vertices varying in two and three directions instead of variation in only one direction. After the point clouds were created, Tetgen was used to create a volume mesh of the cube domain. The resulting Delaunay meshes, which were created without using any quality control features, was used for the graded mesh experiment. See Figure 1(e) for an example of a mesh created with half of its vertices occupying 30% of the space in all three axes and distributed uniformly throughout the rest of the cube volume.

This mesh generation technique results in a structured mesh with heterogeneous elements in terms of volume. In particular, approximately one-fourth, one-half, and one-fourth of the mesh elements can be considered small, medium, and large, respectively. All of the meshes generated contain 8000 vertices and 41,154 tetrahedra.

The results obtained from this experiment are shown in Figure 5. The mesh smoothing results for the graded meshes are similar to those observed in the affine perturbation case. The main difference between the two experiments is the behavior of the conjugate gradient method. For the graded meshes, there is a definite hierarchy among the other four solvers; the rank-ordering is as follows: steepest descent < feasible Newton < trust-region < quasi-Newton. However, the rank of the conjugate gradient method with respect to the other solvers varies as a function of time.

In conclusion, the rank-ordering of the conjugate gradient method varied as a function of time as the graded meshes were smoothed using the aspect ratio mesh quality metric. However, the rank-ordering of the remaining four optimization solvers was as follows: steepest descent < feasible-Newton < trust-region < quasi-Newton.

4.4 Mesh Quality Metric

Our final experiment was designed to investigate the effect of the choice of mesh quality metric on the performance of the optimization methods. For this experiment, we investigated the performance of the various methods on the distduct, foam, gear, hook, and cube meshes by repeating a subset of the above experiments for the inverse mean ratio and vertex condition number quality metrics.

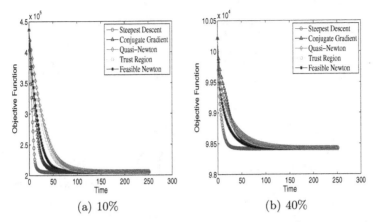

(a) 10%   (b) 40%

Fig. 5. Mesh smoothing results for the graded meshes using the aspect ratio mesh quality metric. The percentages indicate the amount of volume used by the first half of the vertices in a given axis within the cube domain.

The results of performing the scaling experiment for the inverse mean ratio and vertex condition number quality metrics are the same as those for the aspect ratio mesh quality metric described above.

Performing the random perturbation experiment for the inverse mean ratio and vertex condition number quality metrics yielded results that were qualitatively the same, i.e., the results could be classified into one of the above three cases depending upon how large were the perturbations.

The results of performing the affine perturbation experiment for the inverse mean ratio and vertex condition number mesh quality metric yielded results similar to those when the aspect ratio mesh quality metric was used.

Performing the element heterogeneity experiment for the inverse mean ratio mesh quality metric yielded results that were the same as those observed earlier for the aspect ratio mesh quality metric. However, the results are different for the vertex condition number mesh quality metric. When the vertex condition number metric is employed for mesh smoothing in the context of the graded mesh experiment, we observe a small rise in the objective function after a significant initial decrease as seen in Figure 6. The plots in this figure are for the cube meshes with vertices in all three axes distributed nonuniformly to create graded meshes with elements of heterogeneous volume. Although such behavior is rare, it is possible, as local mesh smoothing is being applied with a global objective function. Further investigation into the cause of such behavior for the meshes in this experiment is needed.

In conclusion, the scaling experiment results were insenstivive to the choice of mesh quality metric. However, the perturbation and element heterogeneity results were indeed sensitive to the choice of mesh quality metric. Further research is needed to identify additional contexts where the choice of mesh quality metric influences optimization solver behavior.

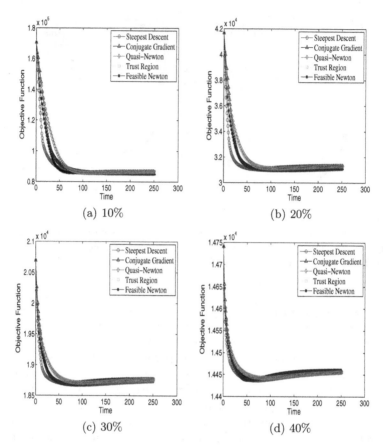

Fig. 6. Mesh smoothing results for the cube meshes with heterogeneous element volumes using the vertex condition number mesh quality metric. The percentages indicate the percentage of volume used by the first half of the vertices in all three axes within the cube domain.

5   Future Work

The results in this study are specific to local mesh quality improvement of unstructured tetrahedral meshes via five optimization solvers, namely, the steepest descent, Fletcher-Reeves conjugate gradient, quasi-Newton, trust-region, and feasible Newton methods, with mesh quality measured according to the three specified quality metrics, namely the aspect ratio, inverse mean ratio, and vertex condition number. The results we obtained may vary dramatically if global mesh quality improvement methods were used instead of the local ones studied here [28, 29]; hence, we plan to investigate global versions of these solvers in future work. In addition, vertex ordering has been shown to play an important role in convergence of the Feasnewt solver when used for local mesh optimization [30]; thus, we will also investigate the effect of vertex ordering in the future. We also plan to investigate the role that other non-shape quality metrics have on the mesh optimization methods with the goal of identifying other contexts where quality metrics influence optimization solver behavior. Finally, we plan to investigate the use of hybrid solvers to improve optimization solver performance.

References

1. Babuska, I., Suri, M.: The p and h-p versions of the finite element method, basic principles, and properties. SIAM Review 35, 579–632 (1994)
2. Berzins, M.: Solution-based mesh quality for triangular and tetrahedral meshes. In: Proceedings of the 6th International Meshing Roundtable, Sandia National Laboratories, pp. 427–436 (1997)
3. Berzins, M.: Mesh quality - Geometry, error estimates, or both? In: Proceedings of the 7th International Meshing Roundtable, Sandia National Laboratories, pp. 229–237 (1998)
4. Babuska, I., Aziz, A.: On the angle condition in the finite element method. SIAM J. Numer. Anal. 13, 214–226 (1976)
5. Fried, E.: Condition of finite element matrices generated from nonuniform meshes. AIAA Journal 10, 219–221 (1972)
6. Shewchuk, J.: What is a good linear element? Interpolation, conditioning, and quality measures. In: Proceedings of the 11th International Meshing Roundtable, Sandia National Laboratories, pp. 115–126 (2002)
7. Freitag, L., Ollivier-Gooch, C.: A cost/benefit analysis for simplicial mesh improvement techniques as measured by solution efficiency. Internat. J. Comput. Geom. Appl. 10, 361–382 (2000)
8. Knupp, P., Freitag, L.: Tetrahedral mesh improvement via optimization of the element condition number. Int. J. Numer. Meth. Eng. 53, 1377–1391 (2002)
9. Freitag, L., Plassmann, P.: Local optimization-based simplicial mesh untangling and improvement. Int. J. Numer. Meth. Eng. 49, 109–125 (2000)
10. Amenta, N., Bern, M., Eppstein, D.: Optimal point placement for mesh smoothing. In: Proceedings of the 8th ACM-SIAM Symposium on Discrete Algorithms, pp. 528–537 (1997)
11. Zavattieri, P.: Optimization strategies in unstructured mesh generation. Int. J. Numer. Meth. Eng. 39, 2055–2071 (1996)
12. Amezua, E., Hormaza, M., Hernandez, A., Ajuria, M.: A method of the improvement of 3D solid finite element meshes. Adv. Eng. Softw. 22, 45–53 (1995)
13. Canann, S., Stephenson, M., Blacker, T.: Optismoothing: An optimization-driven approach to mesh smoothing. Finite Elem. Anal. Des. 13, 185–190 (1993)
14. Parthasarathy, V., Kodiyalam, S.: A constrained optimization approach to finite element mesh smoothing. Finite Elem. Anal. Des. 9, 309–320 (1991)
15. Brewer, M., Freitag Diachin, L., Knupp, P., Leurent, T., Melander, D.: The Mesquite Mesh Quality Improvement Toolkit. In: Proceedings of the 12th International Meshing Roundtable, Sandia National Laboratories, pp. 239–250 (2003)
16. Nocedal, J., Wright, S.: Numerical Optimization, 2nd edn. Springer, Heidelberg (2006)

17. Munson, T.: Mesh Shape-Quality Optimization Using the Inverse Mean-Ratio Metric. Mathematical Programming 110, 561–590 (2007)
18. Cavendish, J., Field, D., Frey, W.: An approach to automatic three-dimensional finite element mesh generation. Int. J. Num. Meth. Eng. 21, 329–347 (1985)
19. Liu, A., Joe, B.: Relationship between tetrahedron quality measures. BIT 34, 268–287 (1994)
20. Knupp, P.: Achieving finite element mesh quality via optimization of the Jacobian matrix norm and associated quantities, Part II - A framework for volume mesh optimization and the condition number of the Jacobian matrix. Int. J. Numer. Meth. Eng. 48, 1165–1185 (2000)
21. Knupp, P.: Matrix norms and the condition number. In: Proceedings of the 8th International Meshing Roundtable, Sandia National Laboratories, pp. 13–22 (1999)
22. Knupp, P.: Algebraic mesh quality metrics. SIAM J. Sci. Comput. 23, 193–218 (2001)
23. Armijo, L.: Minimization of functions having Lipschitz-continuous first partial derivatives. Pacific Journal of Mathematics 16, 1–3 (1966)
24. Kelley, C.T.: Solving Nonlinear Equations with Newton's Method. SIAM, Philadelphia (2003)
25. Sandia National Laboratories, CUBIT Generation and Mesh Generation Toolkit, http://cubit.sandia.gov/
26. Si, H.: TetGen - A Quality Tetrahedral Mesh Generator and Three-Dimensional Delaunay Triangulator, http://tetgen.berlios.de/
27. Knupp, P.: Personal communication (2009)
28. Freitag, L., Knupp, P., Munson, T., Shontz, S.: A comparison of inexact Newton and coordinate descent mesh optimization techniques. In: Proceedings of the 13th International Meshing Roundtable, Sandia National Laboratories, pp. 243–254 (2004)
29. Diachin, L., Knupp, P., Munson, T., Shontz, S.: A comparison of two optimization methods for mesh quality improvement. Eng. Comput. 22, 61–74 (2006)
30. Shontz, S.M., Knupp, P.: The effect of vertex reordering on 2D local mesh optimization efficiency. In: Proceedings of the 17th International Meshing Roundtable, Sandia National Laboratories, pp. 107–124 (2008)

Author Index

Index by Affiliation